Function Field
Arithmetic

Function Field
Arithmetic

by

Dinesh S. Thakur
University of Arizona, USA

World Scientific

N JERSEY • LONDON • SINGAPORE • SHANGHAI • HONG KONG • TAIPEI • CHENNAI

Published by

World Scientific Publishing Co. Pte. Ltd.

5 Toh Tuck Link, Singapore 596224

USA office: 27 Warren Street, Suite 401-402, Hackensack, NJ 07601

UK office: 57 Shelton Street, Covent Garden, London WC2H 9HE

Library of Congress Cataloging-in-Publication Data
Thakur, Dinesh S., 1961–
    Function field arithmetic / by Dinesh S. Thakur.
        p. cm.
    Includes bibliographical references and index.
    ISBN-13 978-981-238-839-1 -- ISBN-10 981-238-839-7
    ISBN-13 978-981-3270-11-4 (pbk) -- ISBN-10 981-3270-11-X (pbk)
    1. Fields, Algebraic. 2. Arithmetic functions. 3. Drinfeld modules. I. Title.

QA247 .T53 2004
512'3--dc22   2004041975

British Library Cataloguing-in-Publication Data
A catalogue record for this book is available from the British Library.

Dedicated to
the fond memories of Bhaikaka,
to Maiatte, Bhai, Aai;
and to my other teachers, family
and friends

Preface

This book, written for those with appreciation for number theory and geometry, gives an exposition of function field arithmetic with emphasis on recent developments having to do with Drinfeld modules, arithmetic of special values of transcendental functions, diophantine approximation and on related interesting open problems. The subject matter is a beautiful blend of arithmetic, algebra, geometry and analysis. We assume a basic background in algebraic number theory and geometry at least at the level of curves.

Number theorists study number fields and function fields together because of various analogies between them in spite of some important differences. Interesting arithmetic of function fields was developed by Kronecker, Dedekind, Weber, Emil Artin, F. K. Schmidt, Witt, Hasse, Weil, Chevalley, Lang, Tate, Serre, Shafarevich, Manin, Parshin, Iwasawa, Grothendieck, Deligne, Arakelov, and on the Drinfeld module side that we write more about by Carlitz, Wade, Wagner, Hayes, Drinfeld, Deligne, Goss, Gekeler, Galovich, Rosen, Jing Yu, Takahashi, Stuhler, Gross, Anderson to name just a few early contributors.

New analogies understood by Carlitz, Drinfeld, Goss, Anderson etc. opened up new territories. The subject is now a mixture of well developed areas, some where analogs to conjectures in the number field case are proved and well-understood, so that there are attempts to bring these ideas in the number fields case in various ways, and some other promising areas, largely unexplored, where the number field case is well-understood and the function field case goes against the naive analogies and is not even conjecturally understood. The good examples of the first type are Riemann hypothesis, Iwasawa theory, Birch and Swinnerton-Dyer conjectures and Langlands conjectures, transcendence theory and special values of Gamma functions, whereas the special values of Zeta functions, Diophantine approximation are of the second type.

Several analogies work in $\mathbb{F}_q[t]$ case, but separate to quite different phenomena in general, unlike earlier theories of phenomena which worked quite smoothly in general, if only with minor changes. This split-up helps us focus on relevant issues. Out of many equivalent ways to look at a particular classical concept, the way which translates well in the function field case may be thought of as a more fundamental one. Hence the analogs help to sharpen our understanding of classical concepts. Further, as we will see, the function field arithmetic is in interesting hybrid analogies with number field case, and its p-adic and q-analogs.

Here is Andre Weil ([Wei79, Vol. II, p. 408] as translated by Mazur) talking about the magic and importance of analogies in mathematics:

Nothing is more fruitful - all mathematicians know it - than those obscure analogies, those disturbing reflections of one theory on another; those furtive caresses, those inexplicable discords; nothing also gives more pleasure to the researcher.

Weil goes on talking about what happens when everything is understood. Fortunately, we are far from that stage here, and the exciting exploration goes on.

In contrast to the earlier function field situation, many areas such as special values of zeta functions, diophantine approximation are not even conjecturally understood. But concrete results that have been obtained suggest a beautiful interesting structures, only partially revealed. Things which do not seem to work as expected are the most interesting things.

In this book, I have tried to tell this interesting story to my fellow students of Mathematics, hoping to share the excitement and the enjoyment I felt working on it. The outline of the book can be best understood by taking a look at the index and introductions to the chapters. One important theme is the algebraic incarnations of gamma and zeta values, and related special functions on product of curves which govern the arithmetic of gamma, zeta values, theta functions and Gauss sums.

For me, it was quite a tough choice between using this occasion to write this book for setting things up in more generality for future reference or going for examples and explanations in the simple motivating cases leaving out the technicalities and the generalizations. Both approaches have their virtues. For the better or for the worse, I have finally opted for the second option. It is indeed true that many of the interesting results and mysteries already show up in the simplest cases. I have not tried to be in the most general setting possible or to give complete proofs each time. I hope that the sketch or sometimes even the catch-words will provide some insight. I have often sacrificed efficiency by dealing with special cases first and spelling out easy consequences. Especially we do $\mathbb{F}_q[t]$ case quite often: Many times it can be done easily explicitly with easier different proofs, and also because many times answers are quite different in that case, many analogies working simultaneously, whereas they break up into different analogies for other A in higher genus.

Simplest way to proceed is usually not the historical one: false attempts, which add to understanding nonetheless, are ignored, and tricks, original ideas or important remarks, at the time, eventually become part of the standard technique. And a lot looks easier and straightforward in retrospect. So in writing a book many times giving proper attributions is hard and referring to the original articles is the best source. I have followed the frequent practice in books of mentioning some relevant references in the notes, rather than in each place as required in research articles. So for attributions original sources need to be consulted. Many times (especially for facts mentioned in passing) I have given only the most convenient reference where you can find the original sources. There is also an extensive bibliography in the book by Goss.

While making comparisons with the literature, note that the notation as well as the normalizations differ in original sources. We use the same notation and name for analogous objects in different context: For example, gamma function Γ stands for the classical gamma function as well as one of the Gamma functions for function fields or its special case for say $\mathbb{F}_q[t]$. The role should be clear from the context and is usually specified. The precise bookkeeping of the notations would have made it harder to read.

This is just a tour of some topics I enjoyed learning and working on, and is by no means exhaustive treatment. Focusing on certain areas and applications, we have omitted or given much shorter treatment (just sketching some ideas and results) than deserved to many topics, such as the moduli questions and topics related to Langlands conjectures, i.e., the original motivations of Drinfeld (see papers of Drinfeld, Deligne-Husemoller, Laumon, Lafforgue and books by Laumon, Gekeler), modular forms (see book by Gekeler), Artin-Hasse-Weil theory and Diophantine geometry (treated in many books).

Only regret I have is that I had to resist the urge to pursue in more detail many things which came up, because of time constraints, and page-limit. I will be very glad if the reader (or readers, let us be optimistic) takes trouble to comment, suggest improvements, points out errors and references etc. I plan to provide updates on the web page.

I thank many people and institutes for their help: Institute for Advanced study, where the writing started and University of Arizona, Tata Institute of Fundamental research and Bhaskaracharya Pratishthan where it continued; my colleagues at these and other places; National Science Foundation, National Security Agency and U. S. Civilian Research and Development Foundation for providing financial support during this period; collaborators Jean Paul Allouche, Greg Anderson, Robert Beals, Minhyong Kim, Felipe Voloch as well as other friends in this area of mathematics such as David Hayes, Ernst Gekeler, Jing Yu, Michael Rosen, Bjorn Poonen, Yuchiro Taguchi, Dale Brownawell, Anatoly Kochubei, Matt Papanikolas.

With his basic contributions as well as tremendous enthusiasm and energy, David Goss has been one of the main leaders and the driving force of the field in the last many years. When I was hesitating to reply positively to the invitation to write this book, as David Goss' book had just been published, David encouraged me to write saying that it would certainly be better to hear different perspectives. I thank him and Greg Anderson for their generosity over the years. I could always count on my friends Greg, David, Ernst, Takahiro Shiota and Jeremy Teitelbaum over the years for trying to answer anything that I would ask.

I dedicate this book to Bhaikaka's memory. He retained informality of conversation in his writings. This is sometimes difficult while communicating mathematical ideas, but I have tried my best, given the constraints, and I would like to think that he would have approved of at least the effort.

This book is also dedicated to my teachers, family and friends, fortunately too many to be listed here. The three categories are also well-mixed. But I must mention John Tate, Barry Mazur, Pierre Deligne and Greg Anderson from whom I learned a lot (though the content and the crude style of this book might prove exactly the opposite!). On the non-mathematical front, I owe a lot to Bhaikaka, Maiatte, Bhai, Aai, Satish, Shubha, Umesh; and to Jyoti and Ashutosh, who kept on nagging me about finishing the book and at the same time provided welcome distraction.

December 2003   Dinesh S. Thakur

Contents

Chapter 1

Number fields and Function fields

We will recall some basic analogies and contrasts in number field and function field situations and then recall some fundamental objects and tools in the study of function fields and the corresponding curves. We will also collect some simple examples for later use.

For the proofs of the facts and the theorems that are not referenced, see eg., the books [Deu73; Che63; Art67; Mor91; Iwa93; Eic66; Sti93; Lor96; Ros02; Gol03]. There are many more good books dealing with general algebraic geometry or Riemann surfaces and curves using algebraic, geometric or analytic perspectives to various degrees. It is good to try to acquire an integrated view and to realize that the distinction between various viewpoints is really artificial and restricting. As we will see in the later chapters, the same can be said about studying Drinfeld modules.

1.1 Global fields: Basic analogies and contrasts

In schools, after we learn how to manipulate (add, subtract, multiply and divide) with integers and rational numbers, we introduce a variable (indeterminate, unknown, transcendent) and learn to manipulate with polynomials and rational functions, learn division algorithms giving remainder of smaller 'size' than that of the divisor. One difference we see is that for polynomials, this concept of size (or absolute value) is non-archimedean. But when we learn about p-adic sizes or absolute values for primes p or irreducible polynomials p, we see that all absolute values for rationals, except the one at 'infinity', are also non-archimedean.

For a rational number or a rational function α, all its absolute values $|\alpha|_v$, suitably normalized, are linked by a **product formula**: $\prod |\alpha|_v = 1$, which for functions basically means that number of zeros is the same as

1

number of poles, when we take into account multiplicities and the point at infinity.

In 1945, Artin and Whaples showed that a field with notions of absolute values satisfying product formula (see [Art67; Art65] for precise statement and simple proofs) is either a number field (i.e., a finite extension of $\mathbb{Q}$) or a function field (i.e., a finite extension of $k(t)$, for some field k). If k is infinite, eg., $\mathbb{C}$, then since we control only the degree of the remainder when we divide, there will be infinitely many residue classes, in contrast to the number field situation. So the function field situation is even closer to the number field situation, when k is a finite field.

For number fields or function fields, for a set S of inequivalent absolute values (including all the archimedean ones, in the number field case), the ring $\mathcal{O}_S$ of S-integers is a Dedekind domain, so that we have unique factorization of ideals into prime ideals. But in the most interesting case of non-empty finite S, we have good control of the passage from the ideals to elements, exactly when k is finite for the function field case: The unit group (which measures the difference between elements and the principal ideals they generate) is finitely generated and, in fact, of rank $|S|-1$, by the Dirichlet unit theorem then (whereas it is not finitely generated otherwise, because $\overline{\mathbb{F}_p}^*$, $\mathbb{Q}^*$ and $\mathbb{F}_p(t)^*$ are not finitely generated). Similarly the class group (which measures the difference between ideals and principal ideals) is finite, when k is finite, but is infinite, for example, when the function field is of positive genus over algebraically closed field k.

Though we will concentrate on finite k, understanding of how things work in complex function fields is very useful and usually learned early through complex analysis, topology, geometry courses. Fortunately, many of those properties carry over with different proofs to other function fields, as was shown by Weil and others.

Unless we relax it explicitly, from now on, by a function field, we would mean a function field of one variable over a finite field, or more precisely, a finite extension of some $\mathbb{F}_q(t)$.

Function fields and number fields are usually studied together as **global fields** (their various completions are the local fields), since the works of Kronecker, Dedekind and Weber. A lot of basic number theory, including class field theory goes parallel for these fields. [Art67; AT68; CF67; Has49; Wei73] are some of the standard books giving such parallel treatment.

Non-archimedean local fields which are finite extensions of $\mathbb{Q}_p$ or $\mathbb{F}((t))$

for a finite field $\mathbb{F}$, can also be characterized as complete non-discrete fields with discrete valuation topology and with finite residue fields or as locally compact non-discrete with respect to non-archimedean absolute values. Together with $\mathbb{R}$ and $\mathbb{C}$, they are called **local fields** and can also be characterized as locally compact, non-discrete topological fields or as completions of global fields.

Despite these strong similarities in the two cases, there are also some basic contrasts in the number field and function field situations. Let us recall them by just key-words: characteristic zero versus finite characteristic, inseparability, having archimedean absolute value and resulting huge difference between topologies and analysis, having differentiation with respect to elements, usual differentiation killing more than constants.

While the completions for different primes are non-isomorphic for $\mathbb{Q}$, and for number fields, we can get isomorphic completions only for conjugate primes, in function fields, the isomorphism class depends only on the degree of the prime.

Another issue is that while $\mathbb{Q}$ is contained canonically in any number field, in characteristic p, the prime field is $\mathbb{F}_p$, and every function field of characteristic p contains infinitely many copies of $\mathbb{F}_p(t)$. The degree over the prime field is infinite now and there is no function field at the 'bottom'. So usually, we **fix a base** function field, and consider its extensions.

Analogs at the simplest level are (i) rational function field $K = \mathbb{F}_q(t)$ and rational number field $\mathbb{Q}$; (ii) polynomial ring $\mathbb{F}_q[t]$ and integer ring $\mathbb{Z}$; (iii) real number field $\mathbb{R}$, which is the completion of $\mathbb{Q}$ at the usual absolute value, and the Laurent series field $K_\infty = \mathbb{F}_q((1/t))$ (here $1/t$ is a parameter at infinity); and (iv) the complex number field $\mathbb{C}$, which is an algebraic closure of $\mathbb{R}$, and C_∞ defined as the completion of an algebraic closure of K_∞, which is 'smallest' extension which is both algebraically closed and complete.

While the only **automorphism** of $\mathbb{Q}$ (or $\mathbb{Z}$ or $\mathbb{R}$ resp.) is trivial, $\mathbb{F}_q(t)$ (or $\mathbb{F}_q[t]$ or $\mathbb{F}_q((1/t))$ resp.) has non-trivial automorphisms, even if we require them to fix $\mathbb{F}_q$. So when we use t we are usually making a choice. The respective groups are those of mobius transformations, affine transformations (or if sign is fixed, just translations), and a group of local units respectively.

Though $\mathbb{Q}$ has no unramified proper extension, $\mathbb{F}_q(t)$ has many such extensions, such as the constant field extensions $\mathbb{F}_{q^r}(t)$ (like some number fields having infinite Hilbert class field tower).

A finite extension L of a function field K is called **geometric extension**, if L and K have the same field of constants, whereas it is called **arithmetic extension** or **constant field extension**, if L is obtained as a compositum of K and a finite extension of a field of constants of K. While arithmetic and geometry are not that easily separated, the terminology, from the days when the (algebraic) geometry was always done over algebraically closed fields, is standard now and we will follow it.

Since there are no singled out archimedean infinite places, if we need, we can **choose some distinguished places to be considered at infinity**. Analog of **totally real** (totally imaginary respectively) is usually taken to mean totally split at infinite places (only one place above each infinite place respectively). There are many more possibilities of **ramification types at infinity** in the function field extensions in contrast to the number field case and there is a phenomenon of wild ramification too. In a relative situation, the extension may be inseparable, though we can always realize any function field as a separable extension of a rational function field.

For the number fields, the ring of integers are the ring of S-integers where S consists of archimedean valuations. Since there are none in function fields, the same definition will make the ring to be the field of constants and would not be that interesting. So we choose some non-empty finite S to consist of valuations at infinity, by decree. For example, singling out the valuation coming from the usual 'degree in t', leads to the ring of integers $\mathbb{F}_q[t]$ in $\mathbb{F}_q(t)$. Hence, in addition to fixing a base, we also fix some places to be considered at infinity and then the ring of integers in an extension would be the integral closure.

In the first few chapters, we will look at $A = \mathcal{O}_S$ with S a singleton set, thought of as consisting of a (single) place at infinity, analogous to the case of $\mathbb{Z}$ or a ring of integers of an imaginary quadratic field. Then the unit group $A^* = \mathbb{F}_q^*$ consists of roots of unity, just as in the classical case. There are $q - 1$ of them now, rather than 2 in the case of $\mathbb{Z}$. Hence by an '**even**' integer (we refer to usual integers, in this context), we mean a multiple of $q - 1$ and by an '**odd**' integer one that is not 'even'. We will see various analogies to justify this terminology. But note that for $q = 2$, every integer is 'even', in general, $s \to 1 - s$ does not switch 'even' and 'odd' and adding two 'odd' integers need not give an 'even' integer etc.

Classically, the elements ± 1 of $\mathbb{Z}^*$, also enter as signs. In fact, there is a unique non-trivial character (continuous homomorphism) (also called sign) of finite order of the multiplicative group $\mathbb{R}^*$ of the completion of $\mathbb{Q}$ at infinity. Its image is $\mathbb{Z}^*$.

If K is a function field, with $\mathbb{F}_q$ as its field of constants, and with a place ∞ of degree d_∞, then its residue field at ∞ is $\mathbb{F}_\infty = \mathbb{F}_{q^{d_\infty}}$ and its completion K_∞ at ∞ is isomorphic to the Laurent series field $\mathbb{F}_\infty((u_\infty))$, where u_∞ is any uniformizer at ∞, i.e., a generator of the maximal ideal of the valuation ring at ∞. The **signs** now live in $\mathbb{F}_{q^{d_\infty}}^*$ rather than $A^* = \mathbb{F}_q^*$.

Let U_1 denote the one units at infinity, i.e., the units of the valuation ring which are congruent to one modulo the maximal ideal. Then we have

$$K_\infty^* \equiv \mathbb{F}_\infty^* \times u_\infty^{\mathbb{Z}} \times U_1.$$

Since q^n-th power of a one-unit tends to one, as n tends to infinity, 'sign' of a one-unit should be one. Hence, it is natural to make the following definition.

Definition 1.1.1   A sign function sgn : $K_\infty^* \to \mathbb{F}_\infty^*$ is a homomorphism which is the identity on $\mathbb{F}_\infty^*$ and trivial on U_1. We also use the conventional extension sgn$(0) = 0$ often. For $\sigma \in \mathrm{Gal}(\mathbb{F}_\infty/\mathbb{F}_q)$, the composite map $\sigma \circ$ sgn is called a twisting of the sign function sgn.

There are exactly $|\mathbb{F}_\infty^*| = q^{d_\infty} - 1$ choices of sign functions, depending on the choices of uniformizers: Any two are then related by sgn$(x) =$ sgn$'(x)\zeta^{\mathrm{ord}_\infty(x)}$, for some $\zeta \in \mathbb{F}_\infty^*$ (which is just the sign of the ratio of the corresponding uniformizers).

Example 1.1.2   The simplest example which we will use often is when the function field is the field of rational functions in a variable t with coefficients in $\mathbb{F}_q$. If we choose the usual place at infinity corresponding to 'degree in t' or equivalently with uniformizer $1/t$, we get $\mathbb{F}_q[t]$ as the ring of integers, and we will consider t as having sign 1, so that the sign of an element of $\mathbb{F}_q[t]$ is the coefficient of the top degree when expressed as a polynomial in t.

Example 1.1.3   For $K = \mathbb{F}_3(t)$, we now choose degree two prime $1 + t^2 = (1 + it)(1 - it)$ (where $i^2 = -1$) as our point at infinity, so that the ring of integers is $= \mathbb{F}_3[x, y]$ with $y^2 = x - x^2$ with $x = 1/(1 + t^2)$ and $y = t/(1 + t^2)$. Here $d_\infty = 2$, so that $\deg(x) = \deg(y) = 2$ and $v_\infty(x) = v_\infty(y) = -1$. The completion at ∞ is $\mathbb{F}_\infty((1 + it))$, with $\mathbb{F}_\infty = \mathbb{F}_3(i)$ being a field of 9 elements. Developing in terms of uniformizer $u := 1 + it$, we see that

$\text{sgn}(x) = -1/\text{sgn}(u)$ and $\text{sgn}(y) = i\,\text{sgn}(x)$. Any choice $\text{sgn}(u) \in \mathbb{F}_\infty^*$ leads to a sign function, and $\text{sgn}(y) = -i\,\text{sgn}(x)$ is a twisting of sign.

If $d_\infty = 1$, the elements with sgn 1 are called **monic** or **positive**. But when $d_\infty > 1$, in addition to such usage, for some purposes, as we will see, it is better to let the generalization of positive element instead to mean an element of sgn in some chosen representatives of $\mathbb{F}_\infty^*/\mathbb{F}_q^*$. We will make clear which usage is made.

We will often focus further on $A = \mathbb{F}_q[t]$ in characteristic p. In this case, many analogies are stronger and more familiar. We then have analogs of division algorithm, principal ideal theorem, unique factorization theorem, and basic congruence theory, with essentially the same proofs.

For example, the order of the multiplicative group $(A/aA)^*$ is given by analog ϕ of Euler's ϕ function $\phi(a) := \prod \text{Norm}\,(\wp_i)^{n_i-1}(\text{Norm}\,(\wp_i) - 1)$, if the ideal aA has prime factorization $\prod \wp_i^{n_i}$.

So we have analog of **Fermat-Lagrange theorem**: $b^{\phi(a)} \equiv 1 \mod a$, for $b \in A$ prime to a.

But we can do much better now: Rather than $a^{\text{Norm}(\wp)(\text{Norm}(\wp)-1)} \equiv 1 \mod \wp^2$ that it implies we have even $a^{p(\text{Norm}(\wp)-1)} \equiv 1 \mod \wp^p$, as follows by taking p-th powers of the congruence obtained modulo $\wp$ by Fermat. So there are no primitive roots modulo $\wp^n$ in general, in contrast to the p^n case for usual (odd) primes p.

In other words, **group structures** of abelian groups A/aA and $(A/aA)^*$ are quite different from those of their counterparts $\mathbb{Z}/n\mathbb{Z}$ and $(\mathbb{Z}/n\mathbb{Z})^*$: While $\mathbb{Z}/n\mathbb{Z}$ is a cyclic group, the cyclic A-module A/aA is a $\mathbb{F}_p$-vector space of dimension $\log_p(q)\deg(a)$ and hence of $(p,\cdots,p)$-type. By the Chinese remainder theorem, both are direct products of their prime power counterparts according to the prime power decomposition of a. Hence it suffices to look at $a = \wp^n$ now, where $\wp$ is a prime of degree d. The multiplicative group of the field $A/\wp A$ is cyclic (i.e., there is a primitive root modulo $\wp$) of order $q^d - 1$ prime to p. The general structure of $(A/\wp^n A)^*$ is quite complicated: The natural homomorphism $(A/\wp^n A)^* \to (A/\wp A)^*$ has kernel a p-group, with typical non-identity element written base $\wp$ as $1 + \sum_{i=s}^{n-1} a_i\wp^i$ (with $a_s \neq 0$) having order p^j, where $p^{j-1}s < n \leq p^j s$. Hence this is a p group of order $q^{d(n-1)}$ and exponent $p^{\lceil \log_p(n)\rceil}$.

In particular, for $n > 1$, $(A/\wp^n A)^*$ is cyclic exactly when $q = p$, $d = 1$, $n = 2$ or when $q = 2$, $d = 1$, $n = 3$.

Note that $A = \mathbb{F}_q[t]$, which is a (countably) infinite dimensional $\mathbb{F}_q$-vector space, is direct sum of countable copies of $\mathbb{Z}/p\mathbb{Z}$, whereas $A_\wp$, which is isomorphic to a power series ring over $\mathbb{F}_{q^{\deg(\wp)}}$ and thus again a vector space, is a direct product of countable copies of $\mathbb{Z}/p\mathbb{Z}$. Each element is thus a p-torsion. For $A_\wp^*$, on the other hand, the only p-torsion is 1. It is a direct product of the cyclic group consisting of $q^{\deg(\wp)} - 1$-th roots of unity and the one unit group, each element of which can be raised to p-adic power. In fact, the one unit group is a countable product of $\mathbb{Z}_p$'s on the basis $1 - w_i u_\wp^j$, where w_i runs through a basis of $A/\wp$ over $\mathbb{F}_p$, $u_\wp$ is a uniformizer at $\wp$ and j runs through positive integers not divisible by p, as we can kill all jp^n-th powers of $u_\wp$ (inductively in j) occurring in the one unit. For details, see for example [GK88]. We leave the description of topology in these identifications to the reader.

Hence we do not have straight analog of $\mathbb{Z}_p^* \cong (\mathbb{Z}/p\mathbb{Z})^* \times \mathbb{Z}_p$ and in Chapter 3, we will naturally get 'cyclotomic' extensions with Galois group $A_\wp^*$ (the fact that such extensions exist is already clear from the class field theory description of the maximal abelian Galois group which we will recall later in this Chapter), whereas by above, $A_\wp$ is a quotient of $A_\wp^*$ of infinite index.

Recall Wilson's theorem that $(p-1)! \equiv -1 \mod p$, for p a prime. The proof and the standard generalization is that, if G is a finite abelian group, then pairing the elements with their inverses, we see that the product of all the elements of G is the product (empty product is 1) of all its elements of order two; and when there is more than one element of order two, grouping all such elements in disjoint triples a_1, a_2, $a_1 a_2$, we see that this product is one. So the product is the unique element of order two, if there is one and if there is none, the product is one.

Let us record for future use **analog of the Wilson theorem** and a generalization in our situation: We see that the product of all elements of $(A/\wp A)^*$ is -1 (which equals 1, if $p = 2$). Now we claim that the product of elements of $(A/\wp^n A)^*$ is -1, unless $q = 2$, $\deg \wp = 1$ and $n = 2$ or 3: This is clear if $p \neq 2$, when $(A/\wp A)^*$ has -1 as its unique element of order two whereas the rest is a p-group with no such element. If $p = 2$, since the elements $1 + \theta\wp^{n-i}$, with θ nonzero mod $\wp$, have order two, if $n \geq 2i > 0$, having unique such element forces us to be in the exceptional cases stated above.

As far as the basic **algorithms** are concerned, not only that naive addition, multiplication is a little simpler because of no carry-overs, but for example, while it is quite hard to calculate the square-free part of a large integer; for a polynomial in characteristic p, the p-th power is visible and once we remove it by just dividing the exponents by appropriate p-power, to get the square-free part, we just divide by the greatest common divisor (obtained by very efficient Euclidean algorithm) of the polynomial with its derivative.

In fact, while there is no polynomial time fast algorithm known for integer factorization, the polynomial **factorization** algorithm of Berlekamp is very efficient: Given $p(t) = p_1(t) \cdots p_r(t) \in \mathbb{F}_q[t]$ of degree d, with p_i distinct irreducible, a non-trivial factor, if any, is obtained in less than (see [Knu69; BS96]) constant times d^3 bit operations, as the greatest common divisor of $p(t)$ with $f_i(t) - \theta$, as θ runs through the elements of $\mathbb{F}_q$ and $f_i(t)$ through linearly independent non-constant vectors in the null-space of the linear transformation 'q-th power minus identity' on $\mathbb{F}_q[t]/p(t)$, obtained by triangularizing the matrix. This works since p_j divides p which divides $f_i^q - f_i = \prod(f_i - \theta)$ and that there are q^r solutions f_i to $f_i^q \equiv f_i \mod p$ of degree less than d, because they can be characterized as chinese remainder solutions to all possible $f_i \equiv \theta_{ij} \mod p_j$, with $\theta_{ij} \in \mathbb{F}_q$.

If all irreducible factors of $p(t)$ are known to have different degrees, complete factorization is more simply achieved with the same complexity bound by computing greatest common divisors of square-free part of $p(t)$ with $t^{q^i} - t$, for $i \leq d/2$, because $t^{q^i} - t$ is the product of all monic irreducibles of degree dividing i.

In the number field situation, we talk about the **class number** of a field to mean the class number of its ring of integers, by abuse of language. For function fields, we may take as class group the full Pic^0, which is the group of degree zero divisors modulo the principal divisors, and its cardinality h as the class number, or we can choose some S as above and look at the divisors supported away from it, to get $\mathrm{Pic}^0(\mathcal{O}_S)$ and h_S.

In function fields, we lose strong **order relation** that we have for real numbers (we will see some implications for multizetas in Chapter 5), lose the uniqueness (up to a constant of integration) of integration, but series convergence is easy because it suffices that terms tend to zero, because of non-archimedean nature of absolute values.

When we consider **local fields**, in addition to analogies, we have **tools for going back and forth** between characteristic zero and nonzero cases for certain kind of problems:

First, there is the important work of J. Ax and S. Kochen proving that truth of first order sentences in the ultraproduct of all $\mathbb{Q}_p$'s or in the ultraproduct of all $\mathbb{F}_p((t))$'s (for any nonprincipal ultrafilter on the set of primes p) is the same. In particular, any first order statement true for all fields $\mathbb{F}_p((t))$ is true for $\mathbb{Q}_p$ for all except possibly finitely many p. For example, consider the statement that 'if $n > d^2$, a homogeneous polynomial of degree d in n variables over K has a non-trivial zero in K'. Artin conjectured this for $K = \mathbb{Q}_p$, Lang proved it for $K = \mathbb{F}_p((t))$, which implied that Artin's conjecture is true for almost all p. Later, in fact, Terjanian found a counterexample to Artin's conjecture, with $p = 2$, $d = 4$ and $n = 18$. See eg., [Che76].

Interestingly enough, though the laurent series fields are easier in some sense because there is no carry-over of digits in the usual operations; from the point of view of logic and model theory, they are still not understood, whereas real and p-adic fields are understood, in the sense that they are known to form a complete, model-complete theory. See [Che76, pa. 65], and [Kuh01].

Second, there is work of Krasner, Deligne, Kazdan etc., showing precise versions (in terms of equivalence of certain categories) of rough principle that for representation of absolute Galois groups or of reductive groups, the theory over a local field of characteristic $p > 0$ is the 'limit' of the theories over local fields in characteristic zero, when their absolute index of ramification tends to infinity. See, for example, de Shalit's paper in [G+92] for references and application proving Artin-Hasse explicit reciprocity laws for p-adic fields from Schmidt-Witt explicit reciprocity laws for local fields in finite characteristic.

1.2 Genus and Riemann-Roch theorem

Now let us recall some classical tools and theorems at function field level, without worrying about distinguished places. Many of the main theorems work in a setting with any field of constants.

Function field is just a birational concept, but usually by the curve X associated to the function field K, we mean the complete, non-singular curve X having K as its function field.

Resolution of singularities of a singular model of a curve is achieved, by a finite number of appropriate quadratic transformations or by normalization: patching the integral closures in quotient field of the affine co-ordinate rings for affine patches.

We can talk about the degree of the function field, only in a relative sense, in contrast to the number field situation. (There is a concept of gonality, which is the smallest positive degree of an element of the function field.) The **genus** of a function field, on the other hand is an absolute concept. For function field over $\mathbb{C}$, the genus is the number of holes or handles in the Riemann surface corresponding to the associated curve. For general function fields, it is defined as the dimension of $H^1(X, \mathcal{O}_X)$, or equivalently, by Serre duality, to be the number of linearly independent holomorphic differentials.

Example 1.2.1   For a plane non-singular curve of degree d in character-istic zero, the genus is $(d-1)(d-2)/2$. For example, if $f(x)$ is a poly-nomial of degree d without multiple roots, then the corresponding Thue curve $F(x, y) = z^d$, where $F(x, y) = y^d f(x/y)$ is homogenization of f, is such a curve: Since $xF_x + yF_y$ is d times F, the affine part $F = 1$ is non-singular, and since we can not have $f = f_x = 0$, the points at infinity are also non-singular. For a 'super-hyper-elliptic' curve $y^k = f(x)$ with f of degree d, prime to k, the genus is $(d-1)(k-1)/2$. A basis of holomorphic differentials is $x^a y^b \, dx/F_y$, $a, b \geq 0$, $a + b \leq d - 3$ in the Thue case; and $x^i dx/y^j$, $0 < j < k$, $(i+1)k + 1 \leq jd$ in the super-hyper-elliptic case. In characteristic p, these calculations are still valid, if p is prime to d (k respectively).

Genus does not change in separable constant field extensions, in par-ticular, for constant field extensions of function fields over finite fields. In the relative situation, for the genus computation, the most useful general formula is the

Theorem 1.2.2   *(Riemann-Hurwitz formula) For a separable, finite, ge-ometric extension L over K,*

$$2g_L - 2 = (2g_K - 2)[L : K] + \deg(R),$$

where the ramification divisor R is just the different of L/K.

When there is only tame ramification, the degree of the ramification divisor is more simply expressed as $\sum(e_P - 1)$, in terms of the local ram-ification indices e_P at the places P of L. The contribution from P is more than $e_P - 1$, if P is wildly ramified. If we have a Galois extension L/K with the Galois group G, then the contribution at P to R is the sum of the contributions for $1 \neq g \in G$ of $\sup\{i : g \in G_{P,i}\} + 1$, where $G_{P,i} := \{g \in G : g \text{ acts trivially modulo } \pi_P^{i+1}\}$ are the usual ramification groups at P.

Castelnuova inequality bounds the genus of the compositum L of K_1, K_2, all with the constant field k: If $d_i = [L : K_i]$, then $(2g_L - 2) \le d_1(2g_{K_1} - 2) + d_2(2g_{K_2} - 2) + 2d_1 d_2$. This gives $g_L \le (d_1 - 1)(d_2 - 1)$ in the case where we have $L = k(x,y)/p(x,y)$ and $K_1 = k(x)$, $K_2 = k(y)$.

Generalizing the well-known fact that there exists a rational function with prescribed zeros and poles as long as number of zeros is the same as the number of poles, and then it is uniquely defined up to a scalar multiple, the Riemann-Roch theorem tells us what happens for general genus function field. Write, as usual, $\mathcal{L}(D)$ for the space of functions f with $(f) + D \ge 0$ (together with zero).

Theorem 1.2.3 *(Riemann-Roch)*

$$l(D) := dim \ \mathcal{L}(D) = \deg(D) + 1 - g + l(K - D),$$

where K is a canonical divisor, i.e., divisor of a differential. In particular, $\deg(K) = 2g - 2$, $l(K) = g$ *and* $l(D) = \deg D + 1 - g$, *if* $\deg D > 2g - 2$.

It is a good exercise to verify the genus calculations in the Example 1.2.1 again by applying each of these theorems.

Associated, in a functorial manner, to a curve X of genus g over a perfect field k, there is a **Jacobian variety** J, which is an abelian variety of dimension g, whose k'-rational points, for $k \subset k' \subset k^{\mathrm{sep}}$, are in bijection with $\mathrm{Pic}^0(X_k\mathrm{sep})^{\mathrm{Gal}_{k'}}$. We have a morphism (**Abel map**) $\alpha : X \to J$ satisfying the universal property for maps of X to abelian varieties. At the level of points, with the above identification, it is an injection sending P to the class of $P - \infty$, for some point $\infty \in X(k)$, at least when $X(k)$ is non-empty. The S-valued points of J are then the isomorphism classes of invertible sheaves $\mathcal{L}$ over $X \times S$, with trivial restriction at $\infty \times S$ and of degree 0 relative to S, and the Abel map $\alpha : X \to J$ is the X-valued point corresponding to the invertible sheaf $\mathcal{O}_{X \times X}(\Delta - X \times \infty - \infty \times X)$, where Δ is the diagonal of $X \times X$.

If $\mathcal{U}$ is the universal line bundle on $X \times J$ of degree 0 trivial at ∞, then the **theta bundle** $\Theta := \wedge^N p_{2*}(\mathcal{U} \otimes p_1^* \mathcal{O}_X((N + g - 1)\infty))^{-1}$ is an invertible sheaf on J independent of $N > g-1$ up to an isomorphism unique up to sign. Let F be algebraically closed. It follows that Θ is the set of isomorphism classes of invertible sheaves $\mathcal{L}$ over X of degree 0, trivial at ∞ such that $h^0(\mathcal{L}((g-1)\infty)) = h^1(\mathcal{L}((g-1)\infty)) > 0$ and since all such $\mathcal{L}$ are of form $\mathcal{O}_X(x_1 + \cdots + x_{g-1} - (g-1)\infty)$, it follows that Θ is the sum of $g-1$ copies of $\alpha(X)$. For classical description and properties of this Abel's map in terms of periods, and relations to theta divisor, theta function, special divisors with the corresponding maps of powers of X to J, we refer to the literature (eg., [And94] and references therein). The discussion of Shtukas in 7.7 and 8.2 can be appreciated better with this knowledge, though we will not use it explicitly.

In several contexts, q^{2g-2} and (the absolute value of) the **discriminant** are good analogs: For example, for unramified extensions, both get raised to the extension degree power. More generally, the fact that the discriminant of the extension is the degree power of the base discriminant times the relative norm of the relative different is just an (multiplicative) analogue of the Riemann-Hurwitz formula. In fact, to transport useful differential descriptions in the function field case, the derivations [Wei79, I, pa. 329] were considered and unified description of the different was achieved. Other important (and related) contexts where $|D|$ and q^{2g-2} occur in parallel fashion are computation of norm of differential idele [Wei73, Chapter VII, Prop. 6] and measure of A_K/K [Wei73, Chapter V, prop. 7 and Chapter VI, Cor. 1]. While consideration of $g = 0$ and $g = 1$ cases show a big contrast to the fact that for number fields discriminant is an integer and that it is (in absolute value) greater than one, except for $\mathbb{Q}$, or it grows with degree; there is (partial) analog of Hermite-Minkowski theorem that there are only a finitely many number fields of a given discriminant: Given $q^{2g-2} \neq 1$, there are only finitely many corresponding q's and g's and given g and q, there are only a finitely many corresponding function fields (see eg., [BP89, pa. 124]).

We will see more examples of this analogy in the next section. Weil [Wei79, I, pa.238] proposes a slight variant of this analogy with genus.

Given a point P, Riemann-Roch shows that there are g **gaps** in the possible orders of poles that functions can have at P and they are between 1 and $2g - 1$. For Riemann surfaces, with $g \geq 2$, we call P a Weierstrass point, if these gaps are other than $1, 2, \cdots, g$ or equivalently, if there is a holomorphic differential with zero of order at least the genus. For example, for hyperelliptic curves $y^2 = f(x)$, of genus at least two, the Weierstrass points correspond to 'branch points' which are infinity and those corresponding to roots of $f(x)$, and have gap sequence of $1, 3, \cdots, 2g - 1$. In characteristic zero, there are only finitely many (between $2g + 2$ to $g^3 - g$ of these, for $g \geq 2$) Weierstrass points, but in finite characteristic, all points can be such, with the definition as above. In 1939, F. K. Schmidt showed that nonetheless, for a given curve, there are g positive integers (between 1 and $2g-1$) such that for all but finitely many points they are the first g non pole-orders. (Closely related is analog, due to Schmidt, of the usual linear independence criterion for $x_1, \cdots, x_n \in K_\infty$ over k as the non-vanishing of their Wronskian obtained by replacing derivatives by Hasse derivatives and replacing successive differentiation by any. See [GV87] for generalization and references.) These finitely many exceptions are then called the **Weierstrass points.**

Example 1.2.4 Schmidt's example: $y^p + y = x^{p+1}$ in odd characteristic p, has generic order sequence $a+bp$, with $a+b \leq p-2$. The $\mathbb{F}_{p^2}$-points are all Weierstrass points and their number is the maximal permitted by the Weil bound.

In this connection, see Stohr-Voloch proof [SV86; Gol03] of Riemann hypothesis for curves.

1.3   Zeta function and class group

Following Artin, we can associate a zeta function to $\mathcal{O}_S$, a special case of the zeta function associated to a scheme, by

$$\zeta_{\mathcal{O}_S}(s) = \prod (1 - \mathrm{Norm}(\wp)^{-s})^{-1} = \sum \mathrm{Norm}(D)^{-s} \in \mathbb{C}, \; Re(s) > 1,$$

where the sum is over positive (effective non-zero) divisors. Here for global fields, as usual, by norm of a non-archimedean place, we just mean the cardinality of the residue class field and we extend norm multiplicatively to divisors. If we take S to be empty, we get (by convention) the zeta function for the function field (i.e., for the complete curve).

Example 1.3.1 Let $A = \mathbb{F}_q[T]$. This is a principal ideal domain, so as there are q^n monic polynomials of degree n, there are q^n ideals of norm q^n, so that

$$\zeta_A(s) = \sum_{I \neq 0} 1/\mathrm{Norm}(I)^s = \sum_{n=0}^{\infty} q^n/q^{ns} = 1/(1 - q^{1-s}) = 1/(1 - qt)$$

with the usual convention that $t = q^{-s}$. Putting back the factor at infinity, which is a place of degree one, so of norm q, we see that the zeta function for $\mathbb{F}_q(T)$ is $1/((1 - qt)(1 - t))$.

This zeta has no zeros in contrast to the Riemann zeta function. There are simple poles at $s = 1 + 2\pi i n/\log(q)$, with residue $\log(q)$ in this case, in contrast to the only pole at $s = 1$ in the classical case. But in fact the zeta function should really be thought of as function of characters (see Tate's thesis in [CF67]): We use the parameter s for $n \to n^{-s}$ in the classical case. If we do the same here to make analogies more transparent, we have to keep in mind that the norm set now is $q^{\mathbb{Z}}$ and the image t of q determines the character, and writing the image as $t = q^{-s}$ gives several values of s corresponding to one character and $\zeta(s)$ is a periodic function of period $2\pi i / \log(q)$.

The prime number theorem, which is closely connected to the nature of zeros and poles on line Re $(s) = 1$, has slightly different analog and it is much simpler to prove. Let us see it in this case:

Since $x^{q^n} - x$ is the product of all monic irreducible polynomials in x of degrees dividing n (a fact, which is immediate from the defining polynomials of finite fields, or equivalently by Fermat's little theorems in this case and multiplicity one calculation using derivatives), we see that eg. if n is prime, there are $(q^n - q)/n$ primes of degree n. For general n, this number is $\sum_{d|n} \mu(n/d)q^d/n$ by the Mobius inversion formula. This is immediately seen to be $q^n/n + O(q^{n/2}/n)$ and this asymptotics (with excellent Riemann hypothesis type error control, and even better if say n is odd) is sometimes stated as analogue of **prime number theorem.**

If we let $\pi(x)$ to be the number of primes of norm not more than x, then, in fact, there is no asymptotic formula, because for the special x's which are q powers, the number jumps wildly. In fact, summing the formula we got above, we get the asymptotic formula

$$\pi(q^{n+1} - 1) = \pi(q^n) \sim (q/(q-1))(q^n/\log_q(q^n)).$$

For a general function field K, with $t = q^{-s}$ substitution, we have

$$\zeta(s) = Z(t) = \prod_P (1 - t^{\deg(P)})^{-1} = \sum_D t^d = h \sum_{[D]} \frac{q^{l(D)} - 1}{q - 1} t^d$$

(where the product is over primes P, the first sum is over effective divisors D of degree d, and the last sum is over divisor classes $[D]$ of such D) by counting in each of the h classes and using the non-trivial fact (which can be deduced eg., by pole counting comparison with zeta for constant field extension) that there is a divisor of degree one. This further simplifies to a rational function

$$Z(t) = L(t)/(1-t)(1-qt)$$

in t by the Riemann-Roch theorem: The series boils down to a geometric series for $d > 2g - 2$ thus giving

$$L(t) = \frac{(1-t)(1-qt)}{q-1}[h(\frac{q^g t^{2g-1}}{1-qt} - \frac{1}{1-t}) + \sum_{0 \leq \deg(C) \leq 2g-2} q^{l(C)} t^{\deg(C)}]$$

which is seen to be a polynomial of degree $\leq 2g$. It has integral coefficients, since $L(t) = (1-t)(1-qt)Z(t)$. Replacing D by $K-D$ for degrees $\leq 2g-2$ and calculating straight with the remaining geometric series part we see the **functional equation** $Z(t) = q^{g-1} t^{2g-2} Z(1/qt)$ (this implies that the degree of $L(t)$ is $2g$), or equivalently $\zeta(s) = q^{(2g-2)(1/2-s)} \zeta(1-s)$, as a consequence of Riemann-Roch (Serre-Poincaré) duality. For the number fields, basically a Poisson formula in adelic setting plays this role (See Tate's thesis [CF67]) and in fact, applied to the characteristic functions of suitable subsets, it implies the geometric Riemann-Roch formula.

Analogy with the functional equation for the Dedekind zeta function becomes apparent after careful analysis of 'local factors at infinity' as in Tate's thesis [CF67, Pa. 346] or [Wei73, Chapter VII, Theorem 3], and we see again that q^{2g-2} is a good analog of the absolute value of the discriminant of the number field.

In the functional equations of L-functions for characters in this theory, the Gauss sums show up, as in the number field case, whereas the gamma function does not show up as there is no archimedean places now.

Recall that in the Artin-Tate-Iwasawa approach [CF67; Wei73; RV99; GGPS90], where we deal with zeta functions as integrals (of characters) over ideles rather than sums over ideals to facilitate the local-global approach, the functional equations of the Zeta functions arise as a result of duality (Poisson formula or Riemann-Roch). Weil [Wei79, III, pa. 159] interpreted the functional equations as product formulas. A different interesting, but as yet non-rigorous, product formula interpretation based on the same ingredients, but with motivations in string theory, is found in [FW87].

From the formula for $L(t)$ above, we get the **class number formula**:

$$L(1) = h = |J\ (\mathbb{F}_q)|.$$

For the Dedekind zeta function, the usual class number formula identifies the leading term of $\zeta(s)$ at $s = 0$ as $-(hR/e)s^{r_1+r_2-1}$. In our case, if we take S consisting of one place of degree δ say, then the only units are $q - 1$ roots of unity. So the replacement for $r_1 + r_2 - 1$ (= the rank of unit group) is 0, and the regulator is 1. Then $\zeta_{\mathcal{O}_S}(t) = \zeta_K(t)(1 - t^\delta)$, so that $\zeta_{\mathcal{O}_S}(0) = -L(1)\delta/(q-1) = -h_S/(q-1)$ gives a good analog of the Dedekind formula.

Weil proved the **Riemann-hypothesis** part of the Weil conjectures in the special case of curves (for hyperelliptic curves this was conjectured by Artin) that the absolute values of reciprocal roots of $L(t)$ are $q^{1/2}$, so that zeros s of $\zeta_K(s)$ have real part 1/2. In the genus one case of elliptic curve E, where we have advantage of natural group structure on points, this was proved by Hasse as follows: Given endomorphism ϕ of degree d, the degree of endomorphism $m\phi + n$ is easily seen to be $dm^2 - \text{Tr}(\phi)mn + n^2$, and since it is non-negative, we have $\text{Tr}(\phi)^2 \leq 4d$. Now $\phi = \text{Frob}_q$ has degree q, so applying this to separable (as is clear from its differential) endomorphism $\phi - 1$, whose kernel corresponds to the $\mathbb{F}_q$-rational points, we see the Riemann hypothesis as stated above or equivalently that $||E(\mathbb{F}_q)| - q - 1| \leq 2q^{1/2}$. In the general case, developing the arithmetic of the Jacobian group-variety, Weil generalized to prove Riemann hypothesis and showed that $L(t)$ is the characteristic polynomial of Frobenius acting on suitable H^1 (or on l-adic Tate modules of the Jacobian) and $1-t$ and $1-qt$ are characteristic polynomials for H^0 and H^2. See the standard books referred above and references there for several proofs, eg., by this Lefschetz fixed point theorem approach or by using Castlenuova inequality for correspondences. Some more elementary proofs by Stepanov, Schmidt, Bombieri or Stohr-Voloch give structurally less, but sometime stronger information in special cases.

Various conjectures in number theory which are known only assuming Riemann hypothesis are thus frequently proved in function field case using this result. For example, for proof by Billharz of Artin's conjecture on primitive roots, see [Ros02, Chapter 10]. There have been various attempts, so far unsuccessful, to transport the ideas of the proofs of the function field Riemann hypothesis to the number field case.

To use good structural geometric results, we take algebraic closure of the finite field of constants of the function field. Since the constant field extensions are just the extensions obtained by adjoining the roots of unity, to get a better analogy between l-power order torsion in the Jacobian on the function field side and the l-Sylow subgroup of the class group on the number field side, Iwasawa [Iwa69] (see also [Wei79, I, pa. 298]) considered the tower of number fields obtained by adjoining l-power roots of unity or more generally $\mathbb{Z}_l$-extensions and developed the Iwasawa theory which, for example, established and conjectured in various contexts, connection between l-adic L-functions with characteristic polynomials arising from Galois modules considered by analogy and formulas for class number growth in such towers. We will come back to such questions later.

Tate gave an analog of the Stickelberger theorem on ideal class annihilators using this machinery as follows: Let K' be a geometric extension over a function field K, with an abelian Galois group G. Let $Cl(K')$ be the class group of K' and let $\theta(T) = \sum_{\sigma \in G} Z(\sigma^{-1}, T)\sigma$ where Z is the partial zeta function for σ. Then the Stickelberger element is $\theta = \theta(1)$. Analog of **Stickelberger theorem** says that

Theorem 1.3.2 $(q-1)\theta \in \mathbb{Z}[G]$ *kills* $Cl(K')$.

Proof. We give a sketch (see [Ros02; GS85] for more details and [Tat84] for details on the general case, due to Tate and Deligne) of the proof: Class group is just the group of $\mathbb{F}_q$-rational points of the Jacobian (of the corresponding curve), i.e., the part of the $\overline{\mathbb{F}_q}$-points of the Jacobian where the Frobenius F acts as the identity. Now by fundamental results of Weil mentioned above, the L-function of a character of G is the characteristic function of the Frobenius on the corresponding component of the Jacobian or rather the Tate module, and hence it kills the component when $T = F$ by the Cayley-Hamilton theorem. Hence $\theta(T)$, which is just a linear combination of L-functions with projection to the components operators, kills the class group when $T = 1$. (One needs $(q-1)$ factor to clear out the denominators to get polynomials, when one makes this sketch precise.) $\square$

Madan [Mad69; Mad70] proved that if F is a finite Galois extension of a function field K, then h_K divides h_F: After reduction step reducing to a simple cyclic extension of degree prime to characteristic, we use the decomposition of the zeta functions into product of L-functions and then use the class number formula and integrality of L-functions etc. in the first proof and class field theory for the second.

Since $J(\mathbb{F}_q)$ is a subgroup of $J(\mathbb{F}_{q^n})$, the divisibility follows for constant field extensions. In fact, in correspondence with Voloch about getting a possible geometric proof without Galois hypothesis, the author learned from him the following nice proof for geometric extensions: If Y and X are the curves corresponding to F and K, then the jacobian of X is (up to isogeny) a factor of that of Y. So the Tate module is a factor, so that the corresponding characteristic polynomial is a divisor, and hence h_K divides h_F.

The Euler product representation implies that zeta has no zeros at s with real part greater than 1, and hence the reciprocal roots α_i of $L(t) =$

$\prod(1 - \alpha_i t)$ have absolute values at most q, so that $h \leq (q+1)^{2g}$. In fact, we know by the Riemann hypothesis analog that $|\alpha_i| = \sqrt{q}$ implying

$$(\sqrt{q} + 1)^{2g} \geq h \geq (\sqrt{q} - 1)^{2g}.$$

This implies, for example, that if $q > 4$, the class number can not be one unless the genus is zero i.e., the case of rational function fields. For a family of function fields of positive genus and with $q \to \infty$, the inequality implies that $\log(h)/g \log(q) \to 1$, analogous to Brauer-Siegel asymptotics, if we note that q^{2g-2} is analog of discriminant here. In [MM80], building on a slightly weaker result by Inaba, analog of Brauer-Siegel theorem is proved, namely $\log(h)/g \log(q) \to 1$, as $m/g \to 0$, where K of genus g and class number h and gonality m (i.e., m is the least integer such that there is x with $[K : \mathbb{F}_q(x)] = m$) runs through function fields over $\mathbb{F}_q$.

More careful analysis of possible distribution of Frobenius eigenvalues has led to the result [LMQ75] that apart from the rational function fields (one for each q) there are only 7 function fields with $h = 1$. (Thus while Honda-Tate results show that there are infinitely many isogeny classes of abelian varieties over $\mathbb{F}_q$ with only one $\mathbb{F}_q$-rational points, this is not true for Jacobians.) While $\mathbb{F}_q[t^{1/n}]$'s and $\mathbb{F}_{q^n}[t]$'s give trivial examples of infinitely many integer rings over $\mathbb{F}_q[t]$ of class number one, and there are many other examples if you allow to vary q, the simplest analogue of Gauss conjecture that there are infinitely many real quadratic quadratic fields with class number one would be that given $\mathbb{F}_q(t)$ and usual ∞, there are infinitely many quadratic extensions in which ∞ splits and the class number of the integral closure of $\mathbb{F}_q[t]$ in them is one. This is still open. See [LV00] for discussion and nice results along this line.

The zeta function also encodes $N_n := |X(\mathbb{F}_{q^n})|$, the number of $\mathbb{F}_{q^n}$-rational points on the curve X corresponding to the function field K with the field of constants $\mathbb{F}_q$:

$$Z(t) = \exp(\sum_{n=1}^{\infty} \frac{N_n t^n}{n}).$$

This is seen as follows: Let a_d be the number of prime divisors of degree d on X. Since each prime of degree d contributes a Galois orbit of size d, we have $N_n = \sum_{d|n} d a_d$, so that the logarithmic derivative of the Euler product $Z(t) = \prod(1 - t^m)^{-a_m}$ is $\sum m a_m t^m/(t(1 - t^m)) = \sum(\sum_{d|n} d a_d)t^{n-1} = \sum N_n t^{n-1}$.

Given $L(t) = \sum a_i t^i = \prod (1 - \alpha_i t)$, we have $N_n = q^n + 1 - \sum \alpha_i^n$. On the other hand, computation of first g values of N_n are sufficient to calculate $L(t)$: By Newton's identities relating the elementary symmetric functions to power functions $S_n := N_n - (q^n + 1)$,

$$L'(t)/L(t) = \sum \frac{-\alpha_i}{1 - \alpha_i t} = -\sum \alpha_i \sum_{n=0}^{\infty} (\alpha_i t)^n = \sum_{n=1}^{\infty} S_n t^{n-1}$$

we get the recursion relations $i a_i = \sum_{j=1}^{i} S_j a_{i-j}$ for $1 \leq i \leq g$. This together with $a_0 = 1$ and $a_{2g-i} = q^{g-i} a_i$ coming from functional equation, determines all the coefficients.

From the Riemann hypothesis, we immediately get $|X(\mathbb{F}_q) - q - 1| \leq \lfloor 2g\sqrt{q} \rfloor$. This is best possible, if we fix X and replace q by q^n letting n tend to infinity. Natural questions in error correcting codes theory, via Goppa codes associated to curves [Sti93] and some data, led to questions of what happens when you let q fixed and let g tend to infinity.

Using other parts of Weil's theorem, such as duality, Serre showed [Sti93] by the following simple nice argument that the bound can be improved to $g\lfloor 2\sqrt{q} \rfloor$: Order α_i's so that $\overline{\alpha_i} = \alpha_{i+g} = q/\alpha_i$, for $1 \leq i \leq g$. Consider the g totally positive algebraic integers $\gamma_i := \lfloor 2\sqrt{q} \rfloor + 1 + \alpha_i + \overline{\alpha_i}$ ($\delta_i := \lfloor 2\sqrt{q} \rfloor + 1 - (\alpha_i + \overline{\alpha_i})$ respectively). The absolute value inequality claimed then follows by the two inequalities that the arithmetic mean of these g numbers is at least their geometric mean, which is at least one, as their product is at least one (as it is in $\mathbb{Z}$ by Galois invariance).

Drinfeld-Vladut [VD83] exploiting beautifully and fully an idea of Ihara, showed that simple high school algebra inequalities, using Weil's theorem applied to all $X(\mathbb{F}_{q^n})$'s together with $|X(\mathbb{F}_{q^n})| \geq |X(\mathbb{F}_q)|$ gives that

$$A_q := \limsup |X(\mathbb{F}_q)|/g \leq \sqrt{q} - 1,$$

as the genus g tends to infinity. This is known to be (using supersingular points, which are defined over quadratic extensions, on elliptic modular or Drinfeld modular curves [Iha81; TVZ82; GS95; Elk01]) the best possible when q is a square. But for any non-square q, A_q is still unknown. There are some bounds known using techniques, for example, from the class field theory.

For the zeta function, there are no trivial zeros and there are $2g$ non-trivial zeros (in t-variable) each leading to (in s variable) infinitely many related by translation by $2\pi i n / \log q$. So there are about $2g \log(q) T/(2\pi)$ of them with imaginary part between 0 and T in contrast to $nT \log(T)/(2\pi)$

asymptotics we get for the Dedekind zeta function for the number field of degree n. On the other hand, there is work of Montgomery, Dyson, Odlyzko, Hejhal, Berry, Katz, Sarnak showing statistics parallel of the distribution of zeros in the global fields case to random matrix theory. In recent work of Katz and Sarnak [KS99], observed and conjectured statistics of the zero distribution of the zeta and L functions in the number field case is understood by using the results for those for curves over function fields (considered as family over finite fields and using Deligne's refined equidistribution results on Weil's conjectures).

1.4   Class field theory and Galois group

Let us start with a very transparent proof by F. K. Schmidt of n-th **power reciprocity** for polynomials, where $n|q-1$. As $|\mathbb{Z}^*| = 2$, whereas $|\mathbb{F}_q[t]^*| = |\mathbb{F}_q^*| = q-1$, this is an analog of quadratic reciprocity for $\mathbb{F}_q[t]$: The n-th power residue symbol $(a/p)_n \in \mathbb{F}_q^*$ is defined as usual. (Those familiar only with Legendre quadratic symbol, can restrict to q odd and $n = 2$.) We have $(a/P)_n \equiv a^{(\mathrm{Norm}(P)-1)/n} \bmod P$ as usual, for prime P not dividing a.

Now let P and P' be irreducible monic polynomials of degree d and d' respectively. Write $P(t) = \prod_0^d (t - \alpha^{q^i})$ and $P'(t) = \prod_0^{d'} (t - \alpha'^{q^{i'}})$.

Now $a \equiv a(\alpha) \bmod (t - \alpha)$. By considering the symbol over $\mathbb{F}_q[t]$ and $\mathbb{F}_{q^d}[t]$, we see that

$$(\frac{a}{P})_n = (\frac{a}{t-\alpha})_n = a(\alpha)^{(q^d-1)/n} = (a(\alpha)a(\alpha^q)\cdots a(\alpha^{q^{d-1}}))^{(q-1)/n}$$

so that

$$(\frac{P}{P'})_n = \prod P'(\alpha^{q^i})^{(q-1)/n} = \prod_{i,i'}(\alpha^{q^i} - \alpha'^{q^{i'}})^{(q-1)/n} = (-1)^{dd'(q-1)/n}(\frac{P}{P'})_n$$

where the last equality follows by symmetry by just counting the number of switches of signs.

More generally, there is **Weil reciprocity**, which gives ratio between evaluations $f((g))$ and $g((f))$ (extended from points to divisors in obvious fashion) in terms of local symbols, for any elements f and g of function fields and says, in particular, that these two quantities are equal if f and g have disjoint support. The last statement can be verified directly for rational function field and then be generalized by covering and norms to general case. For proofs, the connections with power and norm reciprocity and general Artin reciprocity, we refer to [CF67; Ser79].

Global class field theory is the beautiful and well developed theory which explains arithmetic of abelian extensions of a global field in terms of the generalized congruence class groups in the global field.

Now we state a summary (refer to the references below for the details) of some of the main results of global class field theory in a few standard formulations. There are also cohomological or division algebra approaches, useful in many investigations. Some good standard references on class field theory handling global fields are [Art67; AT68; CF67; Wei73; Lan94; Neu86]. See these and also [Ser79; Iwa86] for local class field theory. See [Ser88] for function field class field theory based on geometric approach of Lang using some ideas of Weil and Rosenlicht.

We will not make much explicit use of class field theory, except in explaining explicit class field theory in Chapter 3 and using Lang isogenies in discussions of shtukas and solitons.

We will work in a fixed algebraic closure.

Class field theory in terms of ideal (or divisor) groups:

Let K be a global field. We start by recalling the basic notation. In addition to the symbols denoting usual primes, in the number field case, we formally include a real and complex prime symbol for each real or complex (conjugate pair) embedding of K, as is customary. By a modulus $m = \sum m_v v$, we will mean an effective divisor, i.e, a finite formal positive integral linear combination of primes v, where in the number field case, $m_v = 0$ or 1 for a real prime v and $m_v = 0$ for a complex prime. Write S_m for the set of primes dividing m. For a finite set S of primes of K, denote by $I^S := I_K^S$ the free abelian group generated by finite primes not in S. In the context of extensions, S will also stand for the set of primes above those in S. We write $I := I_K := I^\phi$. Writing (z) for the divisor of $z \in K^*$, let

$$P_m^1 := \{(z) : z \in K^*, \ v(z-1) \geq m_v, \text{if } m_v > 0\}$$

where for real v, with $m_v > 0$ the condition (determining the neighborhood of 1 subgroup) is supposed to mean that the image of z is positive in the corresponding real embedding and for complex v there is no condition. The ray class group modulo m is defined to be $C_m := I^{S_m}/P_m^1$. A congruence subgroup mod m is defined to be a subgroup of finite index (automatic in the number field case) of I^{S_m} containing P_m^1. Recall that for a Galois extension L of K and unramified prime v of K we associate a Frobenius conjugacy class, and hence a Frobenius element $\text{Frob}_v \in \text{Gal}(L/K)$, in case the extension is abelian.

The main theorems of class field theory say that

Theorem 1.4.1 *(1) Given a finite abelian extension L of K, if we let S to be the set of primes of K that ramify in L, then there is a modulus m such that $S_m = S$ and the Artin map (Frobenius maps combined multiplicatively) $\psi : I^S \to \mathrm{Gal}(L/K)$ induces isomorphism of $I_K^{S_m}/P_m^1 N_K^L(I_L^{S_m})$ onto $\mathrm{Gal}(L/K)$.*

(2) Moreover, for any congruence subgroup H mod m, there is a finite abelian extension L of K such that $H = P_m^1 N_K^L(I_L^{S_m})$.

Hence, $\mathrm{Gal}(K^{ab}/K)$ is seen to be isomorphic to the inverse limit of all ray class groups.

In the number field case, the ray class group mod the zero divisor is the class group, which is finite and the corresponding extension is the Hilbert class field, which is the maximal unramified abelian extension. All the principal ideals in the base split in it by the properties of Artin map. In addition, all the ideals in the base become principal in it. This is the principal ideal theorem. In the function field case, the corresponding ray class group is infinite, since by the product formula, the degree of principal divisor is zero, whereas a general divisor can have any degree. If we restrict to degree zero, we get geometric extensions. (Sometimes such a degree zero part is called the ray class group.)

If we take the usual definition of the Hilbert class field as the maximal abelian everywhere unramified extension, then it is of infinite degree over the base, because of the constant field extensions. If we decree that the field of constants should not increase, then there are h such extensions K_i, all with Galois group isomorphic to the class group $Pic^0(K)$. If L is degree h constant field extension, then $LK_i = LK_1 \cdots K_h$, for any i. And this degree h^2 invariantly defined extension is suggested [AT68, pa. 80] by Artin and Tate to be the correct generalization of the Hilbert class field. By Madan's result on class number divisibilities mentioned in previous section, there are no finite 'class field towers' for this notion once the base has class number more than one. Also, Madan [Mad69] shows that if h_K is divisible by p, the characteristic of K, then in no separable extension (eg. Artin-Tate's candidate for Hilbert class field) all degree zero divisors of K become principal.

On the other hand, if you take A to be ring of integers with one point at infinity (i.e., functions with only pole at that chosen point at infinity) and define the Hilbert class field H_A of A (or K and the point at infinity) to be the maximal abelian extension unramified everywhere and split completely at infinity (in other words, the corresponding group being generated by principal ideals and the infinity), we get an abelian extension of degree $h_A = |Pic^0(A)|$. When the point at infinity is a rational point, it is one of the K_i's above, but choosing infinity and working with the corresponding ring of integers gives better analogies. This extension arises naturally in rank one Drinfeld modules theory and the ideals of A become principal in H_A, as we will see in Chapter 3. See also [Ros87].

Class field theory in terms of idele groups:

In this approach, at the expense of dealing with a slightly more complicated objects, we do not have to talk about the modulus and can handle infinite extensions and local-global passage more smoothly.

Let K be a global field and let $\mathcal{A}_K$, $\mathcal{I}_K$ and $C_K = \mathcal{I}_K/K^*$ denote its adele ring, idele group and idele class group respectively. Let $\mathcal{I}_K^1$ denote the subgroup of norm 1 ideles. By the product formula, it contains K^* and $\mathcal{I}_K$ is direct product of $\mathcal{I}_K^1$ and a discrete subgroup isomorphic to $\mathbb{Z}$ in function field case, whereas it is direct product of $\mathcal{I}_K^1$ and a subgroup isomorphic to $\mathbb{R}_+^*$ in the number field case. Let $K_{\infty+}^*$ denote the group of ideles with 1 at each finite place and positive at each real place.

Theorem 1.4.2 *(1) For a finite abelian extension L of K, the Artin map induces isomorphism of $C_K/Norm_K^L C_L = \mathcal{I}_K/K^* Norm\, \mathcal{I}_L$ onto $Gal(L/K)$ and for every open subgroup N of finite index in C_K (in $\mathcal{I}_K$ containing K^* respectively), there exists a unique abelian extension L of K such that $Norm_K^L C_L = N$ ($K^* Norm\, \mathcal{I}_L = N$ respectively).*

(2) For an abelian extension L of K, there is a closed subgroup N of C_K containing the connected component of identity of C_K, such that C_K/N is isomorphic to $Gal(L/K)$.

(3) Conversely, given a closed subgroup of C_K containing the connected component, there is a unique abelian extension with the property above with respect to it.

(4) For a local field K_v and for a finite abelian extension L^v of K_v, the Artin map (which can be obtained by composing the global Artin map of (1) with the canonical inclusion of K_v^* in the ideles) induces isomorphism of $K_v^*/Norm_{K_v}^{L^v}(L^v)^*$ onto $Gal(L^v/K_v)$ and for every open subgroup N of finite index in K_v^*, there exists a unique abelian extension L^v of K_v such that $Norm_{K_v}^{L^v}(L^v)^* = N$.

For a number field K, the Artin map induces an isomorphism of

$\mathcal{I}_K/\overline{K^*K^*_{\infty+}}$ onto $\mathrm{Gal}(K^{ab}/K)$ and for a function field K, it induces an isomorphism of $\mathcal{I}_K/K^*$ onto a dense subgroup of $\mathrm{Gal}(K^{ab}/K)$ and isomorphism of $\mathcal{I}^1_K/K^*$ onto $\mathrm{Gal}(K^{ab}/K^c)$, where K^c is the union of all constant-field extensions of K and K^{ab} is the maximal abelian extension of K.

By the theorem, $\mathrm{Gal}(K^{ab}/K)$ is isomorphic to the inverse limit of C_K modulo its open subgroups of finite index: thus we know the Galois groups of all abelian extensions of K from a knowledge of its idele class group. In the number field case, the Artin map induces an isomorphism between $\mathrm{Gal}(K^{ab}/K)$ and C_K modulo its connected component. In the function field case, the Artin map induces isomorphism between C_K and the dense subgroup of $\mathrm{Gal}(K^{ab}/K)$ consisting of those automorphisms whose restriction to the algebraic closure of the field of constants is an integral power of the Frobenius for the field of constants.

By the usual connection between Frob_v and the splitting properties of v under the Artin map, the filtrations on unit groups and ramification groups correspond. For example, $K^*_\wp$ ($U_\wp$ respectively) is inside the norm group if and only if $\wp$ splits completely (unramified respectively) in L. Further these conditions reduce to local considerations since $\mathrm{Norm}\, C_L \cap K_\wp = \mathrm{Norm}\, L_P$, for $P|\wp$. Another simple, useful corollary is that product (intersection respectively) of the norm groups correspond to intersection (compositum respectively) of the extensions.

In terms of idele groups, the place of C_m is taken by the image (again denoted by C_m) of K^* times group of ideles congruent to 1 modulo v^{m_v} for $m_v > 0$ and with unit components at other places.

In the number field case, C_m is of finite index in C and any open subgroup of C contains image of such for some m. In the function field case, it is contained in the degree 0 (i.e., norm or absolute value 1) part C^0 as a subgroup of finite index. If we let i_1 to be an idele of norm q (corresponds to a divisor of degree one, which is known to always exist), then we can write C as $i_1^{\mathbb{Z}} \times C^0$. The group $i_1^{d\mathbb{Z}} \times C^0$ corresponds to the constant field extension of degree d. The invariant Hilbert class field of Artin-Tate mentioned above corresponds to $i_1^{h\mathbb{Z}} \times C_0$. It is independent of the choice of i_1, whereas K_i's mentioned there correspond to $i_1^{\mathbb{Z}} \times C_0$ and H_A corresponds to $t_\infty \times C_0$, where t_∞ is the local parameter at ∞.

More generally, for an open subgroup H of finite index in C, its intersection H^0 with C^0 is open of finite index. If h is an element of least norm, say q^d, in H, $H = \cup h^k H^0$ and corresponding field has constant field extension of degree d. Any open subgroup of C^0 contains some C_m and hence is of finite index in it.

For a finite abelian extension L of K, its conductor is smallest m such that L is contained in the ray class field for the modulus m.

By the Kronecker-Weber theorem, the maximal abelian extension of $\mathbb{Q}$ is obtained by adjoining all roots of unity to $\mathbb{Q}$, whereas in the function field case, the roots of unity being the constants, we would get K^c in the place of K^{ab}, which is much bigger, as for example, there are many non-constant-field abelian extensions such as Kummer and Artin-Schreier extensions. Carlitz-Drinfeld-Hayes theory of the next two chapters will clarify the matters of this explicit class field theory.

Class field theory via generalized Jacobians:

Here the class field covers are given by geometrical constructions involving generalized Jacobians, which are group varieties, making the class field theory correspondence with generalized ideal class groups more transparent and geometric.

Let K be a function field with the field of constants k, which is either algebraically closed or finite, and let X be a projective, non-singular curve over k having the function field K.

Theorem 1.4.3   *(1) For any rational map $f : X \to G$, to a commutative algebraic group G, regular away from a finite set S, there is a modulus (effective divisor) m with $S_m = S$ such that $f((\phi)) = 0$ for $\phi \in K_m^1$.*

(2) Moreover, for every modulus m, there is a commutative algebraic group J_m and a rational map $f_m : X \to J_m$ having the universal property that, for f as above, there exists unique rational homomorphism $\theta : J_m \to G$ with $f = \theta f_m$.

(3) The map f_m defines by extension to k-rational divisor classes a bijection from C_m^0 to k-points of J_m.

(4) Every geometric abelian finite covering of X is pull-back under a unique separable isogeny for some m, which can be chosen to have support exactly at the ramified points of the covering.

Theorem is first proved for k algebraically closed, then we can descend to finite field case, as the obstruction to descent lies in a Brauer group, which vanishes for a finite field.

For zero divisor m, J_m is just the Jacobian and hence the unramified coverings are pull-backs under isogenies of Jacobians. Suppose ∞ is a degree one point of X. If we embed the curve X into its Jacobian via $P \to P - \infty$ and use the Lang isogeny $x \to x^q - x$, where our field of constants is $\mathbb{F}_q$, the pullback produces an abelian unramified cover of X with Galois group $J(\mathbb{F}_q)$, which is the class group mentioned above. Since the inverse image of 0 consists of just these points, ∞ splits completely. In fact, the cover is the Hilbert cover H_A mentioned above. Choosing ∞ to be any of the h rational points, you get h Hilbert covers corresponding to K_i mentioned above. If ∞ is of degree $d > 1$, then H_A is obtained by first taking a constant field of extension by $\mathbb{F}_\infty$ and then pull-back under Lang isogeny transferred to base change to $\mathbb{F}_\infty$ from $\mathbb{F}_q$ in which we embed using (any) point above ∞.

Class field theory via cohomological approach:

For this we refer to literature except to recall that there the fundamental reciprocity law arises as the following isomorphism obtained by taking cup product with the fundamental class:

$$G^{ab} = H_1(G, \mathbb{Z}) = H^{-2}(G, \mathbb{Z}) \xrightarrow{\sim} H^0(G, C_L) = C_K/\mathrm{Norm}_K^L C_L,$$

where L/K is Galois extension of number fields with Galois group G.

The Galois group:

Thus, class field theory gives good description of the abelianization G_K^{ab} of the absolute galois group G_K of a global field K. Let us compare using group structure differences mentioned in the first section:

For $K = \mathbb{Q}$, the abelian part is $\hat{\mathbb{Z}}^*$, which is product of $\mathbb{Z}_p^*$'s, which is in turn the product of all $\mathbb{Z}_p$'s and all $(\mathbb{Z}/p\mathbb{Z})^*$'s. The usual structure theorem together with Dirichlet theorem on primes in arithmetic progressions (choose $p \equiv 1 + l^n \mod l^{n+1}$), shows that the last product is just the product over primes l and positive integers n of countably many copies of each $\mathbb{Z}/l^n\mathbb{Z}$. For a general number field K, the Leopoldt conjecture on p-adic independence of the basis of units (given by Dirichlet theorem) implies that the free abelian part is $\hat{\mathbb{Z}}^{r_2+1}$. This is known [Was97] for K abelian over $\mathbb{Q}$. Consider now the finite factors: The ray class groups for different prime power towers are linearly disjoint by ramification considerations over the Hilbert class field, so that finite factors come from prime to $\mathrm{Norm}(\wp) = p^f$ factors of $(\mathcal{O}_K/\wp)^*$, which are cyclic groups of order $p^f - 1$. Since f is bounded, $val_l(p - 1) = n$ implies $val_l(p^f - 1) = n$, hence we do get countably many copies of $\mathbb{Z}/l^n\mathbb{Z}$, for each n and sufficiently large primes l. If K is abelian over $\mathbb{Q}$, given l, for sufficiently large n, we can put further consistent congruence condition on p to make it split, thus giving countably many factors $\mathbb{Z}/l^n\mathbb{Z}$. On the other hand, as example $K = \mathbb{Q}(i)$, when $p^f - 1$ is divisible by 4 for odd p, shows we do not get infinitely many $\mathbb{Z}/2\mathbb{Z}$ as factors.

For a function field K of characteristic p, the constant field extensions provide $\hat{\mathbb{Z}} = \prod \mathbb{Z}_p$, and class field theory gives G_K^{ab} as an extension of this by I_K^1/K^* which is just $\prod U_\wp/\mathbb{F}_q^*$ when the class number is one. Now $U_\wp$ is isomorphic to $(A/\wp)^*$ times one-unit group. The one unit group is isomorphic to the product of countably many copies of $\mathbb{Z}_p$. (Dual of profinite is discrete, so we can just count to understand the group.) In contrast to the case above, in this class number one case, we will not get any $\mathbb{Z}/p^n\mathbb{Z}$'s as subgroups and only some of $\mathbb{Z}/l^n\mathbb{Z}$'s for $l \neq p$ will occur as factors in prime power cyclic decomposition of the cyclic group $(A/\wp)^*/\mathbb{F}_q^*$. For example, if $q = 7$, the order of this cyclic group being odd or divisible by 8, we will not get $\mathbb{Z}/2\mathbb{Z}$ or $\mathbb{Z}/4\mathbb{Z}$'s. In the simplest case, $q = 2$, where $\mathbb{F}_q^*$ is trivial, we do not have, for example, a factor $\mathbb{Z}/l\mathbb{Z}$ for primes l, such as 1093 and 3511 (obtained from Brillhart's factorization tables), for which the order of 2 modulo l and l^2 is the same (364 and 1755 respectively). It seems to be expected but not known that there are only finitely many such primes. If the class number is divisible by p, we may get some p-groups coming from the class group.

The question of a good description of the full absolute galois group $G_\mathbb{Q}$ of $\mathbb{Q}$ as a topological group, together with how the various decomposition groups and Frobenius elements sit in it, is a central open problem of number theory. If $G_\mathbb{Q}$ were free profinite (of countable rank), then all finite groups would occur as Galois group over $\mathbb{Q}$, but there are order two (only finite order possible, by Artin's theorem [NSW00, 12.1.7]) 'complex conjugations' in $G_\mathbb{Q}$. It is still conjectured, but not known, that every finite group is a Galois group of a number field over $\mathbb{Q}$.

Let us see what the situation is from some other fields.

The knowledge of full absolute Galois group and of possible extensions is complete and easy for finite fields: We have $\text{Gal}(\overline{\mathbb{F}}_q/\mathbb{F}_q)$ isomorphic to $\hat{\mathbb{Z}}$, the profinite completion of $\mathbb{Z}$. Any finite field $\mathbb{F}_q$ has a unique extension of each degree and it has a cyclic Galois group.

If K is an extension of $\mathbb{Q}_p$ of degree n, and $p \neq 2$, then G_K is explicitly described (see [NSW00, p. 360] and references therein) with $n + 3$ topological generators ($n + 2$ are enough, but make relations complicated) and two relations. This generator-relation description has been used in some Galois representation deformation studies. For a local field K of finite characteristic p, G_K is not (topologically) finitely generated, in fact, its maximal pro-p quotient is free pro-p of countable rank. See [NSW00; Koc67].

The absolute Galois group of a p-adic local field does not determine the local field (up to isomorphism, in the class of p-adic local fields). But interestingly enough, if $G_{\mathbb{Q}_p}$ is isomorphic to G_K, for a field K, then the elementary theory (first order statements in certain natural language of valued fields, in particular, in the usual language of fields) of K and $\mathbb{Q}_p$'s is the same (as Koenigsmann has shown in a recent preprint) in analogy with the Artin-Schreier-Tarski's result that the field with the absolute Galois group $\mathbb{Z}/2\mathbb{Z}$ is really closed and thus elementarily equivalent to $\mathbb{R}$, but in contrast with the obvious case of $\mathbb{C}$ and with the case of local fields of characteristic p, since for every field K of characteristic p, there is a field L of characteristic zero, with G_K and G_L isomorphic. This fact about elementary equivalence does not generalize from $\mathbb{Q}_p$ to all p-adic fields, since that would contradict the first statement of the paragraph, as the elementary equivalence of two p-adic local fields implies their isomorphism. See [Che76; EF99; Jar96] for more and for precise description of the language considered.

As for use of this in understanding of the decomposition groups within the absolute Galois group of a global field, see [NSW00, 12.1.11].

There are only a finitely many extensions of given degree, because of Krasner's lemma [Lan94, Chapter 2, Section 2] implying that 'close' irreducible polynomials give the same extensions and compactness of p-adic integer rings. (For characteristic p local field, the separability assumptions of Krasner's lemma do not hold and in fact, by structure of unit group in Section 1.1, there are infinitely many extensions of degree p for example.) By considerations of ramification group tower, we see that for local fields, all Galois extensions have solvable Galois groups. See [Ser79].

Newton proved that [Ser79] the algebraic closure of $F((t))$ for algebraically closed field F of characteristic zero is the union of $F((t^{1/n}))$, for $n \geq 1$. Thus the absolute Galois group in this case is $\hat{\mathbb{Z}}$. The algebraic closure is much more complicated when F is of finite characteristic p: For example, $x^p - x - t^{-1}$ has no root in the Newton's field, but has root $\sum_{i=1}^{\infty} t^{-1/p^i}$. The algebraic closure of $F((t))$ in this case consists of 'generalized fractional power series' of the form $\sum_{i \in S} f_i t^i$, where $S \subset \mathbb{Q}$ is a well-ordered subset such that for some m, all the elements of mS have denominators power of p (rather than denominator one, in Newton's case) and satisfying further complicated combinatorial conditions, described by Kedlaya in [Ked01], so that they are algebraic over $F((t))$. See [Ked01] also for descriptions of the algebraic closure of $F((t))$, for perfect F and for completion of the algebraic closure etc. If F is a finite field, Kedlaya has given a simple description following Christol's theorem that we will discuss in the last chapter. Such a description for the separable closure is not yet known.

In contrast to the local fields case, for global fields, the absolute Galois group G_K determines K, up to isomorphism in the class of global fields.

See [NSW00, Chapter 12], [Jar96] and references therein, for these and more results. Anabelian geometry conjectures of Grothendieck are vast conjectural generalizations of these facts, and there is a lot of recent progress by Pop, Tamagawa, Mocizuki etc.

Let F be an algebraically closed field and K be a field that is finitely generated and of a transcendence degree one over F. Then the absolute Galois group G_K of K is free profinite on a set of cardinality of F. [Völ96a, pa. 241] and references therein. By analogy, in case $F = \overline{\mathbb{F}_p}$, Shafarevich has conjectured that $\mathrm{Gal}(\overline{\mathbb{Q}}/\mathbb{Q}^{ab})$ is free profinite of countable rank, since by Kronecker-Weber $\mathbb{Q}^{ab}$ is obtained by adjoining all roots of unity, as is the algebraic closure of a finite field. This conjecture would imply that $G_\mathbb{Q}$ is an extension of a profinite group on countable generators by $\hat{\mathbb{Z}}^*$.

As a result of Riemann existence theorem, each finite group is a Galois group [Völ96a, pa. 37] over $K(x)$, where $K = \mathbb{C}$, or algebraically closed subfield of $\mathbb{C}$. Each finite group is a Galois group [Völ96a, pp. 212-213] over $\mathbb{F}_p(x)$ for all but finitely many p and over $K(x)$ where K is a complete field with non-archimedean absolute value (eg., $\mathbb{Q}_p$ or $k((x))$).

Riemann existence theorem also gives a good control on ramification data, so more general questions are naturally raised and partially answered. For example, affine line over algebraically closed field of characteristic zero has no unramified covers, whereas Abhyankar's example $y^{p+1} - xy + 1$ gives an extension of $\overline{\mathbb{F}_p}(x)$ ramified only at $x = \infty$, with the Galois group $PSL_2(\mathbb{F}_p)$. While the full π_1 in this case is not even conjecturally understood, Abhyankar conjectured what all possible finite Galois groups are in this case. This conjecture was proved by Raynaud and Harabater. The situation over $\mathbb{F}_p$ is still unclear.

We refer to Ihara's ICM 1990 talk and MathSciNet for information on the attempt to understand G_Q by considering unramified coverings of projective line minus three points, and various related exciting recent developments.

Another way to understand the absolute Galois group is to try to generalize the class field theory via representation theory: Langlands program.

This envisions natural bijection (matching Frobenius and Hecke eigenvalues) between certain classes of continuous representations of G_K in $Gl_n(\overline{\mathbb{Q}_l})$ ($Gl_n(\mathbb{C})$-representations is a special case, because topological considerations imply that image of G_K is then finite) and certain classes of automorphic representations of $Gl_n(\mathcal{A}_K)$, generalizing the bijection of sets of characters of finite orders of G_K and of C_K provided by the global class field theory.

We will just mention the simplest case of the conjecture to give a flavor of the way it generalizes the class field theory: If $R : G_{\mathbb{Q}} \to Gl_2(\mathbb{C})$ is irreducible representation of Artin conductor N and 'odd' in the sense that determinant of image of complex conjugation is -1, then there is a holomorphic modular new-form $f = \sum a_n q^n$ of weight 1, level N, and character χ such that $\chi(p) = \det R(\mathrm{Frob}_p)$ and $a_p = \mathrm{Tr} R(\mathrm{Frob}_p)$, for all unramified primes p. Since Frob_p tells how the prime p decomposes (in the number field L corresponding to the kernel of R), we see that, for example, $a_p = 2$ is the condition for p to split completely in L, generalizing the 'arithmetic progressions' or generalized congruence conditions of the class field theory. Since these modular forms form finite dimensional spaces (though usual Riemann-Roch technique to find the dimension exactly fails in weight one), these conditions give 'good' descriptions of these primes.

See [Art03] and references therein for detailed introduction giving precise statements and for the recent status. In function field case, Lafforgue [Laf02] has recently settled the conjectures for Gl_n, building on Drinfeld's work on Drinfeld modules and Shtukas and various other lines of work by many mathematicians. We will briefly discuss this in Chapter 6.

Chapter 2

Drinfeld modules

In this chapter, we introduce and develop the basics of Drinfeld modules. These are one dimensional objects of rank any positive integer. In rank one, they are analogs of multiplicative group and reflect analogies with cyclotomic (and complex multiplication) theory, whereas in rank two situation they reflect many analogies with elliptic curves theory. We do not have close cousins in the classical case for the higher rank Drinfeld modules.

We start by motivating the simplest case, that of Carlitz module by analytic approach, introducing analogs of the exponential function. Then we quickly indicate how to get good analogs of e, $2\pi i$, roots of unity, factorial, binomial functions, and zeta functions, Bernoulli numbers, modular forms etc. We will study these in more detail in the later chapters.

Then, after developing the Drinfeld modules for general A, we will see applications to the objects of arithmetic interest. For refined study, in later chapters we generalize the Drinfeld modules, which are one dimensional, to higher dimensions and also use geometric tools in the one dimensional as well as higher dimensional case, in addition to algebraic and analytic tools mentioned in this chapter.

Main general references for this chapter are [Dri74; Car35; DH87; Gos96; G+97; Hay79; Hay].

2.1 Carlitz module and related arithmetical objects

The **exponential function** is one of the most fundamental functions. Let us try to see what its analog, if any, should be in the function field case. We start with the simplest case of $A = \mathbb{F}_q[t]$. By the basic analogies recalled in the first chapter, we would like to have analogue $e(z)$ with range and domain C_∞.

If we say that the exponential is important, because it is its own derivative and try to get a power series solution, solving the usual recursion, we would get $\sum z^n/n!$ as usual, but it does not make sense in characteristic p, as $n! = 0$, if $n \geq p$.

We do not know how to put a polynomial in the exponent, even if we ignore the question of what the base e should be. In fact, the multiplicative functional equation $e(x+y) = e(x)e(y)$ would imply $e(x)^p = e(px) = e(0) = e(0+0) = e(0)^2$, so that $e(0) = 0$ or 1 and so $e(x)$ is identically zero or identically one.

In other words, there is a short supply of such multiplicative functions in finite characteristic, in contrast with the complex case. On the other hand, as a compensation, there is a much larger supply of additive or even $\mathbb{F}_q$-linear functions, in characteristic p. Namely, we can now take $f(z) = \sum f_i z^{q^i}$ instead of just $f(z) = cz$, for some constant c.

So we seek the exponential function to be an entire (given by everywhere convergent power series) among those satisfying $e(x + y) = e(x) + e(y)$. Now for the usual exponential $e^z = 1$ if and only if $z \in 2\pi i \mathbb{Z}$. Since we have switched to additive case from multiplicative, we look at solutions of $e(z) = 0$ and decree simple zeros at $\tilde{\pi} A$, for some $\tilde{\pi} \in C_\infty$ (analog of $2\pi i$) to be fixed.

Now in our setting, meromorphic function is determined up to multiplication by constant by its divisor, i.e., by its zeros and poles with multiplicities, so that

$$e(z) = cz \prod_{\lambda \in \tilde{\pi} A - \{0\}} (1 - z/\lambda).$$

Note that this is additive, because A is $\mathbb{F}_q$-vector space. Now we normalize: First, let the linear term be z, i.e., take $c = 1$. We also have

$$e(tz) = k \prod_{\beta \in (\tilde{\pi}/t)A/\tilde{\pi}A} (e(z) - e(\beta))$$

for some non-zero constant k, since the divisors of both sides are the same. Hence $e(tz)$ is a $\mathbb{F}_q$-linear polynomial of degree q, so of the form $te(z) + ke(z)^q$. We normalize again by choosing $k = 1$.

The resulting analog $e(z) : C_\infty \to C_\infty$ is entire $\mathbb{F}_q$-linear function with linear term z and satisfies

$$e(tz) = te(z) + e(z)^q.$$

It is called **Carlitz exponential** and the map $u \to tu + u^q$ defines **Carlitz module**. In fact, since t generates A over $\mathbb{F}_q$, for any $a \in A$, we can write $e(az)$ as a polynomial $C_a(e(z))$ in $e(z)$. For example, $e(t^2 z) = t^2 e(z) + (t + t^q)e(z)^q + e(z)^{q^2}$ as can be seen by applying the basic functional equation twice. This should be compared with e^{nz} being a monomial $(e^z)^n$ for $n \in \mathbb{Z}$. Simple estimates then show that $e(z) = C_a(e(z/a))$ is, in fact, $\lim C_a(z/a)$ as the degree of a tends to infinity, analogous to classical 'continuous compounding of interest' formula $e^z = (e^{z/n})^n = \lim(1 + z/n)^n$.

The realization, by $a \to C_a$, of A as a subring of (non-commutative) ring of polynomials in Frobenius (q-th power map here) (these are linear functions or endomorphisms of additive group under composition) is the Carlitz module and can be thought of as analogue of $\mathbb{Z}$ (the ring of integers in imaginary quadratic field respectively) sitting in the endomorphism ring of multiplicative group (of complex multiplication elliptic curve respectively). Hence we get $C_a(x)$ as analogs of a-th **cyclotomic polynomials**.

There are many ways to embed A into this non-commutative ring, associating t with a polynomial in Frobenius of degree r will give rank r Drinfeld A-module (so that Carlitz module is rank one) and coefficients of these polynomials, considered as functions of (rank r) A-lattices (which are kernels of their exponential functions), will turn out to be **modular functions**. The space $C_\infty - K_\infty$ will play analog of upper (and lower) half-plane $\mathbb{C} - \mathbb{R}$.

From the kernel $\tilde{\pi}A$ of $e(z)$, now we recover a good analog $\tilde{\pi}$ of $2\pi i$, which is determined up to signs $\mathbb{F}_q^*$ (ambiguity similar to that of ± 1 for $2\pi i$). We will see that adjoining analogs $e(\tilde{\pi}/a)$ of **roots of unity** $\zeta = e^{2\pi i/n}$ (or of $1 - \zeta$, as we compensate for our shift to 0 from 1) give good **cyclotomic extensions**. The Galois group in question is abelian group $(A/a)^*$ in analogy with $(\mathbb{Z}/n)^*$, which is the Galois group of $\mathbb{Q}(\zeta_n)$ over $\mathbb{Q}$. (For example, if we take $a = t$, we are adjoining roots of $tu + u^q = o$, that is $q - 1$-th roots of $-t$, and hence we get Kummer extension of $\mathbb{F}_q(t)$, analogous to quadratic extension of $\mathbb{Q}$, with Galois group cyclic of order $q - 1$, i.e. $(\mathbb{F}_q[t]/(t))^* \cong \mathbb{F}_q^*$.) Note that the roots are now parametrized, not by integers, but by polynomials. So the **Galois representations** you get (from this and higher rank generalizations) are not l-adic as you would get from elliptic curves over function fields (whose points give abelian groups i.e., $\mathbb{Z}$-modules), but $\wp$-adic for prime $\wp$ of A.

Our attempt to imitate $e^z = \sum z^n/n!$ failed, but now that we have $e(z)$ we can try to recover factorials from its coefficients. This leads to an excellent analogue called **Carlitz factorial**. We will study its interpola-

tions, functional equations, special values later. Once we have the exponential $e(z)$ and its inverse function logarithm, say $l(z)$, we can imitate
$\exp(x \log(1+t)) = (1+t)^x = \sum \binom{x}{n} t^n$ by writing

$$e(xl(t)) = \sum \binom{x}{q^i} t^{q^i}$$

thereby defining analogues $\binom{x}{q^i}$ of binomial coefficients.

Similarly, though our attempt to get exponential from differentiation
equation failed, now that we have exponential, we will see in 4.14 a new
analog of differentiation restoring the differential equation property!

Classically, the Bernoulli numbers, which occur in power sums and zeta
values at both negative and positive integers and whose divisibilities give
valuable information on class groups of cyclotomic fields, are defined by the
generating function $z/(e^z - 1) = \sum B_n z^n/n!$. So we can define **Carlitz-
Bernoulli numbers** (which are rational functions in t now) by the generating function $z/e(z) = \sum B_n z^n/n!$, once we have a good factorial as
mentioned above.

In place of the **zeta function** given by $\sum n^{-s} \in \mathbb{C}$ for $Re(s) > 1$, where
we sum over all positive integers, we now sum over all monic polynomials
$a \in A$ to get zeta values $\sum a^{-s} \in \mathbb{F}_q((1/t))$ for $s \in \mathbb{Z}_{>0}$. Just summing
by grouping the terms with the same degree allows one to interpolate to
$s \in \mathbb{Z}$ and by removing the Euler factor to interpolate $\wp$-adically by just
Fermat's little theorem on congruences. We will study its interpolations,
special values later. At the simplest level, we have Carlitz' analog of Euler's
theorem giving the values at positive 'even' integers m (i.e., multiples of
$q - 1$): $\zeta(m) = B_m \tilde{\pi}^m/m!$: Multiplying the logarithmic derivative of the
product formula for $e(z)$ by z, we get

$$\frac{z}{e(z)} = 1 - \sum_{\lambda \in \Lambda - \{0\}} \frac{z/\lambda}{1 - z/\lambda} = 1 - \sum_{n=1}^{\infty} \sum_{\lambda} (\frac{z}{\lambda})^n = 1 + \sum_{n \text{ 'even'}} \frac{\zeta(n)}{\tilde{\pi}^n} z^n$$

since $\sum_{c \in \mathbb{F}_q^\times} c^n = -1$ or 0 according as n is 'even' or not. The comparison
with the generating function of Bernoulli numbers gives the result.

We do not know anything about possibility of a functional equation.

2.2   Drinfeld modules: Basic definitions

Let us try to see how we can generalize the successful construction in the
last section. We might want to replace $\mathbb{F}_q[t]$ by $\mathcal{O}_S$, the ring of S-integers

of a function field K and consider any homomorphism $\mathcal{O}_S \to \mathrm{End}(\mathbb{G}_a)$ or even may want to look at $\mathbb{G}_a^d$ (the fact that $p = 0$ in $\mathcal{O}_S$ essentially rules out other possible groups). Or analytically, we may want to explore arithmetic for any rank lattice in some group. For first six chapters, we restrict to the additive group $\mathbb{G}_a$.

This leads us to the notion of Drinfeld modules, which are these interesting module structures on $\mathbb{G}_a$ of characteristic p.

If F is a field of characteristic p, $\mathrm{End}_F(\mathbb{G}_a)$ can be identified with $F\{\tau_p\}$ defined to be the non-commutative ring of polynomials in variable τ_p and with coefficients in F, with commutation relation $\tau_p f = f^p \tau_p$. This can also be identified with the ring, where the multiplication operation is the composition of functions, of 'additive' polynomial functions $x \to \sum f_i x^{p^i}$. It is a non-commutative integral domain with both left and right division algorithm property.

The derivative map D singles out the linear term coefficient, in characteristic p, i.e., $D(\sum f_i \tau_p^i) = f_0$.

We are forced to have at most one place at infinity, because otherwise we have non-constant units in $\mathcal{O}_S$ and they are mapped by the homomorphism $\mathcal{O}_S^* \to F\{\tau_p\}$ into $F\{\tau_p\}^* = F^*$, and then as elements of $\mathcal{O}_S$ would be algebraic over the ring generated by $\mathcal{O}_S^*$, we get a trivial embedding which does not involve τ_p. In analytic terms, if we have $d > 1$ infinite places in S, we can not embed $\mathcal{O}_S$ discretely as a lattice in C_∞, and will need at least d copies of C_∞.

We will use the following setting and notation throughout the book. Sometimes it will be specialized.

$\mathbb{F}_q$: a finite field of characteristic p having q elements
X: a smooth, complete, geometrically irreducible curve over $\mathbb{F}_q$
K: the function field of X
∞: a closed point of X, i.e., a place of K
d_∞: the degree of the point ∞
A: the ring of elements of K with no pole outside ∞
K_∞: the completion of K at ∞
C_∞: the completion of an algebraic closure of K_∞
$\overline{K}, \overline{K_\infty}$: the algebraic closures of K, K_∞ in C_∞
$\mathbb{F}_\infty$: the residue field at ∞
A_v: the completion of A at a place $v \neq \infty$
g: the genus of X

h: the class number of K

h_A: the class number of A $(= hd_\infty)$

We will consider ∞ to be distinguished place at infinity and call the other places v the finite places. They can be given by a (non-zero) prime ideal $\wp$ of A. As we have not fixed ∞ above, we have already defined K_v, $K_\wp$, $\mathbb{F}_\wp$, C_v, d_v etc.

Then C_∞ is known to be both complete and algebraically closed. We can think of analogs

$$ K \leftrightarrow \mathbb{Q}, \quad A \leftrightarrow \mathbb{Z}, \quad K_\infty \leftrightarrow \mathbb{R}, \quad C_\infty \leftrightarrow \mathbb{C} $$

Instead of $\mathbb{Q}$ we can also have an imaginary quadratic field with its unique infinite place and the corresponding data.

We will see that analogies are even stronger when X is the projective line over $\mathbb{F}_q$ and ∞ is of degree 1, so that $K = \mathbb{F}_q(t)$, $A = \mathbb{F}_q[t]$, $K_\infty = \mathbb{F}_q((1/t))$, namely the situation of the Carlitz module described above.

The Dedekind domain A sits discretely in K_∞ with compact quotient, in analogy with $\mathbb{Z}$ inside $\mathbb{R}$ or the ring of integers of imaginary quadratic field inside $\mathbb{C}$.

We want to define Drinfeld A-module over F (or more generally over a scheme S), but as we discussed in Chapter 1, A is not a canonical base like $\mathbb{Z}$ here, but we decree it to be a base. So we need to be given an A-field structure of F, i.e., a homomorphism $\iota : A \to F$ (or more generally, S should be a scheme over Spec$A$). The kernel of ι is called the **characteristic** of ι, in analogy with the classical situation. The characteristic is a prime ideal $\wp$ of A. If it is the zero ideal, ι is called of generic (or also infinite) characteristic (as it is the generic point) to avoid confusion with field-theoretic zero characteristic. Non-generic characteristic is also called finite characteristic. Unless otherwise stated, when we use extensions such as K, K_∞, C_∞ of K as our F, we assume that ι is the usual embedding.

Definition 2.2.1 Let F be an A-field (also called a field over A) $\iota : A \to F$ of characteristic $\wp$. Then Drinfeld A-module over F (of characteristic $\wp$) is a ring homomorphism $\rho : A \to F\{\tau_p\}$, (we write ρ_a for the image of $a \in A$) with $D\rho = \iota$ and which is non-trivial in the sense that it does not factor through F via ι.

It follows that ρ is injective, since otherwise the map would factor through a finite field (as $F\{\tau_p\}$ is integral domain, the kernel is a prime and hence a maximal ideal of A when non-zero) and will land in F contrary to the assumption.

Hence, Drinfeld A-modules are essentially just non-trivial embeddings of A's into the non-commutative ring $F\{\tau_p\}$ or just non-trivial A-module structures on the additive group in characteristic p. We need not really need to start with a structure map, but non-trivial embedding will lead to such a map when we restrict attention to constant term.

Since $\mathbb{F}_q$ maps via ρ onto a copy of $\mathbb{F}_q$ inside F, using $\rho_a\rho_\theta = \rho_\theta\rho_a$ for a generator θ of $\mathbb{F}_q^*$ shows that in any ρ_a only powers of τ_p that occur with non-zero coefficients are powers of $\tau := \tau_q$. So in fact the image of ρ is inside $F\{\tau\}$ and it is a $\mathbb{F}_q$-algebra homomorphism. We put $\deg(x) := -d_\infty v_\infty(x)$ and $|x|_\infty := q^{\deg(x)}$ as usual. As per the usual convention, $\deg(0) := -\infty$. For $a \in A - \{0\}$, we have $|a|_\infty = |A/a|$.

The function $a \to \deg_\tau(\rho_a)$ clearly extends to a non-trivial valuation on K, and since it is positive for all non-constant $a \in A$, it gives (negative of) valuation at ∞ and so it is $a \to r\deg(a)$ for some positive rational number r. We will see in 2.3.2 that r is a positive integer.

Definition 2.2.2 The positive integer r with the property that $\deg_\tau(\rho_a) = r\deg(a)$ for all $a \in A$ is called the **rank** of ρ.

For ρ of finite characteristic $\wp$, doing the same for the lowest τ exponent $\operatorname{ord}_\tau(a)$ occurring in ρ_a, we get a multiple of valuation $v_\wp$.

Definition 2.2.3 For ρ of finite characteristic $\wp$, the number h such that $\operatorname{ord}_\tau(a) = hv_\wp(a)\deg(\wp)$ for all $a \in A$ is called the **height** of ρ.

As we will see in 2.3.2, the height is again a natural number.

It is sometimes good to think of ρ as the corresponding functor taking F-algebra R to A-module, whose underlying group is $\mathbb{G}_a(R)$ and a multiplication is via ρ_a. A morphism $\phi : \rho \to \rho'$ between two Drinfeld A-module over F is just the corresponding morphism of functors:

Definition 2.2.4 A **morphism** $\phi : \rho \to \rho'$ between two Drinfeld A-module over F is $\phi \in F\{\tau\}$ such that $\phi\rho_a = \rho_a\phi$ for all $a \in A$. A non-zero morphism is called an **isogeny**. Invertible morphism is isomorphism.

A simple count of degrees shows that isogenies can occur only among Drinfeld modules of the same rank and characteristic.

Examples 2.2.5 (1) Let $A = \mathbb{F}_q[t]$. As t generates A over $\mathbb{F}_q$, a general Drinfeld A-module ρ over F, of rank r over F can be fully described by $\rho_t = \sum_{i=0}^{r} f_i\tau^i$, where $f_i \in F$, f_r is non-zero and $f_0 = \iota(t)$. It is of generic characteristic, if f_0 is transcendental over $\mathbb{F}_q$. If $f_0 = 0$, it is of characteristic t. More generally, if $\wp$ is the minimal polynomial of f_0 over $\mathbb{F}_q$, then it is of characteristic $\wp$. So, for example, if $F = A/\wp$ and ι is the canonical map, we get ρ of characteristic $\wp$. Note that $C_t = t + \tau$ is isomorphic to $\rho_t = t + f_1\tau$ by the isomorphism $f_1^{1/(q-1)}$, which in general is defined only over an extension of $\mathbb{F}_q(t, f_1)$. So there is only one isomorphism class of rank one A-modules of generic characteristic in this case.

(2) For general A, we have non-trivial constraints $\rho_{x_i}\rho_{x_j} = \rho_{x_j}\rho_{x_i}$, where x_i's generate A as $\mathbb{F}_q$-algebra. We write this as system of algebraic equations on undetermined coefficients (of τ^k in ρ_{x_i}, for positive $k \leq r \deg x_i$, if we want rank r) and solve by elimination theory. There are many simple tricks to simplify calculations, such as bounds on sizes of coefficients from analytic theory of Section 4, control on field of definition and some coefficients as we will see in next chapter. In particular, 3.5.4 calculates the following example. Let $A = \mathbb{F}_2[x, y]$ with $y^2 + y = x^3 + x + 1$. Then we have $h_A = d_\infty = g = 1$ and $\deg(x) = 2$, $\deg(y) = 3$. Then, as x and y generate A over $\mathbb{F}_2$, $\rho_x = x + (x^2 + x)\tau + \tau^2$ and $\rho_y = y + (y^2 + y)\tau + x(y^2 + y)\tau^2 + \tau^3$ defines a rank one generic characteristic Drinfeld A-module over A (the reader should verify $\rho_x\rho_y = \rho_y\rho_x$).

(3) Let $A = \mathbb{F}_3[x, y]$ with $y^2 = x - x^2$. Then K is just the rational function field, so $h_K = 1$, $h_A = d_\infty = 2$ and $\deg(x) = \deg(y) = 2$. Suppose ρ is a rank one generic characteristic Drinfeld A-module, with $\rho_x = x + x_1\tau + x_2\tau^2$ and $\rho_y = y + y_1\tau + y_2\tau^2$. Then the comparison of coefficients of τ^4 in the equation corresponding to $y^2 = x - x^2$, we see that $(y_2/x_2)^{10} = -1$, and hence the coefficients can not be in A or K. (This argument generalizes to ρ of any rank.) We have $H_A = K(c)$ with $c^2 = -1$ and $\eta = 1 + x + cy$ is the fundamental unit for H_A. We write $\bar{\eta} = 1 + x - cy$ for the conjugate, and leave it to the reader to verify that $\rho_x = x + y\eta\tau + \bar{\eta}\tau^2$, $\rho_y = y + (1 + x)\eta\tau + c\bar{\eta}\tau^2$ (from [Hay79]) now defines a rank one Drinfeld A-module of generic characteristic over $\mathcal{O}_{H_A}$. In fact, we have two Galois conjugate Drinfeld A-modules corresponding to the two choices for c. It is easy to verify that these are non-isomorphic.

(4) If ρ is defined over a finite field F with q elements, then the q-th power Frobenius map is endomorphism of ρ.

We will prove in general in the next chapter, this connection with the field of definition for the rank one module (of generic characteristic) and Hilbert class fields seen in this example. Taking determinant in the higher rank case, we will see that in general, the field of definition contains the Hilbert class field.

As remarked in [DH87, 2.7], the definition of Drinfeld module can be formulated in terms of a $(A, \mathrm{End}_F(\mathbb{G}_a))$-bimodule M which is free of rank one over $\mathrm{End}_F(\mathbb{G}_a)$ and satisfies $|a| = |A/a|^r$, for $0 \neq a \in A$. The choice of a basis element for M identifies it with $\mathrm{End}_F(\mathbb{G}_a)$. We will return to this point of view in Chapter 7.

We started with a Drinfeld modules over fields rather than schemes, since we will mostly deal with those over fields or rings. If we want to define Drinfeld A-modules over a ring R, it is not enough to have just coefficients in R and the rank can not be found just by looking at top degree term, unless we normalize; since now there can be more invertible elements in $R\{\tau\}$ than just non-zero constants.

A simple calculation shows that $\sum r_i \tau^i$ is invertible, if and only if $r_0 \in R^*$ and r_i are nilpotent for $i > 0$ and that conjugating with such automorphisms, we can kill-off any top nilpotent coefficients. Achieving this for one non-constant a automatically achieves this for all.

Hence, we define Drinfeld module of rank r over R similarly, except that we ask for the top coefficient to be unit (rather than just non-nilpotent) in R, so that Drinfeld module of rank r over R gives again a Drinfeld module of rank r by any reductions modulo primes of R. This is in analogy with the usual terminology, which by elliptic curve over R means elliptic curve with good reduction everywhere at points of R. More generally,

Definition 2.2.6 Drinfeld A-module of rank r over a A-scheme S in characteristic p is an invertible sheaf $\mathcal{L}$ and a morphism of rings $\rho : A \to \mathrm{End}_S(\mathcal{L})$ such that locally over open sets where $\mathcal{L}$ is trivial, ρ_a is a polynomial in τ of τ-degree $r \deg(a)$ with top coefficient unit and the constant coefficient being the image of a under the structure map.

Drinfeld modules were called elliptic modules by Drinfeld originally. One reason for the change of terminology is that they give good analogs of elliptic curves in rank two, but not in general.

But there is another sense in which the term elliptic is appropriate: Linear differential operators with invertible top coefficient (symbol) are called elliptic. We will see more about this analogy of these $\mathbb{F}_q$-linear operators with differential operators later, eg., in Chapter 7.

2.3 Torsion points

Structural properties of torsion points, endomorphism rings will be seen to be analogs of what is seen most commonly in elliptic curves theory.

Let $\rho[a] := \{z \in \overline{F} : \rho_a(z) = 0\}$ be the set of a-torsion points of ρ, i.e., the kernel of ρ_a. In fact, this should be considered as a functor, whose L-valued points are $z \in L$ in the kernel of ρ, for an extension L of F. When we drop L, we take all torsion points in an algebraic closure. (Note that $\rho[0] = \overline{F}$, so we usually assume a to be non-zero.) For (non-zero) ideal I of A,

$$\rho[I] := \{z : \rho_i(z) = 0, \text{ for all } i \in I\} = \cap_{i \in I} \rho_i$$

is the set of I-torsion points of ρ.

Clearly, $\rho[(a)] = \rho[a]$. Note that 0 is always a-torsion, for any a, just as 1 is n-th root of unity for any n, in the classical case. The torsion set is $\mathbb{F}_q$-vector space and A-module. As in the elliptic curves case, we can define the $\wp$-adic Tate module $T_\wp(\rho)$, which is $A_\wp$-module and a Galois module.

If ρ is understood, we sometimes write Λ_a for $\rho[a]$ and λ_a for an a-torsion point, in analogy with the classical notations μ_n and ζ_n. We repeat the basic analogies:

$$a \in A \mapsto \rho_a \in \text{End}(\mathbb{G}_a) \leftrightarrow n \in \mathbb{Z} \mapsto P_n := (x \mapsto x^n) \in \text{End}(\mathbb{G}_m)$$

and $\Lambda_a \leftrightarrow \mu_n$, which is the n-torsion of the usual action P of $\mathbb{Z}$ on the multiplicative group $\mathbb{G}_m$.

Let ϕ be a Drinfeld A-module over F. For an ideal I of A, consider the left ideal of $k\{\tau\}$ generated by ϕ_i, $i \in I$. Since we have a division algorithm, this ideal is principal. We denote its monic generator by ϕ_I. Then $\phi[I]$ is clearly the kernel of ϕ_I. We leave it as an exercise to check that $\deg(\phi_I) = r \deg(I)$.

Example 2.3.1   In the example (3) of 2.2.5, the recipe above shows that $\rho_I = (x + cy) + \tau$ for the degree one non-principal ideal $I := (x, y)$. Note that $IA[c] = (x + cy)A[c]$. This is an instance of the explicit principal ideal theorem of Hayes proved in the next chapter.

Note that ρ_I need not commute with ρ_J (see [Hay79] for the full story) and so need not take $\wp$-torsion to itself. We will see in 2.4.3 and 3.5.4 that while ρ_a, for a single non-constant a, determines ρ, ρ_I does not determine ρ.

Theorem 2.3.2   *Let $\wp$ be a non-zero prime ideal of A. Let d be the dimension of $A/\wp$-vector space $\rho[\wp]$. Then $d = r$, if $\wp$ is not the characteristic of ρ and $d = r - h$, otherwise. In particular, the rank and height (when we are in finite characteristic) are natural numbers.*

Proof. Clearly, $|\rho[I]| = q^{\deg_\tau \rho_I - \text{ord}_\tau \rho_I}$. On the other hand, using the exact sequences $0 \to \rho[\wp] \to \rho[\wp^{e+1}] \to \rho[\wp] \to 0$, any $A/\wp$-basis of $\rho[\wp]$ lifts to a basis for $\rho[\wp^e]$ as an $A/\wp^e$ module, by induction and thus $|\rho[\wp^e]| = \text{Norm}(\wp^e)^d = q^{de \deg(\wp)}$.

Let $\wp^e = (a)$, say by taking e to be the class number of A. Then the comparison of the two formulas for $I = \wp^e$ gives that $de \deg(\wp)$ is $re \deg(\wp)$ or $re \deg(\wp) - he \deg(\wp)$ in the two cases respectively. $\qquad\square$

Hence, just as in the elliptic curves case, if we are in rank 2, finite characteristic $\wp$-situation, we have ordinary and supersingular cases depending on whether the number of $\wp$-torsion points is $\text{Norm}(\wp)$ or one, i.e, whether $h = 1$ or 0. We will discuss this more in 6.1.

2.4   Analytic theory

One way to show quickly that Drinfeld modules exist in abundance for any A, and to draw a parallel with more well-known analytic theory of elliptic curves is the analytic description of Drinfeld modules over C_∞ via lattices.

Definition 2.4.1   An A-lattice L is a finitely generated and discrete A-submodule of C_∞, considered A-module with the usual multiplication. Its rank is defined to be the rank of this finitely generated, torsion-free and so projective module. If F is a finite extension of K_∞ in C_∞, then we say that L is an A-lattice over F, if further, L is contained in F^{sep} and invariant under $\text{Gal}(F^{sep}/F)$.

The discreteness is equivalent to having only finitely many elements of L in any finite ball in C_∞. Important thing to notice is that, unlike the classical case over $\mathbb{C}$, where the lattices can have only rank one or two, as $[\mathbb{C} : \mathbb{R}] = 2$, now lattices of arbitrarily high rank do exist.

Similar to Weierstrass theory, we can associate the so-called exponential function $e_L(z) := z \prod'_{l \in L}(1 - z/l)$ to a lattice L. The discreteness immediately implies that the product converges for all z, giving us an entire and thus, by 2.11.4, a surjective function. Further, writing it as a limit as $d \to \infty$ of the product over the $\mathbb{F}_q$-vector space of l's of degree less than d, (which is $\mathbb{F}_q$-linear polynomial), we see that $e_L(z)$ is $\mathbb{F}_q$-linear function. Since e_L vanishes on lattice L by construction, it is periodic with L as a period lattice and gives isomorphism from C_∞/L onto C_∞ because of the surjection. So we always get affine additive algebraic group rather than say various elliptic curves in complex rank 2 case. Note that e_L has coefficients in F, if L is A-lattice over F.

Theorem 2.4.2   *Let F be a finite extension of K_∞ in C_∞. The category of Drinfeld A-modules ρ over F of rank r and of generic characteristic is isomorphic to the category of A-lattices L over F of rank r, with a morphism over F', where $F \subset F' \subset C_\infty$, from L to L' being $z \in F'$ such that $zL \subset L'$.*

Proof.   We show how to get from one data to the other and leave the rest as an exercise. Given L, let e_L be the corresponding exponential. For a non-zero a, put

$$\rho_a^L(x) = ax \prod_{0 \neq l \in a^{-1}L/L} (1 - x/e_L(l)).$$

As A is a Dedekind domain, and L is a projective A-module of rank r, it is isomorphic to $A^{r-1} \oplus I$, for a non-zero ideal I of A. This implies that $a^{-1}L/L$ is isomorphic to the sum of r copies of $A/(a)$. Thus $e_L(a^{-1}L/L)$ is a finite $\mathbb{F}_q$-vector space of dimension $r \deg(a)$ and thus $\rho_a(x)$ is a $\mathbb{F}_q$-linear polynomial of degree $q^{r \deg(a)}$. By comparing the divisors and the derivatives (i.e., the linear terms) of the two sides of the following, we get the equality

$$e_L(ax) = \rho_a^L(e_L(x)).$$

Now $\rho_{ab}^L = \rho_a^L \rho_b^L = \rho_b^L \rho_a^L$ follows from applying it to $e_L(x)$ and we see that ρ_a^L is a Drinfeld module of rank r over F.

Conversely, given ρ, we first show that there exists a corresponding exponential $e(x) = e_\rho(x) = \sum e_i x^{q^i}$, which is, by definition, a $\mathbb{F}_q$-linear entire function satisfying

$$e_\rho(ax) = \rho_a(e_\rho(x)), \quad \text{for all } a \in A$$

and normalized to have $e_0 = 1$:

First we fix a non-constant a and write $\rho_a = \sum_{i=0}^{d} a_i \tau^i$. Solving the functional equation $e(x) = \rho_a(e(a^{-1}x))$ formally by equating coefficients we get a unique solution $e_n(a^{q^n} - a) = \sum_{i=1}^{n} a_i e_{n-i}^{q^i}$, inductively starting with $e_0 = 1$.

Next we show that the functional equation is now automatically satisfied for any $b \in A$: We have $\rho_b e b^{-1} = \rho_b(\rho_a e a^{-1})b^{-1} = \rho_a(\rho_b e b^{-1})a^{-1}$, so that both e and $\rho_b e b^{-1}$ satisfy our functional equation and hence are the same by uniqueness, showing $\rho_b e = eb$, as required.

Next we estimate the coefficient size to show that $e(x)$ converges for all $x \in C_\infty$: Since $a_i = 0$ for $i > d$, for $n \geq d$, we have $e_n(a^{q^n} - a) =$

$\sum_{i=1}^{d} a_i e_{n-i}^{q^i}$. Write $r_n = |e_n|^{q^{-n}}$. It is enough to show that $r_n \to 0$. Taking q^n-th root of the recursion, we see that $|a|r_n \leq \max(|a_i|^{q^{-n}} r_{n-i})$, so that for large n, $r_n \leq \theta \max_{1 \leq i \leq d} r_{n-i}$, for some $\theta < 1$. This implies $r_n \to 0$ as required.

Now define $L = L_\rho$ to be the kernel of e. It is then clearly discrete. By the functional equation of e, it is A-module. Since $e'(x) = e_0 = 1$, it is in separable closure of F and Galois stable. Separability of e also implies that all the zeros are simple and thus $e_L = e\rho$. Hence ρ_a is given by formula above in terms of L. So $|A/a|^r = |a^{-1}L/L|$ and hence its rank is r.

The morphism corresponding to z is given by the τ polynomial corresponding to ρ_z defined by the formula above. $\square$

Remarks 2.4.3 (1) We saw that $\rho_a = \sum a_i \tau^i$ for a single non-constant a fixes e, and thus it fixes ρ. More explicitly, comparing coefficient of τ^k in $\rho_a \rho_b = \rho_b \rho_a$ gives recursive determination

$$(a^{q^k} - a)b_k = \sum_{i=1}^{k} (a_i b_{k-i}^{q^i} - a_i^{q^{k-i}} b_{k-i})$$

of coefficients of $\rho_b = \sum b_j \tau^j$ for any $b \in A$.

(2) Since the exponential is not one to one, there is no inverse function globally defined, but as usual, we mean by a logarithm, choice of an inverse function in some range. We also define $l(z) = \log(z) = \log_\rho(z)$ to be a power series around origin by formally inverting the power series for e. We then have $e(l(z)) = z$ and $l(\rho_a(z)) = al(z)$, when both sides make sense. We will see that it converges for z of degree less than the degree of smallest non-zero period of e. See also [Gos96, 14.4.2].

(3) The Theorem implies that for any A and r, there are Drinfeld A-modules of rank r defined over K_∞^{sep}. In fact, if the rank is one, then by suitable scaling of the given lattice, we get a lattice in K_∞. As all analytic constructions above take place in K_∞, the corresponding Drinfeld module, isomorphic to the one we started with, is defined over K_∞.

(4) From the general description of A-lattices given above, we see that class group acts on isomorphism classes of lattices and thus of Drinfeld modules. In rank one, we see that there are h_A-isomorphism classes of Drinfeld modules over C_∞, one corresponding to each ideal class. Let us see the action directly on the Drinfeld modules: We have seen that the left ideal of $F\{\tau\}$ generated by ρ_i, $i \in I$ is principal generated by monic ρ_I. Since this ideal is carried into itself under right multiplication by ρ_a, we

have $\rho_I \rho_a = \rho'_a \rho_I$, for unique ρ', which is seen to be a Drinfeld module (see [Hay, pa. 7]). We write it as $I * \rho$. If $I = iA$ is a non-zero principal ideal, and l is the leading coefficient of ρ_i, then $(I * \rho)_a = l^{-1} \rho_a l$, thus isomorphic to ρ. We have $I * (J * \rho) = (IJ) * \rho$ and $\rho_{IJ} = (J * \rho)_I \rho_J$. Thus class group acts on isomorphism classes of Drinfeld modules over any F.

(5) For a rank one Drinfeld A-module ρ over an extension of K_∞, with the corresponding period lattice $\tilde{\pi}I$ say, we can give, as in the classical case, an analytic description of the torsion module using the corresponding exponential: $\rho[a] = \{e_\rho(\tilde{\pi}b/a) : b \in I\}$, or more generally, $\rho[J] = \{e_\rho(\tilde{\pi}r) : r \in J^{-1}I\}$.

2.5 Explicit calculations for Carlitz module

It is always useful to have some concrete examples and formulas to play with. Let us first figure out explicitly the exponential, the logarithm, formulas for the coefficients and the fundamental period for the Carlitz module.

Solving for the coefficients of $e(z) = \sum z^{q^i}/D_i$ from the basic equation $e(tz) = te(z) + e(z)^q$ and $D_0 = 1$ gives $D_i = (t^{q^i} - t)D_{i-1}^q$, for $i > 0$. We use the short-form

$$[i] := t^{q^i} - t.$$

Then, with the usual convention that the empty product is 1,

$$D_i = (t^{q^i} - t)(t^{q^i} - t^q) \cdots (t^{q^i} - t^{q^{i-1}}) = [i][i-1]^q \cdots [1]^{q^{i-1}}.$$

Since degree of D_i is iq^i, it follows (again) that $e(z)$ converges for all z.

Since $e(z)$ is periodic, it can not have global inverse, but we can solve for the inverse function $l(z) = \sum(-1)^i z^{q^i}/L_i$, called logarithm again, around origin. Again solving the basic functional equation $tl(z) = l(tz + z^q)$, we get the recursion $L_i = (t^{q^i} - t)L_{i-1}$, which, with $L_0 = 1$, implies

$$L_i = (t^{q^i} - t)(t^{q^{i-1}} - t) \cdots (t^q - t) = [i][i-1] \cdots [1].$$

Since the degree of L_i is $q(q^i - 1)/(q - 1)$, we see that $l(z)$ converges for z of degree less than $q/(q - 1)$, which, as we will see, is the degree of the fundamental period $\tilde{\pi}$ of the Carlitz module.

Now let us see a few ways to get various formulas for the coefficients of the Carlitz module $C_a = \sum C_{a,k} \tau^k$:

Since t generates $\mathbb{F}_q[t]$ over $\mathbb{F}_q$, we immediately get recursive formulas for $C_{a,k}$ from $C_{\sum a_j t^j} = \sum a_j C_{t^j}$ and $C_{t^j} = C_{t^{j-1}}(t + \tau)$.

Another way to get a recursive formula is to use (1) of 2.4.3 for say $a = t$: Equating coefficients in $C_a C_t = C_t C_a = (t + \tau)(\sum C_{a,k}\tau^k) = (\sum C_{a,k}\tau^k)(t + \tau)$, gives recursion relations $C_{a,0} = a$, $C_{a,k} = (C_{a,k-1}^q - C_{a,k-1})/(t^{q^k} - t)$.

To get a direct formula, we note that $e(al(z)) = C_a(e(l(z))) = C_a(z)$. Let us write $e(xl(z)) = \sum_{k=0}^{\infty} \binom{x}{q^k} z^{q^k}$, so plugging in the formulas obtained for the exponential and the logarithm we get

$$\binom{x}{q^k} = \sum_{i=0}^{k}(-1)^{k-i}\frac{x^{q^i}}{D_i L_{k-i}^{q^i}}, \qquad C_{a,k} = \binom{a}{q^k}.$$

Since $C_{a,d} = 0$, if $\deg(a) < d$, this provides q^d zeros a for the polynomial $\binom{x}{q^d}$ of degree q^d. In other words, if we let

$$e_d(x) := \prod_{\deg(a)<d}(x - a)$$

then $\binom{x}{q^d} = e_d(x)/p_d$ for some $p_d \in K$. Since $\binom{t^d}{q^d} = C_{t^d,d} = 1$, we see that $p_d = e_d(t^d)$ is the product of monic polynomials of degree d. Let us try to get a formula for it: The coefficient of x is given by $(-1)^d/L_d$, and also by $\prod(-a)/p_d = (-1)^d(p_0 \cdots p_{d-1})^{q-1}/p_d$. So $p_d/L_d = (p_0 \cdots p_{d-1})^{q-1}$ and hence

$$p_{d+1} = L_{d+1}(p_0 \cdots p_d)^{q-1} = L_{d+1}(p_d/L_d)p_d^{q-1} = (L_{d+1}/L_d)p_d^q = [d]p_d^q.$$

Hence $p_d = D_d$.

Let us summarize these useful facts. Let us first define some convenient **notation** for general A, or more generally for ideals $\mathcal{A}$ of A, even though in the rest of this section, after this paragraph, we will specialize to $A = A = \mathbb{F}_q[t]$. We denote by $\mathcal{A}_d$, $\mathcal{A}_{<d}$, $\mathcal{A}_+$, $\mathcal{A}_{d+}$ etc. the set of elements of $\mathcal{A}$ of degree d, degree less than d, monic, monic of degree d etc. We similarly define $A_{\leq d}$, $\mathbb{Z}_{>0}$ etc. (When $d_\infty > 1$, more useful sign convention is explained in 4.5.) Then,

$$D_d = \text{Product of monic polynomials of degree } d = \prod_{a \in A_{d+}} a,$$

and

$$(-1)^d \prod_{0 \neq a \in A_{<d}} a = (D_0 D_1 \cdots D_{d-1})^{q-1} = D_d/L_d.$$

Carlitz, who discovered these facts, first found these formulas and formulas for $e_d(x)$ using Moore determinants (we explain his approach in 2.11.2) and then went on to get the Carlitz exponential and the Carlitz module.

Yet another way to see this is by using theory of finite fields and careful counting: Note that $[k]$ is the defining polynomial of $\mathbb{F}_{q^k}$ and hence can be interpreted as product of all monic primes of $\mathbb{F}_q[t]$ of degree dividing k.

This implies that $L_d = [d][d-1] \cdots [1]$ is the least common (monic) multiple of the monic polynomials of degree d: for a given prime $\wp$, its multiplicity in both is clearly $\lfloor d/\deg(\wp) \rfloor$.

Another implication is that $D_d = [d][d-1]^q \cdots [1]^{q^{d-1}}$ is the product of monic polynomials of degree d: This again follows by counting a multiplicity of a prime $\wp$ in the product. The counting is exactly analogous to counting of multiplicity in $n!$ classically: For $\wp$ with $\mathrm{Norm}(\wp) \leq q^d$, there are $\lfloor q^d/\mathrm{Norm}(\wp)^e \rfloor$ (we can remove the integral parts here, if the denominator is not greater than the numerator) multiples of $\wp^e$ out of q^d monic polynomials of degree d. (The nice behavior of the Riemann-Roch dimension sequence for $\mathbb{F}_q[t]$ gives a regular supply of divisors.) In other words, the multiplicity of $\wp$ in both sides is easily seen to be $\sum_{e>0} \lfloor q^d/\mathrm{Norm}(\wp)^e \rfloor$. This was first noticed by Sinnott [Gos78].

Using this, yet another way [Gos96] to get the formula for $e_d(x)$ is to prove then Carlitz recursion formula $e_d(x) = e_{d-1}^q(x) - D_{d-1}^{q-1} e_{d-1}(x)$ directly by roots comparisons.

From this formula, we see that for a prime $\wp$, $C_\wp(x)$ is Eisenstein polynomial at $\wp$. In fact, since we know the degrees and the prime factorizations of D_i's and L_i's, it follows from the formula for e_k and $\binom{x}{q^k}$ that if $\wp$ is a monic prime of degree d, then $\binom{\wp}{q^d} = 1$ and for $k < d$, we have

$$v_\wp\left(\binom{\wp}{q^k}\right) = 1, \quad \deg\left(\binom{\wp}{q^k}\right) = (d-k)q^k.$$

Let us now find formulas for the fundamental period $\tilde{\pi}$ in a few ways: First we follow the Carlitz line of thought mentioned above:

$$\sum \frac{\tilde{\pi}^{q^i} z^{q^i}}{D_i} = e(\tilde{\pi}z) = \tilde{\pi}z \prod \left(1 - \frac{z}{a}\right) = \tilde{\pi} \sum (\lim \frac{(-1)^i L_k}{L_{k-i}^{q^i}}) \frac{z^{q^i}}{D_i}$$

where the last equality uses the formula for $e_k(x)$, with the adjusted signs and denominator by $\prod(-a) = D_k/L_k$, as derived above.

So $(q^i - 1)$-th power of $\tilde{\pi}$ is the limit, hence degree count gives

$$\deg(\tilde{\pi}) = q/(q-1).$$

Let us write temporarily $\tilde{\pi}_j := (-[1])^{1/(q-1)}[1]^{(q^j-1)/(q-1)}/L_j$. Then we see that $(-1)^i L_k/L_{k-i}^{q^i} = \tilde{\pi}_{k-i}^{q^i}/\tilde{\pi}_k$. This gives the Carlitz formula

$$\tilde{\pi} = (-[1])^{1/(q-1)} \lim \frac{[1]^{(q^j-1)/(q-1)}}{L_j} = (-[1])^{1/(q-1)} \lim \prod_{i=1}^{j}(1 - \frac{[i-1]}{[i]})$$

where the ambiguity in $\mathbb{F}_q^*$ arising in $(q-1)$-th root is similar to the sign ambiguity in trying to extract $2\pi i$ from $2\pi i \mathbb{Z}$.

Yet another formula for the period, using the same ideas of $e(\tilde{\pi}z) = \tilde{\pi}e_A(z)$ and formulas for $e_A(z)$, is as follows: The constant (or linear, depending on how you view) term t of C_t being the product of non-zero t-torsion points $e(\theta\tilde{\pi}/t)$, $\theta \in \mathbb{F}_q^*$, we get $t = -\tilde{\pi}^{q-1}e_A(1/t)^{q-1}$. We develop this further, for general A, in Chapter 4.

Next we give another formula using a trick (to which we will return in Chapter 8, and in fact, this formula is connected to the previous one with solitons of Chapter 8) of Anderson [AT90, cor. 2.5.8]: Note that $e(\tilde{\pi}/t)$ is a $(q-1)$-th root of $-t$. Consider the generating function

$$E(T) := \sum e(\tilde{\pi}/t^{n+1})T^n$$

of t^n-division points of the Carlitz module. Using the expansion of the exponential, we see that it is a holomorphic function of T plus $\tilde{\pi}/T \sum(T/t)^{n+1} = -\tilde{\pi}/(T-t)$, so that the residue at $T = t$ is $-\tilde{\pi}$. Let us denote $f^{(k)}(T) = \sum f_n^{q^k}T^n$, for $f(T) = \sum f_n T^n$. Then using the Carlitz action, we see the functional equation

$$E^{(1)}(T) = \sum e(\frac{\tilde{\pi}}{t^{n+1}})^q T^n = \sum (e(\frac{\tilde{\pi}}{t^n}) - te(\frac{\tilde{\pi}}{t^{n+1}}))T^n = (T-t)E(T).$$

Clearly $(-t)^{1/(q-1)} \prod_{n=0}^{\infty}(1 - T/t^{q^n})^{-1}$ also satisfies the same functional equation and is thus seen to be the same as $E(T)$. Hence now a straight comparison of the residues gives

$$\tilde{\pi} = (-t)^{1/(q-1)}t \prod_{n=1}^{\infty}(1 - \frac{1}{t^{q^n-1}})^{-1}.$$

Most of the arithmetic for Carlitz module takes place inside much smaller subfield $K_\infty(\tilde{\pi}) = \mathbb{F}_q((-t)^{1/(q-1)})$ (in a sense analog of $\mathbb{R}(2\pi i) = \mathbb{C}$) of C_∞. A natural question is what an analog of π or 2π is, given that now we have a good analog $\tilde{\pi}$ of $2\pi i$. We do not know. There is no circle-diameter relation to help. Classically, $2\pi i$ and the real number 2π have the same absolute value, but degree of $\tilde{\pi}$ is $q/(q-1)$, which is non-integral if $q > 2$, whereas degree is integral for elements of K_∞. (Note that $\tilde{\pi} \in K_\infty$ if $q = 2$.) There are various ways to multiply by $(q-1)$-th root to land up in K_∞. For example, we can take away the terms before the product signs, in the displayed formula above, i.e., take one-unit part using t as in the gamma context of Chapter 4. A naive analogy would be $(-[1])^{1/(q-1)}$ as an analog of i in this context, giving the one-unit $\lim [1]^{(q^j-1)/(q-1)}/L_j \in K_\infty$ as an analog of 2π. It would be nice to have another justification.

Now we turn to torsion points. We will study them in detail in the next chapter. But some useful trivial things to keep in mind are that the a-torsion points form A module, isomorphic to A/a, in our case. The t-torsion consists of 0 and $(-t)^{1/(q-1)}$ ($q-1$ of these). If $q > 2$, and $a \neq 0$, the non-zero torsion points are not rational over $\mathbb{F}_q[t]$. But when $q = 2$, $(q-1)$-th root being rational, by above we get rational non-zero torsion points t, $t+1$ and thus 1, as torsion points for $a = t$, $t+1$ and $t(t+1)$ respectively. This should be compared with the second roots of unity being rational over $\mathbb{Q}$ and the $(q-1)$-st roots of unity being rational over $\mathbb{F}_q(t)$ or K. (See the end of 9.2 where this shows up in an interesting fashion.)

2.6   Reductions

Let ρ be a Drinfeld A-module of rank r over an A-field F, which is equipped with a discrete valuation v and with all the coefficients of ρ_a's integral at v. Let $\mathcal{O}_v$ be the valuation ring at v, with the maximal ideal m_v and the residue field f_v. We denote the naive reduction map obtained by looking at the coefficients modulo m_v by ρ^v. Note that this map might not give a Drinfeld module if all terms containing τ-powers get killed in reduction.

Definition 2.6.1   With the notation as above, ρ is said to have **stable** (**good** respectively) **reduction** at v, if there is isomorphic Drinfeld module ϕ over F with coefficients integral at v such that ϕ^v is a Drinfeld module (of rank r respectively). We say ρ has **potential stable** (**potential good** respectively) reduction, if the corresponding conditions are met over an extension of F and v.

Remarks 2.6.2   (1) If ρ has an isomorphic model over $\mathcal{O}_v$, then ρ has a good reduction.

(2) For $r = 1$, stable is the same as good.

(3) By considering the divisibilities of the top coefficients of images under ρ of generators of A, we see that Drinfeld module over a function field has good reduction at all but finitely many places.

Examples 2.6.3   (1) Any Drinfeld $\mathbb{F}_q[t]$-module ρ over $\mathbb{F}_q[t]$ (so with the top coefficient in constant field and in particular, the Carlitz module) has good reduction at all places of $\mathbb{F}_q[t]$ and ρ itself is a good model.

This should be compared with the fact that there is no elliptic curve over $\mathbb{Z}$. But note that there do exist elliptic curves over some rings of integers. For example, we have Setzer's $y^2 + \sqrt{6}xy - y = x^3 - (2 + \sqrt{6})x^2$ over $\mathbb{Z}[\sqrt{6}]$, with discriminant $(5 + 2\sqrt{6})^3$, which is a unit. Also over function fields, we have constant elliptic curves.

(2) Consider Drinfeld $\mathbb{F}_q[t]$-modules $\rho^{(i)}$ over $\mathbb{F}_q(t)$ given by $\rho_t^{(1)} = t + t\tau + t\tau^2$ and $\rho_t^{(2)} = t + \tau + t\tau^2$ and v corresponding to prime t. Then $\rho^{(1)}$ does not have stable reduction over $\mathbb{F}_q(t)$, but has potential good reduction because conjugation by $t^{1/(q^2-1)}$ gives isomorphic model $\phi_t = t + t^{q/(q+1)}\tau + \tau^2$. On the other hand $\rho^{(2)}$ has a stable, but not good reduction.

Theorem 2.6.4   *Every Drinfeld module ρ, in the setting of the beginning of the section, has potential stable reduction. In particular, we have potential good reduction for ρ of rank one.*

Proof.   Let $m_a = \min_{i>0}(v(a_i)/(q^i - 1))$, where $\rho_a = \sum a_i \tau^i$. Choose generators $a(i)$ of A over $\mathbb{F}_q$ and let $m = \min m_{a(i)}$. In a suitable finite extension of F we have valuation w extending v and an element x with $w(x) = m$. It is easy to see that conjugating by x, we get an isomorphic model over this extension, which has stable reduction. $\square$

Takahashi   [Tak82] proved the following analog of Neron-Ogg-Shafarevich criterion for good reduction of abelian varieties.

Theorem 2.6.5   *Let ρ be as in the beginning of the section. If $\wp$ is a prime different from the characteristic of the residue A-field f_v, then ρ has good reduction at v if and only if the Tate module $T_\wp$ is unramified at v (i.e., the inertia at place above v in F^{sep} acts trivially).*

2.7 Endomorphisms

Let us start with some simple observations and examples.

Example 2.7.1   Since $\tau f = f\tau$ for $f \in \mathbb{F}_q$, we have $\mathbb{F}_q^* \subset \mathrm{Aut}(\rho)$ for any ρ. Similarly, $\mathbb{F}_{q^r}^* \subset \mathrm{Aut}(\rho)$, for $\mathbb{F}_q[t]$-module $\rho_t = t + \tau^r$. Also, if ρ is defined over $\mathbb{F}_q$, then $\mathrm{End}_{\mathbb{F}_q}(\rho)$ which, in general, is a subset of (commutative ring) $\mathbb{F}_q\{\tau\}$, in fact, is equal to it.

Now $\rho(A) \subset \mathrm{End}\,(\rho)$, just as $\mathbb{Z}$ sits in the endomorphism ring of the multiplicative group or of an elliptic curve. The theorem below immediately implies that for Carlitz module, this is the full ring of endomorphisms, while for $\rho_t = t + \tau^r$, the full ring is $\mathbb{F}_{q^r}[t]$ which is A-module of rank r. We consider this as a complex multiplication analog. For $\rho_t = \tau^r$ of characteristic t and rank and height equal to r, we have endomorphism ring $\mathbb{F}_{q^r}\{\tau\}$ of rank r^2 over A.

We have the following basic theorem, as in the elliptic curves case.

Theorem 2.7.2   *Let ρ be a Drinfeld A-module over algebraically closed field F of rank r. Then $\mathrm{End}(\rho)$ is a projective A-module of rank at most r^2. It is commutative of rank at most r, if ρ is of generic characteristic.*

Proof.   (Sketch): If a is prime to the characteristic, we know that $\rho[a]$ is A/a-module of rank r and that $\mathrm{End}(\rho) \otimes A/a \to \mathrm{End}(\rho[a])$ is injective, because the endomorphism vanishing on the a-torsion will be divisible by ρ_a and hence is zero on the left. Hence the first claim follows. In generic characteristic, since we need to deal with only finitely generated field (generated by all coefficients) over K, we can assume, without loss of generality, that we are over C_∞ and then commutativity and the rank assertion is clear from the lattice endomorphisms comparison.   □

Theorem 2.7.3   *Given an isogeny $\phi : \rho \to \rho'$, there exists an isogeny $\psi : \rho' \to \rho$ such that $\psi\phi = \rho_a$, and $\phi\psi = \rho'_a$, for some $a \in A$.*

Proof.   If ϕ is a separable isogeny, then it is of smallest degree killing its kernel, which is a finite A-module and hence killed by ρ_a for some a. By division algorithm in $F\{\tau\}$, we then see that $\rho_a = \psi\phi$. Now $\psi\rho'\phi = \psi\phi\rho = \rho_a\rho = \rho\rho_a = \rho\psi\phi$, so that $\psi\rho' = \rho\psi$ and ψ is isogeny from ρ' to ρ as claimed. Now $\phi\psi\phi = \phi\rho_a = \rho'_a\phi$, so cancellation gives the second equality.

General case is similar [Gos96, 4.7] using scheme theoretic kernel.   □

Now we focus on Drinfeld modules defined over finite A-fields. Again there are many similarities with the elliptic curves over finite fields theory, where theory is simpler because of complex multiplications available. The Riemann hypothesis is now slightly simpler, almost built in the formalism.

Let ρ be a rank r Drinfeld A-module of characteristic $\wp$ over F of cardinality q_1, which is then equal to $q^{m \deg \wp}$, where $m = [F : A/\wp]$. We identify A with its image $\rho(A)$ in the non-commutative ring $F\{\tau\}$. Let π be the (q_1-th power) Frobenius endomorphism of ρ. Since π commutes with ρ, $K(\pi)$ is a commutative field, even though $K\{\tau\}$ inside the quotient division ring of $F\{\tau\}$ is, in general, not commutative.

Under the identification of A with its image, ρ_a (choose a non-constant) has the normalized absolute value (controlled by the top τ-degree term $\tau^{r \deg(a)}$) $q^{\deg(a)}$, so that π has (or rather 'should have', as this passage from absolute value to commutative A to commutative $A(\pi)$ inside non-commutative $\mathbb{F}_q\{\tau\}$ needs some justification) absolute value $q_1^{1/r}$.

Note that the weights in rank r are $1/r$, whereas elliptic curves which are of rank 2 give only weight $1/2$ and classically weights can only be half-integral.

Since we are in characteristic $\wp$, $\rho_\wp$ has factor τ, so its high power is divisible by π. Hence, π lies over $\wp$.

Hence the norm of π from $K(\pi)$ to K is power of $\wp$, and equals $\wp^{m[K(\pi):K]/r}$ by comparing the degrees. Now π being an endomorphism of A-module of degree r, is integral of degree at most r over A, since it satisfies its characteristic polynomial.

Hence we get Gekeler's result [Gek91; Yu95]:

Theorem 2.7.4 *If there is a Drinfeld A-module of characteristic $\wp$ defined over F of degree m over $A/\wp$, then $\wp^m$ is a principal ideal in A.*

So there is no Drinfeld A-module of any rank over A, unless A has class number 1, because otherwise its reduction at a non-principal $\wp$ would lead to a contradiction.

Much more can be proved efficiently and rigorously, as in the elliptic curves case, by using the well-developed theory of skew-fields, division algebras etc. As we will not need it, rather than repeating the arguments in the following references and the basic material referred to there, we will just refer to [Dri74; Dri77b; Gek91; Yu95; Gos96], where the proof of the following main theorem of Honda-Tate theory can also be found.

Call an element π of $\overline{K}$ to be a **Weil number** over F of rank r, if (i) it is integral over A, (ii) there is only one place of $K(\pi)$ which is a zero of π and it lies above $\wp$, (iii) there is only one place of $K(\pi)$ above ∞, (iv) $|\pi| = q_1^{1/r}$, where the absolute value is the unique extension to $K(\pi)$ of the normalized absolute value at ∞ of K, (v) $[K(\pi) : K]$ divides r.

Theorem 2.7.5 *Frobenius endomorphism over F gives a bijective map from the set of isogeny classes of Drinfeld A-modules of rank r over F (as above) onto the set of conjugacy classes of Weil numbers of rank r.*

Remarks 2.7.6 Given a Weil number π, we use its properties to get an embedding of $K(\pi)$ into the quotient division ring of $F\{\tau\}$, which by the restriction to A gives the Drinfeld module with desired properties to show the surjectivity.

2.8   Field of definition

Definition 2.8.1 A subfield of F is called a field of definition for ρ (of any rank and characteristic), if ρ is isomorphic over $\overline{F}$ to a Drinfeld module with all coefficients in this subfield.

As examples, we saw by the analytic theory in 2.4.3 that K_∞ is a field of definition for Drinfeld module of rank one over C_∞, and in 2.2.5, 2.7.4 that K is not a field of definition for a rank one module in general.

Theorem 2.8.2 *Let ρ be a Drinfeld A-module over C_∞ of any rank. There exists a field of definition K_ρ, finitely generated over K, which is contained in every field of definition for ρ.*

Proof. Fix a non-constant $a \in A$ and let $\rho_a = a + \sum a_i \tau^i$. Note that isomorphism (i.e., conjugation) by $c \in C_\infty^*$ gives $c\rho_a c^{-1} = a + \sum c^{1-q^i} a_i \tau^i$.

Let j run through the (finite) subset consisting of i's with $a_i \neq 0$. Write the greatest common divisor, say G, of $q^j - 1$'s as their linear combination $G = \sum e_j(q^j - 1)$. Then the elements

$$I_k := a_k \left(\prod a_j^{e_j}\right)^{(1-q^k)/G}$$

as k runs through the set of j's, are invariant under $a_j \to c^{1-q^j} a_j$ and hence belong to every field of definition for ρ.

We will show that field K_ρ generated over K by I_j's is a field of definition for ρ:

If we choose c with by $c^G = \prod a_j^{e_j}$, then $I_k = c^{1-q^k} a_k$, so that $(c\rho c^{-1})_a$ has coefficients in K_ρ. So by the first remark of 2.4.3, this is true for every $a \in A$. $\square$

The theorem shows how to get the smallest field of definition K_ρ from a single equation ρ_a for some non-constant a for some model ρ. This is analogous to getting the j-invariant (and hence the smallest field of definition) from say the Weierstrass form of the elliptic curve over $\mathbb{C}$.

In the rank one situation, since ρ and $I * \rho$ have the same field of definition, and $I * \rho$ runs through the isomorphism classes, as I runs through the representatives of the ideal class group of A, K_ρ is then independent of choice of ρ. We will see in 3.4 that $K_\rho = H_A$ in this case.

2.9 Points on Drinfeld modules

The analogy with (multiplicative group and) elliptic curves suggests studying structure of the set of F-rational points of Drinfeld module ρ. This is just an additive group of F, considered as A-module via ρ.

Let us first consider the simplest case in finite characteristic. The Carlitz module C is analog of multiplicative group. Indeed, the $\mathbb{F}_{p^n}$-points of multiplicative group give $\mathbb{F}_{p^n}^*$ which is isomorphic to $\mathbb{Z}/(p^n-1)\mathbb{Z}$ as abelian group, i.e., $\mathbb{Z}$-module. Similarly, the A-module structure of $\mathbb{F}_{\wp^n}$ (defined as the unique degree n extension of $\mathbb{F}_\wp = A/\wp$), via reduction of Carlitz action modulo $\wp$, is isomorphic to A-module $A/(\wp^n-1)A$ now with usual A-action: $C_{\wp^n-1}(a) = a^{q^{n\,\deg(\wp)}} - a$ kills it.

For analog of Artin's conjecture on primitive roots for rank one Drinfeld modules (rather than Brillharz function field case mentioned in 1.3), see [HY01].

If we consider, on the other hand, K-points of C, for example, it is easy to see that we get A-module, which is not finitely generated. This is analogous to the fact that the multiplicative group $\mathbb{Q}^*$ is a product of its finite torsion group $\{\pm 1\}$ and a free abelian group of countable rank with basis the prime numbers. The multiplicative group for a number field has a similar structure, as can be shown by using Dirichlet unit theorem and ideal factorization theorem.

But unlike the classical case, where we have Mordell-Weil theorem of finite generation of rational points, for rank two case of elliptic curves, we do not have a nice analog of Mordell-Weil theorem in any rank. In fact, using height functions developed by Denis and himself, Poonen [Poo95] proved the following structure theorem, but now for all ranks:

Theorem 2.9.1 *Let ρ be a Drinfeld A-module over a finite extension F of K. Let $\mathcal{O}_S$ be a ring of S-integers in F. Let M be a non-trivial finite* *separable extension of F and F^{per} be perfection of F. Then each of F, $\mathcal{O}_S$, $F/\mathcal{O}_S$, M/F, F^{per}, considered as A-module via ρ is isomorphic to the direct sum of a free module of countable rank with a finite torsion module.*

We postpone the discussion of the torsion to section 6.1.

2.10   Adjoints and duality

We now describe the notion of adjoint which arose in work of Ore with hints on related duality, which was developed very nicely by Elkies and Poonen as a full-fledged duality theory in $\mathbb{F}_q$-linear calculus and in Drinfeld modules setting. We refer to [Poo96; Elk99; Gos96; Ore33; ABP] for the proofs, as we will not make direct use of unproved assertions from this section.

We will see in 7.1, 7.8 that Drinfeld module setting can be generalized for many purposes by replacing τ by τ^{-1} over a perfect field (or by any automorphism of infinite order).

Let C be algebraically closed field containing $\mathbb{F}_q$ and complete with respect to non-archimedean absolute value $|\ |$. Put

$$\mathcal{F} := \{f = \sum_{i\in\mathbb{Z}} f_i z^{q^i}, f_i \in C, \text{ convergent for all } z \in C\}.$$

Note that $f \in \mathcal{F}$ if and only if $|f_i|^{q^{-i}}$, $|f_{-i}| \to 0$ as $i \to \infty$, and that $\mathcal{F}$ is closed under addition and composition (denoted by $\circ$).

Theorem 2.10.1   *If $f \in \mathcal{F}$ is non-zero, then the kernel of f is discrete if and only if $f_i = 0$ for $i << 0$, and it is compact if and only if $f_i = 0$ for $i >> 0$.*

Definition 2.10.2   Define the adjoint of $f = \sum f_i \tau^i$ to be $f^* := \sum \tau^{-i} f_i = \sum f_i^{q^{-i}} \tau^{-i}$.

We have $f^{**} = f$, $(f \circ g)^* = g^* \circ f^*$, and if $f \in \mathcal{F}$, then $f^* \in \mathcal{F}$.

Let ρ be a Drinfeld module with the corresponding exponential e, logarithm l. Taking adjoints of usual

$$e(az) = \rho_a(e(z)), \quad al(z) = l(\rho_a(l(z))), \quad \rho_{ab}(z) = \rho_a(\rho_b(z))$$

we get

$$l^*(az) = \rho_a^*(l^*(z)), \quad ae^*(z) = e^*(\rho_a^*(z)), \quad \rho_{ab}^*(z) = \rho_a^*(\rho_b^*(z)), \quad a, b \in A,$$

so that adjoint of exponential is 'logarithm' of adjoint of Drinfeld module and vice versa, except adjoint of logarithm is formal (whereas adjoint of exponential is entire by above). The author had looked at the notion of adjoint independently and had noticed these curious connections before he became aware of Ore's work through David Goss. See also 4.14, 6.5, where we see again how D_i's and L_i's get connected. Note also that e^* appears in [Car60].

Ore gave a nice formula connecting the root spaces of adjoints:

Theorem 2.10.3 *Let $f = \sum_{i=0}^{d} f_i \tau^i$, with $f_0, f_d \neq 0$, so that f is separable. Let $\{w_1, \cdots, w_d\}$ be a $\mathbb{F}_q$-basis of the kernel W of f. Put*

$$\overline{w_i} := M(w_1, \cdots, \hat{w_i}, \cdots, w_d)/M(w_1, \cdots, w_d),$$

where M is the Moore determinant of Section 2.11.2 and the entry with hat is supposed to be omitted. Then $\overline{w_i}^q$'s give a $\mathbb{F}_q$-basis for the kernel of f^*.

Proof. Without loss of generality, we assume that f is monic. If β is a root of $f^*(x)$, then $f(\tau) = (\tau - \beta^{(1-q)/q}) P(\tau)$, for some P. (This is an easy exercise on adjoints and manipulations in linear polynomials using division algorithms in Section 2.11.1, noting that β is a root of $x^{q^{-1}} - \beta^{(1-q)/q} x$.)

If $v_1, \cdots, v_{d-1}$ is a basis of the kernel $W_1 \subset W$ of P, then we have (1) $f(\tau) = (\tau - P(w)^{q-1}) P(\tau)$, for any $w \in W - W_1$ (by degree counting and checking vanishing of the right side by the substitution of the roots $cw + \sum c_i v_i$, with $c, c_i \in \mathbb{F}_q$ of f), (2) $P(w) = M(v_1, \cdots, v_{d-1}, w)/M(v_1, \cdots, v_{d-1})$ (by the Moore determinant identity of Section 2.11.2).

Hence $\beta = c P(w)^{-q}$, with $c \in \mathbb{F}_q^*$, is in the span of the Moore determinant quotients claimed, by the properties of the determinants under a change of a basis. □

Let us now look at the pairing on the root spaces defined by Elkies and Poonen:

If $f = \sum f_i \tau^i \in \mathcal{F}$ is such that $f(1) = 0$, then we can divide by $1 - \tau$: Let $b_n := \sum_{-\infty}^{n} f_i = -\sum_{n+1}^{\infty} f_i$ and $b := \sum b_i \tau^i$. Then $b \in \mathcal{F}$, using the two expressions for coefficients to check convergence on left and right. Also,

$$f = \sum f_i \tau^i = \sum (b_i - b_{i-1}) \tau^i = b \circ (1 - \tau).$$

For $\alpha \in \ker f$ and $\beta \in \ker f^*$, put

$$\langle \alpha, \beta \rangle_f := g_\alpha^*(\beta),$$

where $g_\alpha \in \mathcal{F}$ such that $f \circ \alpha = g_\alpha \circ (1 - \tau)$ is obtained by applying the remark above to $f \circ \alpha$.

Theorem 2.10.4   *If $f \in \mathcal{F}$ is non-zero, then the pairing*

$$\langle \, , \, \rangle_f : ker \, f \times \, ker \, f^* \to \mathbb{F}_q$$

exhibits the two kernels as Pontryagin duals as topological $\mathbb{F}_q$-vector spaces.

In context of Drinfeld module torsion, we get Poonen-Taguchi's

Theorem 2.10.5   *Let ρ be a Drinfeld module over a field F. Fix a nonzero $a \in A$. Let $\rho[a]$ and $\rho^*[a]$ denote the kernels of ρ_a and ρ_a^* respectively on $\overline{L}$. Then*

$$[\, , \,]_a : \rho[a] \times \rho^*[a] \to \widehat{(A/a)} := Hom_{\mathbb{F}_q}(A, \mathbb{F}_q), \alpha, \beta \to (\overline{a} \to \langle \rho_{\overline{a}}(\alpha), \beta \rangle_a)$$

is a Galois-equivariant (Galois acts trivially on right) perfect pairing of finite A-modules.

For further properties analogous to those of Weil pairing for abelian varieties, see [Poo96]. (Analog of Weil pairing duality, more along classical lines, was defined by Anderson (unpublished) and later considered by Taguchi, using exterior power / determinant construction (see eg., 7.1.3 (4)) in category of shtukas or t-motives of Chapter 7. For a detailed treatment and applications, see University of Groningen thesis of 2003 of van der Heiden).

Theorem 2.10.6   *Let ρ be a Drinfeld A-module over C_∞, with $e(z) \in C_\infty[[z]]$ being the associated exponential. Then there is a natural pairing $ker \, e \times ker \, e^* \to \hat{A} := Hom_{\mathbb{F}_q}(A, \mathbb{F}_q)$ which exhibits $ker \, e^*$ as the Pontryagin dual of $ker \, e$ as an A-module.*

If the lattice $ker \, e$ is isomorphic to the direct sum $I_1 \oplus \cdots \oplus I_r$, then $ker \, e^*$ is isomorphic to $K_\infty/(I_1^{-1}J) \oplus \cdots \oplus K_\infty/(I_r^{-1}J)$ as a topological A-module, where J is an ideal of A isomorphic to Ω_A (differentials on X regular away from ∞) as A-module. Further, $ker \, e^*$ is the closure of union of $\rho^*[a]$'s.

2.11   Useful tools in non-archimedean or finite field setting

In order to keep the flow unhindered, we collect here some algebraic and analytic tools that we need elsewhere.

We will start with some tools to deal with $\mathbb{F}_q$-linear functions. Then we look at some basic calculus facts in the non-archimedean domain, including Newton polygons. Some of these we have already used or referred in this chapter. The Dwork's formula will be used in Chapter 5.

(a) Properties of $k\{\tau\}$

We recall here some basic algebraic facts, leaving the easy proofs as an exercise. See [Gos96] for details, if necessary.

Let q be a power of the characteristic p in which we operate. We call $P(\tau) = \sum p_i \tau^i$ (we write coefficients to the left) the τ-polynomials (or power or Laurent series, as the case may be). The corresponding $P_x(x) := \sum p_i x^{q^i}$ are $\mathbb{F}_q$-linear (polynomials or series), also called q-linear.

Once the additivity property is understood at the variable level or over infinite field extension such as algebraic closure of the base (eg., we do not call $x \to x + (x^q - x)^2$ $\mathbb{F}_q$-linear (for $q > 2$), though, as a function it is the same as $x \to x$ over $\mathbb{F}_q$), then the naive notion of a linear function agrees with the terminology: If we have additive $P(x)$ in this sense, then its derivative is easily seen to be some constant c, so $P(x) = cx + \sum_{j>0} c_i x^{pj}$. Subtracting off any $\sum a_i x^{p^i}$ part, we are left with additive $Q(x)$, which if non-zero, will give non-zero term with exponent greater than one and not divisible by p, when we take its maximal possible p-power root, contradicting its zero derivative property.

The roots of $\mathbb{F}_q$-linear, separable polynomials form an $\mathbb{F}_q$-vector space (τ-polynomials are analogous to differential d/dx-polynomials in this respect) and conversely a separable polynomial whose roots form an $\mathbb{F}_q$-vector space is $\mathbb{F}_q$-linear. So given a polynomial we can always find its $\mathbb{F}_q$-linear multiple, by taking a polynomial with roots $\mathbb{F}_q$-span with appropriate q-power multiplicity.

As in the classical Euclidean algorithm for polynomial rings over a field, we have right division algorithm for τ-polynomials: Given such P_1 and $P_2 \neq 0$, we have $P_1 = QP_2 + R$, for unique τ-polynomials, Q and R, with $\deg_\tau R < \deg_\tau P_2$. So every left ideal of $k\{\tau\}$ is principal.

If you want to divide by P_2 from left, we have similar theorems when k is perfect, i.e., $\tau(k) = k$. This property is needed in taking q-th roots while commuting τ variable pass the coefficients. For example, if we divide $t\tau$ by τ from left, it is $t\tau = \tau t^{1/q}$, whereas from right, it is just $t\tau = (t)(\tau)$.

(b)   Moore determinant

The τ-version (i.e., q-linearized version) of the more familiar Vandermonde determinant $|x_j^{i-1}| = \prod_{i>j}(x_i - x_j)$ is the Moore determinant

$$M(x_i) := |\tau^{i-1}(x_j)| = |x_j^{q^{i-1}}| = \prod_i \prod_{f_j \in \mathbb{F}_q} (x_i + f_{i-1}x_{i-1} + \cdots + f_1 x_1).$$

The proof of the last equality is similar to that of the Vandermonde identity: By row, column manipulations and linearization, we see that the right side divides the determinant (assuming independence of variables) and then degree and sign comparison gives the equality. We can also use an induction on the number of variables to simplify the matters:

$$M(x, x_1, \cdots, x_d)/M(x_1, \cdots, x_d) = \prod (x + \sum f_i x_i).$$

Now let us describe calculations of Carlitz: Note that $M(1, t, \cdots, t^d)$ is determinant of Moore as well as Vandermonde, so that the two evaluations give us that the product of all monic polynomials of degree $\le d$ is $\prod_{i=0}^{d} D_i$, so that taking ratios we see that D_d is the product of monic polynomials of degree d. More generally, expanding by the last column gives

$$e_d(x) = M(1, t, \cdots, t^{d-1}, x) = \sum_{j=0}^{d}(-1)^{d-j}x^{q^j} M_j = \sum \frac{(-1)^{d-j}x^{q^j} D_d}{D_j L_{d-j}^{q^j}},$$

because the minor M_j is of Vandermonde type $M_j = \prod_{k>i}(t^{q^k} - t^{q^i})$ (where k or i is not allowed to be j).

As an example of analogy between differential linear operators and our linear operators, we see a property analogous to Wronskians: If $x_1, \cdots, x_d$ are linearly independent in $\mathbb{F}_q$-vector space, if and only if $M(x_1, \cdots, x_d) \ne 0$.

(c) q-resultants

One of the basic tools in the explicit elimination theory is the resultant $R(A, B)$ of two polynomials $A(x) = \sum_{i=0}^{a} a_i x^i = a_a \prod(x - \alpha_j)$ and $B(x) = \sum_{i=0}^{b} b_i x^i = b_b \prod(x - \beta_j)$. It is defined to be the determinant of the $a + b$ order square matrix whose i-th row consists of $i - 1$ zeros followed by $a_a, \cdots, a_0$ followed by zeros in the remaining places again, if $1 \le i \le b$. If $b < i \le a+b$, it consists of $i - b - 1$ zeros followed by $b_b, \cdots, b_0$ followed by zeros in the remaining places. The main formula is $R(A, B) = a_a^b b_b^a \prod(\alpha_i -$

β_j). So, if the leading coefficients a_a and b_b are nonzero, then $R(A,B) = 0$ if and only if A and B have a common root.

Hence, for example, from two polynomial equations in two variables, treating only one of them as variable, we can eliminate the other by getting one variable equation 'resultant equal to zero'.

The following $\mathbb{F}_q$-linearized or τ- version was considered by Ore [Ore33] (in fact, Ore did not write down this formula explicitly): For $A(\tau) = \sum a_i \tau^i$ similar to above, the q-resultant $R_\tau(A(\tau), B(\tau))$ is defined to be the determinant of the $a + b$ order square matrix whose i-th row consists of $i - 1$ zeroes followed by $a_a^{q^{b-i}}, \cdots, a_0^{q^{b-i}}$ followed by zeros in the remaining places again, if $1 \le i \le b$. If $b < i \le a+b$, it consists of $i - b - 1$ zeros followed by $b_b^{q^{b+a-i}}, \cdots, b_0^{q^{b+a-i}}$ followed by zeros in the remaining places.

Let us for simplicity assume that A and B are monic and separable. Let α_i and β_i be the basis of the $\mathbb{F}_q$-vector space of the roots of A and B respectively. Then we see that the above resultant is

$$R_\tau = \frac{M(\alpha_1, \cdots, \alpha_a, \beta_1, \cdots, \beta_b)}{M(\alpha_1, \cdots, \alpha_a)M(\beta_1, \cdots, \beta_b)}.$$

If we denote the corresponding polynomials by $A_x(x) = \sum a_i x^{q^i}$ and B_x respectively, then $R_\tau^{q-1} = R(A_x(x)/x, B_x(x)/x)$. The point is thus the great reduction in size, from order $q^a + q^b - 2$ to $a + b$, and the parametric form with respect to q. This implies that if the resultant is zero, the two bases α_i, β_j together form a linearly dependent set over $\mathbb{F}_q$.

(d)   Non-archimedean calculus

In non-archimedean setting such as ours, where we work with a field F complete with respect to a valuation, we have some simplifications such as series converges if and only if the terms tend to zero, but there are also complications such as that there are many locally constant functions, such as valuation, making integration difficult or that the derivative is zero for any $\mathbb{F}_q$-linear power series having no x-term.

We will just mention tool of rigid analysis developed by Tate and others, which is important in analysis here, for example in study of modular forms of Chapter 6 and in general for studying analytic properties in detail. We will not use it and will just refer the interested reader to the standard books on the subject.

Let us see how Newton polygons tell us about the zeros of a power series: Consider $f(x) = \sum_{i=0}^{\infty} f_i x^i \in F[[x]]$. In $\mathbb{R}^2$, we plot points $(i, v(f_i))$

and call the lower convex hull of these points, the Newton polygon of f. In other words, we start with a vertical line through origin and rotate it counter-clockwise, making a polygonal line with corners the plotted points it hits in turn. For polynomials, we stop at the last non-zero coefficient.

First we assume without loss of generality that $f_0 = 1$, as it is trivial to account the zeros at the origin. The recipe is that f has exactly m zeros with valuation $-s$, if the Newton polygon has a side with slope s with horizontal projection of length m. See, for example, [Kob77; Kob80].

Sketch of the proof: If there is a zero z with $v(z) = -s$, there are at least two terms $f_i z^i$ and $f_j z^j$ of minimum valuation and joining them would give (part of) side of slope s. Conversely, having a side as in the statement gives two such terms with indices m apart. Taking out the valuation part and reducing modulo the maximal ideal we get a polynomial of degree m and so m zeros which lift by Hensel's lemma.

As another immediate corollary to this prescription of the zeros, the version of Picard's theorem in our case is stronger: Non-constant entire function (i.e., given by everywhere convergent power series) $C_\infty \to C_\infty$ is surjective.

In other words, we do not have an entire function like exponential missing the value zero. As a corollary, we have stronger Weierstrass product theorem, without a need for exponential convergence factors: if f is entire function with zeros z with multiplicities m_z, then $f(x) = cx^{m_0} \prod_{z \neq 0}(1-x/z)^{m_z}$, where c is a constant.

Similarly, a meromorphic function is determined up to a multiplication by constant, by its divisor, i.e., by the specification of its zeros and poles with multiplicities.

Another way to see this is Weierstrass preparation type theorem: By changing the function and the variable by scalar factors, if necessary, we can assume that our non-constant entire function f has the form $\sum f_i x^i$, with $|f_i| \leq 1$, $|f_i| < 1$ for $i < d$ and $|a_d| = 1$. By solving congruences modulo powers of the maximal ideal of integer ring, we factor $f(x) = (x^d + c_1 x^{d-1} + \cdots + c_d)(b_0 + b_1 x + \cdots)$. Since our field is algebraically closed, we get a zero of the polynomial in front, showing that an entire non-constant function always has a zero and hence is surjective, from which it is straightforward to deduce the result. Note that the zeros obtained via Newton polygons/ Weierstrass preparation are algebraic over our Laurent series field (or over p-adic fields in analogous situation). Also, if the horizontal projection $m = 1$, then the corresponding zero is in F.

The completion C_∞ of an algebraic closure of K_∞ is algebraically closed.

(e) Dwork's trace formula

We follow the treatment in [Wan96a; Wan93]. We first work in the p-adic setting.

Let $\mathbb{C}_p$ be the completion of an algebraic closure of $\mathbb{Q}_p$ and $\mathcal{O}_p$ be the ring of integers in a finite extension of $\mathbb{Q}_p$ in $\mathbb{C}_p$. For $u = (u_1, \cdots, u_n) \in \mathbb{Z}_{\geq 0}^n$, write X^u for the column vector $(X_1^{u_1}, \cdots, X_n^{u_n})$. Let $B(X) = \sum b_u X^u$ be a power series convergent over $\mathcal{O}_p$, with coefficients b_u being $r \times r$ matrices with entries in $\mathcal{O}_p$. (For the simplest application in Chapter 5 to the Carlitz zeta function, the case $n = r = 1$ is sufficient.) Associated Frobenius matrix F_B is infinite matrix indexed by $\mathbb{Z}_{\geq 0}^n \times \mathbb{Z}_{\geq 0}^n$ with (u, v)-th entry being the block b_{qu-v}. (The formula below will not depend on how u's are ordered.)

Theorem 2.11.1 *(Dwork's trace formula) For each integer $k \geq 1$,*

$$S_k(B) := \sum_{x^{q^k-1}=1, x \in \mathbb{C}_p^n} Tr(B(x^{q^{k-1}}) \cdots B(x^q) B(X)) = (q^k - 1)^n \, Tr(F_B^k).$$

Proof. Since $\sum x_i^{u_i}$, where x_i run over $(q-1)$-th roots of unity, is $q-1$ or zero according as whether $q-1$ divides u_i or not,

$$S_1(B) = \sum Tr(b_u) \sum x^u = (q-1)^n \sum Tr(b_{(q-1)u}) = (q-1)^n Tr(F_B).$$

For $k > 1$, similarly,

$$S_k(B) = \sum Tr(b_{u^{(k-1)}} \cdots b_{u^{(0)}}) \sum_{x^{q^k-1}=1} x^{q^{k-1}u^{(k-1)}+\cdots+qu^{(1)}+u^{(0)}}$$

$$= (q^k - 1)^n \sum_{q^{k-1}u^{(k-1)}+\cdots+qu^{(1)}+u^{(0)} \equiv 0 \mod q^k-1} Tr(b_{u^{(k-1)}} \cdots b_{u^{(0)}}).$$

Noting that solutions of $q^{k-1}u^{(k-1)} + \cdots + qu^{(1)} + u^{(0)} = (q^k - 1)v^{(k-1)}$, with $u^{(i)}, v^{(k-1)} \in \mathbb{Z}_{\geq 0}^n$ are just $u^{(k-1)} = qv^{(k-1)} - v^{(k-2)}, \cdots, u^{(0)} = qv^{(0)} - v^{(k-1)}$, with all $v^{(i)} \in \mathbb{Z}_{\geq 0}^n$, we see that $S_k(B)$ is then

$$(q^k-1)^n \sum_{v^{(k-1)}, \cdots, v^{(0)} \in \mathbb{N}^n} Tr(b_{qv^{(k-1)} - v^{(k-2)}} \cdots b_{qv^{(0)} - v^{(k-1)}}) = (q^k-1)^n Tr(F_B^k)$$

as claimed. $\square$

For nice accounts of how this was used by Dwork to give a proof of rationality of L-functions for varieties over finite fields, see [Kob77; Wan96a]. For more general versions and applications, and analytical subtleties, see [Wan96a].

For later applications, let us also note the multiplicative form of the theorem above. If V over $\mathbb{F}_q$ is an affine variety in $\mathbb{A}^n$, we define

$$L(B/V,T) := \prod_{\text{closed points } \bar{x} \text{ of } V} \text{Det}(I - T^{d(x)}B(x^{q^{d(x)-1}}) \cdots B(x^q)B(x))^{-1},$$

where x is the Teichmüller lifting of $\bar{x}$ and $d(x)$ is the degree of $\bar{x}$ over $\mathbb{F}_q$.

The theorem above shows that Fredholm determinant $\text{Det}(I - TF_B)$ is well-defined and equals $\exp(-\sum Tr(F_B^k)T^k/k)$.

Now $L(B/\mathbb{G}_m^n, T) = \exp(\sum S_k(B)T^k/k)$, hence expanding $(q^k-1)^n$, we see that

Theorem 2.11.2   *(Dwork's formula in multiplicative form)*

$$L(B/\mathbb{G}_m^n, T)^{(-1)^{n-1}} = \prod_{i=0}^{n} \text{Det}(I - q^i F_B T)^{(-1)^i \binom{n}{i}}.$$

Remarks 2.11.3   (1) Wan noted that for characteristic p applications, we do not need Teichmüller lifting and $x = \bar{x}$, but as we can lift coefficient b_u to characteristic zero, the passage from the additive to the multiplicative form, which involves denominators, is still valid. In fact, when q is zero, the right hand simplifies to $\text{Det}(I - F_B T)$.

(2) For a subset S of $\{1, 2, \cdots, n\}$, write F_B^S for the submatrix (b_{qu-v}) with u, v such that $u_i \geq 1$, $v_i \geq 1$ for all $i \in U$. Then using combinatorial argument on torus decomposition, Wan shows that

$$L(B/\mathbb{A}^n, T)^{(-1)^{n-1}} = \prod_{i=0}^{n} \prod_{|S|=i} \text{Det}(I - q^{n-i}TF_B^S)^{(-1)^{n-i}}.$$

(3) For affine V defined by equations $g_1(X) = \cdots = g_k(X) = 0$, we have

$$L(B/V,T) = L(B\phi(X, Z_1, \cdots, Z_k)/\mathbb{A}^{n+k}, T/q^k),$$

where, for example, the over-convergent power series $\phi(X, Z)$ is given by $\exp(\pi_p H(X, Z) - \pi_p H(X^q, Z^q))$ with Dwork's π_p is a $(p-1)$-th root of $-p$ and $H(X, Z) = \sum Z_i g_i(X)$ with $g_i(X)$ lifted to characteristic zero, say by Teichmüller liftings of coefficients.

(4) If F_B is compact operator, then the Fredholm determinant is entire. In general, its study gives good information on meromorphicity.

Chapter 3

Explicit class field theory

By the Kronecker-Weber theorem, the cyclotomic theory (i.e., the theory of extensions obtained by adjoining roots of unity (torsion of $\mathbb{G}_m$) to the base field $\mathbb{Q}$) is the same as the theory of abelian extensions over $\mathbb{Q}$ and gives thus the class fields explicitly.

For a function field K of one variable over a finite field, in addition to the usual cyclotomic extensions $K(\mu_n)$, which are just constant field extensions now, there are, of course, many more abelian extensions of K, like Kummer or Artin-Schreier extensions. Thus the two theories seem to diverge.

But Carlitz in 1930's developed another analog for the cyclotomic theory (using torsion of $\mathbb{G}_a$ via the Carlitz module now) for $A = \mathbb{F}_q[t]$ case. But somehow (most probably because of his several papers, especially with the titles which conveyed a little) this global explicit class field theory was forgotten. The similarity of Lubin and Tate's approach to explicit local class field theory to that of Carlitz global theory was noticed by Carlitz student Hayes, who then developed it more fully by handling all the abelian extensions in the $\mathbb{F}_q[t]$ case, and then in general, using the theory of Drinfeld modules. He also developed singular theory for general A, i.e., replacing A by orders in A. A very elegant detailed treatment can be found in [Hay74] for $\mathbb{F}_q[t]$ theory, [Ha2] for general A as well as the singular theory (with some restrictions and different proofs), and [Hay85; Hay] for the full general theory.

This explicit class field theory uses the theory of rank one Drinfeld modules to give explicit constructions and uses the structural theorems of the class field theory to show that what we construct explicitly are the correct objects predicted to exist by the class field theory. The results and techniques should be compared with those of cyclotomic theory [Lan90;

Was97] and complex multiplication theory [BCH$^+$66]. The constructions
are technically easier and give good control on equations compared to the
geometrical explicit theory.

3.1 Torsion of rank one Drinfeld modules

Adjoining roots of unity gives an abelian extension of any field. Similarly,
we have

Theorem 3.1.1 *If ρ is Drinfeld A-module of rank one over F, and if $a \in A - \{0\}$, then $F(\rho[a])$ is abelian extension of F, with the Galois group isomorphic to a subgroup of $(A/(a))^*$.*

Proof. If a is prime to the characteristic, by 2.3, $\rho[a]$ is naturally isomorphic to $A/(a)$ as an A-module, and if we choose a generator λ for it, then clearly any Galois conjugate $\sigma(\lambda)$ can be expressed uniquely as $\rho_{a_\sigma}(\lambda)$ for some $a_\sigma \in (A/(a))^*$. (More generally, any other torsion point is also a polynomial in λ via ρ-action and so $F(\Lambda_a) = F(\lambda)$.) Straightforward tracing through the basic definitions shows that a_σ does not depend on the choice of generator λ and that $\sigma \to a_\sigma$ is injective homomorphism of the Galois group of $F(\rho[a])$ over F into the abelian group $(A/(a))^*$.

If $a = \wp$, the characteristic of ρ, then $\rho[a] = \{0\}$ and we get a trivial extension, in analogy with adjoining p-th roots of unity (there is only 1) in characteristic p.

More generally, $\rho_\wp = f\tau^{\deg \wp}$, for some $f \in F^*$, as we are in rank and height one situation. Hence $a\wp^n$-torsion is just a-torsion and we reduce to the first case. Now $(A/(a))^*$ is isomorphic to a subgroup of $(A/(a\wp^n))^*$. $\square$

We can say much more, if $F = K$, as is the case for the Carlitz module and the following examples [Hay79]. We will see in Section 5 how to construct such examples. But as an exercise, the reader should verify the recipe and calculate ρ_y from ρ_x, as explained in (1) of 2.4.3.

Examples 3.1.2 (i) $A = \mathbb{F}_2[x, y]$ with $y^2 + y = x^3 + x + 1$,

$$\rho_x = x + (x^2 + x)\tau + \tau^2, \quad \rho_y = y + (y^2 + y)\tau + x(y^2 + y)\tau^2 + \tau^3.$$

(ii) $A = \mathbb{F}_4[x, y]$ with $y^2 + y = x^3 + w$, $w^2 + w + 1 = 0$,

$$\rho_x = x + (x^8 + x^2)\tau + \tau^2, \quad \rho_y = y + (x^{10} + x)\tau + (x^{32} + x^8 + x^2)\tau^2 + \tau^3.$$

(iii) $A = \mathbb{F}_3[x, y]$ with $y^2 = x^3 - x - 1$,

$$\rho_x = x + y(x^3 - x)\tau + \tau^2, \quad \rho_y = y + y(y^3 - y)\tau + (y^9 + y^3 + y)\tau^2 + \tau^3.$$

(iv) $A = \mathbb{F}_2[x, y]$ with $y^2 + y = x^5 + x^3 + 1$, and $\rho_x = x + (x^2 + x)^2\tau + \tau^2$.

The first 3 are of genus 1 and the last of genus 2. The last model is singular at infinity, but there is only one place at infinity. In the last example, y being of degree 5, the expression for ρ_y is quite huge.

We will soon see that existence of a Drinfeld A-module of rank one over K is equivalent to existence of such over A and which in turn is equivalent to $h_A = 1$. Examining all 7 examples of non-rational function fields with $h = 1$ in [LMQ75], only 4 of them have place of degree one, and it is unique. Calling it ∞, we get the 4 examples above with $h_A = hd_\infty = 1$. Hence these are all possible examples. So the first hypothesis of the following theorem will be seen to be redundant. In any case, the theorem immediately applies to these 4 examples in addition to the Carlitz module.

Theorem 3.1.3 *Let A be of class number one. Let ρ be a Drinfeld A-module of rank one over A. Then the extension $K(\Lambda_a)$ is abelian over K with its Galois group isomorphic to $(A/(a))^*$. A prime $\wp$ of A not dividing a is unramified and we have*

$$\lambda^{\mathrm{Frob}(\wp)} = \rho_{(\wp)}(\lambda) = \rho_\wp(\lambda)$$

for $\lambda \in \Lambda_a$, where we choose the sign of the generator $\wp$ of its ideal in such a way (eg., monic for Carlitz module) that the top coefficient of $\rho_\wp$ is 1.

Proof. For the first part, by the theorem above, it only remains to compare the cardinality of the Galois group with that $\phi(a) := |(A/(a))^*|$.

We will first consider the prime power case. Let $\wp$ be a prime (in fact we can choose a generator of its ideal, as $h_A = 1$) of A of degree d. Let λ be a generator of the A-module $\Lambda_{\wp^n}$. Since

$$\rho_{\wp^n}(x) = \rho_\wp(\rho_{\wp^{n-1}}(x)) = \wp\rho_{\wp^{n-1}}(x) + \rho_{\wp^{n-1}}(x)^2 P(x)$$

for some polynomial $P(x) \in K[x]$, we have

$$\wp = \rho_{\wp^n}(0)/\rho_{\wp^{n-1}}(0) = \prod \rho_a(\lambda)$$

where the product is over representatives $a \in A$ of $(A/\wp^n)^*$.

As the coefficients of ρ_a are in A, $\rho_a(\lambda)/\lambda$ is integral over A and so is its inverse, because it is of the same form, as a is prime to $\wp$. So it is a unit in the integral closure $\mathcal{O}$ of A in $K(\lambda)$. (These are analogs of **cyclotomic**

units $(1 - \zeta_n^a)/(1 - \zeta_n)$, where $(a, n) = 1$ and will be studied in more detail below. Hence λ plays a role similar to $1 - \zeta$.) Hence $\wp$ is $\phi(\wp^n)$-th power of λ times a unit. Hence the degree of the extension is at least the ramification index $\phi(\wp^n)$ of $\wp$ in it. Hence the Galois group, which is by the theorem above isomorphic to a subgroup of $(A/\wp^n)^*$ is in fact isomorphic to the full group.

We also see that $\wp$ is totally ramified in it.

On the other hand, all other finite primes are unramified in it: The discriminant of $\mathcal{O}$ divides that of (a priori a subring, in fact equal to it, as we will later see) $A[\lambda]$ which in turn divides norm of $\rho'_{\wp^n}(\lambda)$. But because of linearity of polynomial $\rho_a(x)$, $\rho'_{\wp^m}(x)$ is its constant term $\wp^n$.

The multiplicativity of $\phi(a)$ now allows us to show that for any a the degree of the extension is $\phi(a)$, by taking compositums, since the linear disjointness is guaranteed at each stage by the ramification description.

To make the Galois action explicit, we choose $\wp$ not dividing a, as in the statement of the theorem.

Claim: The monic polynomial $\rho_\wp(x)/x$ and more generally, $f(x) := \rho_{\wp^n}(x)/\rho_{\wp^{n-1}}(x)$ is Eisenstein at $\wp$.

Since $\rho_{\wp^n}(x) = \rho_\wp(\rho_{\wp^{n-1}}(x))$, the statement for f follows immediately from that for $\rho_\wp$. For the latter, we have already seen it for the Carlitz module using the explicit formulas in Section 2.5. We prove it now, in general, without explicit formulas: Since $f(x) = \prod(x - \rho_a(\lambda))$ and since each $\rho_a(\lambda)$ is divisible by λ, all except the top coefficients are divisible by λ and hence by $\wp$. The constant coefficient is $\wp$ as we are in generic characteristic. This proves the claim.

So $\rho_\wp(\lambda) \equiv \lambda^{q^d} \mod P$, where P is the prime above $\wp$. Hence $\rho_\wp(\lambda) \equiv \mathrm{Frob}_\wp(\lambda) \mod P$. To show that they are equal and not just congruent, we take the derivative of $\rho_a(x) = \prod(x - \rho_b(\lambda))$, where $b \in A$ runs through mod a. Noting that the derivative is just a, we get

$$a = \prod_{b \ \mathrm{mod}\ a, b \not\equiv c} (\rho_b(\lambda) - \rho_c(\lambda))$$

for every $c \in A$. Now P does not divide a, so that all these conjugates have distinct images modulo P. Hence we get the Galois action as claimed. $\square$

In other words, the Frobenius at $\wp$ acts as $\wp$ (chosen of sign described in the theorem) through its image in $(A/(a))^*$ (via ρ) on the generator λ. So the splitting of primes is exactly parallel to the cyclotomic case.

Now we turn to the decomposition of the infinite place in this extension. Since elements of $A^* = \mathbb{F}_q^* \hookrightarrow (A/aA)^*$, act just by multiplication, $\mathbb{F}_q^*$ is contained in the decomposition and the inertia groups at infinity. In fact, both these groups are equal to $\mathbb{F}_q^*$. (The torsion points are $\tilde{\pi} e_A(b/a)$ and we have seen that $C_a(x)/x$ is polynomial in x^{q-1} and $\tilde{\pi}^{q-1} \in K_\infty$. This shows that infinity ramifies to degree $q - 1$.)

The fixed field of the decomposition group

$$K(\Lambda_a)^+ = K(\lambda^{q-1}) = K(\prod_{\theta \in A^*} (\theta\lambda))$$

is analogous to the totally real subfield $\mathbb{Q}(\zeta_n)^+ = \mathbb{Q}(\sum_{\theta \in \mathbb{Z}^*} \zeta_n^\theta)$ of the full cyclotomic field.

Another way to see this decomposition, eg. for the Carlitz module, since we have good explicit equation for C_a, is to use Newton polygon method to analyze its decomposition in C_∞: We sketch simple model case of $a = \wp$, a prime of degree d. We see that the cyclotomic polynomial $\rho_a(x)/x$ can be written as $f(x^q - 1)$, with $f(u) = \sum f_i(t) u^{(q^i-1)/(q-1)}$. The explicit formulas in 2.5 show that $\deg(f_i) = (d-i)q^i$. The points corresponding to these non-zero terms are exactly the vertices of the Newton polygon. The first slope has then horizontal projection one giving a root in K_∞. Since the extension is Galois, all the roots are in K_∞ and ∞ splits completely. Then $x^{q-1} = u$ accounts for the total ramification of degree $q - 1$ from $K(\Lambda_a)^+$ to $K(\Lambda_a)$ (see [Hay74] for details).

We repeat the discriminant calculation in the theorem in a slightly different way: If we index the a-torsion points as λ_i, we have

$$\text{Disc}\,(\rho_a) = \prod_{i<j}(\lambda_i - \lambda_j)^2 = \prod_{\lambda_i \neq 0} \lambda_i^{q^{\deg(a)}-1} = \pm(\iota(a)/a_{r\,\deg(a)})^{q^{\deg(a)}-1}$$

where the first equality follows by definition, the second by additivity of ρ_a and the third by comparison of coefficients of ρ_a.

Let us try to generalize to the general A setting of 2.2. The discussion above suggests that we need to understand the field of definition of Drinfeld A-module, and possibility of a model over its ring of integers. Recall that such a model would have unit top coefficients and we can ask whether the top coefficients can be chosen to be constants.

As Hayes remarks in [Hay], rather than controlling the field of definition first [Hay79], it turns out to be more efficient to control the top coefficient first:

3.2 Sign normalization of the top coefficient

Let ρ be a Drinfeld A-module of rank r over F. In this section, we will write $\mu(a)$ for the top coefficient of ρ_a. (Here $\mu(0) = 0$ by the usual conventions.) Then we have

$$\mu(a_1 a_2) = \mu(a_1)\mu(a_2)^{q^{r\,\deg(a_1)}} = \mu(a_1)^{q^{r\,\deg(a_2)}}\mu(a_2)$$

and if a_i are of the same degree, then also $\mu(a_1 + a_2) = \mu(a_1) + \mu(a_2)$.

Given a non-constant a, conjugating by a suitable element, we can always choose an isomorphic model ρ with $\mu(a)$ in a finite field, say even $\mu(a) = 1$. The last equality of the first equation above then shows that $\mu(a)$ is in a finite field for all a.

Let us study these normalizations in more detail:

Since by the Riemann-Roch, for every sufficiently large n, there are elements in A of order $-n$ at infinity, using them and the equation above, we can extend the definition of μ uniquely (satisfying the equation above) from A to the domain K_∞, using the decomposition of K_∞ in 1.1. In other words, we can extend the definition with $\mu(U_1) = 1$ and uniquely determined $\mu(\pi)$ by using the decompositions $a = \text{sgn}(a)\pi^{ord_\infty(a)}u_1$, where u_1 is a one-unit at infinity: To define μ of some element, we just have to multiply by a suitable element of A to get large enough negative order to have $a \in A$ of that order, so that we can define the μ of the product as $\mu(a)$ and using the equation above, define μ of the element we started with. It is a simple exercise to check that this process is independent of the choices made. Restricting μ on the subfield $\mathbb{F}_\infty$ (identified with the residue field) of K_∞, we get an automorphism i_ρ of $\mathbb{F}_\infty$ fixing $\mathbb{F}_q$. (In particular, $\mathbb{F}_\infty$ embeds in the field of definition K_ρ.)

Fixing a sign function sgn, our base object becomes (K, ∞, sgn).

Definition 3.2.1 We say that ρ is sgn-normalized, if μ_ρ is a twisting of sgn, as defined in 1.1.1.

Theorem 3.2.2 *Every Drinfeld A-module ρ' over C_∞ is isomorphic to a sgn-normalized A-module ρ.*

Proof. Let π be a uniformizer at ∞ with $\text{sgn}(\pi) = 1$. If we let $\rho = c\rho'c^{-1}$, with $c \in C_\infty$ such that $c^{1-q^{rd_\infty}} = \mu_{\rho'}(\pi^{-1})$, then $\mu_\rho(\pi) = 1$. Hence, for $x = \sum_{i \geq n} x_i \pi^i \in K_\infty = \mathbb{F}_\infty((\pi))$, with $x_i \in \mathbb{F}_\infty$, $x_n \neq 0$, we have $\mu(x) = \mu(x_n) = \mu(\text{sgn}(x)) = i_\rho(\text{sgn}(x))$. $\square$

If ρ is of rank one, then, by Theorem 2.6.4, it has potential good reduction, so that this model will have integral coefficients. We restrict to rank one from now on.

Proposition 3.2.3 *If ρ is a Drinfeld A-module of rank one over C_∞, with $\mu_\rho(A) \subset \mathbb{F}_\infty$, then ρ is sgn-normalized.*

Proof. By Theorem 3.2.2, there is $\theta \in C_\infty$, such that $\theta\rho\theta^{-1}$ is sgn-normalized. The hypothesis then implies that $\theta^{q^{\deg(a)}-1} \in \mathbb{F}_\infty^*$ for all non-zero $a \in A$. But the greatest common divisors of $\deg(a)$'s is d_∞ and so the greatest common divisor of $q^{\deg(a)} - 1$'s is $q^{d_\infty} - 1$. Writing $\theta^{q^{d_\infty}-1}$ as $\zeta \in \mathbb{F}_\infty^*$, we have $\theta^{q^{\deg(a)}-1} = \zeta^{\deg(a)/d_\infty}$. This proves the claim by the remark after 1.1.1. $\square$

Proposition 3.2.4 *If ρ and $\rho' = c\rho c^{-1}$ are sgn-normalized, then $c \in \mathbb{F}_\infty^*$ and $\mu_\rho = \mu_{\rho'}$.*

Proof. We have $\mu_{\rho'}(x) = c^{1-q^{\deg(a)}}\mu_\rho(x)$, for $x \in K_\infty$. Hence the first claim follows from $\mu_\rho(\pi^{-1}) = 1$ and $\mu_{\rho'}(\pi^{-1}) = 1$. Then the second claim follows since d_∞ divides $\deg(a)$. $\square$

Corollary 3.2.5 *Each isomorphism class of Drinfeld A-modules of rank 1 over C_∞ contains exactly $(q^{d_\infty} - 1)/(q - 1)$ sgn-normalized ones.*

Proof. By Theorem 3.2.2, each class contains a sgn-normalized ρ. For each $c \in \mathbb{F}_\infty^*$, $c\rho c^{-1}$ is sgn-normalized, and since $\mathrm{Aut}(\rho) = \mathbb{F}_q^*$, there are exactly $(q^{d_\infty} - 1)/(q - 1)$ distinct ones. That these are all the possibilities follows from the proposition. $\square$

We call a sgn-normalized rank one Drinfeld A-module over C_∞ a **Hayes module**.

Carlitz modules (one for each $\mathbb{F}_q[t]$) and the four examples (all of class number one) in 3.1.2 are examples of Hayes modules, all with $d_\infty = 1$. For the example (3) of 2.2.5, with $d_\infty = 2$, conjugating by the top coefficient gives you a Hayes module.

Let X be the set of Hayes A-modules. It is easy to verify that X is preserved under the * action (see 2.4.3) of ideals of A and the action of $\mathrm{Aut}(C_\infty/K)$ and that these actions commute.

A fractional ideal I stabilizes $\rho \in X$, if and only if I is a principal ideal say iA (since it preserves the isomorphism class) and $\mu_\rho(i) \in \mathrm{Aut}(\rho) = \mathbb{F}_q^*$, by (4) of 2.4.3. In other words, I is principal generated by a sgn one element (sometimes also called a positive element).

Definition 3.2.6 Let $I(A)$ be the group of non-zero fractional ideals of A, $P(A)$ ($P_1(A)$ respectively) be the subgroup of principal (generated by sgn one element respectively) fractional ideals. Let $Pic_1(A) = I(A)/P_1(A)$. It is called the narrow class group of A (relative to sgn). Let $Pic(A) := I(A)/P(A)$. It is the usual class group.

Hence we have $h_1 := |Pic_1(A)| = |X| = h_A(q^{d_\infty} - 1)/(q - 1)$ and

Theorem 3.2.7 *The set X of sgn-normalized A-modules is a principal homogeneous space for (i.e., has a transitive, faithful action of) $Pic_1(A)$ under the $*$ action.*

3.3 Normalizing Field as a class field

Now we return to the question of identifying the fields of definition.

Choose a non-constant $a \in A$ and $\rho \in X$ and let H_1 be the field generated over K by the coefficients of ρ_a. By (1) of 2.4.3, H_1 is independent of the choice of a and since all $I * \rho$ are defined over H_1, by (4) of 2.4.3, it is also independent of the choice of ρ and just depends on the base object (K, ∞, sgn). The field H_1 is called the **normalizing field** for rank one Drinfeld A-modules over this base triple.

Proposition 3.3.1 *The field H_1 is an abelian extension of K with Galois group naturally isomorphic to a subgroup of $Pic_1(A)$.*

Proof. Let $\rho \in X$. For any $\sigma \in \text{Aut}(C_\infty/K)$, $\sigma\rho$ is sgn-normalized and hence is defined over H_1. Hence H_1 contains all the conjugates over K of its finite set of generators and hence is a finite, normal extension of K.

Let $c \in C_\infty$ be an isomorphism taking ρ to $\rho' = c\rho c^{-1}$ defined over the smallest field of definition K_ρ of ρ. Then for a non-constant $a \in A$ of sgn one, we have $c^{1-q^{\deg(a)}} = \mu_{\rho'}(a) \in K_\rho$, so that $K_\rho(c)/K_\rho$ is separable. Hence H_1 is separable over K_ρ. Since $K_\rho \subset K_\infty$, by (3) of 2.4.3, K_ρ is separable over K. So H_1 is separable and hence Galois extension of K.

Since the action of $\text{Gal}(H_1/K)$ on X commutes with the $*$ action of $Pic_1(A)$, we get a homomorphism from the Galois group into $Pic_1(A)$ by sending σ to I, where $\sigma\rho = I * \rho$. It is injective by Theorem 3.2.7, since by definition of H_1, $\sigma\rho \neq \rho$, for non-trivial σ. $\square$

Let B_1 be the integral closure of A in H_1. Since a sgn-normalized ρ is defined over B_1, by Theorem 2.6.4, we can reduce it modulo any prime w

of B_1. Let $r_w : B_1 \to B_1/w$ be the reduction map and let $\wp$ be a prime of A under w.

Lemma 3.3.2 *The reduction map $\rho \to r_w \circ \rho$ is one-to-one on X.*

Proof. Suppose ϕ and ϕ' in X reduce to the same module modulo w. By Theorem 3.2.7, we can write $\phi' = I * \phi$, for some ideal I of A. We may assume that I is prime to w: By weak approximation, choose $x \in K$ congruent to one modulo ∞ and with valuation at w negative of that of I. Then writing $xI = I_1/I_2$, with I_1 and I_2 relatively prime ideals of A, $I_1 I_2^{h_1-1}$ is prime to w and belongs to the same narrow ideal class as I.

Now for any $a \in A$, we have $\phi_I \phi_a = \phi'_a \phi_I$, which is the same modulo w as $\phi_a \phi_I$. So $r_w(\phi_I) \in \text{End}(r_w \circ \phi) = A$, since we are in rank 1, by Theorem 2.7.2. Hence there is $i \in A$, such that $\phi_I \equiv \phi_i \mod w$. Hence comparison of the leading coefficient implies $\mu_\phi(i) = 1$, so that i is positive. It is enough to show now that $I = iA$.

The congruence above shows that I, iA or $I + iA$-torsion of $r_w \circ \phi$ is the same in the algebraic closure of B_1/w and hence the degrees of these three ideals are the same and the claim follows. □

Proposition 3.3.3 *The extension H_1 over K is unramified at every finite place $\wp$ of A.*

Proof. If σ belongs to the inertia group at $\wp$, then $\sigma\rho \equiv \rho \mod \wp$, so that $\sigma\rho = \rho$ by the lemma. But this implies $\sigma = 1$, as H_1 is generated by the coefficients. □

Theorem 3.3.4 *If $\sigma_I \in \text{Gal}(H_1/K)$ denotes the Artin automorphism associated to I, then for every $\rho \in X$,*

$$\sigma_I \rho = I * \rho.$$

In particular, the natural map from $\text{Gal}(H_1/K)$ into $\text{Pic}_1(A)$ is onto isomorphism.

Proof. It is enough to prove this for $I = \wp$, a prime. Then $\sigma_\wp$ is the Frobenius at $\wp$. By the lemma, it is enough to prove the equality modulo w above $\wp$. Since rank and height are one, in characteristic $\wp$, $\rho_\wp$ reduces modulo w to $\tau^{\deg(\wp)}$. Hence the defining equation $\rho_\wp \rho = (\wp * \rho)\rho_\wp$ reduces modulo w to what we want. □

Combining with (4) of 2.4.3, we get

Corollary 3.3.5   *If for $k \in K^*$, σ_k denotes the Artin automorphism for fractional ideal kA, then*

$$\sigma_k \rho = \mu_\rho(k)^{-1} \rho \mu_\rho(k).$$

Since all the primes of A are unramified in H_1, the following theorem applies to division polynomials of Hayes modules:

Theorem 3.3.6   *Let w be a valuation corresponding to an unramified prime above $\wp$ in a finite extension of K. If ϕ is a Drinfeld A-module of rank one defined over a valuation ring $\mathcal{O}_w$, then $\phi_{\wp^{e-1}}(x)$ divides $\phi_{\wp^e}(x)$ in $\mathcal{O}_w$ and the quotient is Eisenstein at $\wp$.*

Proof.   First let $e = 1$. Since the reduction modulo the maximal ideal has characteristic $\wp$, all except the top coefficient belong to the maximal ideal, as the rank and height are one. So it suffices to show that $w(D(\phi_\wp)) \le 1$:

Let $a \in A$ be such that $w(a) = 1$ and let $(a) = \wp I$. By (4) of 2.4.3, $\phi_a = \mu(a)(\wp * \phi)_I \phi_\wp$, so that $1 = w(D(\phi_a)) \ge w(D(\phi_\wp))$.

For $e > 1$, $\phi_{\wp^e}(x) = f(\phi_{\wp^{e-1}}(x))\phi_{\wp^{e-1}}(x)$, where $f(x) = (\wp^{e-1} * \phi)_\wp(x)/x$. So f is Eisenstein and $\phi_{\wp^{e-1}}(x) \equiv x^{q^{(e-1)\deg(\wp)}}$ modulo the maximal ideal, as the rank and height is one for the reduction.   □

Here is an explicit principal ideal theorem for H_1:

Theorem 3.3.7   *For a Hayes A-module ϕ and an ideal I of A, we have*

$$IB_1 = D(\phi_I)B_1$$

where $D(\phi_I)$ denotes the constant coefficient of ϕ_I.

Proof.   By (4) of 2.4.3, we may assume that $I = \wp$ is a prime ideal. We know from the previous theorem that the valuation of $D(\phi_\wp)$ is one, for any prime of B_1 above $\wp$. It is enough to show that the valuation w for prime not above $\wp$ is zero:

Choose $e \ge 1$, so that $\wp^e = aA$ is principal and put $J = \wp^{e-1}$. By (4) of 2.4.3,

$$w(D((\wp * \phi)_J)) + w(D(\phi_\wp)) = w(\mu(a)^{-1}a) = 0,$$

as $\mu(a)$ is a unit of B_1. Now since both ϕ and $\wp * \phi$ are over B_1, both the summands on the left are non-negative, so that both are zero.   □

3.4 Smallest field of definition as a class field

Now we identify the smallest field of definition K_ρ for a rank one ρ (over C_∞), which can be assumed to be sgn-normalized by Theorem 3.2.2.

By the remark before 3.2.1, $K\mathbb{F}_\infty \subset K_\rho \subset H_1$. Let $c \in C_\infty$ be an isomorphism such that $\rho' = c\rho c^{-1}$ is defined over K_ρ. Because $\mathrm{Aut}(\rho) = \mathbb{F}_q^*$, the G in the proof of Theorem 2.8.2 is just $q - 1$. Hence that proof shows that $c_0 := c^{q-1} \in H_1$ and $H_1 = K_\rho(c_0)$.

Proposition 3.4.1 *The subfield K_ρ of H_1 is the fixed field of the automorphisms σ_k, $k \in K^*$. The extension H_1 of K_ρ is a Kummer extension of degree $d := (q^{d_\infty} - 1)/(q - 1)$ of K_ρ. For $k \in K^*$, we have*

$$c_0^{\sigma_k} = \mu_\rho(k)^{q-1} c_0.$$

Proof. By Corollary 3.3.5, σ_k's fix all the invariants I_j of 2.8.2, which generate K_ρ. Let σ_k also denote some extension of σ_k to an isomorphism from $H_1(c)$ to C_∞. By Corollary 3.3.5,

$$\rho' = \sigma_k \rho' = c^{\sigma_k} \sigma_k \rho c^{-\sigma_k} = (c^{\sigma_k - 1} \mu_\rho(k)^{-1}) \rho' (c^{\sigma_k - 1} \mu_\rho(k)^{-1})^{-1}.$$

Thus $c^{\sigma_k - 1} \mu(k)^{-1} \in \mathrm{Aut}(\rho') = \mathbb{F}_q^*$, which implies the displayed formula of the proposition. This implies $d \leq [H_1 : K_\rho]$.

Since $\mu_\rho(\pi^{-1}) = 1$, $c_0^{(q^{d_\infty} - 1)/(q-1)} = c_0^{q^{d_\infty} - 1} = 1/\mu_{\rho'}(\pi^{-1}) \in K_\rho$, so that we get the reverse inequality and thus the degree of the extension is d as claimed.

Since d is also the order of $P(A)/P_1(A)$, which is the kernel of the natural surjective homomorphism from $Pic_1(A)$ to $Pic(A)$, K_ρ is the fixed field of $P(A)/P_1(A)$ (under the natural identifications) as claimed. □

Hence K_ρ is independent of sgn and of ρ. We will now denote it by H and show that it is indeed the Hilbert class field H_A of A:

Theorem 3.4.2 *The smallest field of definition H of Drinfeld A-modules of rank one over C_∞ is the Hilbert class field of A: It is the maximal abelian extension of K, unramified at every finite place and completely split at ∞. It is extension of degree h_A of K and has Galois group isomorphic to $Pic(A)$ under the Artin map.*

If ρ is defined over H, then $\sigma_I \rho = I * \rho$, for any non-zero fractional ideal I of A.

Proof. Since $H \subset H_1$, it is unramified at all finite places and abelian, by 3.3.1 and 3.3.3. Since $H \subset K_\infty$, by (3) of 2.4.3, it is completely split at ∞. By the proposition, the Galois group is isomorphic to $Pic(A)$ and thus by the class field theory, it is the Hilbert class field. The action of Artin map follows from 3.3.4 by straight calculation relating ρ to conjugate Hayes module. $\square$

Remarks 3.4.3   (1) The action of Artin map here is the analog of classical $j(\mathcal{L})^{\sigma_I} = j(I^{-1}L)$ for j-invariant of the elliptic curve with complex multiplication with the corresponding lattice $\mathcal{L}$.

(2) Just as an elliptic curve with complex multiplication (see [BCH$^+$66] for details) by imaginary quadratic $\mathcal{O}_K$ has the Hilbert class field as its smallest field of definition (over K, so that even for $y^2 = x^3 - x$ we think of $\mathbb{Q}(i)$ as the field of definition), we see that the rank one Drinfeld A-modules in generic characteristic has the Hilbert class field $H := H_A$ of A as its smallest field of definition.

(3) But in contrast to the elliptic curves case [ST68, pa. 507 cor. 2 thm. 9], we will see in the next section that they can even be defined over its ring of integer $B := \mathcal{O}_H$ (just as the multiplicative group (the rank one situation) is defined over $\mathbb{Z}$). Further, if $d_\infty = 1$, then $H = H_1$ as we saw. In addition, the top coefficients can be arranged to be constants. In general, to arrange this, we needed a slightly bigger class field H_1 (corresponding to the narrow class group) as the field of definition.

3.5   Ring of definition

Theorem 3.5.1   *Every Drinfeld A-module ρ of rank one over C_∞ is isomorphic to ϕ over $B := \mathcal{O}_H$, i.e., with the coefficients in B and with the top coefficients of ϕ_a (for $a \neq 0$) units in B.*

Proof.   By 3.2.2, we may assume that ρ is sgn-normalized. It is enough to show that there is w such that w^{q-1} is a unit of B_1 such that $\phi = w^{-1}\rho w$ has coefficients in B:

Let $\wp$ be a prime ideal of A of degree congruent to one modulo d_∞. By 3.3.7, $\wp B_1 = D(\rho_\wp)B_1$ and by the principal ideal theorem for the Hilbert class field of A from the class field theory (see eg., [Ros87]), $\wp B = \theta B$ say. Then $w^{q-1} := D(\rho_\wp)/\theta$ is clearly a unit. It is enough to show that $F_v(\phi) = \phi$ for any finite place v of B, where $F_v \in \mathrm{Gal}(H_1/H)$ is Frobenius at v.

By the properties of the Hilbert class field, the norm of v to A is principal, generated by a_v say. We write μ for $\mu_\rho(a_v)$. Then by 3.3.5,

$$F_v(\rho) = F_{(a_v)}(\rho) = \mu^{-1}\rho\mu.$$

Hence $F_v(D(\rho_\wp)) = D(\rho_\wp)\mu^{1-\text{Norm}\wp} = D(\rho_\wp)\mu^{1-q}$, as $\deg(\wp) \equiv 1$ mod d_∞. So $F_v(w)\mu/w \in \mathbb{F}_q^*$.

Hence $F_v(\phi) = F_v(w^{-1}\rho w) = \mu w^{-1}F_v(\rho)w\mu^{-1} = \phi$. $\qquad \square$

Remarks 3.5.2   If $d_\infty = 1$, then $B = B_1$ and we choose ϕ sgn-normalized. Hayes [Hay79] had proved it in more generality under the hypothesis that K has a place of degree one. Takahashi proved the general version, using his criterion 2.6.5 for the good reduction and theory of Hecke characters. The following proof is due to Hayes given in correspondence (about removing the hypothesis, unaware of Takahashi result) with the author in 1988. Learning of this, Deligne then gave an interpretation and a variant (see 3.7) in a letter to Hayes. In [Hay], answering a Deligne's suggestion, Hayes gives a proof which does not use the principal ideal theorem for the Hilbert class field.

This theorem proved by using the principal ideal theorem can be used to give an explicit version of the principal ideal theorem:

Theorem 3.5.3   *For ϕ a Drinfeld A-module of rank one and generic characteristic over B and I and ideal of A, we have $IB = D(\phi_I)B$.*

Proof.   As all finite primes of A are unramified in B and the top coefficient of ϕ is a unit, the proof of 3.3.7 carries over. $\qquad \square$

Example 3.5.4   Now let us see how this theory helps in simplifying calculation of rank one examples, eg. (1) of 3.1.2: From what we have seen, a sgn-normalized example exists in this case over A, so that putting top coefficients to be one, we can write $\rho_x = x + x_1\tau + \tau^2$ and $\rho_y = y + y_1\tau + y_2\tau^2 + \tau^3$. Then comparison of the τ coefficient in $\rho_x\rho_y = \rho_y\rho_x$ gives $y_1 = x_1(y^2 + y)/(x^2 + x)$. Now we know that $y_1 \in A$, so we can put $x_1 = a(x^2 + x)$, for some $a \in A$. Now τ^2 coefficients comparison shows that degree of y_2 is $8 + 3\deg a$. The τ^3 coefficients comparison gives $x_1^8 + x_1 + y_2^4 + y_2 = 0$, implying $32 + 8\deg a = 32 + 12\deg a$. So $\deg a = 0$, hence $a = 1$ and $x_1 = x^2 + x$.

More generally [Hay91], for A with $g = d_\infty = 1$, we use x of degree 2 and y of degree d, odd say and use the commutation relation $\rho_x\rho_y = \rho_y\rho_x$. Comparing coefficients of τ^j, for $j = 1$ to $d - 1$, we solve for y_j recursively in terms of x_1. Finally x_1 is a root of the greatest common divisor of two polynomials you obtain by comparing the coefficients of τ^d and τ^{d+1}. This gcd is thus a power of the minimal polynomial, which can be obtained as usual by gcd with derivatives method to get rid of multiplicities. By the theory above, this also constructs the Hilbert class field and the degree of the minimal polynomial is the class number.

In 8.2 (c), we see another method to write such examples.

Finally, here is the example, due to Lingsueh Shu, promised in Section 2.3:

Example 3.5.5   Let A be with $h_A > 1$ and $d_\infty = 1$ and having an ideal I of degree one. (eg, $A = \mathbb{F}_3[x, y]/(y^2 = x^3 - x + 1)$, and $I = (x, y - 1)$.) Let ρ be a Hayes A-module, so that $\rho_I = c + \tau$, with $c \in H_A$. If we define μ by $\mu^{q-1} = c/c^\sigma$, for a non-trivial $\sigma \in \mathrm{Gal}(H_A/K)$, then we have the same I-th isogeny $\rho_I^\sigma = (\mu^{-1}\rho\mu)_I = c^\sigma + \tau$ for non-isomorphic Hayes modules ρ^σ and $\mu^{-1}\rho\mu$.

3.6   Cyclotomic fields

Now we look at the same theory, but with a conductor or equivalently with a level structure. We will not use the last section.

Let the modulus m be a non-zero effective divisor with support on A. We denote the corresponding ideal of A also as m.

We now treat ∞ as a place at infinity, analogous to the real place. Fix a sign function sgn. In this section, positive will mean sgn 1. Denote by X the set of corresponding Hayes modules. We consider modifications of the generalized class groups, by using corresponding sign conditions as in the real case. Compare with Section 1.4.

Definition 3.6.1   Let I_m be the group of all fractional ideals prime to A and P_{m1} be the subgroup of principal (fractional) ideals generated by positive $k \in K^*$ congruent to one modulo m. The **narrow ray class group modulo m (relative to sgn)** is defined as $Pic_{m1}(A) := I_m/P_{m1}$.

Let X_m be the set of pairs (ϕ, λ), where $\phi \in X$ and λ is a generator of the m-torsion $\phi[m]$ of ϕ.

The *-action of ideals I of A in I_m extends to X_m by $I * (\phi, \lambda) := (I * \phi, \phi_I(\lambda))$. The stabilizer of any point is P_{m1}. Since the cardinalities of X_m and $Pic_{m1}(A)$ are the same, we see that X_m is a principal homogeneous space for *-action of of $Pic_{m1}(A)$.

Put $K_m := H_1(\phi[m])$. Exactly as in 3.3, we see that K_m is Galois extension of K unramified away from m and ∞. By Theorem 3.3.6, we see that support of m is totally ramified in it. For ideal I of A in I_m, we have $\lambda^{\sigma_I} = \phi_I(\lambda)$. Hence K_m is independent of ϕ. Further it is abelian extension with the Galois group naturally isomorphic to $Pic_{m1}(A)$.

Since the positive elements of A generate A/m, the map $a \to \sigma_a := \sigma_{(a)}$, for $a \in A$ prime to m sets up an isomorphism between $(A/m)^*$ and $\mathrm{Gal}(K_m/H_1)$. From the Galois action described above, for any $k \in K^*$ congruent to one modulo m, $\lambda^{\sigma_k} = \mu_\phi(k)^{-1}\lambda$. Therefore $\mathrm{Gal}(K_m/K)$ contains a natural subgroup isomorphic to $\mathbb{F}_\infty^*$ which is the decomposition and inertia group at ∞. Hence ∞ splits completely in its fixed field denoted by K_m^+: analog of maximal real subfield of the m-th cyclotomic field.

Theorem 3.6.2 *If λ is a generator of m-torsion, then $H_1(\Lambda_m) = K(\lambda)$.*

Proof. It is enough to show that the only automorphism of $H_1(\Lambda_m)$ over K fixing λ is identity, or equivalently that for an ideal I of A prime to m, $\phi_I(\lambda) = \lambda$ implies $I = (i)$, with positive i congruent to one modulo m.

Since ϕ_I maps m-torsion of ϕ onto that of $I * \phi$, $\phi_I(\lambda) = \lambda$ implies that ϕ and $I * \phi$ have the same m-division points. So $\phi_m = (I * \phi)_m$.

Now choosing $m^e = (m_1)$, by repeated use of

$$\phi_{m^j} = (m^{j-1} * \phi)_m \phi_{m^{j-1}} = \mathrm{Frob}_{m^{j-1}}(\phi_m)\phi_{m^{j-1}},$$

we can express ϕ_{m_1} solely in terms of ϕ_m. Thus $\phi_{m_1} = (I * \phi)_{m_1}$ and so by (1) of 2.4.3 we have $\phi = I * \phi$. (Note that this implies that sgn-normalized ϕ is determined by a single ϕ_I for a non-trivial I. Compare with 3.5.5.)

But by (4) of 2.4.3 this implies that $I = (i)$ where we can assume i positive (i.e., of sgn 1), since the automorphism group is just $\mathbb{F}_q^*$. Then $Frob_I(\lambda) = \phi_i(\lambda) = \lambda$ implies that i is congruent to 1 modulo m. □

This proof is from a letter of Hayes to the author in 1987.

3.7 Moduli approach

This section just sketches some ideas and will not be used elsewhere. The point to be made here is only that for those who have already made an initial investment through learning of the moduli schemes, the questions in the preceding sections, strategies and techniques for solving them are more straightforward.

We will see in 8.7 that the coarse moduli scheme of Drinfeld A-modules (of rank one) is the integral closure B of A in the Hilbert class field for A, since it is pullback of A under the Lang isogeny Frob -1. (Note that also by analytic theory in 2.4, we can see that it is a scheme over $\mathrm{Spec}(A)$ with the class group of A acting on it.)

But since Drinfeld modules have automorphisms, it is not a fine moduli scheme and so it does not follow that Drinfeld modules can be defined over their 'field of definition' which is H_A. (Since the automorphism group in the rank one case is $\mathbb{F}_q^*$, only for $q = 2$ the approach above works.) As noted above, Deligne defined an extra structure killing the automorphisms and producing a moduli problem with the fine moduli scheme giving thus another proof that Drinfeld modules exist over B. (It is somewhat curious that the approach in the previous section led to an easier identification of the field of definition than that of the ring of definition, whereas in the moduli approach, both seem to be of the same difficulty.)

We now just record here the moduli problem he proposed, without going into the techniques and details of the proof using moduli schemes:

First note that $D(\rho_I)$ for a Drinfeld module ρ of rank one with Lie algebra L (over R) can be thought of as an invertible element of $L^{\otimes(NI-1)} \otimes_A I$, with the tensor power keeping track of how it changes under reparametrization $\rho \to w\rho w^{-1}$. Pick ideals I_j of A and integers n_j such that $\sum n_j(NI_j - 1) = q - 1$. Then we get $D(\rho) := \otimes D(\rho_{I_j})^{\otimes n_j} \otimes_A \otimes I_j^{\otimes n_j}$.

Here is the moduli problem: Use a generator of $\prod_j I_j^{n_j} \otimes_A B$ provided by the principal ideal theorem to view $D(\rho)$ as an invertible element of $L^{\otimes(q-1)}$ and for rings R over B consider the moduli of the data 'ρ over R with $\lambda \in Lie(\rho)$ such that ρ is in the prescribed geometric isomorphism class and $\lambda^{\otimes(q-1)} = D(\rho)$'. Then the data of λ destroys the automorphisms, so that the fine moduli space is just $\mathrm{Spec}(B)$ and we get the required Drinfeld module over B.

Standard way to kill automorphisms is to introduce level structure. Drinfeld's version of the cyclotomic theory is then the special case $r = 1$ of

Theorem 3.7.1   *Let I be an ideal of A divisible by at least two primes, then there is a fine moduli scheme M_I^r for Drinfeld modules with level I and rank r. It is affine and of finite type over $\mathrm{Spec}(A)$, r-dimensional and regular, with structure map $M_I^r \to \mathrm{Spec}(A)$ flat with $r - 1$-dimensional fibers and smooth outside support of I. For $I \subset J$, the natural morphism $M_I^r \to M_J^r$ is finite and flat.*

Theorem 3.7.2   *The scheme $(M_I^1)_K$ is just (spectrum of) the maximal*

abelian extension of K unramified outside I and totally split at ∞.

The scheme M^1 defined to be the inverse limit of M_I^1's is isomorphic to the spectrum of the ring of integers in the maximal abelian extension of K totally split at ∞. The action of finite idele class group A_f^*/K^* is that of class field theory.

Drinfeld [Dri77b] also constructed a covering using 'level ∞-structure' which gives integers in maximal abelian extension.

3.8 Summary

Let us summarize the main statements scattered in the proofs so far and some of the easy implications (left to the reader).

(1) Let $H = H_A$ denote the Hilbert class field of A, i.e., the maximal abelian unramified extension of K where ∞ splits completely. Its field of constants has degree d_∞ over $\mathbb{F}_q$ and $[H : K] = hd_\infty = h_A$. The Galois group $\mathrm{Gal}(H/K)$ is naturally isomorphic to the class group of A via the Artin map. Hence, for example, a prime P of A splits completely if and only if it is principal.

The corresponding idele group is $K^* \times K_\infty^* \times U_f$, where U_f consists of ideles which have unit components at all finite primes.

Let B be the integral closure of A in H.

(2) There are $h(A) = hd_\infty$ isomorphism classes of Drinfeld A-modules of rank one, generic characteristic over $\overline{K}$. Each isomorphism class has a model over H and models in other isomorphism classes can be obtained as $\mathrm{Gal}(H/K)$ conjugates of such a model. In fact, H can be characterized as minimal or common field of definition.

Further, the model ϕ can be chosen over B, i.e., with the leading coefficient of ϕ_a a unit in B. So it has Good reduction everywhere.

(3) There are $q^{d_\infty} - 1$ sign functions sgn, depending on choice of uniformizers. For $\sigma \in \mathrm{Gal}(\mathbb{F}_\infty/\mathbb{F}_q)$, $\sigma \circ \mathrm{sgn}$ is called a twisted sign function and a Drinfeld module (always assumed below to be of rank one and generic characteristic) is called sgn-normalized if the leading coefficient of D is twisting of sign.

Any Drinfeld module (of rank one and generic characteristic) is isomorphic to a sgn-normalized one, for a given sgn. Given sgn, let H_1 be the field generated by coefficients of such a sgn-normalized ϕ. It is abelian of degree $hd_\infty(q^{d_\infty} - 1)/(q - 1)$ over K, unramified except at ∞ over K and H_1/H is totally ramified at ∞. In particular, if $d_\infty = 1$, $H_1 = H$.

H_1 is the class field corresponding to $K^* \times \pi^{\mathbb{Z}} \times U_+$ where π is a uniformizer at ∞ with $\mathrm{sgn}(\pi) = 1$ and U_+ is the subgroup of the idele group consisting of those ideles with unit components at finite places and with the component at ∞ of sgn 1. In other words, a finite place v of K splits completely in H_1 if and only if v corresponds to a principal ideal xA with $\mathrm{sgn}(x) \in \mathbb{F}_q^*$.

Let B_1 be the integral closure of A in H_1.

(4) Let i be the coefficient of τ^0 in ϕ_I for a ϕ over B, then $iB = IB$. This is the explicit version of the usual principal ideal theorem.

(5) Let ϕ be a sgn-normalized A-module over H_1. Let m be a modulus not containing ∞, so that it can be identified with an ideal of A. Then $K_m := H_1(\Lambda_m) = K(\Lambda_m) = K(\lambda_m)$ (where λ_m is any generator) is an abelian extension over H_1 with Galois group naturally isomorphic by the Artin map to $(A/m)^*$.

The subgroup of the idele group corresponding to the extension K_m of K is $K^* \pi^{\mathbb{Z}} U^*(m)$, where π is a positive (i.e. $\mathrm{sgn}(\pi) = 1$) uniformizer at ∞, $U^*(m)$ consists of those ideles i with $i_P \equiv 1 \mod m$, i_Q a Q-unit for primes Q distinct from P and $\mathrm{sgn}(i_\infty) = 1$.

In particular, K_m is independent of the choice of a sgn-normalized ϕ. But it depends on the choice of sgn. $K_m(\mu_{(q^{d_\infty}-1)^2/(q-1)})$ is independent of the choice of sgn also.

In K_m, the primes in the support of m are ramified, the other finite primes are unramified and the ramification index of ∞ is $q^{d_\infty} - 1$, with the decomposition and the inertia groups same.

(6) The fixed field of this group is K_m^+, called the plus part or the maximal real subfield: ∞ splits completely in it.

The corresponding idele group is $K^* K_\infty^* U^*(m)$.

A nice consequence of the splitting criteria above is that, if one selects sgn functions appropriately and if v_1 and v_2 are distinct places of K and if Λ_{ij} ($i,j = 1,2; i \neq j$) denotes v_i-torsion for a sgn-normalized ϕ for A with v_j as the infinite place; then

$$K(\Lambda_{ij})(\mu_{q^{\deg v_i}-1}) = K(\Lambda_{ji})(\mu_{q^{\deg v_j}-1}).$$

3.9 Maximal abelian extension

We work in a fixed algebraic closure as usual.

For $\mathbb{Q}$, we get the maximal abelian extension by adjoining all roots of unity, i.e., taking compositum of all m-th cyclotomic fields (this is the famous Kronecker-Weber theorem).

For a quadratic imaginary field K, we similarly get [BCH$^+$66] it by adjoining to the Hilbert class field $K(j(E))$ co-ordinates in (genus zero curve) $E/\mathrm{Aut}(E)$ of all the torsion of the corresponding elliptic curve E over it with $\mathcal{O}_K$ (or a non-principal ideal class) multiplication.

Note that the compositum of all K_m's, as in cyclotomic theory over $\mathbb{Q}$, even together with the constant field extensions, can not give the maximal abelian extension, as from (5) of the section above, it is still tamely ramified at ∞, in fact with ramification index dividing $q^{d_\infty} - 1$ at any finite level.

By the idele group description in (6) of the previous section, the idele group corresponding to the compositum of all K_m^+'s is $K^* K_\infty^*$ and hence the compositum is nothing but the maximal abelian extension of K in which ∞ splits completely. Since K_m is a subfield of K_a for $a \in m$, so that we can just take compositum of all $K(\Lambda_a)^+$'s instead, for all non-zero a's.

Now let ∞_1 and ∞_2 be any two places of K, with corresponding A_1 and A_2. Then taking the compositum of the two corresponding extensions we get the maximal abelian extension of K, as the intersection of the idele subgroups $K_{\infty_1}^*$ and $K_{\infty_2}^*$ is $\{1\}$. This can be considered as the Kronecker-Weber theorem for function fields.

In the case of rational function field $K = \mathbb{F}_q(t)$, more economical description [Ha] is given as follows: Let $A_1 := \mathbb{F}_q[t]$ and $A_2 := \mathbb{F}_q[1/t]$.

(i) Consider $\overline{\mathbb{F}_q}(t)$, which is the extension obtained by adjoining all the roots of unity, i.e., union of all constant field extensions. (ii) Consider the compositum of all $K(\Lambda_a)$'s for $a \in A_1 - \{0\}$ for the Carlitz module for A_1, i.e., the maximal (Drinfeld type) cyclotomic extension relative to ∞. (iii) Consider the compositum of all the extensions obtained by taking the totally real part of those extensions obtained by adjoining $1/t^k$-torsion for $k \geq 2$ of the Carlitz module corresponding to A_2.

Then these three extensions are linearly disjoint and their compositum is the maximal abelian extension of K.

There is another way to get a good family of abelian extensions using the singular theory of orders as developed by Deuring for elliptic curves with complex multiplication. We just sketch the results. For details see, [Hay79]:

A subring R of A which contains 1 and has K as its field of fraction is called an order in A. The conductor of R is the largest ideal in A which is also an ideal in R. Developing theory of R modules similarly, the field of definition H_R is a class field of A: The corresponding class group is the group of principal ideals of A generated by xA, with $x \in R$ prime to the conductor of R. (Only primes dividing the conductor can ramify.) The compositum of all H_R's for all orders is again the maximal abelian extension of K in which ∞ splits completely.

This is analogous in the classical case, to adjoining j-invariant for all orders or adjoining $j(K)$ to the imaginary quadratic field K. To get the maximal abelian extension, classically [BCH+66] we adjoin to K all roots of unity, all $j(z)$ for $z \in K^*$ with positive imaginary part and square roots of all elements thus obtained.

3.10 Cyclotomic theory of $\mathbb{F}_q[t]$

Let us go a little further into the strong analogies between the usual cyclotomic theory of $\mathbb{Q}(\zeta_n)$, where ζ_n is a primitive root of $x^n - 1 = 0$ and of $K(\lambda_a)$, when $K = \mathbb{F}_q(t)$ and λ_a is a primitive root of the $C_a(z) = 0$ for the Carlitz module C for $A = \mathbb{F}_q[t]$. In perfect analogy with the classical case [Lan90; Was97], we have

Theorem 3.10.1 *The extension $K(\lambda_a)$ as above is a Galois extension of K with the Galois group naturally isomorphic to $(A/aA)^*$:*

The action of the Frobenius element $\sigma_\wp$ for prime $\wp$, relatively prime to a is (where we use $\wp$ to be a monic representative): $\sigma_\wp(\lambda_a) = C_\wp(\lambda_a)$.

Its ring of integers (i.e. the integral closure of A in it) $\mathcal{O}$ is $A[\lambda_a]$.

If $\wp$ is an irreducible monic polynomial in A which does not divide a, then it splits in the extension into g conjugate distinct prime ideals each of residue degree f over $\wp$, where $fg = \phi(a) := |(A/aA)^*|$ and f is the order of $\wp$ modulo a.

The infinite prime ∞ of K splits completely into ∞_i in the sub-extension $K(\lambda_a)^+ := K(\prod_{\theta \in \mathbb{F}_q^*} \theta \lambda_a) = K(\lambda_a^{q-1})$ of index $q - 1$ and then each ∞_i is totally ramified in $K(\Lambda_a)$.

To describe how the primes in A ramify in the extension, we specialize, for simplicity, to the prime-power case $a = \wp^n$ where $\wp$ is an irreducible polynomial of degree d. Then as in the number field case, $\wp$ is totally ramified and (λ_a) is the prime ideal above $\wp$ (analogous to $(1 - \zeta_p)$ being the prime ideal above p in $\mathbb{Q}(\zeta_{p^n})$). In fact, the discriminant in the number field case is (up to signs) $p^{n-1}(pn - n - 1)$-th power of p, whereas here it is (up to signs) $\text{Norm}(\wp)^{n-1}(\text{Norm}(\wp) - n - 1)$-th power of $\wp$. Analog q^{2g-2} of the discriminant mentioned in Chapter one, on the other hand, is

$\text{Norm}(\wp)^{n-1}(n\text{Norm}(\wp) - n - 1 - q(\text{Norm}(\wp) - 1)/(d(q-1)))$-th power of $\text{Norm}(\wp)$.

The proofs [GR81a; GR82; Hay74] of the statements above are parallel to the usual proofs [Lan90; Was97] and are good exercises for the reader interested in getting more familiar with the cyclotomic issues.

3.11 Cyclotomic units and conjectures of Brumer and Stark

In the cyclotomic situation, there is interesting connection due to Kummer: the class number of $Q(\zeta_{p^n})^+$ is the index of the group of cyclotomic units in it to the total unit group. (This does not arise from the isomorphism of the corresponding groups.) The corresponding theorem was proved in [GR81a]. (For the slightly more complicated case of general cyclotomic fields in classical, as well as for general A situation, we refer to [GR82; Was97; And96a; Shu94a].)

Let $L = K(\Lambda_a)$ and $F = K(\Lambda_a)^+$, where $a \in A$ is nonconstant. The intersection of the subgroup of L^* generated by the elements of $\Lambda_a - \{0\}$ with $\mathcal{O}_L^*$ is the group E of cyclotomic units. Then $\mathcal{O}_L^* = \mathcal{O}_F^*$ just as in the classical case. We then have the following [GR81a] analog of Kummer's theorem.

Theorem 3.11.1 (GALOVICH-ROSEN): *Let $A = \mathbb{F}_q[T]$ and $a = \wp^n$, where $\wp$ is a prime of A. Then $h(\mathcal{O}_F) = [\mathcal{O}_F^* : E]$.*

Proof. (Sketch) Let $S = \{\infty_i\}$ denote the set of the infinite primes of F. Let $\text{Div}^0(S) \supset P(S) \supset \mathcal{E}$ be the groups of divisors of degree zero supported on S, divisors of elements of $\mathcal{O}_F^*$ and the divisors of cyclotomic units respectively. It is elementary to see that the index in the theorem is equal to that of $\mathcal{E}$ in $\text{Div}^0(S)$. The calculation of the divisors of cyclotomic units using the basis $\infty_i - \infty_0$ of $\text{Div}^0(S)$ allows us to express this index as determinant, which by the Dedekind determinant formula, can be expressed as a product of certain character sums. Finally, the analytic class number formula for the Artin-Weil zeta and L-functions for the function fields identifies this product as the class number $h(F)$ of F. The theorem follows by noticing that $[\text{Div}^0(S) : P(S)] = h(F)/h(\mathcal{O}_F)$. □

In fact, Tate pointed out that the calculation of the divisor of cyclotomic units easily implies (just as in the classical case worked out by Stark [Tat84, Chapter III.5]) the corresponding special case of Brumer-Stark conjectures, which we now recall.

For a finite abelian extension L of a global field K with Galois group G and a non-empty set T of places of K containing at least all those ramify in L, by character/Fourier theory, there is a unique $\theta = \theta_{T,L} \in \mathbb{C}[G]$ such that

$$\psi(\theta) = L_T(0, \overline{\psi})$$

for all complex valued characters ψ of G (extended linearly to group algebra), where the L function

$$L_T(s, \psi) := \prod_{\wp \notin T} (1 - \psi(\mathrm{Frob}_\wp)\mathrm{Norm}(\wp)^{-s})^{-1}, Re(s) > 1$$

is a rational function in q^{-s} and is finite at 0. Let μ be the number of roots of unity in L and $\omega := \mu\theta$.

The following theorem of Deligne and Tate is analog of Brumer-Stark conjectures (giving ideal class annihilators and abelian extensions via analytic processes in the ground field) for function fields.

Theorem 3.11.2   *Let K be a function field and P any prime divisor of L. We have*

(1) $\omega \in \mathbb{Z}[G]$,

(2) If $|T| \geq 2$, then P^ω is a principal divisor (ℓ) of some $\ell \in L$,

(3) If $T = \{v\}$, then P^ω is $(\ell) + nv_L$, for some $\alpha \in L$ and $n \in \mathbb{Z}$ and where v_L denotes the simple sum of places of L above v.

(4) If $\lambda^\mu = \ell$, then $L(\lambda)$ is abelian over K.

Explicit proof of this theorem was given by Hayes [Hay85] by first using the functorial properties to reduce the general abelian extension case to L being the totally real cyclotomic case of conductor M with the place at infinity taken to be the place below P (note that we can choose infinity arbitrary, in contrast to the number field case) and then using the M-division points λ of Hayes sign-normalized Drinfeld module for A corresponding to K and ∞ and carrying out an analog of the calculation above.

A classical analogy would be

$$\ell = (1 - \zeta_m)(1 - \zeta_m^{-1}) = (\zeta_4(\zeta_{2m} - \zeta_{2m}^{-1}))^2$$

which is a unit if m is divisible by at least 2 distinct primes. See [Tat84, Chapter III.5], and [Hay85] for details.

3.12   Some contrasts and open questions

As we saw in Chapter 1, the Iwasawa theory was motivated [Was97; Iwa69] by the function field situation. But if we try to take naive analogs of various aspects of Iwasawa theory back into the function field situation, we see several crucial differences.

First of all, the class field theory description of the abelianization of the absolute Galois group in Chapter 1 immediately shows that a function field has unique $\mathbb{Z}_p$-extension coming from constant field extensions for p different from its characteristic, whereas infinitely many independent $\mathbb{Z}_p$ extensions for p equal to the characteristic. This itself is in great contrast with the number field situation. Gold and Kisilevsky [GK88] show that in fact the class number growth can be made arbitrarily large by suitably choosing geometric $\mathbb{Z}_p$ extension, in stark contrast to Iwasawa's theorem. For more, see the survey [LZ98]. We might want to focus on some particular $\mathbb{Z}_p$ extensions, such as those obtained (Igusa theory) inside extension of $\mathbb{F}_q(t)$ by adjoining all the p^n-torsion of elliptic curve with j-invariant t, in some analogy with cyclotomic $\mathbb{Z}_p$ extensions.

Consider 'cyclotomic' extensions now, where we take instead of constant field cyclotomic towers, the Drinfeld cyclotomic towers of this chapter. For simplicity, let $A = \mathbb{F}_q[t]$ and consider just the Carlitz tower. While the full cyclotomic $\wp^n$-tower has the Galois group $A_\wp^*$, analogous to the situation over $\mathbb{Q}$; as we saw in Section 1.1, it is much wilder group containing infinitely many copies of $\mathbb{Z}_p$ and having quotient isomorphic to $A_\wp$ of infinite index in it. Such extensions are not explored much (but see recent works of De Jong, Böckle, Khare, Angles), as we do not know any good structure theory.

For Carlitz module and a prime $\wp$ of degree d, by the Riemann-Hurwitz theorem, the genus g_n of $K(\Lambda_{\wp^n})$ is asymptotically $d(q^d - 1)nq^{d(n-1)}$ as n tends to infinity (see [Hay74, Thm. 4.1]). So the discussion in Section 1.3 on Brauer-Siegel analogs implies that, if $q > 4$, *log* h_n is of the order (in the sense that the ratio of the two sides is bounded between two positive constants) $n(\text{Norm } \wp)^n$, as n tends to infinity. Does the same estimate hold in the classical case? For the minus part of the class number such estimate holds in the classical case (see [Was97, Thm. 4.20]) and also in our case (if $q > 4$), by similar considerations. But one should note that for $q = 2$ the minus part of h_n is just 1 in our case and also that there is a difference between the notions of class numbers of fields and those of their 'rings of integers'.

Carlitz-Kummer extensions of field F containing torsion $C[a]$, obtained by adjoining roots z of $C_a(z) = f$, for $f \in F$ to F have been studied, and analogy with the Galois theory in classical case is studied in [CL01] and decomposition laws have been studied in [Sch90] and J. Zhao in Chinese science bulletin 1994.

Kisilevsky [Kis93] proved analog of strong form of Leopoldt's conjecture that for a place v of a function field K, multiplicatively independent set of v-adic one units is also independent over $\mathbb{Z}_p$. Of course, one of the main uses of this in Iwasawa theory in determining $\mathbb{Z}_p$-extensions of a number field fails here because of the totally different situation we have noted.

Ki-Seng Tan and others have used Iwasawa theoretic methods for $\mathbb{Z}_p$-extensions for function fields of characteristic p in their attacks on refined conjectures of Mazur and Tate of Birch, Swinnerton-Dyer type.

As for analog of another famous question related to Iwasawa theory, namely Kummer-Vandiver conjecture, see 5.2 and 8.9.

Chapter 4

Gauss sums and Gamma functions

Gauss and Jacobi sums are important algebraic numbers in cyclotomic theory via connection with power reciprocity laws, ideal class relations in cyclotomic fields via Stickelberger theorem, coherence properties via Hecke characters, Euler systems, basic occurrence in character theory and special values of L-functions, root numbers (local constants occurring in the functional equations of L-functions), finite field theory.

These are discrete versions (both finite sums, instead of integrals, of multiplicative characters, analogous to x^s, multiplied by additive characters, analogous to e^x) of Gamma and Beta functions, which are transcendental functions important in cyclotomic theory via special values occurrence as periods in complex multiplication theory, as local factors at infinity in L-function theory and again through their connections with interpolations and their relation to Gauss and Jacobi sums via the Gross-Koblitz formula. Euler introduced gamma function in an attempt to extend the domain for the factorial function important in combinatorics. See [Kob80; GK79; CF67; GGPS90; Lan90; Was97; DMOS82; Del79; Art64] etc. for the classical material, especially for the statements and proofs of the theorems whose analogs we discuss.

We will see that at the simplest level, when we fix an infinite place, there are two kinds of factorials/gamma functions, the first one will turn out to be closely linked with cyclotomic theory obtained by taking constant field extensions and hence following the standard practice will be called 'arithmetic gamma'. The second is linked with the Drinfeld cyclotomic case, which does not increase the constants (unless $d_\infty > 1$, when we have to expand from monic to other signs) and hence following the standard practice it will be called 'geometric gamma'. As mentioned before, it should be understood though that arithmetic and geometry are not really as separate as the names indicate.

The analogies work the best, as is usual, for the $\mathbb{F}_q[t]$ case. Often the proofs are also simple in this case, even though the phenomenon is general. So often we will deal with this case separately.

To keep the analogies more apparent and to keep the notation less, we will use the same notation for classical objects such as Gauss sums and Gamma functions and their function field counterparts (some times many for one!).

From 4.9 onwards we restrict to $d_\infty = 1$, and to principal ideal class, having described in detail how to handle signs when $d_\infty > 1$ and general ideal classes, especially in 4.7.

Apart from the references provided, see [Tha91a; Tha] for 4.5-4.12, [Tha92b] for 4.15.

4.1 Gauss and Jacobi sums: Definitions

Classically, a Gauss sum g_χ is defined to be

$$ g_\chi = - \sum_{x \in \mathbb{F}_{p^m}^*} \chi(x)\psi(\mathrm{Tr}\,x) $$

for a non-trivial multiplicative character $\chi : \mathbb{F}_{p^m}^* \to \mathbb{C}^*$, a non-trivial additive character $\psi : \mathbb{F}_p \to \mathbb{C}^*$ and the trace Tr from $\mathbb{F}_{p^m}$ to $\mathbb{F}_p$.

If we use the usual characters, say C_∞-valued, we end up in the field of constants. We view ψ rather as an isomorphism of $\mathbb{Z}$-modules $\mathbb{Z}/p \to \mu_p$ and replace it by an isomorphism of A-modules $\psi : A/\wp \to \Lambda_\wp$, where $\wp$ is a prime of degree d of A.

This analog ψ of additive character is no longer a character in usual sense, but for $A = \mathbb{F}_q[t]$ say, it has analogous analytic description: $\psi(a) = e(aa_0\bar{\pi}/\wp)$, where $a \in A$ is a representative of the corresponding class modulo $\wp$ and a_0 is prime to $\wp$. It satisfies $\psi(az) = \rho_a(\psi(z))$, for $a \in A$.

To obtain non-vanishing Gauss sums, we restrict the class of multiplicative characters to those giving $\mathbb{F}_q$-homomorphisms $\phi : k \to M$, where M is a field containing $K(\Lambda_\wp)$ and k is a finite field of 'characteristic $\wp$' i.e. a finite extension of $\mathbb{F}_\wp = A/\wp$. In other words, we restrict to $\mathbb{F}_q$-homomorphisms $\chi_j : A/\wp \to M$, $(j \mod d)$ indexed so that $\chi_j^q = \chi_{j+1}$ (special multiplicative characters which are q^j-powers of 'Teichmuller character').

Then we define [Tha87; Tha88] basic Gauss sums by

$$g_j := g(\chi_j) := - \sum_{z \in (A/\wp)^*} \chi_j(z^{-1}) \psi(z).$$

Example 4.1.1   For $\wp = t$, the basic Gauss sum is just a t-torsion point.

We have thus defined Gauss sums taking values in the function fields $K(\Lambda_\wp)(\mu_{q^d-1})$, by combining the Drinfeld cyclotomic theory with the classical cyclotomic theory of constant field extensions i.e. the extensions obtained by adjoining the usual roots of unity to your field.

The fact the we do not get anything different from these d basic Gauss sums (once ψ is fixed) by taking the trace from a bigger extension follows immediately from a simple lemma:

Lemma 4.1.2   *If h is a function on $\mathbb{F}_{q^f}$ with a values in a ring containing $\mathbb{F}_{q^f}$ and with $h(0) = 0$, then*

$$\sum_{x \in \mathbb{F}_{q^f}^*} x^{-q^j} h(Tr(x)) = \sum_{y \in \mathbb{F}_q^*} y^{-q^j} h(y).$$

Proof.   This follows from $\sum_{Tr(x)=y} x^{-q^j} = y^{-q^j}$, which is [Tha88] an easy exercise.   □

The Fourier coefficients of the additive character ψ in terms of all characters of $(A/\wp)^*$ are exactly the basic Gauss sums:

Proposition 4.1.3 $\psi(z) = \sum g_j \chi_j(z).$

Proof.   The coefficient of $\chi(z)$ is zero for $\chi \neq \chi_j$, because of $\mathbb{F}_q$-linearity of ψ and is g_j for $\chi = \chi_j$, by Fourier inversion.   □

This should be compared with the classical $\psi(z) = \sum g_\chi \chi(z)$. It also shows why we restricted the class of multiplicative characters to get the non-vanishing sums.

Proposition 4.1.4   *The basic Gauss sums g_j are non-zero.*

Proof.   As ψ is nonzero, g_{j_0} is nonzero for some j_0 and hence using the Galois action $g_{j_0+kd_\infty}$ is nonzero for any integer k. In particular, all g_j's are nonzero, when $(d, d_\infty) = 1$ (eg. when $d_\infty = 1$). In the general case, we need only a slightly more involved argument: put $\psi_\mu(z) := \psi(\mu z)$, for $\mu \in A/\wp$. Since the pairing $A/\wp \times A/\wp \to \Lambda_P$ defined by $(x, y) \to \psi(xy)$ is non-degenerate bilinear pairing over $\mathbb{F}_q$, any $\mathbb{F}_q$-linear function from $A/\wp$, which is a d-dimensional vector space over $\mathbb{F}_q$, to M is a linear combination of ψ_μ's. Now $\psi_\mu = \sum(\chi_j(\mu)g_j)\chi_j$. If g_{j_0} were zero, then ψ_μ's would be in a vector space generated by less than d of χ_j's which is a contradiction. $\square$

The general Gauss and Jacobi sums are defined as monomials in the basic ones: Let $0 < n < q^d$ with $n = \sum n_j q^j$, $0 \le n_j < q$. Then the gauss sum corresponding to the n-th power of the Teichmüller character is defined as $\prod g_j^{n_j}$. More generally, we call $\prod_{j=0}^{d-1} g_j^{m_j}$, with $m_j \in \mathbb{Z}$, as general Gauss sums and those with $\sum m_j q^j$ divisible by $q^d - 1$ are considered also as Jacobi sums.

To compare with the classical references as well as the gamma monomials we will study later, we generalize the notation slightly to accommodate primes in constant field extensions lying above $\wp$ as follows:

Let P be a prime of $K\mathbb{F}_{q^r}$ above $\wp$, with relative residue class degree f. Choose a prime $\tilde{P}$ above P in $K\mathbb{F}_{q^{df}}$. We may assume that χ_0 is Teichmüller character for $\tilde{P}$. This fixes the numbering of g_j's.

Given an element $\underline{a} = \oplus\, m(i)\frac{n(i)}{1-q^r}$ in the free abelian group with basis $(1 - q^r)^{-1}\mathbb{Z}/\mathbb{Z}$, choosing representatives with $0 < n(i) \le q^r - 1$ for the basis elements and expanding $n(i) = \sum n(i)_j q^j$ in base q digits as above, by abuse of notation we write $\underline{a}$ as $\oplus\, m_j \frac{q^j}{1-q^r}$, with $m_j = \sum m(i)n(i)_j$.

Definition 4.1.5 For an element of the form $\underline{a} = \oplus_{j=0}^{r-1} m_j(q^j/(1-q^r))$ as above, define Gauss sum

$$g(\underline{a}, P) = \prod_j (g_j g_{j+r} \cdots g_{j+df-r})^{m_j}.$$

We let $g(\underline{a})$ stand for $g(\underline{a}, P)$ when $P = \wp$ and $r = d$.

Also, for $k \ge 0$, put $\underline{a}^{(q^k)} = \oplus m_j(q^{j+k}/(1 - q^r))$ and $n(\underline{a}) = \sum m_j \langle q^j/(1 - q^r)\rangle$, where $\langle x \rangle$ is representative of x mod $\mathbb{Z}$ such that $0 \le \langle x \rangle < 1$. If $n(\underline{a})$ is an integer, we call the sum Jacobi sum. For example, $J_j = g_{j-1}^q/g_j$ are Jacobi sums. Note also that $g(q^j/(1 - q^{df}), \tilde{P}) = g_j$.

In contrast with the classical case, the Jacobi sum is not a character sum built out of multiplicative characters, since the multiplicative characters take values in a finite field. But we will see below that these monomials in the basic gauss sums have the right Galois invariance and algebraic properties to be called Jacobi sums.

4.2 Gauss and Jacobi sums: $\mathbb{F}_q[t]$ case

For this section, we restrict to $A = \mathbb{F}_q[t]$ and the Carlitz module. We will see many analogies [Tha87; Tha88] with the classical case.

The situation of the factorization, signs, absolute values etc. is quite different in the general case, but the Galois theoretic properties are similar, but we stick to $\mathbb{F}_q[t]$ case anyway for simplicity.

Let $\wp$ be a monic prime of A of degree d. Clearly $g_j \in L :=$ $K(\Lambda_\wp)(\mu_{q^d-1})$. The Galois group of L over K is canonically isomorphic to the product of the Galois groups for $K(\Lambda_\wp)$ and $K(\mu_{q^d-1})$ over K. So the powers of q-th power Frobenius σ for the constant field extension and elements $\mu \in (A/\wp)^* = \mathrm{Gal}(K(\Lambda_\wp)/K)$ can be thought of as elements of $\mathrm{Gal}(L/K)$. We will denote the image of $l \in L$ under g in the $\mathrm{Gal}(L/K)$ as l^g. The following two theorems follow easily and the proofs are left as an exercise.

Theorem 4.2.1 (1) $g_j^\sigma = g_{j+1}$ (j mod d).
  (2) $g_j^\mu = \chi_j(\mu)g_j$.

Theorem 4.2.2 (1) $g(\underline{a}, P)$ is independent of the choice of $\tilde{P}$.
  (2) $g(\underline{a}, P)$ belongs to $\mathbb{F}_{q^r}(\Lambda_\wp)$.
  (3) If $n(\underline{a})$ is an integer, then the Jacobi sum $g(\underline{a}, P)$ belongs to $\mathbb{F}_{q^r}(t)$.
  (4) $g(\underline{a}, p)$ depends on the choice of ψ only up to multiplication by q^d-1-th root of unity, and is independent of ψ, if $n(\underline{a})$ is an integer.
  (5) $g(\underline{a}^{(q^d)}, P) = g(\underline{a}, P)$.

Theorem 4.2.3 (Analog of the Hasse-Davenport theorem) If P' is a prime (in $\mathbb{F}_{q^{rt}}[t]$ for some t) of relative residue class degree s over P, then $g(\underline{a}, P') = g(\underline{a}, P)^s$.

Proof. Follows from Lemma 4.1.2, and the part (5) of the Theorem above using that $q^j/(1-q^d) = q^j(1+q^d+\cdots+q^{d(s-1)})/(1-q^{ds})$. $\qquad\square$

See remark 4.6.6 for an analog of the second Hasse-Davenport theorem. Now we turn to the factorization of Gauss sums:

Theorem 4.2.4 We have

$$\prod_{j \;\; \mathrm{mod}\; d} g_j^{q-1} = (-1)^d \wp.$$

In particular, g_j lie above $\wp$.

Proof. Since $\psi(tz) = t\psi(z) + \psi(z)^q$, we have $g_j\chi_j(t) = g_j t + g_{j-1}^q$. As $g_j \neq 0$, we get

$$J_j = g_{j-1}^q/g_j = -(t - \chi_j(t)) = -\wp_{1-j}$$

for j mod d. Multiplying these equations, we get the theorem. □

This can be restated as

$$g(y)g(-y) = g(y)g(-1-y) = g(-1) = \prod g_j^{q^j} = (-1)^d\wp$$

analogous to the **reflection formula** $g(\chi)g(\overline{\chi}) = \chi(-1)q$ for classical Gauss sums as well as the one for the factorial function. In particular, $g(1/2)^2 = (-1)^d\wp$, if $p \neq 2$. See 4.6 for more along these lines.

Set $\wp_j := t - \chi_{j-1}(t)$ (j mod d). These are monic representatives of primes above $\wp$ in $\mathbb{F}_{q^d}[t]$. Observe that χ_{1-j} is then the Teichmüller character of $\wp_j$ and $\wp_j^\sigma = \wp_{j-1}$. Let $\overline{\wp_j}$ be the unique prime above $\wp_j$ in $\mathcal{O}_L$. Put $\overline{\wp} = \overline{\wp_1}$.

Lemma 4.2.5 *There exists a unique $\lambda \in K_\wp(\psi(1)) \subset L_{\overline{\wp}}$ such that $\lambda^{\mathbb{N}\wp-1} = -\wp$ and $\lambda \equiv \psi(1) \mod \psi(1)^2$.*

Proof. Uniqueness is obvious. Now $C_\wp(u)/u$ is an Eisenstein polynomial $u^{\mathbb{N}\wp-1} + \cdots + \wp$, so its root $\psi(1)$ generates totally tamely ramified abelian extension of $K_\wp$ of degree $\mathbb{N}\wp - 1$, but any $\mathbb{N}\wp - 1$-th root λ also generates such an extension. By local class field theory such extensions are obtained by adjoining $\mathbb{N}\wp - 1$-th root of prime elements, so we have $K_\wp(\psi(1)) = K_\wp((-\tilde{\wp})^{1/(\mathbb{N}\wp-1)})$ for some prime element $\tilde{\wp}$. Now from the equations satisfied by these generating elements, we see that $\wp$ and $\tilde{\wp}$ are norms from these extension. Thus their ratio is one-unit and hence a $\mathbb{N}\wp-1$-th power. So the extension is the same as $K_\wp(\lambda)$. □

Theorem 4.2.6 *(Analog of the Stickelberger theorem)* $g_j/\lambda^{q^j} \equiv 1/d_j$ *mod $\overline{\wp}$ for $0 \leq j < d$, where $d_0 = 1$ and $d_j = -\wp_{1-j}d_{j-1}^q$.*

Proof. By the lemma above,

$$\psi(\text{Tr}(x)) = C_{\text{Tr}(x)}(\psi(1)) \equiv \text{Tr}(x)\psi(1) \equiv \text{Tr}(x)\lambda \mod \psi(1)^2.$$

As $\chi_0(y) \equiv y \mod \overline{\wp}$, it follows that

$$g_0 \equiv -\lambda \sum \chi_0(x^{-1})\text{Tr}(x) \equiv \lambda \mod \overline{\wp}^2.$$

Theorem follows by combining the evaluation of J_j in the proof of Theorem 4.2.4. □

Theorem 4.2.7 *(1)* $(g_j) = \overline{\wp}_{1-j}\overline{\wp}_{2-j}^{q}\cdots\overline{\wp}_{d-j}^{q^{d-1}}$.

(2) The valuation, normalized in the usual fashion, so that the valuation of t is -1, of g_j at any infinite place of L is $-1/(q-1)$.

Proof. Part (1) follows from theorem above. Raising both sides of (1) to the q^d-1-th power, part (2) follows from the known splitting of infinite place in the cyclotomic extension and the fact that $\overline{\wp}_j^{q^{d-1}} = \wp_j = t - \chi_{1-j}(t)$, and $(1 + q + \cdots + q^{d-1})/(q^d - 1) = 1/(q-1)$. $\qquad\square$

Part (2) is analog of the classical absolute value evaluation for the Gauss sums, with $1/(q-1)$ analogous to $1/2$ as usual, as explained in 1.1.

Theorem 4.2.8 *(Analog of Weil's theorem on Jacobi sums Hecke characters) If $n(\underline{a})$ is an integer, then the Jacobi sums $g(\underline{a}, P)$ extended multiplicatively to a function on ideals, give an algebraic Hecke character χ of F of conductor 1 (or rather supported at ∞) with the algebraic part $\theta = \sum n(\underline{a}^{(q^j)})\sigma^{-j}$ where the sum runs over j in the Galois group $\mathbb{Z}/(r)$. More precisely, for monic $\alpha \in \mathbb{F}_{q^r}[t]$, we have $\chi((\alpha)) = \alpha^\theta b^{\deg\alpha}$, where $b = (-1)^s$, $s = \sum_0^{r-1} n(\underline{a}^{(q^i)})$.*

Proof. The evaluation of J_j in the proof of Theorem 4.2.4 shows that $g_j = \eta_{1-j}\eta_{2-j}^{q}\cdots\eta_{df-j}^{q^{df-1}}$ for some choice $\eta_i = (-\wp_i)^{1/(q^{df}-1)}$. Put $\beta_i = (\eta_i\eta_{i+r}\cdots\eta_{i+df-r})^{(q^{df}-1)/(q^r-1)}$ $(i \bmod r)$. Then $g(q^j/(q^r - 1), P)$ equals $\beta_{1-j}\beta_{2-j}^{q}\cdots\beta_{r-j}^{q^{r-1}}$.

The hypothesis implies that there are integers t_i (i from 0 to $r-1$) such that $\sum t_i q^i = n(\underline{a})(q^r - 1)$ and

$$
\begin{aligned}
\chi((\wp)) &= \prod g(q^j/(q^f - 1), P)^{t_j} \\
&= (\beta_1\beta_2^{q}\cdots\beta_r^{q^{r-1}})^{t_0}(\beta_r\cdots)^{t_1}\cdots(\beta_2\cdots\beta_1^{q^{r-1}})^{t_{r-1}} \\
&= \beta_1^{n(\underline{a})(q^r-1)}\beta_2^{n(\underline{a}^{(q)})(q^r-1)}\cdots \\
&= (-1)^{df(n(\underline{a})+\cdots+n(\underline{a}^{q^{(q^{r-1})}})))/r}\wp_1^{n(\underline{a})}\cdots\wp_r^{n(\underline{a}^{(q^{(q^{r-1})})})}.
\end{aligned}
$$

$\qquad\square$

Remark 4.2.9 Using Stickelberger elements, Hayes and Anderson constructed Hecke characters with much more general conductors. See [Hay93; And] for this and references to earlier work of Takahashi and Gross.

4.3   Gauss and Jacobi sums: General A

We now show that the factorizations [Tha91b] and absolute values [Tha93a] situation for general A is quite interesting, but different from the $\mathbb{F}_q[t]$ case above. The geometry behind the factorization in general will be explained in 8.2.

Let $\wp$ be a prime of degree d as before. Let us take ρ sgn-normalized. Then the basic Gauss sums will take values in $H_1(\Lambda_\wp, \mu_{q^d-1})$.

We focus on the 'norm' $G_\wp =: \prod g_j^{q^{-1}}$ (where the product runs over j modulo d). We will see a scenario quite different from Theorem 4.2.4, in general.

First note that Eisenstein property of $\rho_\wp(u)/u$ at $\wp$ implies that $\wp$ divides $G_\wp$.

Examples 4.3.1   (i) If $\wp$ is a prime of degree one, then explicit principal ideal theorem implies easily that $G_\wp B = \wp B$.

(ii) Let $q = 2$ and $\wp \in A$ be a prime of degree two. Let $D_\wp = \wp + p_1\tau + p_2\tau^2$ and $\Lambda_\wp = \{\lambda_i\}$. Then a direct calculation shows

$$G_P = g_0 g_1 = \sum \lambda_i^2 + \sum_{i \neq j} \lambda_i \lambda_j = 0 + p_1/p_2 = p_1/p_2.$$

Since p_2 is a unit by our assumption, $G_\wp B = p_1 B$. Example of (i) of 3.1.2, with $\wp = x$, now shows that $G_\wp = x(x+1)$, so we see that primes other than $\wp$ can also enter into the factorizations of Gauss sums.

To study this phenomenon in detail, we first focus on the case $h_A = h = d_\infty = 1$ (i.e., examples of 3.1.2), when the cyclotomic theory is the simplest and analogous to the classical case, but still leads to quite different but interesting factorizations.

Theorem 4.3.2   *Let $h = d_\infty = 1$, and $\wp \in A$ be a prime degree d and sgn 1. Then $G_\wp$ is given by the following recipe.*

(1) For Example 3.1.2 (i), $G_\wp = \wp \wp^\sigma$ where σ is order four automorphism of A given by $\sigma(x) = x + 1$, $\sigma(y) = y + x + 1$.

(2) For Example 3.1.2 (ii), $G_\wp = \wp(\wp^\sigma)^3$, where σ is the order two automorphism of A given by $\sigma(x) = x$, $\sigma(y) = y + 1$ and for a given sign function such that x and y have sign 1.

(3) For Example 3.1.2 (iii), $G_\wp = (-1)^d \wp(\wp^\sigma)^2$, where σ is the order three automorphism of A given by $\sigma(x) = x + 1$, $\sigma(y) = y$.

(4) For Example 3.1.2 (iv), $G_\wp = \wp(\wp^\sigma)^2 \wp^{\sigma^2}$, where σ is the order four automorphism of A given by $\sigma(x) = x + 1$, $\sigma(y) = y + x^2$.

Proof. Note that (1)-(3) are of genus 1, while (4) is of genus 2.

When A has degree 2 element (x in our cases), one can obtain a simple expression for g^q_{j-1}/g_j as follows. $\rho_x = x + x_1\tau + x_2\tau^2$ gives, as in the proof of Theorem 4.2.4, $\chi_j(x)g_j = xg_j + x_1g^q_{j-1} + x_2g^{q^2}_{j-2}$. One uses this relation to successively eliminate higher powers from a similar relation obtained from ρ_y with suitable element, eg., y in our cases, of higher degree.

Write $\chi_0(x) = \theta$ and $\chi_0(y) = \beta$, so that $\chi_j(x) = \theta^{q^j}$ and $\chi_j(y) = \beta^{q^j}$. In our four cases one gets the following expressions for g^q_{-1}/g_0 respectively:

$$\frac{x(x + \theta) + y + \beta}{x + \theta + 1}$$

$$\frac{x^2(x + \theta) + y + \beta}{x + \theta}$$

$$\frac{-y(x - \theta) + y - \beta}{x - \theta + 1}$$

$$\frac{(x + \theta)(x^4 + x^3 + (1 + \theta)x^2) + y + \beta}{x^3 + \theta x^2 + (1 + \theta)x + \theta^2 + \theta}$$

Now $G_\wp = \prod(g^q_{j-1}/g_j)$. (Claim about the signs immediately follows from this.) Let the product of numerators of the expression (as written, without reducing) be N and product of denominators be D. Then we claim that in our four cases we have respectively:

$$N = \wp(\wp^\sigma)^2\wp^{\sigma^3}, \quad D = \wp^\sigma\wp^{\sigma^3}$$

$$N = \wp^2(\wp^\sigma)^4, \quad D = \wp\wp^\sigma$$

$$N = \wp(\wp^\sigma)^3(\wp^\sigma)^\mu, \quad D = \wp^\sigma(\wp^\sigma)^\mu \quad \mu(y) = -y, \mu(x) = x$$

$$N = (\wp\wp^{\sigma^2}\wp^{\sigma^3})^2(\wp^\sigma)^4 \quad D = \wp\wp^{\sigma^2}(\wp^\sigma\wp^{\sigma^3})^2$$

Theorem immediately follows if we establish this claim. We will give details only for (1), others being similar. Let $\wp_1, \ldots, \wp_d$ be the primes above $\wp$, in the integral closure of A in $K(\mu_{q^d-1})$, numbered so that $x \equiv \theta, y \equiv \beta$ mod $\wp_1$. We want to show (a) $\prod(x + \theta^{2^j} + 1) = \wp^\sigma\wp^{\sigma^3}$ and (b) $\prod(x(x + \theta^{2^j}) + y + \beta^{2^j}) = \wp(\wp^\sigma)^2\wp^{\sigma^3}$.

Case I: $\wp \neq \wp^{\sigma^2}$: Proof of (a): $\wp_1$ divides $x + \theta$, hence $\wp$ divides $\prod(x + \theta^{2^j})$. But this product being invariant by σ^2, $\wp^{\sigma^2}$ also divides it. Count of degrees then shows that $\prod(x + \theta^{2^j}) = \wp\wp^{\sigma^2}$. Applying σ to both sides we get (a).

Proof of (b): (a) and the fact that $\wp$ divides $G_\wp$, imply that $\wp\wp^\sigma\wp^{\sigma^3}$ divides N. It is enough to show that $\wp^2$ divides N^{σ^3} or that $\wp_1^2$ divides $n_1^{\sigma^3}$, where we have put $n_j =: x(x + \theta^{2^j}) + y + \beta^{2^j}$. Now, since $x^2 \equiv \theta^2$ and $y^2 \equiv \beta^2$ modulo $\wp_1^2$, we have modulo $\wp_1^2$,

$$n_1^{\sigma^3} \equiv (x+1)(x+1+\theta^2) + y + x + \beta^2 \equiv x^3 + x + 1 + y^2 + y \equiv 0$$

as required.

Case II: $\wp = \wp^{\sigma^2}$: Proof of (a): Write $\wp = a(x)y + b(x)$, $a(x), b(x) \in \mathbb{F}_2[x]$. Invariance by σ^2 implies that $a(x) = 0$. As degree x is 2, the equation $\wp(\theta) \equiv 0$ (modulo $\wp_1$) is of degree $d/2$ in θ, $\theta = \theta^{2^{d/2}}$, which implies $\wp_1$ divides $x + \theta$ and $x + \theta^{2^{d/2}}$ and hence $\wp^2$ divides $\prod(x + \theta^{2^j})$. (a) now follows as in case I.

Proof of (b): In case I , we saw $(\wp_1^\sigma)^2$ divides n_1. Similarly, it is easy to see that $\wp_1$ and $\wp_1^{\sigma^3}$ divide n_0. $\wp_1^2$ divides $n_1^{\sigma^3} = n_1^\sigma + 1$. Now $\wp_1$ does not divide n_1^σ, hence $\wp_1^{\sigma^3}$ does not divide n_1 and so $\wp_1^\sigma \neq \wp_1^{\sigma^3}$; so that we can still conclude $N = \wp(\wp^\sigma)^2\wp^{\sigma^3}$. $\qquad\square$

Remarks 4.3.3   (1) The factorizations of g_j's can be easily deduced from these factorizations of Jacobi sums achieved in the proof, as in the previous section.

(2) We will reprove and generalize this theorem in Chapter 8, where we also discuss the case of infinite place of higher degree in rational function fields. See [Tha91b] for earlier approach.

Now we turn from factorization question to the question of valuation at infinity of Gauss sums, and again we will see behavior for general A quite different from the classical or that of $\mathbb{F}_q[t]$ in 4.2.7. We will restrict to A of $d_\infty = 1$ for simplicity.

Let ρ be a Drinfeld module over K_∞ with the corresponding lattice $\tilde\pi A$, with and ideal $\mathcal{A}$ of A. Let $\wp$ be a prime of A of degree d. Note that $\Lambda_\wp = \{e(\tilde\pi r) : r \in \wp^{-1}A\}$ and that $g_j \in \tilde\pi\mathbb{F}_{q^d}((t^{-1}))$.

Theorem 4.3.4   *The degree of the Gauss sum is same as the maximum possible degree of a $\wp$-torsion element.*

Proof.   Let $R \subset \wp^{-1}A$ be a set of representatives modulo $\mathcal{A}$ of the lowest possible degrees. It is easy to see that there is a $\mathbb{F}_q$- basis $\{r_1, \cdots r_d\}$ of R

such that $\{r_1 + r_2\theta_1 + \cdots + r_d\theta_d : \theta_i \in \mathbb{F}_q\}$ is exactly the subset of monic (i.e. of sgn 1) elements of R of maximal degree.

We can assume that the torsion points $\psi(r)$ are just $e(\tilde{\pi}r)$, for $r \in R$. The product formula for the exponential shows that the degree of $\psi(r)$ is maximal, when the degree of r is maximal. Now ψ being additive, the maximal degree is degree of $\psi(r_1)$. It is enough to show that this top degree does not get canceled in the summation. Since both ψ and χ_j are $\mathbb{F}_q$-linear, and $q - 1 = -1$ in characteristic p, we have $g_j = \sum \chi_j(z^{-1})\psi(z)$, where now the sum is taken over the monic representatives of $\wp^{-1}A/A$. If we note that $\chi_j(r_i)$ is a basis of $\mathbb{F}_{q^d}$ over $\mathbb{F}_q$, the theorem then follows from the following lemma. $\square$

Lemma 4.3.5 *If* $f_1, \cdots, f_d$ *is a basis of* $\mathbb{F}_{q^d}$ *over* $\mathbb{F}_q$, *then*

$$\sum := \sum_{\theta_i \in \mathbb{F}_q} \frac{1}{f_1 + f_2\theta_2 + \cdots + f_d\theta_d} \neq 0.$$

Proof. Consider the Moore determinant of 2.11 (b)

$$M(x_1, \cdots, x_k) := \prod_{j=1}^{k} \prod_{\theta_i \in \mathbb{F}_q} (x_j + x_{j+1}\theta_{j+1} + \cdots + x_k\theta_k).$$

Then with $P(t) := \prod(t + f_2\theta_2 + \cdots + f_d\theta_d)$, where the product is over all $\theta_i \in \mathbb{F}_q$, we have $\sum = P'(f_1)/P(f_1)$ and $P(f_1) = M(f_1, \cdots, f_d)/M(f_2, \cdots, f_d)$. As $P(t)$ is an $\mathbb{F}_q$-linear polynomial, $P'(t)$ is just the coefficient of t in $P(t)$ and hence equals $\prod(f_2\theta_2 + \cdots + f_d\theta_d)$, where now the product runs through $\theta_i \in \mathbb{F}_q$ not all zero. But this is just $(-1)^{d-1}M(f_2, \cdots, f_d)^{q-1}$, because $\prod_{\theta \in \mathbb{F}_q^*} \theta = -1$ and $(-1)^{(q^{d-1}-1)/(q-1)} = (-1)^{d-1}$. Hence $\sum = (-1)^{d-1}M(f_2, \cdots, f_d)^q/M(f_1, \cdots, f_d)$ is nonzero, as it is product of terms which are nonzero because f_i are linearly independent over $\mathbb{F}_q$. $\square$

Definition 4.3.6 Let $0 \le n_1 < n_2 \cdots < n_g$ be the integers n ('gaps of $\mathcal{A}$') so that there are no elements of $\mathcal{A}$ of degree n. Call $\wp$ exceptional with respect to $\mathcal{A}$ (or rather its ideal class) if n_g is a gap for the fractional ideal $\wp^{-1}\mathcal{A}$ (i.e. there is no element of degree n_g in $\wp^{-1}\mathcal{A}$). We say that $\wp$ is exceptional, if it is exceptional with respect to some $\mathcal{A}$.

Remarks 4.3.7 (1) When $\mathcal{A} = A$, the number of gaps g is just the genus g, by the Riemann-Roch.

(2) The number $n_g - g$ is an ideal class invariant. We take $n_0 = -1$ to retain this property when $g = 0$.

Theorem 4.3.8 *(1) Principal primes are not exceptional; (2) Primes of degree more than g are not exceptional. In particular, there are at most finitely many exceptional primes. (3) Primes of degree one which are not principal are exceptional.*

Proof. Let $\wp$ be a prime of degree d. Since n_g is the largest gap of $\mathcal{A}$, there is an element, say $e \in \mathcal{A}$ of degree $n_g + d$. If $\wp$ is principal, $\wp = (P)$ say, then $e/P \in \wp^{-1}\mathcal{A}$ has degree n_g and (1) follows.

In general, given $\wp$, let d' be the degree of the smallest degree element in the smallest degree ideal $\overline{\wp}$ in the ideal class inverse to that of $\wp$. Counting the gaps of $\overline{\wp}$, we see that $d' \leq \deg \overline{\wp} + g$. If $d > d' - \deg \overline{\wp}$, then $\wp^{-1}$ has an element of negative degree and just as above we see that $\wp$ in that case can not be exceptional. Thus (2) follows.

Now let $\wp$ be of degree one and not principal (equivalently $g \neq 0$). Then 0 is a gap for $\wp$, but not for $\mathcal{A}$. Hence the count of gaps shows that the largest gap for $\mathcal{A}$ is also the largest gap for $\wp$ and hence $\wp$ is exceptional for $\mathcal{A} = \wp$. □

Remarks 4.3.9 (i) If $\wp$ is a non-principal prime of the lowest possible degree for $\mathcal{A}$ for a hyper-elliptic K, then $\wp$ has $2g - 1$ as a gap and hence is exceptional for $\mathcal{A} = \wp$.

(ii) When $g = 1$, the theorem shows that the exceptional primes are exactly the primes of degree one and are hence $h - 1$ in number.

(ii) By the first part of the theorem, there are no exceptional primes when $h_A = 1$. On the other hand, in appendix to [Tha93a] Voloch gives a nice characterization of exceptional primes and proves that they do exist when $h > 1$.

Lemma 4.3.10 *If $\wp$ is (resp. is not) exceptional, the highest degree element in R has degree less than (resp. equal to) n_g.*

Proof. If $\wp$ is not exceptional, $\wp^{-1}\mathcal{A}$ has an element of degree n_g, which is not congruent to any element of lower degree modulo $\mathcal{A}$, since n_g is a gap for $\mathcal{A}$. On the other hand, any element of $\wp^{-1}\mathcal{A}$ of degree more than n_g is congruent modulo $\mathcal{A}$ to one of the lower degree as can be seen by subtracting an element of the same degree (which exists, as n_g is largest gap) and opposite sign. □

Theorem 4.3.11 *Let ρ be a Drinfeld module over H. Let $i_k : H \hookrightarrow K_\infty$ be the embedding corresponding to an infinite place ∞_k of H and $\tilde{\pi}\mathcal{A}$ be the corresponding lattice. Then the degree of the Gauss sum is the same at any* *prime above* ∞_k *and is less than or equal to* $q^{n_{\underline{g}}-\underline{g}+1}/(q-1)$, *with equality if and only if* $\wp$ *is not exceptional with respect to* $\mathcal{A}$.

Proof. Let $n(i)$ be the number of monic elements of $\mathcal{A}$ of degree i. Then it is easy to see that $n(i)$ is q^{i-j}, if $n_j < i < n_{j+1}$, where for convenience we take $n_{g+1} = \infty$. We will see in Section 4.7 that the degree of $\tilde{\pi}$ is given by p-adic sum $\sum (q-1)in(i) = \Sigma_1 + \Sigma_2$, where Σ_1 is the sum over $i \leq n_{\underline{g}}$ and Σ_2 over $i > n_{\underline{g}}$. Then

$$\Sigma_2 = (q-1)\sum_{k=1}^{\infty}(n_{\underline{g}}+k)q^{n_{\underline{g}}+k-\underline{g}} = -(n_{\underline{g}}+1)q^{n_{\underline{g}}-\underline{g}+1} + \frac{q^{n_{\underline{g}}-\underline{g}+2}}{q-1}.$$

By Theorem 4.3.4, the degree of the Gauss sum is the degree of $e(\tilde{\pi}r_1) = \tilde{\pi}r_1 \prod(1 + r_1/a)$, where the product runs over nonzero $a \in \mathcal{A}$. Let us compute the degree of $r_1 \prod(1 + r_1/a)$. Note that there are no terms of negative degree in the product by the choice of R. It is clearly sufficient to consider only the case where $\wp$ is not exceptional. Then the degree is easily seen to be

$$n_{\underline{g}} + \sum_{i \neq n_j, i \leq n_{\underline{g}}} (n_{\underline{g}} - i)(q-1)n(i) = n_{\underline{g}}[1 + \sum(q-1)n(i)] - \sum(q-1)in(i).$$

Now the sum in the bracket telescopes to $q^{n_{\underline{g}}-\underline{g}+1}$ by the determination of $n(i)$ above. Combining with the formula for the degree of $\tilde{\pi}$, the degree of the Gauss sum then turns out to be $-q^{n_{\underline{g}}-\underline{g}+1} + q^{n_{\underline{g}}-\underline{g}+2}/(q-1) = q^{n_{\underline{g}}-\underline{g}+1}/(q-1)$ as claimed. $\qquad \square$

Remarks 4.3.12 Let D be the degree of a non-principal prime of A of smallest possible degree. If the largest gap for A is smaller than $D+g-1$ (for example, $D > 1$ and A with gaps 1 to g), then $\wp$ is not exceptional for $\mathcal{A} = \wp$ and hence for such A's, every prime has 'generic' infinite valuation for at least one ∞_k by the theorem. The situation when A has a gap $2g-1$ gives another example of this, since then there are no exceptional primes for $\mathcal{A} = A$: The largest gap for $\wp^{h-1}$ is $\leq 2g-1+(h-1)d$ and hence the largest gap for $\wp^{-1}$ is $\leq 2g-1-d$. In fact, for general $\mathcal{A}$, in appendix to [Tha93a] Voloch shows how to find $\mathcal{A}$ with no exceptional primes with respect to it.

4.4 Sign of Gauss sums for $\mathbb{F}_q[t]$

In this section, we restrict to the case $A = \mathbb{F}_q[t]$ and the Carlitz module.

By an m-th order Gauss sum, when m divides $q^d - 1$, we mean $g(y)$ with y having reduced denominator m. Note that $g(q^j/(q^d-1)) = g_j$ corresponds to the multiplicative characters χ_j of order $q^d - 1$. In particular, if $p \neq 2$, we can talk about the 'quadratic Gauss sum' $g(1/2) = \prod g_j^{(q-1)/2}$.

We can consider g_j as an element of $\tilde{\pi} K_\infty(\zeta_{q^d-1})$ and can talk about its 'sign' sgn. By formulas for $\tilde{\pi}$ in 2.5, $\tilde{\pi}^{q-1} \in K_\infty$ and $\mathrm{sgn}(\tilde{\pi}^{q-1}) = -1$. We first give a formula for $\mathrm{sgn}(g_j)$:

Theorem 4.4.1   *Write* $\epsilon := \mathrm{sgn}(\tilde{\pi})$ *and* $t_j := \chi_j(T)$. *Then*

$$\mathrm{sgn}(g_j) = (-1)^{d-1} \epsilon \prod_{k=1}^{d-1} (t_k - t_0)^{-q^j}.$$

Proof.   Consider the $\mathbb{F}_q$-basis $r_i := t^{d-i}$, $0 < i \leq d$ for the representatives of $A/\wp$. Then 4.3.5 shows that $\mathrm{sgn}(g_j)$ is ϵ times $\chi_j(\sum)$, where $\sum :=$ $\sum_{\theta_i \in \mathbb{F}_q} (r_1 + r_2\theta_2 + \cdots + r_d\theta_d)^{-1}$. By the last calculation in 4.3.5, the Moore determinant identity of 2.11 (b), and formula in 2.5 relating D_i and L_i, we see that

$$\sum = (-1)^{d-1} \frac{(D_{d-2}\cdots D_0)^q}{(D_{d-1}\cdots D_0)} = (-1)^{d-1} \frac{(D_{d-2}\cdots D_0)^{q-1}}{D_{d-1}} = \frac{(-1)^{d-1}}{L_{d-1}}.$$ □

So the dependence of the $\mathrm{sgn}(g_j)$ or even $\mathrm{sgn}(g(1/m))$ on $\wp$ is quite complicated and not just through d. But when $m = 2$, i.e., the case of quadratic Gauss sums, we have the following analog of Gauss' theorem.

Theorem 4.4.2   *Let* $i := \epsilon^{(q-1)/2}$. *(Note* $i^2 = -1$.) *Then* $s_2 :=$ $\mathrm{sgn}(g(1/2))$ *depends only on the congruence class of* q *and* d *modulo 4. In fact, we have*

$$s_2 = (-i)^{d+2} \quad (q \equiv 1 \mod 4) \quad s_2 = (-i)^{d^2+2} \quad (q \equiv 3 \mod 4).$$

In other words, s_2 *is* i, $(-1)^{(q-1)/2}$, $(-i)(-1)^{(q-1)/2}$ *or* -1 *according as* d *is congruent to* $1, 2, 3$ *or* 4 *modulo 4.*

Proof.   We know that $g(1/2)^2 = (-1)^d \wp$ and hence s_2 is a fourth root of unity, a priori. By the theorem above, since $g(1/2) = \prod g_j^{(q-1)/2}$, we see that $s_2 = i^d \prod (t_k - t_0)^{-(q^d-1)/2}$, with $0 < k < d$. Now $(q^d - 1)/2 =$ $(q^{d-1} + \cdots + q + 1)(q - 1)/2$. Since $t_k^{q^r} = t_{k+r}$, we have

$$\prod (t_k - t_0)^{q^{d-1}+\cdots+1} = \prod (t_j - t_i) = (-1)^{d(d-1)/2} \prod (t_j - t_i)^2$$

where $d > j \neq i \geq 0$ in the second product and $d > j > i \geq 0$ in the third. On the other hand,

$$\prod_{j>i}(t_j - t_i)^{q-1} = \frac{\prod_{j>i}(t_{j+1} - t_{i+1})}{\prod_{j>i}(t_j - t_i)} = (-1)^{d-1}$$

where the last equality follows from the fact that there are $d - 1$ reversals of the sign, namely when $j = d - 1$. (Another way: the 'discriminant' is square exactly when d is odd.) Putting this together, we see $s_2 = i^d(-1)^{d(d-1)(q-1)/4}(-1)^{d-1}$, which is equivalent to the formulas claimed. □

For formulas and analogies for signs of higher order Gauss sums see [Tha93a].

4.5 Arithmetic Factorial and Gamma: Definitions

Euler's Gamma function arose via interpolation of the factorial function, which is very useful in combinatorics, elementary number theory. We will see how in case of $A = \mathbb{F}_q[t]$, Carlitz factorial, with good divisibility properties, gives rise to Goss interpolations at all primes in a fashion analogous to Morita's interpolations in the non-archimedean p-adic case and how the values relate to periods and Gauss sums. For general A, while the last two aspects are retained after suitable interpretations, the divisibility properties are not that good. For geometric gamma, even for $A = \mathbb{F}_q[t]$, its values at positive integers ($a \in A+$) are not integral! Still it will be seen to be a good analog.

In this section, for a given factorial function the corresponding gamma function is defined by variable shift of one, as in the classical case.

(a) $\mathbb{F}_q[t]$ case

As it seems quite often to be the case, several viewpoints coincide in $\mathbb{F}_q[t]$ case to give the same notion of factorial, the factorial introduced by Carlitz [Car35]. We restrict to this case first and explain several analogies, only some will generalize to general A and retain good properties.

The simplest way to introduce it by analogy may be by Sinnott's observation [Gos78] of its prime factorization: For $n \in \mathbb{Z}_{\geq 0}$, let

$$\Pi(n) := n! := \prod_{\wp \text{ monic prime}} \wp^{n_\wp} \in \mathbb{F}_q[t], \quad n_\wp := \sum_{e \geq 1} \lfloor \frac{n}{\text{Norm}\wp^e} \rfloor.$$

We use the Gauss' notation Π for the factorial when it is more convenient.

If we let $\wp$ run through (positive) primes, then we get the prime factorization of the usual factorial. In $\mathbb{F}_q[t]$ case, a more useful formula, but which does not make analogies immediately apparent, is Carlitz' formula:

$$\Pi(n) = \prod_i D_i^{n_i}, \quad \text{for } n = \sum n_i q^i, \text{ with } 0 \le n_i < q.$$

We have already seen a proof in 2.5 that both are the same, in case $n = q^i$:

$$D_i = (t^{q^i} - t)(t^{q^i} - t^q) \cdots (t^{q^i} - t^{q^i - 1}) = \text{Product of monic polynomials of degree}$$

The general case follows from $\sum \lfloor n/\text{Norm}(\wp)^e \rfloor = \sum n_i \lfloor q^i/\text{Norm}(\wp)^e \rfloor$, where $n = \sum n_i q^i$ is the base q expansion of n, i.e., $0 \le n_i < q$.

Examples 4.5.1 For $q = 3$, $\Pi(3) = \Pi(4) = \Pi(5) = t^3 - t$, $\Pi(6) = (t^3 - t)^2$ and $\Pi(9) = t^9 - t$.

Note that $D_i = (q^i)!$ fits in with the analogy:

$$e^z = \sum z^n/n!, \quad e_{\mathbb{F}_q[t]}(z) = \sum z^{q^n}/D_n.$$

The general factorial is obtained then by these basic building blocks by digit expansion, which we will see again and again in various contexts. It can also be motivated by the desire to have integral binomial coefficients: For example, $q^{n+1}!/q^n! = D_{n+1}/D_n = [n+1]D_n^{q-1} = [n+1]q^n!^{q-1}$, which suggests that the factorial of $q^{n+1} - q^n = (q-1)q^n$ should be D_n^{q-1}.

We do not have Euler integral formula analog, but Hankel's integral formula can be interpreted as residue consistent with the expansion of the exponential above.

Goss [Gos80c, appendix] made interpolations of the factorial at all places of $\mathbb{F}_q[t]$ as follows:

Since $D_i = t^{iq^i} - t^{(i-1)q^i + q^{i-1}} + \text{lower degree terms}$, the unit part

$$\overline{D_i} =: D_i/t^{\deg D_i} = 1 - 1/t^{(q-1)q^{i-1}} + \cdots$$

tends to 1 in $\mathbb{F}_q((1/t))$ as i tends to ∞. So the unit part of $\Pi(n)$ interpolates to a continuous function called ∞-adic factorial, $\overline{\Pi}(n)$:

$$\overline{\Pi} : \mathbb{Z}_p \to \mathbb{F}_q((1/t)), \quad \sum n_i q^i \to \prod \overline{D_i}^{n_i}.$$

Let v (sometimes we use symbol $\wp$) be a prime of A of degree d. Since D_i is the product of all monic elements of degree i, we have a Morita-style v-adic factorial $\Pi_v : \mathbb{Z}_p \to \mathbb{F}_q[t]_v$ for finite primes v of $\mathbb{F}_q[t]$ given by

$$\Pi_v(n) = \prod(-D_{i,v})^{n_i}$$

where $D_{i,v}$ is the product of all monic elements of degree i, which are relatively prime to v, and n_i are the digits in the q-adic expansion of n. This makes sense since $-D_{i,v} \to 1$, v-adically, as $i \to \infty$, if $n = \lfloor i/d \rfloor$, $D_{i,v}$ is a q-power power of all elements in $(A/v^n A)^*$ and hence is -1, for large i, by Wilson theorem analog in 1.1.

Let us give a direct proof, in the case where v is of degree one. Using the automorphism sending t to $t + \theta$, $\theta \in \mathbb{F}_q$, we can assume without loss of generality that $v = t$. Now, for a general monic prime v of degree d, we have $D_{i,v} = D_i / v^w D_{i-d}$, where w is such that $D_{i,v}$ is a unit at v. So in our case,

$$-D_{n,t} = \left(\prod_{i=0}^{n-1}(1 - t^{q^n - q^i})\right) / \left(\prod_{i=0}^{n-2}(1 - t^{q^{n-1} - q^i})\right) \to 1, \text{ as } n \to \infty.$$

In fact, it is easy to evaluate the value:

$$\Pi_t\left(\frac{1}{1-q}\right) = \lim \prod_{i=0}^{N} -D_{i,t} = -\lim \prod_{j=0}^{N-1}(1 - t^{q^N - q^j}) = -1$$

because the product telescopes after the first term $-D_{0,t} = -1$. We will see much more general result later.

(b)   General A

The different analogies that we have discussed for $\mathbb{F}_q[t]$ diverge for general A, giving different possible generalizations and we have to choose the ones with best properties. Here we give such definitions and prove some good properties, leaving a detailed comparison between the alternatives for later.

Let $\mathcal{A}$ be an ideal of A. First notice that, the field of constants being $\mathbb{F}_q$, if $x \in \mathcal{A}_{\leq i}$ (see notation in 2.5), then $ax \in \mathcal{A}_{\leq i}$, $a \in \mathbb{F}_q^\times$, so the signs (elements of $\mathbb{F}_q^\times$) appear in $\mathbb{F}_q^\times$-equivalence classes. So choose representatives for $\mathbb{F}_\infty^\times / \mathbb{F}_q^\times$ and let D_i be the product of all elements a of $\mathcal{A}$ of degree id_∞ with $\text{sgn}(a)$ being one of these representatives. (Note that even for $h = d_\infty = 1$ cases, now D_i need not divide D_{i+1}, unlike the $\mathbb{F}_q[t]$ case.) So $D_i \in \mathcal{A} \subset A$. Also let d_i be the number of these elements. We choose a uniformizer $u = u_\infty$ at ∞. The one-unit part $\overline{D}_i$ with respect to u is obviously independent of the choice of representatives. (Notice that for $d_\infty = 1$ and $1 \in \mathbb{F}_q^\times / \mathbb{F}_q^\times$ as the representative $D_i =$ product of all monic (i.e. with sign 1) elements in A of degree i.) We will prove in 4.7.2 that $\overline{D_i} \to 1$ as $i \to \infty$. Then we can define $\overline{\Pi}$ and $\overline{\Gamma}$ similarly.

By Riemann-Roch theorem, $d_i = (q^{d_\infty} - 1)q^{id_\infty+c}/(q-1) \to 0$ q-adically as i tends to infinity and (following a suggestion by Gekeler) we can recover the degree of the Gamma as follows:

The map $\mathbb{N} \to \mathbb{Z}$ given by $z \to \deg\Pi(z)$ interpolates to a continuous function $\deg\Pi : \mathbb{Z}_p \to \mathbb{Z}_p$ given by $\sum z_i q^i \to \sum id_\infty z_i d_i$.

Hence we p-adically complete $K_\infty^\times$, i.e. define $\hat{K}_\infty^\times =: \varprojlim K_\infty^\times/K_\infty^{\times p^n}$.

Since finite fields are perfect, signs in $K_\infty^\times$ project to 1 in $\hat{K}_\infty^\times$. If $d_\infty > 1$, one can do slightly better. Since d_i, for large i, is not only divisible by large power of q, but also by $(q^{d_\infty} - 1)/(q - 1)$, we can put $\tilde{K}_\infty^\times =:$ $\varprojlim K_\infty^\times/K_\infty^{\times(q^{d_\infty}-1)p^n/(q-1)}$ and take it as the range, in the evident fashion. The signs in $\mathbb{F}_\infty^\times$ survive now in $\mathbb{F}_\infty^\times/\mathbb{F}_q^\times$.

Then we define the ∞-adic interpolation $\Pi = \Pi_\infty : \mathbb{Z}_p \to \tilde{K}_\infty^\times$ with

$$\Pi(z) = \overline{\Pi}(z)u^{-\deg\Pi(z)/d_\infty}.$$

We use the symbol Π again, as we have recovered the degree part. For first reading, the reader may want to just ignore the complications with signs, by restricting to $d_\infty = 1$ case and ignore the degrees by just focusing on $\overline{\Pi}$ instead. We will do the same except in the next couple of sections.

Let v be a finite place of A relatively prime to $\mathcal{A}$, and of degree d. We form $\tilde{D}_i =: D_{i,v}$ as usual by removing the factors divisible by v.

Definition 4.5.2 Let $\tilde{D}_i$ be the product of elements a of degree id_∞ with $\text{sgn}(a)$ one of the chosen representatives and $v(a) = 0$. Let S_i be the set of these elements.

We will prove $(-1)^{d_\infty}\tilde{D}_i \to 1$, so put

Definition 4.5.3

$$\Pi_v(\sum z_i q^i) := \prod((-1)^{d_\infty}\tilde{D}_i)^{z_i}$$

so that $\Pi_v : \mathbb{Z}_p \to K_v$.

Even though D_i depends on a choice of representatives for $\mathbb{F}_\infty^\times/\mathbb{F}_q^\times$, $\tilde{D}_i$ for large i does not, because the number of elements in S_i of given sign is a multiple of $q^d - 1$ and $q^d - 1$-th power kills the choice. Similarly it is independent of the choice of sgn for large i. So Γ_v is again unique on a smaller disc of $\mathbb{Z}_p$. In any case, a value $\Gamma_v(z)$ is determined up to multiplication by an element in $\mathbb{F}_q^{\times}$. (So there is a unique function for $q = 2$.)

Lemma 4.5.4 $(-1)^{d_{\infty}} \tilde{D}_i \to 1.$

Proof. It is enough to prove that there is an integer l such that for $i >> 0$, $(-1)^{d_{\infty}} \tilde{D}_i \equiv 1 \bmod v^w$ with $w = \lfloor id_{\infty}/d \rfloor - l$.

Assume that i is sufficiently large. Then product of elements of $(A/Av^w)^{\times}$ is -1 mod v^w, by generalization in 1.1 of Wilson's theorem analog. (Notice that once i and w are large, we escape the exceptions in 1.1.) Now the elements of S_i are equidistributed among the cosets, since elements of S_i with fixed sign are. (This is because, for large i, all members of one coset can be transferred to any other by subtracting coset-representatives of degree less than id_{∞}, leaving the sign unchanged.) This takes care of the case $p = 2$, so now assume that the characteristic is odd. It is enough to prove that the number of elements of S_i in any coset is $\equiv d_{\infty} \bmod 2$. Now $\#A_{\leq kd_{\infty}} = q^{kd_{\infty}+c}$, by Riemann's theorem, so that $\#\{a \in A : \deg a = id_{\infty} \text{ with } \operatorname{sgn}(a) \text{ one of the chosen}\} = (q^{d_{\infty}} - 1)/(q-1)q^{(i-1)d_{\infty}+c}$. So $\#S_i = q^{(i-1)d_{\infty}-d+c}(q^d - 1)(q^{d_{\infty}} - 1)/(q-1)$, but the number of cosets is $(q^d - 1)q^{d_w}$, so that the required number is $q^r(q^{d_{\infty}} - 1)/(q-1) \equiv (q^{d_{\infty}} - 1)/(q-1) \equiv d_{\infty} \bmod 2$ as claimed. (Here l is chosen so that the equidistribution works and r is positive.) $\square$

4.6 Functional equations in arithmetic case

In this section, we show how the structure of the functional equations for the factorial functions, for all places and all A's (and for another factorial defined later in Chapter 8), follow just from manipulation of p-adic digits of the arguments. So the proofs of functional equations reduce to this plus calculation of one single value: the value of gamma at 0, which we take up in the later section. The same type of functional equations are also satisfied by Gauss sums and the proof is the same.

After the more familiar reflection and multiplication formula, we will prove a general functional equation directly. We will see later in Section 4.12 how the cyclotomy and the Galois groups play a role in this structure.

Recall that the classical gamma function satisfies (1) Reflection formula: $\Gamma(z)\Gamma(1 - z) = \pi/\sin \pi z$ and (2) Multiplication formula:

$$\Gamma(z)\Gamma(z + \frac{1}{n}) \cdots \Gamma(z + \frac{n-1}{n})/\Gamma(nz) = (2\pi)^{(n-1)/2}n^{1/2-nz}.$$

We will prove analogs of these and also their p-adic counterparts in the function field case, by first proving relations in the abstract setting below. Consider a function f defined on $\mathbb{Z}_p$ by

$$f(\sum a_j q^j) =: \prod f_j^{a_j}$$

for some f_j's. You can think of f_j's as independent variables with the evident manipulation rules. Put $g(z) = f(z - 1)$. The various factorial functions ('f') and gamma functions ('g') introduced above are all of this form.

We want to get formal relations satisfied by f. In particular, we would like to know when $\prod f(x_i)^{n_i} = 1$ formally i.e. independently of f_i's.

First we have a reflection formula:

Theorem 4.6.1 $g(z)g(1 - z) = g(0)$ *or equivalently* $f(z)f(-1 - z) = f(-1)$.

Proof.   Let the digit expansion of z be $z = \sum z_j q^j$. Since $-1 = \sum(q - 1)q^j$, and $0 \le q - 1 - z_j < q$, $-1 - z = \sum(q - 1 - z_j)q^j$ is a digit expansion. Hence the relation with f's follows. The one with g is the same, once we replace z by $z + 1$.   □

Another way to prove this is as follows:

Lemma 4.6.2   *For* $z \ne 0$, $g(z + 1)/g(z)$ *depends only on* $r := \text{ord}_q(z)$.

Proof.   In fact, the definition imply $g(z + 1)/g(z) = f(z)/f(z - 1) = f_r/(f_0 \cdots f_{r-1})^{q-1}$.   □

Now $g(1) = f(0) = 1$, so it is true for $z = 0$. And since $\text{ord}_q(z) = \text{ord}_q(-z)$, by the Lemma, we have $g(z + 1)/g(z) = g(1 - z)/g(-z)$, hence by induction it is true for all integers z. Integers being dense in $\mathbb{Z}_p$, (and $f_n \to 1$ as $n \to \infty$ for the product to make 'sense'), the Theorem again follows easily.

Next we have multiplication formula:

Theorem 4.6.3   *For* $z \in \mathbb{Z}_p$ *and* $(n, q) = 1$,

$$g(z)g(z + \frac{1}{n}) \cdots g(z + \frac{n - 1}{n})/g(nz) = g(0)^{(n-1)/2}$$

(here, if n is even, so that q is odd, then we mean by $g(0)^{1/2}$ the element $\prod_{j=0}^{\infty} f_j^{(q-1)/2}$ whose square is $g(0)$).

Proof. Again we will prove in two different ways. First, by well-known results (eg. Hardy and Wright, chapter 9) on digit expansions: If $(n, q) = 1$, $-1/n$ has a purely recurring expansion of r recurring digits where r is minimal such that n divides $q^r - 1$. Essentially, the recurring digits for $-a/n$ are just obtained by permutations of those of $-1/n$, so that the sum of the i-th digits of all of them is constant. This constant is easily seen to be $(q - 1)(n - 1)/2$, as $-1/n + \cdots + -(n - 1)/n = -(n - 1)/2 = ((n - 1)/2) \sum (q - 1)q^j$.

We can also proceed as in the second proof: Since $\mathrm{ord}_q(z) = \mathrm{ord}_q(nz)$, the Lemma implies that

$$\frac{g(z)g(z + \frac{1}{n})\cdots g(z + \frac{n-1}{n})}{g(nz)} \cdot \frac{g(nz + 1)}{g(z + \frac{1}{n})g(z + \frac{2}{n})\cdots g(z + 1)}$$

$$= \frac{g(z)}{g(z + 1)} \cdot \frac{g(nz + 1)}{g(nz)} = 1$$

and again as before it is enough to prove the claim for a single z, say $z = 1/n$. So we want to prove $g(1/n)\cdots g((n - 1)/n) = g(0)^{(n-1)/2}$, which follows from the reflection formula above, by pairing $g(a/n)$ with $g((n - a)/n)$. $\qquad\square$

We now give a more general functional equation. For the motivation and more discussion of this result, see Section 4.12.

Let N be a positive integer prime to p. For $x \in \mathbb{Q}$, define $< x >$ by $x \equiv< x >$ modulo $\mathbb{Z}$, $0 \leq< x >< 1$. If $\underline{a} = \sum m_i[a_i]$ ($m_i \in \mathbb{Z}, a_i \in \frac{1}{N}\mathbb{Z} - \{0\}$) is an element of the free abelian group with basis $\frac{1}{N}\mathbb{Z} - \{0\}$, put $n(\underline{a}) =: \sum m_i\langle a_i\rangle$. Also, for $u \in (\mathbb{Z}/N\mathbb{Z})^\times$, let $\underline{a}^{(u)} =: \sum m_i[ua_i]$.

Theorem 4.6.4 *If*

$$n(\underline{a}^{(q^j)}) \text{ is an integer independent of } j, \qquad\qquad (**)$$

then

$$\prod f(-\langle a_i\rangle)^{m_i} = f(-1)^{n(\underline{a})}.$$

Proof. Multiplying the denominator if necessary, we may assume that $N = q^r - 1$.

The idea of the proof is to 'disentangle' the monomial on left in terms of 'basis' elements $f(q^j/(1 - q^r))$'s. The hypothesis says that each basis element occurs with power $n(\underline{a})(q - 1)$. Since the product of the basis elements is $f(1/(1 - q))$, we get the right side as claimed. Let us see how this works out.

Put $\underline{a} = \sum m_i a_i$ and

$$N a_i = b_i = b_{r-1,i} q^{r-1} + b_{r-2,i} q^{r-2} + \cdots + b_{0,i}.$$

So the given monomial is equal to

$$\prod f(\frac{-q^j}{N})^{\sum m_i b_{j,i}} = f(\frac{1}{1-q})^{\sum m_i b_{r-1,i}} \prod f(\frac{-q^j}{N})^{\sum m_i (b_{j,i} - b_{r-1,i})}.$$

For $1 \leq j \leq r - 1$, (**) is equivalent to

$$\sum m_i b_i / N = \sum m_i b_i q^j / N - \sum m_i b_{r-1,i} q^{j-1} - \cdots - \sum m_i b_{r-j,i}.$$

Claim : $\sum m_i (b_{j,i} - b_{r-1,i}) = 0$.

This is obvious if $j = r - 1$, suppose it is true for $j = r - 1, \cdots, r - t + 1$. Then (**) for $j = t$ says that $\sum m_i b_i (q^t - 1)/N$ is

$$[(\sum m_i b_{r-1,i})(q^{t-1} + q^{t-2} + \cdots + 1)] - [\sum m_i b_{r-1,i} - \sum m_i b_{r-t,i}].$$

Now (**) for $j = 1$ gives $\sum m_i b_i (q - 1)/N = \sum m_i b_{r-1,i}$ and hence $\sum m_i b_i (q^t - 1)/N$ is equal to the first $[\cdots]$ and hence the second $[\cdots]$ is zero, proving the claim for $j = r - t$. Induction completes the proof. $\square$

Remarks 4.6.5 (1) This general functional equation is better explained together with its relations to cyclotomic theory and with the multiplication and reflection formulas in Section 4.12.

(2) By the definitions in 4.1, the Gauss sums $g(a)$'s that we have defined for a, a fraction which is integral at p, also fit the framework of functions f here (where we restrict to p-adically integral fractions only) and thus we get the same functional equations for them. (The letter f we used for factorial (and not g for gamma) works for Gauss sums.) We have already remarked after Theorem 4.2.4 the analog of formal reflection formula, together with the calculation of $g(-1) = g((q^l - 1)/(1 - q^l))$. The multiplication formula gives an analog of the second Hasse-Davenport relation $\prod g(\chi_1 \chi_2)/g(\chi_1) = g(\chi_2)^m \chi_2(m^{-m})$, with χ_i being multiplicative characters of $\mathbb{F}_q^*$ and the product running over χ_1's of order m, a fixed divisor of $q - 1$, giving a 'distribution' relation.

4.7 Special values for arithmetic Γ_∞

(a) Periods: $\mathbb{F}_q[t]$ case

Now we relate the special values of $\overline{\Gamma}$ to the period $\tilde{\pi}$ of the Carlitz module for $A = \mathbb{F}_q[t]$. In the next sub-section, we will derive more general results in different fashion.

In the rest of this section, for $0 \neq f \in \mathbb{F}_q((1/t))$, $f/t^{\deg f}$ will be denoted by $\overline{f}$. We saw in 2.5 that

$$\tilde{\pi} = (-1)^{1/(q-1)} \lim [1]^{q^k/(q-1)}/[1] \cdots [k]$$

so $\tilde{\pi}^{q-1} \in \mathbb{F}_q((1/t))$ and $\overline{\tilde{\pi}^{q-1}}$ makes sense. By $\overline{\tilde{\pi}}$ we will denote its unique $q-1$-th root which is a one unit in $\mathbb{F}_q((1/t))$. A similar remark applies to $\tilde{\pi}^{1/(q-1)}$.

Theorem 4.7.1 *For $0 \leq a \leq q-1$, we have*

$$\overline{\Gamma}(1 - \frac{a}{q-1}) = (\overline{\tilde{\pi}})^{a/(q-1)}.$$

In particular, We have $\overline{\Gamma}(0) = \overline{\tilde{\pi}}$, and if $q \neq 2^n$, then

$$\overline{\Gamma}(1/2) = \sqrt{\overline{\tilde{\pi}}}.$$

Proof. Since $-1 = \sum (q-1)q^i$, we have

$$\overline{\Gamma}(0) = \overline{\Pi}(-1) = \lim \overline{(D_0 \cdots D_n)^{q-1}}.$$

Now

$$(D_0 \cdots D_n)^{q-1} = D_{n+1}/[1] \cdots [n+1].$$

Hence

$$\overline{\Gamma}(0)^{q-1}/\overline{\tilde{\pi}^{q-1}} = \lim \overline{D_{n+1}^{q-1}}/\overline{[1]}^{q^{n+1}} = 1$$

since we have already seen that $\overline{D_i} \to 1$, and since $\overline{[1]}^{q^n} \to 1$, because any one unit raised to the q^n-th power tends to 1 as $n \to \infty$.

Hence we have proved $\overline{\Gamma}(0) = \overline{\tilde{\pi}}$. Observing that $a/(1-q) = \sum aq^i$ for $0 \leq a \leq q-1$, we get the theorem. $\qquad \square$

Corollary 4.7.2 *For $A = \mathbb{F}_q[t]$, we have $\Gamma(0)^{q-1} = -\tilde{\pi}^{q-1}$.*

Proof. Both sides have degree $q/(q-1)$ and $\tilde{\pi}^{q-1} \in \mathbb{F}_q((1/t))$ has sign -1 as we see from any of the formulas for it in 2.5, whereas any $\Gamma(z)$ has sign one by construction. $\qquad \square$

To investigate the nature of gamma values at all fractions, it is sufficient to look at all $\overline{\Pi}(q^j/(1-q^k))$ for $0 \le j < k$, since a general value is up to a harmless translation a monomial in these basic ones. They can be related to the periods $\tilde{\pi}_k$'s of Carlitz modules for $\mathbb{F}_{q^k}[t]$. For example,

Theorem 4.7.3   *We have* $\overline{\tilde{\pi}_k} = \frac{\overline{\Pi}(q^{k-1}/(1-q^k))^q}{\overline{\Pi}(1/(1-q^k))}$.

Proof.   We have

$$\frac{\overline{\Pi}(1/(1-q^k))}{\overline{\Pi}(q^{k-1}/(1-q^k))^q} = \lim \frac{\overline{D_{kn}D_{k(n-1)}\cdots D_0}}{\overline{D_{kn-1}{}^q \cdots D_{k-1}{}^q}}$$

$$= \lim \overline{[kn][k(n-1)]\cdots[k]}$$

$$= (\overline{\tilde{\pi}_k})^{-1}.$$

$\square$

This is the Chowla-Selberg formula for constant field extensions, as will be explained in section 4.12. Similarly, it can be shown, for example, that

$$\overline{\Pi}(1/(1-q^2))^{q^2-1} = \overline{\tilde{\pi}^q \tilde{\pi}_2}^{-(q-1)}$$

$$\overline{\Pi}(q/(1-q^2))^{q^2-1} = \overline{\tilde{\pi}\tilde{\pi}_2}^{q-1}.$$

See Chapter 10 for applications to transcendence of these results.

(b)   Periods: General A

Recall the fact from 2.2 that the degree is always a multiple of d_∞ and recall the notation $\mathcal{A}_i$ from 2.5. The Riemann part of the Riemann-Roch theorem shows that $\#\mathcal{A}_{\le id_\infty} = q^{id_\infty+c}$ for $i \in \mathbb{N}$, $i >> 0$, c some constant.

Fix a sgn function (see 1.1.1) and let ρ be a corresponding sign-normalized (see 3.2.1) rank one Drinfeld A-module with corresponding rank one lattice Λ and exponential $e_\rho = e_\Lambda$. Let $\tilde{\pi} \in C_\infty$ be a corresponding 'period' defined up to an element in $\mathbb{F}_q^\times$ by the equation $\Lambda = \tilde{\pi}A$. This period $\tilde{\pi}$ of ρ is an analog of period $2\pi i$ (up to ± 1) in the situation $\mathbb{Z} \hookrightarrow \mathrm{End}G_m$.

Choose a uniformizer u at ∞ of sgn 1. (For example, $u = 1/t$ for $A = \mathbb{F}_q[t]$.) As in 1.1, z in $K_\infty^\times$ can be written uniquely as $z = \mathrm{sgn}(z) \times \bar{z} \times u^n$, with $\mathrm{sgn}(z) \in \mathbb{F}_\infty^\times, \bar{z} \in U_1, n \in \mathbb{Z}$.

We will use Gekeler-Hayes formula ((*) below) to prove the identity $\Gamma(0) = \tilde{\pi}$ multiplied by a root of unity.

Let x be an element of A of degree > 0, say of degree d and with $\mathrm{sgn}(x) = 1$. The coefficient of linear term of ρ_x is x, and ρ is sgn-normalized.

Hence x is the product of its nonzero roots:

$$x = \prod_{a \in A/Ax} {}' \tilde{\pi} e_A(a/x) = \tilde{\pi}^{q^d-1} \prod {}' e_A(a/x).$$

So

$$\tilde{\pi}^{1-q^d} = \frac{1}{x} \prod_{a \in A/Ax} {}' e_A(a/x).$$

This implies $\tilde{\pi}^{q^d-1} \in K_\infty$. We will see later that $\tilde{\pi}^{q^{d_\infty}-1} \in K_\infty$.
Hence

$$\tilde{\pi}^{1-q^d} = 1/x \prod_{a \in A/Ax} (a/x) \prod_{b \in A} (1 - a/xb) = x^{-q^d} \left(\prod a \prod (xb-a)/xb \right) \quad (*)$$

which is the limit of the same expression with $b \in A$ replaced $b \in A_{\leq Nd_\infty}$
as N tends to infinity. (This follows, since the exponential itself is such
a limit.) Now $(xb-0)/xb = 1$, so if N is large, then the numerator of the
right hand side of $(*)$, with $xb - 0$ allowed, is just the product of nonzero
elements of $A_{\leq Nd_\infty + d}$, whereas the denominator is $x^{(\#A_{\leq Nd_\infty}-1)*q^d}$ times
the q^d-th power of the product of nonzero elements of $A_{\leq Nd_\infty}$ (as there are
q^d a's).

Take the one-unit part of both sides and notice that

$$\overline{x}^{-q^d} * \overline{x}^{-(\#A_{\leq Nd_\infty}-1)*q^d} = \overline{x}^{-\#A_{\leq Nd_\infty}} = \overline{x}^{-q^{Nd_\infty+c}} \to 1$$

since a one-unit raised to the q^r-th power tends to 1 as $r \to \infty$. Hence,

$$\tilde{\pi}^{1-q^d} = \left(\prod_{a \in A_{\leq Nd_\infty+d}} {}' a \right) / \left(\prod_{b \in A_{\leq Nd_\infty}} {}' b \right)^{q^d}.$$

With the definitions of 4.5.2, our equation becomes

$$\tilde{\pi}^{1-q^d} = \lim_{N \to \infty} (\overline{D_0} \cdots \overline{D_{N+d/d_\infty}})^{q-1} / ((\overline{D_0} \cdots \overline{D_N})^{q-1})^{q^d}.$$

So

$$\lim (\overline{D_{N+d/d_\infty}}^{q-1} / (\overline{D_N}^{q-1})^{q^d} = 1.$$

But in characteristic p, q^d power spreads out the power series expansion,
so $\overline{D_i}$ being a one-unit, we get $\overline{D_i}^{q-1} \to 1$, so $\overline{D_i} \to 1$, as promised in 4.5.2.
So $\overline{\Gamma}$ is well-defined and we get

$$\tilde{\pi}^{1-q^d} = \overline{\Gamma}(0)^{1-q^d}.$$

Lemma 4.7.4

$$\gcd\{q^d - 1 : d = \deg(x),\ x \in A,\ sgn(x) = 1\} = q^{d_\infty} - 1.$$

Proof. It is enough to prove that gcd of d's is d_∞. First of all, by Riemann's theorem we know that gcd of degrees of elements of A is d_∞. Next, since K is dense in K_∞, there is an element of degree 0 in K of any given sign. Multiplying by elements in A of high degree, clearing the denominators we see that A has, for some large i, elements of degree i of all signs. Now, choose elements x_k's in A of degrees d_k such that gcd of d_k's is d_∞. Then multiplying by powers of x_k, we get all signs in degree $i + nd_k$ for all positive n, so that the gcd in question divides the gcd of $i + nd_k$ which is d_∞. □

Hence

$$\bar{\tilde{\pi}}^{1-q^{d_\infty}} = \overline{\Gamma}(0)^{1-q^{d_\infty}}.$$

If $\bar{\tilde{\pi}}$ is that $1 - q^{d_\infty}$-th root of $\bar{\tilde{\pi}}^{1-q^{d_\infty}}$ which is one-unit then $\overline{\Gamma}(0) = \bar{\tilde{\pi}}$.

Analyzing the degrees on both sides of the equation, we get

$$\deg \tilde{\pi}^{q^d-1} = \lim\left(q^d \sum_{b \in A_{Nd_\infty}}{}'(\deg b) - \sum_{a \in A_{Nd_\infty+d}}{}'(\deg a) + d\#A_{Nd_\infty}\right)$$

$$= (q^d - 1)\sum (q-1)id_\infty d_i.$$

Since $-1 = \sum(q-1)q^i$, we get $\deg \Pi(-1) = \sum(q-1)id_\infty d_i$. So $\deg \Pi(-1) = \deg \tilde{\pi}$ and hence putting the degree part back, we see that $\Pi(-1) = \tilde{\pi}$ times a root of unity.

Now we analyze the signs in (*). The sign of the right hand side is the limit as $N \to \infty$ of the sign of the product of nonzero elements of $A_{Nd_\infty+d}$ divided by the product of nonzero elements of A_{Nd_∞} (as q^d-th power is identity on $\mathbb{F}_\infty$). Varying d, a simple gcd argument then shows that $\tilde{\pi}^{1-q^{d_\infty}} = \epsilon\Gamma(0)^{1-q^{d_\infty}}$ where the sign ϵ is the stationary limiting sign (we have shown that limit exists, for a more direct proof see [Gek86, pa. 30]) of the product of elements of A of degree Nd_∞ as $N \to \infty$. If $d_\infty = 1$, as $\prod_{a \in \mathbb{F}_q^\times} a = -1$, and as $(-1)^{q^d} = -1$ it follows by straightforward counting using Riemann's theorem that $\epsilon = -1$. Hence,

$$\Gamma(0)^{q-1} = -\tilde{\pi}^{q-1} \in (K_\infty^\times)_{sgn=1} \subset \hat{K}_\infty^\times$$

as $(q-1)\deg\Gamma(0) \in \mathbb{Z}$ rather than just in $\mathbb{Z}_p$.

We summarize the discussion in

Theorem 4.7.5

$$\Gamma(0) = \mu\tilde{\pi}$$

(where μ is at most $(q^{d_\infty} - 1)^2$-th root of unity. If $d_\infty = 1$, it is $(q - 1)$-th root of -1), in the sense that '$q^{d_\infty} - 1$-th powers of the both sides' are the same, considered in $K_\infty^\times$.

For transcendence applications, see Chapter 10.

Remarks 4.7.6 (1) If $q = 2$ and $d_\infty = 1$, the question of signs disappears.

(2) The Γ function is independent of the choices of coset representatives of $\mathbb{F}_\infty^\times/\mathbb{F}_q^\times$, and it depends on the uniformizer only through its sign. If t and t' are uniformizers with $t = at'$, $a \in \mathbb{F}_\infty^\times$, then $a^{\deg \Gamma/d_\infty}\Gamma = \Gamma'$, so there are $(q^{d_\infty} - 1)/(q - 1)$ gamma functions (hence unique if $d_\infty = 1$) and the gamma function is independent of the sgn choice on a smaller disc of $\mathbb{Z}_p$.

(3) All rank 1 normalized Drinfeld A-modules are isogenous, so periods for different choices of A are algebraic multiples of each other.

4.8 Special values of arithmetic Γ_v

We will prove strong results for $\mathbb{F}_q[t]$ case, but only weak results for general A. We will prove comparable strong results only in 8.3 and 8.6.

(a) $\mathbb{F}_q[t]$ case: Analog of Gross-Koblitz

The Gross-Koblitz formula, based on crucial earlier work by Honda, Dwork and Katz, expresses Gauss sums lying above a prime p in terms of values of Morita's p-adic gamma function at appropriate fractions (see [Kob80; GK79]).

Honda conjectured and Katz proved a formula for Gauss sums made up from p-th roots of unity in terms of p-adic limits involving factorials, combining two different calculations of Frobenius eigenvalues on p-adic (Crystalline or Washnitzer-Monsky) cohomology of Fermat and Artin-Schreier curves. Gross and Koblitz interpreted this as special value of Morita's p-adic interpolation of factorial, defined around that time.

Theorem 4.8.1 *('Analog of the Gross-Koblitz formula'): Let $A = \mathbb{F}_q[t]$ and $\wp$ is a monic prime of A of degree d. Let λ be the $q^d - 1$-th root of $-\wp$*

as in 4.2.5. Then for $0 \le j < d$, we have

$$g_j = -\lambda^{q^j} / \Pi_\wp\left(\frac{q^j}{1 - q^d}\right).$$

Proof. We have $\tilde{D}_a = D_a / D_{a-d\wp^l}$, where l is such that $\tilde{D}_a$ is a unit at $\wp$. Hence, using the base q expansion $q^j / (1 - q^d) = \sum q^{j+id}$ we get,

$$\Pi_\wp\left(\frac{q^j}{1 - q^d}\right) = \lim(-1)^{m+1}\tilde{D}_j \cdots \tilde{D}_{j+md} = \lim(-1)^{m+1}D_{j+md}/\wp^w$$

where $w = \mathrm{ord}_\wp D_{j+md}$. Now the recursion formula for D_i gives

$$D_{j+md} = [j+md][j-1+md]^q \cdots [j+1+(m-1)d]^{q^{d-1}} D_{j+(m-1)d}^{q^d}.$$

Let $t = au$ be the decomposition in $K_\wp$ of t, a unit at $\wp$ (without loss of generality, $\wp \neq t$), as the product of its 'Teichmuller representative' a and its one unit part u. As $a^{q^{md}} = a$ and $u^{q^n} \to 1$ as $n \to \infty$, we have, as $m \to \infty$, $[l+md] = ((au)^{q^{md+l}} - t) \to (a^{q^l} - t)$, which is just (negative of) one of the $\wp_j$'s of 4.2. Using this in the limit above and counting powers of $\wp$, using the description of $[i]$ given above, we see that

$$\Pi_\wp\left(\frac{q^j}{1 - q^d}\right)^{1-q^d} = (-\wp_{1-j})(-\wp_{2-j})^q \cdots (-\wp_{-j})^{q^{d-1}} / \wp^{q^j}.$$

Comparing with the Stickelberger factorization 4.2.7 we see that factorizations are the same and we fix the root of unity by comparing the congruence 4.2.6 with $\Pi_\wp\left(\frac{q^j}{1-q^d}\right) \equiv -\tilde{D}_j \equiv -D_j \bmod \wp$, which follows from $-\tilde{D}_i \equiv 1 \bmod \wp$ for $i \ge d$, which in turn follows from Wilson theorem analog in 1.1 since the product expression in 2.5 for D_i means $\tilde{D}_i$ is the product of all reduced representatives modulo $\wp$, each taken q^{i-d} times. $\square$

Note that this proof is quite direct and does not need a lot of machinery, unlike the proof in the classical case.

(b)   General A

Theorem 4.8.2 $\Gamma_v(0) = (-1)^{\deg v - 1}$ *for all v prime to A. For $0 \le a \le q - 1$ $\Gamma_v(1 - \frac{a}{q-1})$ are roots of unity and $\Gamma_v(\frac{b}{q-1})$ is algebraic for $b \in \mathbb{Z}$.*

Proof.   The first statement of the theorem implies the rest and the first statement will follow, if we show that

$$\left(\prod m\right)^{q-1} \equiv (-1)^{d-1}$$

mod v^{l_i} where m runs through monic polynomials prime to v (of degree d) and of degree not more than t_i and with $l_i, t_i \to \infty$ as $i \to \infty$. Given l_i, choose t_i so that $\{am\}$ as a runs through $\mathbb{F}_q^\times$ spans the reduced residue class system mod v^{l_i} (for example, in $\mathbb{F}_q[t]$ case, $t_i = dl_i - 1$ works). Then it is easy to see that $\{am\}$ covers each reduced residue class equal number (which is a power of q) of times. Hence we have

$$-1 \equiv \left(\prod a\right)^{\#\{m\}}(\cdots)^{q-1}.$$

But $\prod a = -1$, so we are done if $p = 2$. Assume p is not two, then we have to show that $\#\{m\} \equiv d \bmod 2$. But for some c we have

$$\#\{m\} = (q^{t_i} - q^{t_i - d})/(q - 1) \equiv (q^d - 1)/(q - 1) \equiv d \bmod 2. \qquad \square$$

Using congruences to decide the root, one can pin down the roots of unity mentioned in the theorem.

Now we prove an analog of theorem of Deligne [DMOS82, pa. 91], in our situation. Let $A = \mathbb{F}_q[t]$. Let N be a positive integer prime to p. Choose r such that N divides $q^r - 1$ and let $F = \mathbb{F}_{q^r}(t)$. Consider prime P of $\mathbb{F}_{q^r}[t]$. If $\underline{a}$ satisfies the hypothesis of Theorem 4.6.4, then by 4.2.8, $g(\underline{a}, P)/(\mathrm{N}P^{n(\underline{a})})$ is a Hecke character $\chi_{\underline{a}}(P)$ for F of finite order. More precisely $\chi_{\underline{a}}(P) = (-1)^{r \deg Pn(\underline{a})}$.

Theorem 4.8.3 *Under the hypothesis of Theorem 4.6.4, if we put* $\Omega_{\underline{a}} := \prod \Pi(-\langle a_i \rangle)^{m_i}/\tilde{\pi}^{n(\underline{a})}$, *then* $\Omega_{\underline{a}}^\tau/\Omega_{\underline{a}} = \chi_{\underline{a}}(\tau)$ *for any* $\tau \in \mathrm{Gal}(F^{sep}/F)$.

Proof. Theorem 4.7.4 shows that $M := \Gamma(0)/\tilde{\pi} = (-1)^{1/(q-1)}$, if Frob_P denotes the Frobenius ($q^{r \deg P}$-th power), then $\mathrm{Frob}_P M/M = (-1)^{r \deg P}$.

It is sufficient to look at action of $\tau = \mathrm{Frob}_P$'s. By theorem 2.8 and by 4.2.8, we see that both sides are $(-1)^{r \deg Pn(\underline{a})}$. $\qquad \square$

Remarks 4.8.4 (1) If one takes a general A, with $d_\infty = 1$ and chooses a sgn-normalized Drinfeld module and corresponding period $\tilde{\pi}$, this theorem and its proof carry over word to word, when one takes $\chi_{\underline{a}}(P) = (-1)^{r \deg Pn(\underline{a})}$ as a definition of $\chi_{\underline{a}}$ (as a character of $F =: K(\mu_{q^r - 1})$ in this case). But in this general case, one loses the connection with the gauss sums of 4.1.

(2) As we will see in 4.12.3, we handle 'more' relations than classical case, at the same time, we can prove the full result and not just up to sign (see [GK79, remark after theorem 4.5]) and also that in relations generated by multiplication and reflection, we can use $\Gamma(\underline{a})$ instead of $\Pi(-\underline{a})$ in the definition of $\Omega_{\underline{a}}$.

(3) Instead of using 4.2.8 to get an expression for $\chi_{\underline{a}}$, we could have used 4.8.1 to give another [Tha91a] proof.

We prove analog of Gross-Koblitz theorem for general A in Chapter 8.
For the rest of the chapter, unless noted otherwise, we restrict to $d_\infty = 1$, for simplicity.

4.9   Geometric Factorial and Gamma: Definitions

Gamma function we studied so far had domain in characteristic zero, even though the values were in characteristic p. This is connected to the fact that as we have seen, for example in $A = \mathbb{F}_q[t]$ case, its arithmetic is linked up with cyclotomic extensions $\mathbb{F}_q(t)(\mu_n)$ of $\mathbb{F}_q(t)$ which are just constant field extensions. More precisely, Gross-Koblitz analog 4.8.1 or Chowla-Selberg analog 4.7.3 are related to Stickelberger elements of these constant field extensions. The fractions we handle there are p-integral, and so are of the form $m/(q^r - 1)$. The values of gamma function at these fractions are connected to extension $\mathbb{F}_q(t)(\mu_{q^r-1})$ which is a general constant field extension of finite degree.

But we have another nice family (one for each infinite place) of cyclotomic extensions $\mathbb{F}_q(t)(\Lambda_a)$ of Carlitz-Drinfeld-Hayes, so one would expect another gamma function with characteristic p domain such that its special values at fractions with denominator a are related to arithmetic of $\mathbb{F}_q(T)(\Lambda_a)$.

With all its nice analogies, the gamma function of first part has one feature strikingly different than classical gamma function. It has no poles. Usual gamma function has no zeros and has simple poles exactly at 0 and negative integers (negative of positive integers). Monicity is an analog of positivity, but positivity is closed under both addition and multiplication, while monicity only under multiplication; also, for $p = 2$, positive is the same as negative and for $q = 2$ all integers are negative! Having thus decided upon the location of the poles, note that in our non-archimedean case, divisor determines function up to multiplicative constant. The simplest choice of constant we choose below also seems to be the best for the analogies we describe later.

Hence we define geometric gamma function as a meromorphic function defined as

Definition 4.9.1

$$\Gamma(x) =: \frac{1}{x} \prod_{n \text{ monic}} (1 + \frac{x}{n})^{-1} \in C_\infty \cup \{\infty\}, \quad x \in C_\infty.$$

From this point of view of divisors, the factorial Π should be defined as $\Pi(x) =: x\Gamma(x)$.

Classically, we have $x\Gamma(x) = \Pi(x) = \Gamma(x+1)$, whereas in this game, the first equality is natural for the gamma and factorial defined here and the second equality holds for the ones considered previously in Sections $4.5 - 4.8$. Consequently, gamma and factorial now differ by more than just a harmless change of variable. Also, in characteristic p, addition of p brings you back, so giving value at $x + 1$ in terms of that at x will not cover all integers by recursion anyway.

Remarks 4.9.2 Definitions can easily be modified, as in Section 7, so as to make good sense for general A (or ideal $\mathcal{A}$), by choosing sign representatives if $d_\infty > 1$.

We have

$$\frac{\Gamma'}{\Gamma}(x) = -\sum_{n \text{ monic}} \frac{1}{x+n} - \frac{1}{x} = -\zeta(x, 1), \frac{\Pi'}{\Pi}(0) = -\zeta(1)$$

where $\zeta(1)$ is the Carlitz zeta value of 2.1 or Chapter 5 and $\zeta(x, s)$ is analog of Hurwitz's partial zeta function defined analogously. But classically value at zero of the logarithmic derivative of factorial is $-\gamma$, where γ is Euler's constant. Hence in our case, analog, denoted by γ again, of Euler's constant gamma is

$$\gamma = \zeta(1) = -\frac{\Pi'}{\Pi}(0)$$

(and this is equal to $\log(1)$, where $\log$ is the inverse function (with $\log(0) = 0$) to the Carlitz exponential, if $A = \mathbb{F}_q[t]$ and also equal to $\tilde{\pi}/(t^2 + t)$ if further, $q = 2$, as we will see in the next chapter.

Remarks 4.9.3 (1) In fact, interestingly enough, for $a \in A = \mathbb{F}_q[t]$, $1/\Gamma(a) \in A$, and even $2/\Gamma(a) \in A_+ \cup \{0\}$ when $a \in A_+$. To see this, it is enough to see that $\prod_{n \in k+} n$ divides $\prod_{n \in k+}(a + n)$: Choosing $n_{\wp, e}$, appropriately, of degree less than $e \deg \wp$ one sees that if $\wp^e$ divides n then it also divides $a + (n + n_{\wp, e})$. Another way to see this is to note that from

2.5, we have

$$\prod_{a \in A_j+} (1 + \frac{x}{a}) = \frac{e_j(t^j + x)}{D_j} = 1 + \binom{x}{q^j} \in \mathbb{F}_q[t].$$

(2) For general A, if $j > \deg x$, $\prod_{n \in j+} \frac{n}{n+a} = 1$ and so if $a \in A - A_-$, then $\Gamma(a) \in K$. But the integrality of reciprocal as above fails in general because of the irregular behavior in Riemann-Roch for higher genus. For example, if $A = \mathbb{F}_3[x, y]/(y^2 = x^3 - x - 1)$, then neither $\Gamma(y)$ nor $1/\Gamma(y)$ is integral.

(3) A simple example that $1/\Gamma(t) = 2t(t+1)$, when $A = \mathbb{F}_q[t]$, and $p \neq 2$, shows that the prime factorization does not depend only on the degree of the prime in contrast to the arithmetic gamma function case.

This rationality of values at integers gives hope for interpolation a la Morita.

Definition 4.9.4   For $a \in A_v$, let $\bar{a} =:\ a$ or 1 according as whether $v(a) = 0$ or $v(a) > 0$ respectively, and when $x \in A$, put

$$\Pi_v(x) =: \prod_{j=0}^{\infty} (\prod_{n \in A_j+} \frac{\bar{n}}{x+n}).$$

Again terms are 1 for large j. Hence $\Pi_v(a) \in K$ for $a \in A$.

Lemma 4.9.5 Π_v *interpolates to* $\Pi_v : A_v \to A_v^*$ *and is given by the same formula, as in the definition, even if* $x \in A_v$. *Similarly,* $\Gamma_v(x) =: \Pi_v(x)/\bar{x}$ *interpolates.*

Proof.   In fact, it is easy to see that if $x \equiv y \mod v^l$, then $\Pi_v(x) \equiv \Pi_v(y)$ mod v^l.   □

Examples 4.9.6   Let $A = \mathbb{F}_q[t]$. Then $\Pi_t(t^2)$ is $(t + 1)/\{(t^2 + 1)(t - 1)(t^2 + t - 1)\}$ if $q = 3$ and is 1 if $q = 2$. When $q = 3$, $\Pi_v(t)$ is $-1/(t+1)$, 1, or $1/(t+1)$ according as $\deg(v) \neq 1$, $v = t+1$ or v is of degree one, but not $t + 1$ respectively.

Remarks 4.9.7 $1/\Pi_v(x) = \prod_{j=0}^{\infty} - \prod_{n \in j+} (\bar{x} + n)$ up to multiplication by a fixed root of unity by Theorem 4.7.5.

4.10   Functional equations in geometric case

Reflection formula for Π:

For $q = 2$, all nonzero elements are monic, so

$$\Gamma(x) = \frac{1}{e_A(x)} = \frac{\tilde{\pi}}{e(\tilde{\pi}x)}$$

where e_A is the exponential corresponding to lattice A and e is the exponential corresponding to sgn-normalized Drinfeld module with period lattice $\tilde{\pi}A$. Hence, for $x \in K - A$, $\Gamma(x)$ has algebraic ratio with $\tilde{\pi}$ and so, as we will see in Chapter 10, $\Gamma(x)$ is transcendental. From the point of view of their divisors, $e(\tilde{\pi}x)$ being analogous to $\sin(\pi x)$ (i.e. both have simple zeros at integers and no poles), this observation suggests relation between Γ and sine. We state these reflection relations as follows to make the analogy more visible.

Classical reflection formula can be stated as $\prod_{\theta \in \mathbb{Z}^\times} \Pi(\theta x) = \frac{\pi x}{\sin \pi x}$ and here we clearly have

Theorem 4.10.1

$$\prod_{\theta \in A^\times} \Pi(\theta x) = \frac{\tilde{\pi}x}{e(\tilde{\pi}x)}.$$

Hence the classical name 'half sine' for Γ should be replaced by '$\frac{1}{q-1}$-th exponential' in our case! Notice also that interesting cyclotomic part missing from the reflection formula for the arithmetic gamma is present now. One interesting difference is that for x a fraction, the ratio (denote it by $\Omega_{\underline{a}}$ again to stress analogy with situation in Theorem 4.8.3) obtained by dividing the left-hand side by first power of the period $2\pi i$ ($\tilde{\pi}$ resp.) lies in $\mathbb{Q}(\mu_{2n})$ ($K(\Lambda_a)$ resp.) if the denominator of x is n (a resp.). Consequently, if τ is an element of relevant Galois group fixing appropriate roots of unity (i.e. n-th in the classical case, a-th torsion in our case (and $q^r - 1$-th roots of unity in case of Theorem 4.8.3), then from what we have just said, $\Omega_{\underline{a}}{}^\tau / \Omega_{\underline{a}}$ is 1. Hence the character $\chi_{\underline{a}}$ occurring as in the analog of Deligne's theorem (take the formula in Theorem 4.8.3 as the definition of $\chi_{\underline{a}}$) is trivial in our case; whereas it can be nontrivial in the classical case [DMOS82, pa. 91] or [GK79, pa. 577]). On the other hand, if we look at $\Omega_{\underline{a}}{}^\tau / \Omega_{\underline{a}}$ for τ not necessarily fixing the appropriate roots of unity, one gets cyclotomic units in both cases.

Multiplication formula for Π:

Theorem 4.10.2 *Let $g \in A$ be monic of degree d and let α run through* *full system of representatives modulo g. Then*

$$\prod_\alpha \Pi\left(\frac{x+\alpha}{g}\right) = \Pi(x)\tilde{\pi}^{(q^d-1)/(q-1)}((-1)^d g)^{q^d/(1-q)} R(x)$$

where t is any integer larger than max(deg $\alpha, 2g_K) + d$ and

$$R(x) = \frac{\prod_{\beta \in A_{\leq t}+} \beta + x}{\prod_\alpha \prod_{a \in A_{\leq t+d}+} ga + \alpha + x}.$$

Proof. Using Theorem 4.7.5 and the fact, proved in Section 4.7.2, that $\overline{D_j} \to 1$ as $j \to \infty$, straight manipulations give

$$\prod_\alpha \Pi\left(\frac{x+\alpha}{g}\right) = \prod_j \prod_{a \in A_j+} \prod_\alpha \frac{ga}{(ga+\alpha)+x}$$

$$= R(x) \lim_{j \to \infty} \left(\prod_{k=0}^{j+d} \prod_{a \in A_k+} \frac{(ga)^{q^d}}{a+x}\right)\left(\prod_{k=j+1}^{j+d} \prod_{a \in A_k+} \frac{1}{(ga)^{q^d}}\right)$$

$$= R(x)\Pi(x) \lim \frac{g^{q^d(q^{j-c}-1)/(q-1)} \prod_{k=0}^{j} \prod_{a \in A_k+} a^{q^d-1}}{D_{j+1} \cdots D_{j+d}}$$

$$= R(x)\Pi(x)g^{q^d/(1-q)}\Pi_0(-1)^{(q^d-1)/(q-1)}$$

$$= R(x)\Pi(x)\tilde{\pi}^{(q^d-1)/(q-1)}((-1)^d g)^{q^d/(1-q)}$$

where in the last but one line we temporarily denote the arithmetic factorial by Π_0. □

Remarks 4.10.3 (1) Here $R(x)$ and c in the proof take care of beginning irregularity in Riemann-Roch. eg. $R(x) = \prod_{\alpha \text{ monic}} (x + \alpha)$ and $c = -1$ in case $A = \mathbb{F}_q[T]$ and $\{\alpha\}$ being set of all polynomials of degree not more than d.

(2) Analogy with usual multiplication formula in 4.6 is quite visible in that an analogous combination of factorials is a nonzero algebraic quantity times analogous power of period: Instead of $(n-1)/2$ here one has $(\text{Norm}(g) - 1)/(q-1)$, but we have seen that 2 and $q-1$ represent choices of signs in respective situations, whereas n and $\mathbb{N}g$ are number of residue classes that are relevant. For $x \in K - A$, call the algebraic part $\Omega_{\underline{a}}$ again. Now $\underline{a}$ corresponds to the multiplication formula. Note that $\Omega_{\underline{a}} \in K(\Lambda_g)$, which makes, just as in remarks above, the corresponding character $\chi_{\underline{a}}$ trivial again, in contrast with the classical situation.

Multiplication formula for Π_v:

Theorem 4.10.4 *Let α, g be as in the previous Theorem and with $(g, v) = 1$. Then*

$$\prod_\alpha \Pi_v(\frac{x+\alpha}{g})/\Pi_v(x) \in K(x)^\times.$$

Proof. By similar manipulations as above, We have

$$\prod_\alpha \Pi_v(\frac{x+\alpha}{g}) = \prod_\alpha \prod_j \prod_{a \in A_j+} \frac{\tilde{a}}{(\frac{ga+\alpha+x}{g})} = Q(x)\Pi_v(x) \lim_{j \to \infty} \frac{g^{m_j}(\tilde{D}_0 \cdots \tilde{D}_j)^{q^d-1}}{\tilde{D}_{j+1} \cdots \tilde{D}_{j+d}}$$

where $Q(x)$ is a rational function, which just as above, takes care of beginning irregularity in Riemann-Roch. Straight Riemann-Roch counting shows that m_j tends to an integer. By Lemma 4.5.4 and Theorem 4.8.2, (1) $-\tilde{D}_k \to 1$ and (2)$(\tilde{D}_0 \cdots \tilde{D}_j)^{q-1} \to (-1)^{\deg(v)-1}$. Hence, up to elements in $K(x)^\times$, the left-hand side of the equation above is $\Pi_v(x)$. $\qquad\square$

Reflection formula for Π_v:

Theorem 4.10.5 *Given $a \in A_v$, write a_v for its mod v representative which is of degree less than $\deg(v)$. Then*

$$\mu := \prod_{\theta \in \mathbb{F}_q^\times} \Pi_v(\theta a)/\bar{a} \in \mathbb{F}_q^*$$

and $\mu = -\text{sgn}(a_v)^{-1}$, if $a_v \neq 0$ and $\mu = 1$, otherwise.

Proof. Consider first a nonzero $a \in A$ of degree d. If n runs through monics of some fixed degree more than d, then so does $n+a$. So to calculate $\Pi_v(a)$, we need the product in the definition of Π_v only for $j \leq d$. Suppose first that $\deg(v) > d$, then in the product for $j \leq d$, there is a single cancellation: $(a/\text{sgn}(a))+\theta a = 0$ term, for $\theta = -\text{sgn}(a)^{-1}$. Writing $n/(n+\theta a)$ as $\theta^{-1}n/(\theta^{-1}n+a)$, we see that this product has numerator the product of all nonzero integers of degree $\leq d$ and denominator the product of these translated by a (canceling the zero term). So the product would have been a except for the correction for the adjusted sign for the canceled term. Hence in this case μ is as claimed.

In general, there can be several cancellations, leading to several sign adjustments for θ's, so the answer is the product of certain elements of $\mathbb{F}_q^*$ times $\bar{a}$ as claimed.

But this sign can be determined using the congruence property of Π_v noted in 4.9.5, which shows that the sign is as claimed, when $a_v \neq 0$. If

v divides a, $\Pi_v(a) \equiv \Pi_v(0) = 1$ implies the second claim. Finally v-adic continuity allows us to extend the result to $a \in A_v$. $\qquad\square$

Remarks 4.10.6   (1) In particular, if $q = 2$, $\Pi_v(a) = \bar{a}$, for $a \in A$, so that $\Pi_v(x) = \bar{x}$ for $x \in A_v$ by continuity. Hence $\Gamma_v(x) = 1$.

(2) When $A = \mathbb{F}_q[t]$, and $\{\alpha\}$ the representatives of degree less than d, the rational function of 4.10.4 is $\prod_{\alpha \in A_+} \overline{(x + \alpha)}$ times $g^m g_v^{(q^h-1)/(q-1)}(-1)^{hd}$, where $h = \deg(v)$, g_v is the Teichmüller representative of g and $-m$ is the number of $n \in A_{<d}+$, with n not congruent to $-x$ modulo v. In general, it can be given explicitly at expense of more complicated formula in terms of beginning irregularity in Riemann-Roch.

4.11   Special values of geometric Γ and Γ_v: $\mathbb{F}_q[t]$ case

We saw that for $q = 2$, all geometric gamma values at proper fraction are algebraic multiples of $\tilde{\pi}$ and all v-adic gamma values are algebraic.

Now we look at special results in the simplest case, that of $A = \mathbb{F}_q[t]$ and v a prime of degree one, to give a flavor of what can be done. More such results can be found [Tha91a]. Full results will be derived in Chapter 8, using Anderson's solitons.

We restrict now to $A = \mathbb{F}_q[t]$.

Lemma 4.11.1   *Let $D_{r,\eta,t}$ denote the product of monic polynomials of degree r, which are congruent to $\eta \in \mathbb{F}_q^*$ modulo t. Then*

$$D_{r+1,\eta,t} = D_r t^{q^r} (1 - \eta/(-t)^{(q^{r+1}-1)/(q-1)}).$$

Proof.   Let $x_0 = t^{r+1} + \eta$ and $x_i = t^{r-i+1}$, for $i > 0$, then the right side of Moore determinant identity is $D_{r+1,\eta,t}(t^{q^{r-1}} D_{r-1}) \cdots (t D_0)$, whereas splitting the first row in two terms by elementary operations, the determinant on the left is seen to be $t^{(q^{r+1}-1)/(q-1)} + (-1)^r \eta$ times the determinant of the matrix with the (i,j)-th entry being $(t^{r-i})^{q^{r-j}}$. The last determinant is $D_r \cdots D_0$ by the Moore identity again. $\qquad\square$

Now notice that we are dealing with denominator t, and t-th torsion of Carlitz module is $\lambda_t = (-t)^{1/q-1}$. So the formula in the lemma can be rewritten as

$$\prod_{n \in A_d+} (1 + \eta/nt) = 1 - \eta\lambda_t/\lambda_t^{q^{d+1}}.$$

Consider the Carlitz module over $B = \mathbb{F}_q[\lambda_t]$, the integral closure of $A = \mathbb{F}_q[t]$ in the t-th cyclotomic field. This is a Drinfeld module of rank $q - 1$ over A and of rank one over B. Hence this can be considered as t-th Fermat motive, having complex multiplications by t-th cyclotomic field. Denote its period by π_B.

Theorem 4.11.2   *With this notation, we have $\pi_B = \lambda_t^{q/(q-1)}\Pi(1/t)$.*

Proof.   We have

$$\Pi(1/t) = \prod_{d=1}^{\infty}(1 - \lambda_t/\lambda_t^{q^d})^{-1}.$$

On the other hand using the formula in 2.5 for the Carlitz period in the case of base B, we see that

$$\pi_B/\Pi(1/t) = \lim(\lambda_t^q - \lambda_t)^{q^n/(q-1)}/(\lambda_t^q)^{(q^n-1)/(q-1)} = \lambda_t^{q/(q-1)},$$

as $\lim(1 - \lambda_t^{1-q})^{q^n/(q-1)} = 1$. $\square$

For expression of $\Pi(-1/t)$ as a 'quasi-period', see Chapters 6 and 8.

Now let us look at the v-adic values. First we set up some preliminary notation. The Galois group of $K(\lambda_t)$ over K can be identified with $\mathbb{F}_q^*$, with $\eta \in \mathbb{F}_q^*$ acting as $\sigma_\eta(\lambda_t) = \eta\lambda_t$. If v is a prime of degree d, then the Galois group of $K\mathbb{F}_v$ over K can be identified with $\mathbb{Z}/d\mathbb{Z}$, with $j \in \mathbb{Z}/d\mathbb{Z}$ acting as $\tau_j(\zeta_{q^d-1}) = \zeta_{q^d-1}^{q^j}$. Let $\bar{\lambda}$ be the Teichmüller representative of λ_t, so that $\lambda_t^{q^{di}} \to \bar{\lambda} \in \mathbb{F}_v \subset K_v$ as $i \to \infty$. Write $\wp = \lambda_t - \bar{\lambda}$. Then $\wp^\eta$'s are the primes of $K(\lambda_t)$ above v.

Theorem 4.11.3   *Let $A = \mathbb{F}_q[t]$, v be a monic prime of degree d of A which is congruent to 1 modulo t and $\eta \in \mathbb{F}_q^*$. Then*

$$\Pi_v(\eta/t) = (-\bar{\lambda})^{(q^d-1)/(q-1)}(\wp^{\sigma_\eta})^{-\sum \tau_j}, \text{ for } \eta \neq 1, \quad \Pi_v(1/t) = \frac{v}{t}(\wp)^{-\sum \tau_j}.$$

In particular, $\Pi_v(a/t)$ is algebraic, for any $a \in A$.

Proof.   Write $w_j := \prod_{n \in A_{j+}} \overline{n/n + \eta/t}$ and $x_j := \prod_{n \in A_{j+}} n/(n + \eta/t)$. Since v is monic congruent to 1 modulo t, for $j \geq d$, we have $w_j = x_j/x_{j-d}v^{r_j}$, where r_j is such that the right side is a unit at v.

For $0 \leq j < d$, $w_j = x_j$, unless $j = d - 1$ and $\eta = 1$, in which case, they differ in $n = (v - 1)/t$ term giving $w_{d-1}t/v = x_{d-1}$. Hence, for $\eta \neq 1$, the product telescopes and we get

$$\Pi_v(\eta/t) = \prod w_j = \lim_{} \prod_{i=0}^{d-1} (1 - \eta\lambda_t/\lambda_t^{q^{nd+i}})^{-1} = \prod_{i=0}^{d-1} (1 - \eta\lambda_t/\overline{\lambda}^{q^i}).$$

Similarly, we calculate $\eta = 1$ case, where the answer is a unit at v. $\quad\square$

We leave it to the reader to express the values in terms of the corresponding Stickelberger elements and refer to Chapter 8, where the result is generalized to any denominator.

4.12   Comparisons and uniform framework

For simplicity, we restrict to $d_\infty = 1$. In fact, for the geometric case, most of the evidence is for the base $A = \mathbb{F}_q[t]$ and the hypotheses (Hi) below have not been scrutinized much in general.

Here we compare and explain analogies by giving unified treatment for the gamma functions in the three cases: classical, arithmetic (of Section 4.5) and geometric (of Section 4.7).

Classical	Arithmetic	Geometric
$\Gamma(\mathbb{Z}_+) \subset \mathbb{Z}_+$	$\Gamma(\mathbb{Z}_+) \subset A_{>0}^*$	$\Gamma(A - A_{\leq 0}) \subset K^*$
$\Gamma : \mathbb{C} - \mathbb{Z}_{\leq 0} \to \mathbb{C}^*$	$\Gamma : \mathbb{Z}_p \to \hat{K}_\infty^*$	$\Gamma : C_\infty - A_{\leq 0} \to C_\infty^*$
$\Gamma_p : \mathbb{Z}_p \to \mathbb{Z}_p^*$	$\Gamma_v : \mathbb{Z}_p \to K_v^*$	$\Gamma_v : A_v \to A_v^*$

We look at the questions of algebraicity, transcendence, relations to the periods of the special values of gamma functions at fractional arguments. First note how special value combinations occurring in the reflection and multiplication formulas, for z a fraction, introduce cyclotomic ('sin $\pi a/b$') and Kummer ('$n^{1/2-nz}$') extensions and keep in mind the vague connection $\Gamma(1/a) \leftrightarrow e(\tilde{\pi}/a) \leftrightarrow a$-torsion of ρ.

A unified treatment requires some unified notation and identification of similar objects:

	Classical	Arithmetic	Geometric
I: integers in domain	$\mathbb{Z}$	$\mathbb{Z}$	A
F: fractions in domain	$\mathbb{Q}$	$\{a/b \in \mathbb{Q} : (p,b)=1\}$ $= \{a/(q^m-1)\}$	K
$\bar{A}$: algebraic nos. in range	$\overline{\mathbb{Q}}$	$\overline{K}$	$\overline{K}$
underlying objects	$\mathbb{G}_m$, CM elliptic curves	$\mathbb{G}_m$, CM ρ's	CM ρ's
exponential e	classical e	classical for domain, e_ρ for range	e_ρ
period θ	$2\pi i$	$2\pi i$ for domain, $\tilde{\pi}$ for range	$\tilde{\pi}$
B: base field	$\mathbb{Q}$	K	K
$\mathcal{O}$: base ring	$\mathbb{Z}$	A	A
extension $B(e(\theta f))$	usual cyclotomic	constant field	Drinfeld cyclotomic
Galois group G identified with	$\mathrm{Gal}(\mathbb{Q}(\mu_N)/\mathbb{Q})$ $=(\mathbb{Z}/N\mathbb{Z})^*$	$\mathrm{Gal}(K(\mu_N)/K)$ $=q^{(\mathbb{Z}/r\mathbb{Z})}$	$\mathrm{Gal}(K(\Lambda_N)/K)$ $=(A/NA)^*$

In this table, $f = a/N$, with $a, N \in I$ and r was defined by $\mathbb{F}_{q^r} = \mathbb{F}_q(\mu_N)$ and in the geometric case we restricted to $h = d_\infty = 1$ or even to $A = \mathbb{F}_q[t]$ first for simplicity. See below for more general situation.

We restrict our attention to gamma values at proper fractions. In all the three cases, if $f \in F - I$, $i \in I$, then $\Gamma(f+i)/\Gamma(f) \in B \subset \overline{A}$. So for our question of algebraicity, it is enough to look at $f \in (F - I)/I$. To extract this information, we define for $f \in F - I$, a rational number $\langle f \rangle \in \mathbb{Q}$ such that $\langle f \rangle$ depends only on $f \mod I$:

Definition 4.12.1 (1) For classical and arithmetic gamma: For unique $n \in \mathbb{Z}$, $0 < f - n < 1$. Put $\langle f \rangle := f - n$.

(2) For geometric gamma: Let $A = \mathbb{F}_q[t]$. For unique $a \in A$, $f - a = -a_1/a_2$, where $a_1, a_2 \in A$, $\deg a_1 < \deg a_2$, a_2 monic. Put $\langle f \rangle := 1$ or 0, according as a_1 is monic or not.

Motivation for this definition and the definition in general case will be provided later. Note that for $q = 2$, $\langle f \rangle = 1$, in the geometric case.

Definition 4.12.2 Consider a finite formal sum $\underline{f} = \oplus m_i[f_i]$, $m_i \in \mathbb{Z}$, $f_i \in F - I$. This is nothing but an integral linear combination of symbols $[f_i]$, i.e. a divisor. Let N be a common denominator for the f_i's. Put $\Gamma(\underline{f}) := \prod \Gamma(f_i)^{m_i}$ and make similar definitions for Π, Γ_v, Π_v etc. Also put $m(\underline{f}) := \sum m_i \langle -f_i \rangle \in \mathbb{Q}$.

For $\sigma \in G$, let $\underline{f}^{(\sigma)} := \oplus m_i[f_i^{(\sigma)}]$, where $f_i^{(\sigma)}$ is just multiplication of f_i by σ as an element of the identification of the Galois group given in the table.

For example, in the case of arithmetic gamma, if σ corresponds to q^j, then $f_i^{(\sigma)} = q^j f_i$.

For a finite place v, let $\mathrm{Frob}_v \in G$ be the Frobenius, so that it makes sense to talk of Frob_v-orbits of f_i or of $\underline{f}$.

Consider the hypothesis/recipe/conjecture:

(H1): If $m(\underline{f}^{(\sigma)})$ is independent of $\sigma \in G$, then $\Gamma(\underline{f})/\theta^{m(\underline{f})}$ and $\Gamma_v(\underline{f})$ belong to $\overline{\mathcal{A}}$.

(H2): If $\underline{f}$ is a linear combination of Frob_v-orbits, then $\Gamma_v(\underline{f}) \in \overline{\mathcal{A}}$ (or rather algebraic in appropriate v-adic context).

Examples 4.12.3 With $\underline{f} = \langle f \rangle + \langle 1 - f \rangle$, $m(\underline{f}) = 1$ and with $\underline{f} = \oplus_{j=0}^{n-1} \langle z + j/n \rangle - \langle nz \rangle$, $m(\underline{f}) = (n-1)/2$ the hypothesis of H1 is satisfied in the first two cases. These correspond to the reflection and multiplication formulas, classical and of Section 4.6 respectively. In case II, Theorems of Section 4.6 show that $\Gamma(-\underline{f})/\tilde{\pi}^{n(\underline{f})} \in \overline{\mathcal{A}}$. This implies that $\Gamma(\underline{f})/\tilde{\pi}^{\sum m_i - n(\underline{f})} \in \overline{\mathcal{A}}$, but in case of reflection and multiplication, $\sum m_i = 2n(\underline{f})$. Hence $\Gamma(\underline{f})/\tilde{\pi}^{\sum m_i \langle -f_i \rangle}$ works in both cases under the independence hypothesis.

What is known about H1 and H2 and are they best possible?: H1 is true in all the three cases, but still we have written it as a hypothesis, because we will consider more situations later, in 8.3 and 8.7.2, which fit in this framework: For 8.3 gamma, it is true with $\theta = \Gamma(0)$, but for gamma in 8.7.2, it is not known.

In the classical case, $\underline{f}$ satisfying the condition in H1 is a linear combination, with rational coefficients, of the examples above (i.e. the conclusion follows from the known reflection and the multiplication formulas) by a result of Koblitz and Ogus. With a proof analogous to that of Koblitz and Ogus, the same can be proved for the geometric case. (We do need rational coefficients, so that we get some non-trivial χ_a's for a's which are not integral linear combinations of our functional equations. See [Sin97a] for analysis of rational versus integral linear combinations of functional equations and Anderson's 'ϵ-generalization' of Kronecker-Weber theorem [And02] in the classical case, inspired by this analysis.) But, since the independence condition in arithmetic case is much weaker than the one in the classical case, the functional equations of Theorem 4.6.4 (which together with Theorems 4.7.5, 4.8.2 prove H1 in this case directly) are more general than those generated by multiplication and reflection formulas.

H2 follows from the Gross-Koblitz theorem in classical case, and for the arithmetic case with $A = \mathbb{F}_q[t]$ by 4.8.1. We will see a full proof for geometric case, when $A = \mathbb{F}_q[t]$ in Chapter 8. There is some evidence in other cases.

In the complex multiplication situation, we also have a Chowla-Selberg type formula (up to multiplication by element of $\overline{\mathcal{A}}$) for the period, in terms of the gamma values at appropriate fractions. This formula is predicted via a simple calculation involving the 'brackets' introduced above. It works as follows:

Let E be a Drinfeld A-module (or elliptic curve for the case I) over B^{sep} with complex multiplication by the integral closure of $\mathcal{O}$ in the appropriate abelian extension L of B as in the table, eg., a constant field extension for arithmetic and Drinfeld cyclotomic extension for geometric case etc. (In fact, we can get much more flexibility using higher dimensional A-motives of Anderson considered in Chapter 7, eg., the solitons give rise to higher dimensional A-motives with Drinfeld cyclotomic CM whose periods are values of $\Gamma(z)$, for $z \in K$, in the geometric case with $A = \mathbb{F}_q[t]$. But here we will be content with this simple case.)

For $f \in F - I$, let $h(f) : \mathrm{Gal}(B^{sep}/B) \to \mathbb{Q}$ be defined by $h(f)(\sigma) := \langle -f' \rangle$, where $e(\theta f)^\sigma = e(\theta f')$. (Note that in arithmetic case, the exponential used here is the classical exponential according to the table.)

Let $\chi_{L/B}$ be the characteristic function of $\mathrm{Gal}(L/B)$. In other words, $\chi_{L/B} : \mathrm{Gal}(B^{sep}/B) \to \mathbb{Z}$ is such that $\chi_{L/B}(\sigma)$ is 1, if σ is identity on L and is 0 otherwise.

We then have the hypothesis/recipe/conjecture:

(H3): If $\chi_{L/B} = \sum m_f h(f)$, $m_f \in \mathbb{Z}$, then $\mathrm{Period}_E / \prod \Gamma(f)^{m_f} \in \overline{\mathcal{A}}$.

This is known in classical case. In function fields, there is only some evidence.

Example 4.12.4 $\mathcal{O} = A = \mathbb{F}_q[t]$, $K = B = \mathbb{F}_q(t)$, $L = \mathbb{F}_{q^k}(t)$. Then the rank k Drinfeld A-module: $u \mapsto tu + u^{q^k}$ is the rank one Carlitz module over $\mathbb{F}_{q^k}[t]$. Hence we want to express its period $\tilde{\pi}_k$ in terms of the gamma values at fractions with the gamma function built up from the base. If τ is the q-power Frobenius, then $\chi_{L/B}(\tau^n)$ is 1, if k divides n and is 0 otherwise. The same is true for

$$[qh(\frac{q^{k-1}}{1-q^k}) - h(\frac{1}{1-q^k})](\tau^n) = q\langle \frac{q^{k+n-1}}{q^k-1} \rangle - \langle \frac{q^n}{q^k-1} \rangle.$$

Hence H3 predicts that $\Gamma(q^{k-1}/(1-q^k))^q/\Gamma(1/(1-q^k))$ is an algebraic multiple of $\tilde{\pi}_k$. We know this from 4.7.3.

Remarks 4.12.5 (1) For transcendence implications and questions of algebraic independence etc. we refer to Chapters 10 and 11. In particular, we discuss whether the Hi's are best possible there.

(2) In case I, the Γ function is closely connected with ζ and L functions, it is in some sense a factor at infinity for the ζ function and hence appears in functional equations. Such connections are missing in the other cases. We only have $\Pi'(x)/\Pi(x) = \zeta(x, 1)$ in the third case.

(3) The Hurwitz formula $\zeta(x, s) = \langle -x \rangle - 1/2 + s \log \Gamma(x) + o(s^2)$ around $s = 0$ connects partial zeta, 'brackets' and logarithm of gamma. For case III, the partial zeta value for $x = a_1/a_2$ (with $\deg a_1 < \deg a_2$ and $a_2 \in A_+$) is obtained by summing $q^{-s \deg a}$ over monic a congruent to a_1 modulo a_2. Hence at $s = 0$, it equals $1 - 1/(q-1)$ or $-1/(q-1)$ according as a_1 is monic or not. This is the motivation for the definition, due to Anderson, of the brackets as above in the geometric case

(4) Consider general A with $d_\infty = 1$, for simplicity. If we have a representative a_1 modulo a_2 with $d_1 := \deg a_1 < d_2 := \deg a_2$, we have

$$\zeta(s, a_1, a_2) = q^{-sd_1} + q^{-sd_2} + qq^{-s(d_2+n_1)} + \cdots + q^k q^{-s(d_2+n_k)} + \frac{q^{k+1}q^{-s(d_2+n_{k+1})}}{1 - q^{1-s}},$$

where the last term is just sum of the geometric series obtained when the regularity in Riemann-Roch counting kicks in. Thus putting $s = 0$, we get the same formula as above, even with irregularity of Riemann-Roch in the beginning. In particular, the formula is independent of such representative, even though it is no longer unique as in $\mathbb{F}_q[t]$ case. Now suppose that the degree d_1 of a smallest degree representative is more than d_2 (d_1 is independent of such representative), then $a_1 + aa_2$ is monic for any a (not necessarily monic) of degree less than $d_1 - d_2$, so we have to add (q −

2)$(q^{\ell((d_1-d-2)\infty)} - 1)/(q-1)$ to the partial zeta value earlier in the monic case, and subtract $(q^{\ell((d_1-d-2)\infty)} - 1)/(q-1)$ in the non-monic case.

In $q = 2$ case, there is no change, as we would expect from triviality of bracket being consistent with our special values results in 4.10.

Since there are $q-1$ signs, one monic and $q-2$ non-monic, these corrections add to zero, and thus for functional equations, it does not matter whether we use the correct partial zeta values as brackets, or just use 1 and 0 according as minus of smallest degree representative is monic or not. But in Chowla-Selberg hypothesis H3, it will matter. For $d_\infty > 1$, do we need to subtract $1/(q^{d_\infty} - 1)$? We have not done calculations to probe these issues any further.

When $h > 1$, there are of course h geometric gamma's depending on $\mathcal{A}$, and Galois action permutes them and criterion talks about such collection. But we can just choose action of smaller group generated by Frobenius of principal ideals and then get theory for single gamma, say the one corresponding to $\mathcal{A}$.

We prove v-adic algebraicity for new gamma in Chapter 8, but the special values situation at ∞ for this gamma is still unclear.

(5) Many of our examples so far can be considered as simple cases of Fermat motives: Drinfeld modules (or their higher dimensional generalizations we will study in Chapters 7 and 8) with complex multiplication by ring of integers of 'cyclotomic fields': For $\underline{a}$ satisfying the 'functional equations' $n(\underline{a}^{(\sigma)}) = m$, the motive is m-th tensor power (see Chapter 7) of the Carlitz module, which is rank one over $\mathbb{F}_q[t]$. The other examples in 4.7 and 4.11, such as $\underline{a} = -1/t$ corresponding to the Carlitz module over $\mathbb{F}_q[\Lambda_t]$, or $\underline{a} = q * (q^{j-1}/(1 - q^j)) + (-1) * (1/(1 - q^j))$ (both with $n(\underline{a}) = 1$ and $n(\underline{a}^{(\sigma)}) = 0$, for σ not the identity) corresponding to the Carlitz module over $\mathbb{F}_{q^j}[t]$. In each case, periods were essentially $\Gamma(\underline{a})$, and Frobenius eigenvalues for $\wp$, the Gauss sums $g(\underline{a})$ (Stickelberger elements for $\sum n(\underline{a}^{(\sigma)})\sigma^{-j}$) also corresponding to $\Gamma_\wp(\underline{a})$. This should fit with analog (not yet developed fully) of v-adic periods and framework of Chowla-Selberg-Ogus formula. These phenomena will be generalized for cyclotomic extensions of $\mathbb{F}_q[t]$ in Chapter 8.

(6) To stress the analogies, we did not deal with important **two variable gamma function of Goss** [Gos88], which is essentially

$$\Gamma(x, y) := \frac{1}{x} \prod_{j=0}^{\infty} \prod_{n \in A_{j+}} (1 + x/n)^{-y_j} \Gamma(y), \quad (x, y) \in C_\infty \times \mathbb{Z}_p,$$

where y_j are base q digits of y. It handles both constant field and Drinfeld cyclotomic extensions at once and arithmetic and geometric gamma functions occur as its specializations. Analogs of Hi can be formulated (see [Tha91a] for this, for some examples and variants) for it and they are open. Unfortunately, the two variable domains of this gamma function and of the zeta functions of Goss which we mention in the next chapter are not compatible, and the zeta-gamma connection is an open question.

(7) Anderson has some ideas on adelic gamma and Barnes gamma analogs.

(8) For work on gamma monomials and distributions, see [BGY01; BGKY03].

4.13 More analogies for $\mathbb{F}_q[t]$: Divisibilities

Function field situation presents a curious hybrid of objects and identities in analogies with number field case, and its p-adic and q-analog or finite differences counterparts: We can do analysis for global fields by working in various completions, real or p-adic or by introducing deformation parameters such as q, or generic points.

We will assume that the reader is familiar with the p-adic situation, though some facts are reviewed in the last section of the next chapter.

In the last section, we review quickly the q-analogs situation for a comparison.

The defining polynomial $[n] = t^{q^n} - t$ of $\mathbb{F}_{q^n}$, which can also be described as the product of monic irreducible polynomials of degree dividing n, does sometimes play some role analogous to the usual n, or rather q^n.

We have $q^n! = D_n = ([n] - [0])([n] - [1]) \cdots ([n] - [n-1])$ in this context looking like a factorial of $[n]$. There is also q-twisted recursion analog of $n! = n(n-1)!$: $D_n = [n]D_{n-1}^q$. One has twisted recursions

$$\Pi(q^{n+1}) = [n+1]\Pi(q^n)^q, \quad [n+1] = [n]^q + [1]$$

in place of the usual $(n+1)! = (n+1)n!$ and $(n+1) = n+1$ respectively.

In this vein, note that $[n][n-1] \cdots [1] = L_n$. In fact, we can also think of 'factorial' of $n-k$ by product of differences as above, but in a backward manner: $([k+1] - [k])([k+2] - [k]) \cdots ([n] - [k]) = L_{n-k}^{q^k}$. We will come back to this in Section 6.5.

These vague analogies are made much more precise by a definition, due to Manjul Bhargava [Bha97; Bha00] of factorial with very good combinatorial and divisibility properties in a very general context. We describe some good properties referring to the original nice papers for details and more.

See also papers by Shashikant Mulay [Mul95; Mul99] where he defines a double factorial, whose ratios give factorials, in even more general setting.

Let X be arbitrary nonempty subset of a Dedekind ring (i.e., noetherian, locally principal and with all primes maximal) R. Special cases would be Dedekind domains $\mathcal{O}_S$ coming from global fields or their quotients. Bhargava associates to a natural number k an ideal $k!_X := \prod \wp^{v_k(X,\wp)}$ of R, with the exponents v_k of the primes $\wp$ of R defined as follows: Let a_0 be any element of X. Choose a_k to be an element of X which minimizes the exponent of the highest power of a prime $\wp$ dividing $(a_k - a_0)(a_k - a_1)\cdots(a_k - a_{k-1})$ and $v_k(X,\wp)$ be this exponent. It can be proved that it is well-defined, independent of the choices involved.

The sequence a_i is called a $\wp$-ordering. If a_i is $\wp$-ordering for all $\wp$, then $k! = (a_k - a_0)\cdots(a_k - a_{k-1})$.

Examples 4.13.1 (1) The sequence $0, 1, 2, \cdots$ in $X = R = \mathbb{Z}$ gives simultaneous p-ordering, for all p and leads to the usual factorial. For X consisting of q-powers for integer $q > 1$ in $R = \mathbb{Z}$, we have $k!_X = (q^k - 1)\cdots(q^k - q^{k-1})$. For the set of $(q^j - 1)/(q - 1)$'s, we get the q-factorial. For one more interesting special case, see 4.16.6.

(2) For $X = R = \mathbb{F}_q[t]$, we have the following simultaneous $\wp$-ordering: Let $0 = a_0, a_1, \cdots, a_{q-1}$ be the elements in $\mathbb{F}_q$ and put $a_n = \sum a_{n_i} t^i$ where $n = \sum n_i q^i$ is the base q expansion of n. Hence $n! = (a_n - a_0)\cdots(a_n - a_{n-1}) = \prod [i]^{n_i + n_{i+1}q + \cdots + n_h q^{h-i}}$, which is the Carlitz factorial of n.

If q is a prime, then as a nice mnemonic we can think of associating to base q expansion $n = n(q) = \sum n_i q^i$ a polynomial $a_n = n(t) = \sum n_i t^i$ and with this ordering the Carlitz factorial can be described by the usual formula $n! = (n-0)(n-1)\cdots(n-(n-1))$. (But keep in mind that addition of n's is like integers, with carry-overs and not like polynomials!) The same works for general q except we have to identify n_i between 0 and $q - 1$ with elements of $\mathbb{F}_q$ by force then.

(3) Since the definition of this factorial is local in $\wp$, it leads to Goss-Sinnott ideal valued generalization (see [Gos80b] and also [Gos96, Sec. 9.3]) for general A, for $X = R = A$. But it seems that we do not have a simultaneous $\wp$-ordering in non-$\mathbb{F}_q[t]$ case.

(4) Let $R = \mathbb{F}_q[t]$. We saw that if $X = R$, then $D_i = (q^i)!_X$. We also have $D_i = i!_X$, for $X = \{t^{q^j} : j \geq 0\}$ or for $X = \{[j] : j \geq 0\}$ justifying the naive analogies mentioned above.

Bhargava shows that generalized factorial retains the most important divisibility properties of the usual and Carlitz factorial, such as (1) $k!l!$ divides $(k + l)!$ (integrality of binomial coefficients), (2) $0!1! \cdots n!$ divides $\prod_{i<j}(x_i - x_j)$ for any $n + 1$ elements $x_i \in X$ etc.

As far as the divisibilities and combinatorial questions are concerned, this seems to be the right notion. In our study, we have focused more on interpolations and special values relations to cyclotomy: periods and Stickelberger elements, and in that respect, our earlier definitions seem more suitable. In general, for arithmetic factorial, even D_i need not divide D_{i+1}, and for geometric factorial, which has quite different domain than the usual integers, the values at integers are not integers. For the case of $\mathbb{F}_q[t]$, all these many points of view come together, still leading to the two notions of factorials that we have seen and a few notions of binomial coefficients, as we will see next.

4.14 Binomial coefficients

Once we have factorials $\Pi(n)$'s for n in $\mathbb{Z}_{\geq 0}$ or in $A - A_-$, we can consider the binomial coefficients $\Pi(n)/(\Pi(k)\Pi(n-k))$. As we have mentioned, they are integral in the arithmetic case (when $A = \mathbb{F}_q[t]$ for our factorials or more generally for Goss-Sinnott-Bhargava factorials), but not in the geometric case, where even the factorial is not integral.

From digit expansion formula for factorial it follows that $n!/(m!(n - m)!) = 1$, if there is no carry over base q in adding m to $n - m$. This also follows from Lucas theorem analog, since unit modulo all primes implies constant and monic implies one.

More useful for our purposes are the binomial polynomials analogs $\binom{x}{q^n} = e_n(x)/D_n \in K[x]$ of the usual $x(x-1)\cdots(x-n+1)/n! \in \mathbb{Q}[x]$. We have already encountered these basic objects for n a power of q in 2.1, 2.5. For a general n, we will define them by digit expansion as in the factorial case: If $n = \sum n_i q^i$ is the base q expansion of n, i.e., $0 \leq n_i < q$, then we put $\binom{x}{n} := \prod \binom{x}{q^i}^{n_i}$.

Why are these good analogs of binomial coefficient polynomials? Firstly, we present a curious mix of analogies:

$$\binom{x}{q^i} = e_i(x)/D_i \leftrightarrow \frac{x(x-1)\cdots\left(x - (q^i - 1)\right)}{q^i!}.$$

Here $0, 1, \ldots, q^i - 1$ is a natural full residue class system for $\mathbb{Z}/q^i\mathbb{Z}$ whereas

$n \in A$, deg $n < i$ is a natural full residue class system for A/aA with a such that $Na = q^i$.

Hence by the definition in 2.5 of $e_i(x)$, the numerators are analogous. We've seen that the denominator $D_i = g_{q^i}$ is an analog of $q^i!$. Now $q^i! = q^i(q^i - 1)\ldots 1$, but D_i is a product of monic elements of A degree i. That's why we called this analogy a curious mixture.

With the ordering mentioned above $\binom{x}{q^k} = \prod_{i=0}^{q^k-1}(x - a_i)/(a_{q^k} - a_i)$. If we replace q^k by n on right side, what we get does not agree with Carlitz binomial coefficient considered above, and is not as intrinsic as it depends on this arbitrary ordering. But Bhargava shows that they do give you good basis. Again for general A, these diverge from the analogs we considered, and to get basis we can not use simultaneous ordering, so that there is no nice formula.

Secondly, classically the binomial coefficient $\binom{x}{k}$ is coefficient of t^k in the expansion of $(1 + t)^x = e^{x\log(1+t)}$, whereas $\binom{x}{q^i}$ is coefficient of t^{q^i} in $e(x\log t)$ (where log is inverse function of e of section I and coefficients of $t^m, m \neq q^i$ are zero). Translating from 'analytic' to 'algebraic' language, multiplication by $x \in \mathbb{Z}$ for $\mathbb{G}_m$ is represented by $(1 + t)^x - 1 = \sum \binom{x}{k}t^k$ whereas 'Carlitz action' of $x \in A$ on $\mathbb{G}_a$ is represented by $\sum \binom{x}{q^i}t^{q^i}$.

Thirdly, Carlitz [Car40b] showed that $\binom{x}{k}$ takes A-values on A and in fact form a basis of A-module of polynomials in and over K which map A into A.

Bhargava's binomial coefficients mentioned above also have this property. It is curious that Bhargava's definition automatically leads to the digit expansion definition from the basic objects, in case of factorials, but not in the case of these binomial coefficients.

Before we move on to the general A, let us mention in passing some other interesting analogs of binomial coefficients: First, Carlitz [Car35] mentioned the following as having some properties analogous to binomial coefficients:

For $n, k \in \mathbb{Z}_+ \cup \{0\}$, put $[n, k] =: D_n/(D_k L_{n-k}^{q^k})$ if $n \geq k$ and 0 otherwise. As we have seen that these are just the non-vanishing coefficients of $e_n(x)$, i.e., non-vanishing elementary symmetric functions of the set $A_{<n}$, hence clearly integral. Carlitz mentions easily verifiable recursion $[n, k] = [n - 1, k - 1]^q[n]/[k]$ analogous to $\binom{n}{k} = \binom{n-1}{k-1}n/k$ and parallel to the q-twisted recursion $D_n = [n]D_{n-1}^q$ thought analogous to $n! = n(n - 1)!$ as mentioned above. Also $[n, k] = [n - 1, k - 1]^q + D_{n-1}^{q-1}[n - 1, k]$ is somewhat like $\binom{n}{k} = \binom{n-1}{k-1} + \binom{n-1}{k}$ for the usual binomial coefficients. We note here

that $[n,k]$ are just like $n!/(k!(n-k)!)$, in view of analogies in 4.13, where for $(n-k)!$, we use the 'backwards' differences analogy mentioned there. Note though that $[n,k]$ and $[n,n-k]$ need not be the same because of the asymmetry.

As we saw above these naive analogies are somewhat justified by Bhargava with his p-orderings. This suggests looking at similar backwards differences binomial coefficients for general p-orderings. In fact, Bhargava [Bha97] has considered these objects in the generality of his setting as ratio of Vandermonde products, which in our case of simultaneous $\wp$-orderings becomes,

$$\prod_{0\leq i<j<n, i\neq k\neq j} (a_j - a_i)/ \prod_{0\leq i<j<n-1} (a_j - a_i).$$

He shows that this is integral and gives [Bha97, Theorem 6, 7] some analogies.

In this book, we will only focus on binomial coefficient polynomials $\binom{x}{k}$, from now on.

(a)   Binomial coefficients as nice basis

The usual binomial coefficients take $\mathbb{Z}$-values on $\mathbb{Z}$ and hence by continuity assume $\mathbb{Z}_p$-values on $\mathbb{Z}_p$. In fact, they form a basis of the $\mathbb{Z}$-module of polynomials (in x) over $\mathbb{Q}$ which map $\mathbb{Z}$ into $\mathbb{Z}$ (and also of the $\mathbb{Z}_p$-module of polynomials over $\mathbb{Q}_p$ that map $\mathbb{Z}_p$ into $\mathbb{Z}_p$).

Theorem 4.14.1 *(Mahler [Mah58]) Any $f \in Conti\ (\mathbb{Z}_p, \mathbb{Z}_p)$ can be uniquely written as*

$$f(x) = \sum_{k\geq 0} f_k \binom{x}{k}, \qquad f_k \in \mathbb{Z}_p, \qquad a_k \to 0.$$

The f_k may be recovered as the higher differences of f:

$$f_k = (\Delta^k f)(0) = \sum_{i=0}^{k} (-1)^i \binom{k}{i} f(k-i),$$

where $\Delta g(x) := g(x+1) - g(x)$ is the usual finite difference operator.
Conversely, any series $\sum_{k\geq 0} f_k \binom{x}{k}$, $f_k \in \mathbb{Z}_p$, $f_k \to 0$ converges to an element of Conti $(\mathbb{Z}_p, \mathbb{Z}_p)$.

Carlitz' student Wagner showed that (note we use the same notation for binomial coefficients in function fields)

Theorem 4.14.2 *(Wagner [Wag71a; Wag71b]) Any* $f \in Conti$ *$(A_\wp, A_\wp)$ can be uniquely written as*

$$f(x) = \sum_{k \geq 0} f_k \binom{x}{k}, \qquad f_k \in A_\wp, \qquad f_k \to 0.$$

The f_k may be recovered as

$$f_k = (-1)^m \sum_{\deg a < m} \frac{G'_{a^m - 1 - k}(a)}{\Pi(q^m - 1 - k)} f(a) \quad \text{where} \quad q^m > k$$

and for $n = \sum n_i q^i$, $0 \leq n_i < q$,

$$G'_n(x) := (\prod_{n_i < q-1} e_i(x)^{n_i})(\prod_{n_i = q-1} (e_i(x)^{q-1} - D_i^{q-1})).$$

Conversely, any series $\sum_{k \geq 0} f_k \binom{x}{k}$, $f_k \in A_\wp$, $f_k \to 0$ converges to an element of Conti $(A_\wp, A_\wp)$.

Remarks 4.14.3 (1) Higher differences interpretation for coefficients, as in previous theorem, is given in the next subsection.

(2) The sum in the formula for f_k can also be written as a sum over a with $0 \leq$ Norm $a \leq k$. Also if none of the q-digit for $q^m - 1 - k$ is $(q-1)$, then $\frac{G'_{q^m-1-k}(a)}{\Pi(q^m-1-k)} = \binom{a}{q^m-1-k}$, a 'binomial coefficient'. Carlitz shows that $G'_n(x)/\Pi(n)$ also satisfy the integral basis property of the binomial coefficients.

(3) Classically, as well as in our case of linear or general functions, the corresponding binomial coefficient basis is orthonormal (see [Koc98] for the corresponding Banach space of continuous functions, in the usual sense that the supremum norm $\|f\|$ for continuous function f on the power series ring coincides with $\sup |f_i|$, for f and f_i related as in the theorem in the general case.

(4) Wagner, Goss, Kochubei, Yang [Wag71b; Wag74; Gos89; Koc99; Yan98] related degrees of smoothness of f to orders of growth of sizes of coefficients of f_k.

(5) For more general results on integrality and interpolation, see [BK99; Tat99] and references therein for work of Amice etc. Voloch and Jeong [Vol98; Jeo01; Jeo00a; Jeo00b] studied and compared with Hasse hyperdifferential orthonormal basis.

(6) Keith Conrad [Con00] showed that for getting basis for continuous functions from that of continuous linear functions, the construction used for binomial coefficients using digit expansions is quite general phenomenon.

We will use these theorems in calculating zeta measures in the next chapter.

(b)   Difference and differentiation operators

Carlitz [Car35] considered a linear operator Δ given by

$$\Delta f(x) = f(tx) - tf(x)$$

on linear functions. (Recall that in characteristic p, one has a wide variety of linear functions.) Then $\Delta x^{q^n} = [n]x^{q^n}$. Consider also

$$d_\tau := \Delta^{1/q}.$$

Then we have $d_\tau\binom{x}{q^i} = \binom{x}{q^{i-1}}$ as an analog of classical difference operator identity $\Delta\binom{x}{i} = \binom{x}{i-1}$, where just here we write $\Delta F(x) = f(x+1) - f(x)$. Instead of iteration, Carlitz defined higher difference operators $\Delta^{(i)}$ for $i \geq 0$, by $\Delta^{(0)} f(x) = f(x)$, and

$$\Delta^{(i+1)} f(x) = \Delta^{(i)} f(tx) - t^{q^i} \Delta^{(i)} f(x), \quad (i \geq 0),$$

so that $\Delta^{(1)} = \Delta$. Then

$$\Delta^{(n)} f(x) = f(t^n x) + \sum_{k=0}^{n-1} (-1)^{n-k} \Big(\sum_{0 \leq i_1 < \cdots < i_{n-k} \leq n-1} t^{q^{i_1} + \cdots + q^{i_{n-k}}} \Big) f(t^k x).$$

We have [Car35] analog of classical higher difference formula in 4.14.1: For $\mathbb{F}_q$-linear f, we have

$$f(x) = \sum \Delta^{(i)} f(1) \binom{x}{q^i}.$$

In fact, Δ is just a commutator with t (or any generator $t + \alpha$, $\alpha \in \mathbb{F}_q^*$, of $\mathbb{F}_q[t]$ over $\mathbb{F}_q$). We have the derivation rule

$$\Delta(f \circ g) = f \circ (\Delta g) + (\Delta f) \circ g,$$

where $\circ$ denotes composition. Then d_τ is linear and satisfies derivation rule for composition if one restricts to 'linear' power series with coefficients in $\mathbb{F}_q$.

The functional equation $e_C(tx) = te_C(x) + e_C(x)^q$ is equivalent to

$$d_\tau e_C(x) = e_C(x)$$

just like $d/dx(e^x) = e^x$. So d_τ is also a good analog of differentiation operator. This is in contrast to the usual derivative leading to $d/dx(e_C(x)) = 1$

and killing all except the x term in the $\mathbb{F}_q$-linear series. Note also that with the original Carlitz normalizations, we would have obtained a minus sign.

In fact, this holds term by term, and one has

$$d_\tau(x^{q^n}/\Pi(q^n)) = x^{q^{n-1}}/\Pi(q^{n-1}), \quad d_\tau(x^{q^n}/[n]) = d_\tau(x^{q^{n-1}})$$

analogous to $d/dx(x^n/n!) = x^{n-1}/(n-1)!$ and $d/dx(x^n/n) = x^{n-1}$ resp.

The (non-commutative) ring of differential operators for us is $\mathcal{D} := C_\infty\{\tau, d_\tau\}$, with commutation relations

$$\tau w = w^q \tau, \quad d_\tau w = w^{q^{-1}} d_\tau, \quad d_\tau \tau - \tau d_\tau = -[-1].$$

The action of $\mathcal{D}$ on the functions is the obvious one, except perhaps for $d_\tau(\sum a_n \tau^n) := \sum a_n^{1/q}[n]^{1/q}\tau^{n-1}$. We note that $d_\tau(f+g) = d_\tau(f) + d_\tau(g)$ and $d_\tau(wf) = w^{1/q}d_\tau(f)$ and $d_\tau(fw) = d_\tau(f)w$, for $w \in C_\infty$.

We will see more development and applications of this viewpoint in the sections on Bernoulli polynomials, DeRham cohomology and Hypergeometric functions.

Kochubei [Koc98] has studied related operators in $\wp$-adic settings and pointed out analogies with quantum mechanics commutation relations of creation and annihilation operators. Let us briefly recall these referring to the paper for more. (The classical and p-adic situation is briefly recalled in the last section for comparison.)

In $\mathbb{F}_q[t]$ case, for Schrödinger representation, consider $K_t = \mathbb{F}_q((t))$ and the corresponding 'complex numbers' C_t say. Let X be a Banach space, over C_t, of all continuous $\mathbb{F}_q$-linear C_t-valued functions with the supremum norm. Write b_i for binomial coefficient $\binom{x}{q^i}$. Then $a^+ := \tau - 1$ and $a^- := d_\tau$ are continuous $\mathbb{F}_q$-linear operators on X with $a^- a^+ - a^+ a^- = [1]^{1/q}$ (which is also $-[-1]$). The operator $a^+ a^-$ possesses the orthonormal eigenbasis b_i, with eigenvalues $[i]$. In fact, $a^+ b_{i-1} = [i]b_i$ and $a^- b_i = b_{i-1}$ for $i \geq 1$ and $a^- b_0 = 0$.

For Bargmann-Fock representation [Koc99], we look at the space of $\sum a_n t^{q^n}/D_n$, with $|a_n|_t \to 0$, with supremum norm, and orthogonal basis $\tilde{b}_i = t^{q^i}/D_i$, with $\tilde{a}^+ = \tau$ and $\tilde{a}^- = d_\tau$ and we have the same identities as before with $\tilde{\ }$ counterparts.

Finally, we should mention Hasse derivative $H^{(n)}$ of order n: $H^{(n)}(z^m) = \binom{m}{n}z^{m-n}$, which is often a good substitute of $(d/dz)^n/n!$ in finite characteristic. In a recent preprint of Jeong, Kim, Park, Son various related formulas are given with applications to function field Bernoulli identities (see 4.16).

4.15   Relations between the two notions of binomials

As we have mentioned, several analogies, which work for $\mathbb{F}_q[t]$, diverge for general A, leading to different notions of binomial coefficients. We focus and relate two of these notions for general A. We will see the arithmetic applications of this relation to the zeta values in the next chapter.

We now work with general A, but restrict to $d_\infty = 1$ for simplicity. We choose a sgn-normalized ρ for A, and let $e(z) = \sum z^{q^i}/d_i$ and $l(z) = \sum z^{q^i}/l_i$ be the corresponding exponential and logarithms.

Define one version of binomial coefficients $\{{}^{\,t}_{q^i}\}$ by

$$e(tl(z)) = \sum_{i=0}^{\infty} \{{}^{\,t}_{q^i}\} z^{q^i} \quad \text{Then} \quad \{{}^{\,t}_{q^i}\} = \sum_{k=0}^{i} t^{q^k}/d_k l_{i-k}^{q^k}.$$

Only for this section, we will put

$$\frac{1}{L_i} := S_i := \sum_{a \in A_{i+}} \frac{1}{a}.$$

Note that L_i defined here is $(-1)^i L_i$ for the $\mathbb{F}_q[t]$ case. Put

$$e_i(t) := \prod_{a \in A_{<i}} (t - a) \in A[t], \quad \text{and} \quad D_i := \prod_{a \in A_{i+}} a.$$

The second version of binomial coefficients is

$$\binom{t}{q^i} := \frac{e_i(t)}{D_i}.$$

Then Carlitz results in 2.5 imply

Theorem 4.15.1   *For $A = \mathbb{F}_q[t]$ and ρ as in our example,*

$$\{{}^{\,t}_{q^i}\} = \binom{t}{q^i}, \quad D_i = d_i, \quad L_i = l_i.$$

For general A, $d_i, l_i \in H$, $\{{}^{\,t}_{q^i}\} \in H[t]$ and $D_i, L_i \in K$, $\binom{t}{q^i} \in K[t]$. (In fact, $D_i \in A$.) We want to generalize Theorem 4.15.1 by relating d_i's to D_i's, l_i's to L_i's and $\{\}$'s to $()$'s. Applications will be provided in the next chapter.

If A has a monic element, say p_i, of degree i (e.g. if $i > 2g - 1$), then $e_{i+1}(t) = \prod_{c \in \mathbb{F}_q} e_i(t + cp_i) = e_i(t)^q - D_i^{q-1} e_i(t)$, since $e_i(p_i) = D_i$. So

$$e_{i+1}(t) = e_i(t)^q - D_i^{q-1} e_i(t),$$

if p_i exists. If p_i does not exist, $e_{i+1}(t) = e_i(t)$. Let

$$\left\{{t \atop q^i}\right\} = \sum_{k=0}^{i} a_{ik} t^{q^k}, \qquad \left({t \atop q^i}\right) = \sum A_{ik} t^{q^k}.$$

By the Riemann-Roch theorem, there are g Weierstrass gaps. So if $i \geq 2g$, then $A_{ik} = 0$ for $k > i - g$, and

$$A_{i0} = (-1)^{i-g} \frac{(D_0 D_1 \cdots D_{i-1})^{q-1}}{D_i}.$$

On the other hand, the definition of $\left\{{t \atop q^i}\right\}$ implies that for any i,

$$a_{i0} = 1/l_i, \qquad a_{ii} = 1/d_i \quad A_{i(i-g)} = 1/D_i.$$

Let $L_i(n)^{-1} := S_i(n) := \sum_{a \in A_{i+}} 1/a^n$.
Then for i for which p_i exists (e.g. $i > g - 1$),

$$1 - \left({t \atop q^i}\right) = -\left({t - p_i \atop q^i}\right) = -\prod_{a \in A_{i+}} \left(\frac{t}{a} - 1\right).$$

Taking negative of logarithmic derivative of both sides, we get

$$\frac{A_{i0}}{1 - \sum_{k=0}^{i} A_{ik} t^{q^k}} = -\sum_{a \in A_{i+}} \frac{1}{t - a} = \sum_{n=0}^{\infty} S_i(n+1) t^n.$$

Remarks 4.15.2 Hence $S_i(n)$ is a homogeneous polynomial in A_{ik}'s, with coefficients in $\mathbb{F}_p$, of weight n, if A_{ik} is given weight q^k. In particular,

$$S_i(1) = S_i = \frac{1}{L_i} = A_{i0}.$$

This quantity plays a crucial role in the arithmetic of gamma functions, periods, zeta functions, as we will see.

Theorem 4.15.3 *For* $i \geq 2g$,

$$\left\{{t \atop q^i}\right\} = \sum_{k=0}^{g} c_{ik} \left({t \atop q^i}\right)^{q^k}, \qquad c_{ik} \in H$$

$$\sum_{k=0}^{g} c_{ik} = 1, \quad c_{i0} = \frac{L_i}{l_i}, \quad c_{ig} = \frac{D_i^{q^g}}{d_i}.$$

Proof. If $t \in A$, then $e(tl(z)) = \rho_t(e(l(z))) = \rho_t(z)$. If, further, $\deg t < i$, then the degree of polynomial $\rho_t(z)$ is less than q^i, so that immediately from the definition of this binomial, $\{ {t \atop q^i} \} = 0$. Hence the $\mathbb{F}_q$-linear polynomial $\{ {t \atop q^i} \}$ of degree q^i in t has elements of A of degree less than i as its zeros. Now $\binom{t}{q^i}$ has precisely these zeros. By the Riemann-Roch theorem, this accounts for q^{i-g} zeros of $\{\}$. Since the zeros of a $\mathbb{F}_q$-linear polynomial form a vector space over $\mathbb{F}_q$, choosing a basis and using the fact that the top degree coefficient of $\{ {t \atop q^i} \}$ is $1/d_i$, we see that

$$d_i \left\{ {t \atop q^i} \right\} = \prod_{c_1, \cdots, c_g \in \mathbb{F}_q} e_i(t + c_1 t_{i1} + \cdots + c_g t_{ig}) = \sum_{k=0}^{g} r_{ik} e_i(t)^{q^k}$$

for suitable t_{ij}, r_{ij}'s. This implies the first statement of the theorem. As ρ is sgn-normalized, $\{ {p_i \atop q^i} \} = 1$. Hence the second statement of the theorem follows by putting $t = p_i$ in the first statement. The rest follows by comparison of the lowest (resp. top) degree coefficients in the first statement using our evaluations of corresponding a_{ij} and A_{ij}'s above. □

Note that this Theorem implies the previous Theorem. Our goal now is to understand c_{ik}'s.

Theorem 4.15.4   *When $g = 1$, $i \geq 2g$, we have*

$$\left\{ {t \atop q^i} \right\} = (1 - \mu_i) \binom{t}{q^i}^q + \mu_i \binom{t}{q^i}$$

$$D_i^q = (1 - \mu_i) d_i, \quad S_i = \frac{1}{L_i} = \frac{1}{\mu_i l_i}$$

with $f_i := d_i / d_{i-1}^q$, $g_i := l_i / l_{i-1}$ and

$$\mu_i = \frac{(f_{i+1} - f_i + l_1^{q^i}) g_{i+1}}{(f_{i+1} - f_i) g_{i+1} - f_{i+1} l_1^{q^i}}.$$

Proof.   Comparison of the two formulas for A_{i0} gives

$$\frac{D_{i+1}}{L_{i+1}} = -\frac{D_i^q}{L_i}.$$

Let b_i be the coefficient of $t^{q^{i-2}}$ in $e_i(t)$. Comparing the coefficients of $t^{q^{i-1}}$ in the functional equation for $e_i(t)$, we get $b_{i+1} = b_i^q - D_i^{q-1}$.

Let $\mu_i = L_i / l_i$, then by the previous Theorem, we get the first two formulas claimed. We want to show the claimed formula for μ_i. Now comparing the coefficients of $t^{q^{i-1}}$ in the first formula of the Theorem, using the second formula and expressions for the two binomial coefficients above, we get

$$\frac{d_i}{d_{i-1}l_1^{q^{i-1}}} = b_i^q + \frac{d_i L_i}{D_i l_i}.$$

Subtracting q-th power of this equation from the same equation, but with i replaced by $i+1$, and using the recursion above for b_i, we get

$$\frac{1}{l_1^{q^i}}\left(\frac{d_{i+1}}{d_i} - \frac{d_i^q}{d_{i-1}^q}\right) = -D_i^{q(q-1)} + \left(\frac{d_{i+1}L_{i+1}}{l_{i+1}D_{i+1}} - \frac{d_i^q L_i^q}{l_i^q D_i^q}\right)$$

which equals to

$$-D_i^{q(q-1)} - \frac{1}{D_i^q}\left(\frac{d_{i+1}L_i}{l_{i+1}} + \frac{d_i^q L_i^q}{l_i^q}\right).$$

By definitions of f_i and g_i and the first two statements of the Theorem, the above equation reduces, after canceling the common factor d_i^{q-1}, to

$$\left(\frac{f_{i+1}}{f_i} - 1\right)\frac{f_i}{l_1^{q^i}} = -(1-\mu_i)^{q-1} - \frac{1}{1-\mu_i}\left(\frac{f_{i+1}\mu_i}{g_{i+1}} + \mu_i^q\right).$$

Simple algebraic manipulation leads to the claimed formula for μ_i. $\quad\square$

Finally, we give an explicit formula for f_i and g_i, when $g = 1$. (In Section 8.2 (c), there are formulas for sgn-normalized Drinfeld modules also, so for x_1 etc. below.) By the Riemann-Roch theorem, we can find monic $x, y \in A$ of degree 2 and 3 respectively. Put $[i]_x := x^{q^i} - x$ and $[i]_y := y^{q^i} - y$. Let $\rho_x = x + x_1\tau + \tau^2$ and $\rho_y = y + y_1\tau + y_2\tau^2 + \tau^3$. (In fact ρ_x determines ρ.)

Theorem 4.15.5   *When g=1, we have*

$$f_i = \frac{[i]_y - (y_2 - x_1^q)[i]_x}{[i-1]_x^q + y_1 - (y_2 - x_1^q)x_1}$$

$$g_i = \frac{-[i]_y + (y_2 - x_1)^{q^{i-2}}[i]_x}{y_1^{q^{i-1}} - [i-1]_x - (y_2 - x_1)^{q^{i-2}}x_1^{q^{i-1}}}.$$

Proof. Comparing the coefficients of z^{q^i} in the functional equation $e(az) = \rho_a(e(z))$, with $a = x$ and with $a = y$, we get

$$0 = -\frac{[i]_x}{d_i} + \frac{x_1}{d_{i-1}^q} + \frac{1}{d_{i-2}^{q^2}}, \quad 0 = -\frac{[i]_y}{d_i} + \frac{y_1}{d_{i-1}^q} + \frac{y_2}{d_{i-2}^{q^2}} + \frac{1}{d_{i-3}^{q^3}}.$$

Denote the right-hand sides of these equations by $x[i]$ and $y[i]$ respectively. Then the equation $y[i] - x[i-1]^q - (y_2 - x_1^q)x[i] = 0$ simplifies to the first claim.

Similar method, using functional equation of the logarithm instead of the exponential gives the second claim. $\square$

Remark 4.15.6 By these theorems, we see that (see also [Tha92b]) when $g \leq 1$ and $d_\infty = 1$, we can explicitly calculate each of the arithmetically interesting functions (of i) $D_i^{q^g}/d_i$, L_i/l_i, D_i^q/D_{i+1}, $A_{ik}l_i^{q^k}$, $S_i(n)l_i^n$, $S_i(n)/S_i^n$, and each function is of the form (at least for large i) $f(i) = h(x_1^{q^{i-k}}, \cdots, x_n^{q^{i-k}}, x_1, \cdots, x_n)$, for some $h(X_1, \cdots, X_{2n}) \in \mathbb{F}_q(X_1, \cdots, X_{2n})$, where x_i are generators of H over $\mathbb{F}_q$. In other words, basic arithmetically interested quantities such as D_i, L_i, $S_i(n)$ are linked to corresponding quantities coming from the theory of Drinfeld modules, via functions on the Hilbert cover of X cross itself specialized to graphs of (i-th) powers of Frobenius. This has been generalized in principle, but not explicitly, to any A with $d_\infty = 1$ in [And94], and we will also see similar 'interpolation' phenomena, in the context of Shtukas, solitons etc.

4.16 Bernoulli numbers and polynomials

Bernoulli numbers B_n and their close cousins B_n/n occur in many investigations: Power sums, theory of finite differences and sums via the Euler-Maclaurin sum formula; factorials, divergent series, cyclotomic theory, distributions, measures and special values of zeta and L-functions, K-theory, topology. Some good references are the appendix to Milnor's book on K-theory, Borevich-Shafarevich book on Number theory, Lehmer's American Mathematical Monthly paper of December 1988 and references therein.

There is no unanimity over the sign and indexing conventions, but we will mean the ones defined via the generating function identity $z/(e^z - 1) = \sum z^n B_n/n!$. It is easy to see that $B_0 = 1$, $B_1 = -1/2$, $B_{2n+1} = 0$ for $n > 0$.

Let us first recall some classical formulas:

Define Bernoulli polynomials $B_n(x)$ via

$$\sum_{n=0}^{\infty} \frac{B_n(x)z^n}{n!} = \frac{ze^{xz}}{e^z - 1},$$

so that comparison with the definition above gives immediately that $B_n(0) = B_n$ and that

$$B_n(x) = \sum_{k=0}^{n} x^k \binom{n}{k} B_{n-k},$$

which is suggestively written (remembered) as $(B + x)^n$. It follows that

$$B_n'(x) = nB_{n-1}(x).$$

This implies, by straight induction on $n > 0$ and integration using the identity above that

$$B_n(x + 1) - B_n(x) = nx^{n-1}.$$

Adding for $x = 0$ to $x = m - 1$, we get telescoping sum

$$\sum_{x=0}^{m} x^n = \frac{B_{n+1}(m + 1) - B_{n+1}(0)}{n + 1} = \frac{B_{n+1}(m) + (n + 1)m^n - B_{n+1}}{n + 1},$$

which was the main sum formula for which Bernoulli numbers were invented. Since, as a polynomial in m, the right-hand side has lowest degree term $B_n m$, we get a p-adic limit formula for B_n:

$$B_n = \lim_{k \to \infty} \left(\sum_{a=1}^{p^k - 1} a^n\right)/p^k.$$

We will use this in the next Chapter.

The $B_n(x)$ can also be characterized as the unique monic polynomial of degree n satisfying the distribution relation (leading to measures)

$$\frac{1}{m} \sum_{k=0}^{m-1} B_n\left(x + \frac{k}{m}\right) = \frac{B_n(mx)}{m^n}.$$

It also satisfies

$$B_n(1 - x) = (-1)^n B_n(x).$$

Following Carlitz, we can similarly attach Bernoulli numbers and polynomials (see [Car35; Car40a; Car41] and [Gos78; Gek89c]), for example to a Hayes module (sgn-normalized rank one Drinfeld module) via $z/e(z) = \sum z^n B_n/n!$, and $e(zx)/xe(z) = \sum z^n B_n(x)/n!$, if we decide on the notion of factorials to be used. Let us consider another variant of Bernoulli polynomials $\overline{B}_n(x)$ defined, more in analogy with the classical case, by $ze(zx)/e(z) = \sum z^n \overline{B}_n(x)/n!$.

Here we concentrate only on the Carlitz module i.e., the $\mathbb{F}_q[t]$ case, where all the analogies point to the Carlitz factorial as we have seen.

After a few remarks on Bernoulli polynomials, we will return to more details on B_n's: With Carlitz definition, we have

$$B_n(x) = \sum_{q^i \leq n+1} \frac{n!}{q^i!(n - q^i + 1)!} B_{n-q^i+1} x^{q^i-1}$$

and $B_n(0) = B_n$ and that $B_n(x)$ and B_n are identically zero, unless n is 'even'. On the other hand, while $\overline{B}_n(x)$ does not have these properties, unlike $B_n(x)$, it is $\mathbb{F}_q$-linear polynomial given by

$$\overline{B}_n(x) = \sum_{q^i \leq n} \frac{n!}{q^i!(n - q^i)!} B_{n-q^i} x^{q^i},$$

in mixed analogies with classical one at least when $n = q^k$. The two variants are immediately seen to be related by $\overline{B}_n(x) = xB_{n-1}(x)n!/(n - 1)!$. This implies that $\overline{B}_n(x)$ is the same as $xB_{n-1}(x)$ unless $n \equiv q \mod q(q-1)$ and also that $\overline{B}_{q^k} = L_k(xB_{q^k-1})$. The real advantage of $\overline{B}_n(x)$ will be clear in 4.16.2 with its **differential equation**.

Since $e(z)$ is linear, we also get **reflection formula** $\overline{B}_n(1-x) = -\overline{B}_n(x)$ for $n \neq 1$, compared to the classical $B_n(1 - x) = (-1)^n B_n(x)$.

If we ask for the **distribution relation** analog

$$\frac{1}{a} \sum_{\deg a' < \deg a} B_n(x + \frac{a'}{a}) = \frac{B_n(ax)}{a^n},$$

then linear polynomials like $\overline{B}_n(x)$ will certainly not satisfy them, since linearity makes x dependence drop out as number of terms is multiple of characteristic. But in fact, there is no such polynomial B_n because the right-hand side vanishes for large degree of a, because once you expand in powers, the power sums $\sum (a')^j$ vanish eg. by 5.3.1 when j is 'even' and by replacing a' by $\theta a'$ with θ being a generator of $\mathbb{F}_q^*$, if j is not 'even'. So we do not have such characteristic p-valued distribution polynomials, but

Rosen, Galovich, Yin [GR81b; Yin00] etc. have looked at characteristic zero valued distributions in function field contexts.

Finally, we record the recursion formulas that follow from comparison of coefficients in $(ze(xz)/e(z))e(z) = ze(xz)$:

$$\overline{B}_{q^m}(x) = x^{q^m} - \sum_{j=1}^{m} \text{Binom}(q^m, q^j)\overline{B}_{q^m+1-q^j}(x),$$

$$\overline{B}_n(x) = -\frac{n!}{(n+1)!} \sum_{1<q^j\leq n+1} \text{Binom}(n+1, q^j)\overline{B}_{n+1-q^j},$$

where n is not a q-power and $\text{Binom}(a, b)$ is temporary notation for $a!/(b!(a-b)!)$.

Let us now focus on Bernoulli numbers: Equating coefficients of z^n in $(z/e(z))e(z) = z$ gives $\sum B_m \delta_{n-m}/(m!(n-m)!) = 0$, for $n > 1$ where we temporarily write $\delta_i = 1$ if i is a power of q and 0 otherwise. This gives recursion formula $B_n/n! = -\sum B_{n-(q^i-1)}/(q^i!(n-(q^i-1))!)$. Another way to see this is just (for $n \neq 1$) 'sum of coefficients of $\overline{B}_n(x)$ is zero', as follows from $\overline{B}_n(x+1) - \overline{B}_n(x) = 0$. Note that we saw that classical $B_n(x+1) - B_n(x) = nx^{n-1}$ responsible for power sum formula not only fails, but would have been useless anyway, as adding 1 a few times brings you back in characteristic p and you do not get telescoping power sums anyway.

Alternately, solving this system of equations expresses $(-1)^n B_n/n!$ as the determinant of the n cross n matrix whose k-th row is the vector $(\delta_{k+1}/(k+1)!, \delta_k/k!, \cdots, \delta_2/2!, 1, 0, \cdots, 0)$. (The classical counterpart is the same except all the delta's are always one.)

In a recent preprint of Jeong, Kim, Park and Son

$$B_n = n! \sum_{j=1}^{n} (-1)^j \sum_{q^{e_i}>1, \sum(q^{e_i}-1)=n} \frac{1}{q^{e_1}! \cdots q^{e_j}!},$$

which is more compact formula of similar flavor, is proved using Hasse derivatives, as analog of classical

$$B_n = n! \sum_{j=1}^{n} (-1)^j \sum_{e_i>1, \sum(e_i-1)=n} \frac{1}{e_1! \cdots e_j!}.$$

We see $B_{p^m k}/(p^m k)! = (B_k/k!)^{p^m}$, by taking p-th power of the generating function and comparing with substituting z with z^p.

The **von-Staudt theorem** identifies the fractional part of the Bernoulli number B_n for even $n > 0$: $B_n - \sum 1/p$ is integer, where the sum is over the primes p with $(p-1)$ dividing n. Hence the denominator of B_n is the product of such primes and is thus always divisible by 6, for example. The **Lipschitz-Sylvester theorem** says $a^n(a^n - 1)B_n/n$ is integer for any integer a. It follows that $p-1$ divides n when p divides denominator of B_n. It also follows easily that primes dividing denominators of B_n or of B_n/n are exactly the same and hence gives formula for the denominator of B_n/n recalled in 4.16.6.

Let us turn to $\mathbb{F}_q[t]$ case: Again $B_n = 0$ unless n is 'even'. So we will assume below that n is even. Now B_n is a rational function in $\mathbb{F}_q(t)$ and in general B_n/n will not make good sense. We think of $B_n(n-1)!/n!$ as an analog of B_n/n instead. We use short-form P_h for the product of monic primes of degree h. Since $[n] = \prod_{d|n} P_d$, by Mobius inversion, $P_h = \prod_{d|h} [d]^{\mu(h/d)}$.

Theorem 4.16.1   *(Carlitz)* $B_{q^h - q^i} = (-1)^{h-i}(q^h - q^i)!/L_{h-i}^{q^i}$.

Proof.   We prove this by induction on $q^h - q^i$, the case $h = 0$ being clear. Equating coefficients in $z/e(z)e(z) = z$ and using p-action gives

$$\frac{B_{q^h - 1}}{(q^h - 1)!} = -\sum_{i=1}^{h} \frac{B_{q^h - q^i}}{((q^h - q^i)!(q^i)!)} = -\sum \frac{(-1)^{h-i}}{(L_{h-i}^{q^i}(q^i)!)} = \frac{(-1)^h}{L_h}$$

where the last equality follows by equating coefficients in $e(l(z)) = z$.   □

Remarks 4.16.2   (1) Hence, putting these values in the formula for Bernoulli polynomials in terms of the Bernoulli numbers, we see that $\overline{B}_{q^n}$ is just the monic polynomial $q^n!\{ \frac{x}{q^n} \}$. Hence,

$$\Delta \overline{B}_{q^n}(x) = [n]\tau \overline{B}_{q^{n-1}}.$$

This can be considered as an analog of the classical identity $(xd/dx)B_n(x) = nxB_{n-1}(x)$ and shows why (in this respect) linear $\overline{B}_n(x)$ is better behaved than Carlitz $B_n(x)$. (See 6.5 and 4.14 for more on these analogies.)

This nice behavior and simple recursion for q^n's is similar to that of factorials, binomials, Gauss sums, but we do not see any 'digit principle' building all $\overline{B}_n(x)$'s from these 'basic' ones.

(2) $B_{q^h - 1} = (q^h - 1)!/L_h = D_h/L_h = [h - 1]^{q-2} \cdots [1]^{q^{h-2}-2}/[h]$, so that the denominator is P_h unless $q = 2$ and

$h = 2$, when we have $B_3 = 1/[2]$ and hence in that case, the denominator is $P_1 P_2$.

(3) We see that unlike in the classical case, n being divisible by Norm$(\wp)-1$ is not sufficient to imply that $\wp$ divides the denominator of B_n. Otherwise, for example [1] would have always divided these denominators.

(4) Taking $q = 2$, $h = 3$ and $i = 1$, we see that B_6 has denominator $P_2 = [2]/[1] = t^2 + t + 1$, whereas $B_6 5!/6!$ has denominator [2], divisible by degree one primes also. So again unlike the classical case, at least for $q = 2$, the primes occurring in the denominators of B_n and $B_n(n-1)!/n!$ are not always the same. We leave it to the reader to verify the same conclusion for $n = 10$, 18 and to find additional such examples.

Theorem 4.16.3 *For any $a \in A = \mathbb{F}_q[t]$, $a^n(a^n - 1)B_n(n-1)!/n!$ is integers, i.e., is in A.*

Proof. We have $e(az) = ae(z) + \sum a_i e(z)^{q^i}$, with a_i integral. So

$$\sum (a^n - 1)B_n \frac{z^n}{n!} = \frac{az}{e(az)} - \frac{z}{e(z)} = z(-\sum a_i e(z)^{q^i - 2})/(a + \sum a_i e(z)^{q^i - 1}).$$

Following classical analogy, let us call Hurwitz series any power series of the form $\sum c_n z^n/n!$ with c_n integral, for example $e(z)$. The numerator and the denominator of the right side are integral linear combinations of Hurwitz series. The divisibility properties (namely integrality of binomial coefficients $(a+b)!/(a!b!))$ of factorial implies that product of Hurwitz series is also a Hurwitz series and the same for ratio if the denominator has constant coefficient 1 (since then we can develop in geometric series). To apply the last property we replace z by az to get Hurwitz series $\sum (a^n - 1)a^n(B_n(n-1)!/n!)z^{n-1}/(n-1)!$, implying the claim, i.e., an analog of Lipschitz-Sylvester theorem. $\qquad\square$

By taking a to be a primitive root modulo $\wp$, we see that if $\wp$ divides the denominator of B_n (or of $B_n(n-1)!/n!$), then Norm$(\wp) - 1$ divides n. We have already seen that the converse is false in our case. Instead Carlitz proved following more complicated **analog of von-Staudt Clausen theorem**:

Theorem 4.16.4 *Let $A = \mathbb{F}_q[t]$, with $q = p^n$, p a prime. Let $m = \sum m_i p^i$ $(0 \le m_i < p)$ be 'even' and let $d(m)$ denote the (monic) denominator of B_m for A. Let P_k denote the product of monic primes of degree k. Put $h := (\sum m_i)/(n(p-1))$.*

(1) Let $q > 2$. If h is an integer and $q^h - 1$ divides m, then $d(m) = P_h$, otherwise $d(m) = 1$.

(2) Let $q = 2$. If $h \neq 2$ and $q^h - 1$ divides m, then $d(m) = P_h$. If $h = 2$ and $q^h - 1$ divides m, then $d(m) = P_2$ for m a multiple of 2 and $d(m) = [2] = P_2 P_1$ otherwise. If $q^h - 1$ does not divide m, then for m not a multiple of 2 and with $h = 2$ (i.e., of the form $2^k + 1$), $d(m) = P_1$ and otherwise $d(m) = 1$.

Note that if q is prime, then h is clearly an integer, but in general, it may not be. For $q = 2$, $h = l(m)$ is just the number of base 2 digits.

Theorem 4.16.5   *Let $q > 2$. Then, the primes dividing the denominators of B_n or $B_n(n-1)!/n!$ are the same. In fact, if q^k is the maximum power of q dividing n, and if the denominator of B_n is P_h, then the denominator of $B_n(n-1)!/n!$ is $P_h^{\lfloor k/h \rfloor + 1}$. When $q = 2$, the primes dividing the denominators of B_n or $B_n(n-1)!/n!$ are the same, unless n is of the form $2^i + 2^k$, with $k > i > 0$.*

Proof.   Let $\wp$ divide the denominator of $B_n(n-1)!/n!$. For the first part, it is enough to prove that it divides $d(n)$.

(Classically, this follows immediately from von-Staudt and Lipschitz-Sylvester as follows: By choosing a to be a primitive root modulo $\wp$, we see that $\mathrm{Norm}(\wp) - 1$ divides n, which is sufficient by classical von Staudt statement to imply $\wp$ divides $d(n)$. This argument fails here.)

Note that $(n-1)!/n! = 1/L_k$, which is 1 if $k = 0$. So there is nothing to prove in the case $k = 0$. Now we do induction on k.

Note $(qn)!/n!^q = \prod [i+1]^{n_i}$ is integral. Let $\wp$ divide the denominator of $B_{qn}(qn-1)!/(qn)!$. Then by

$$\frac{(qn-1)! B_{qn}}{(qn)!} = \frac{(qn-1)!}{(qn-q)!} \frac{(q(n-1))!}{(n-1)!^q} \left(\frac{(n-1)! B_n}{n!}\right)^q$$

we see that $\wp$ divides the denominator of $B_n(n-1)!/n!$, which by induction hypothesis implies that $\wp$ divides the denominator of B_n. Now when $q > 2$, or under the hypothesis when $q = 2$, the comparison of the von-Staudt statement conditions for n as for qn then implies that $\wp$ divides the denominator of B_{qn}. The second part follows from noting that $L_k = [k] \cdots [1]$ and that $[i]$ is a product of P_j's with j dividing i. $\square$

Remarks 4.16.6   (1) Analogous statement for the classical case is an easy implication of von-Staudt.

(2) We will see denominators of B_n/n coming up again in zeta values and tensor power torsions.

(3) In the classical case the von-Staudt theorems imply that the denominator of B_n/n is given by $\prod_{(p-1)|n,\,p^m||n} p^{m+1}$. Using this, Bhargava has made a very nice calculation that when X is the set of primes in $R = \mathbb{Z}$,

$$n!_X = \prod_p p^{\sum_{e=0}^{\infty} \lfloor (n-1)/(p^e(p-1))\rfloor}$$

is $2^{\lfloor n/2 \rfloor}$ times the product of denominators of first $\lceil n/2 \rceil$ B_k/k's. In $\mathbb{F}_q[t]$ case, however the corresponding factorial is given with a similar formula, with p replaced by Norm($\wp$) in the exponents, but because of the different nature of von-Staudt, there seems to be no easy connection to Bernoulli denominators.

(4) The Kummer congruences (even at the lowest level: $B_{n+p-1}/(n+p-1) \equiv B_n/n \mod p$, for $p-1$ not dividing n) do not hold for $B_n(n-1)!/n!$'s as can be seen from direct calculations with low values. We will see the zeta connection as well as another set of Bernoullis satisfying Kummer congruences in the next chapter.

4.17 Note on finite differences and q-analogs

This section or its notation will not be used anywhere else in this book. In fact, it clashes with notation elsewhere.

We will not try to set up or explain a dictionary, but here are some common q-analogs, which we have taken from [KC02], omitting (usually) easy proofs.

Classical difference calculus deals with difference operator $\Delta f(x) := f(x+1) - f(x)$ or more generally with

$$d_h f(x) := f(x+h) - f(x), \quad D_h f(x) := d_h f(x)/d_h x$$

(without limit $h \to 0$). In q-analogs, (where q is thought of as e^h, $q \to 1$ would be the classical limits), we look at

$$d_q f(x) := f(qx) - f(x), \quad D_q f(x) := d_q f(x)/d_q(x) = (f(qx) - f(x))/((q-1)x).$$

Then D_q and D_h are linear operators and we have $d_q(f(x)g(x)) = f(qx)d_q g(x) + g(x)d_q f(x)$ and similar formula for D_q. There is no nice chain rule in general.

Put $[n] := [n]_q := (q^n - 1)/(q - 1)$, so that $[n]_q \to n$ as $q \to 1$ and

$$D_q x^n = [n]x^{n-1}, \quad D_q x^n/[n]! = x^{n-1}/[n-1]!.$$

Introduce factorials as $[n]! := [1][2] \cdots [n]$, q-binomial coefficients as $[n, j] := [n]!/([j]![n-j]!)$ and q analogs of $(x - a)^n$ as

$$(x - a)^n_q := (x - a)(x - qa) \cdots (x - q^{n-1}a), \quad (x - a)^{-n}_q := 1/(x - a)^n_q.$$

Then

$$(x - a)^{m+n}_q = (x - a)^m_q (x - q^m a)^n_q, \quad D_q(x - a)^n_q = [n](x - a)^{n-1}_q.$$

We have q-Taylor formula for polynomials f:

$$f(x) = \sum (D^j_q f)(c) \frac{(x - c)^j_q}{[j]!}.$$

This gives Gauss formula $(x + a)^n_q = \sum_{j=0}^{n} q^{j(j-1)/2}[n, j]a^j x^{n-j}$.

When q is a prime power, $[n, j]$ is the number of j-dimensional subspaces in the n-dimensional vector space $\mathbb{F}^n_q$.

We have $[n, j] = [n-1, j-1] + q^j[n-1, j] = q^{n-j}[n-1, j-1] + [n-1, j]$, for $1 \le j \le n - 1$. Also, $[m + n, k] = \sum_{j=0}^{k} q^{(k-j)(m-j)}[m, j][n, k - j]$.

We have q-exponential

$$e^x_q := \sum_{j=0}^{\infty} x^j/[j]! = 1/(1 - (1 - q)x)^{\infty}_q, \quad D_q e^x_q = e^x_q.$$

We have $e^x_q e^y_q = e^{x+y}_q$, if $yx = qxy$, but not in general. Under this commutation condition, we also have $(x + y)^n = \sum[n, j]x^j y^{n-j}$.

For $|q| < 1$, the limit of $[n]$ as $n \to \infty$ is $1/(1 - q)$.

There are Euler, Jacobi, Ramanujan identities and connections with partition functions.

We have q-hypergeometric series $\sum c_n x^n$, where c_{n+1}/c_n is a rational function $R(q^n)$ of q^n. With $R(t) = (a_1 - t^{-1}) \cdots (a_r - t^{-1})/((b_1 - t^{-1}) \cdots (b_s - t^{-1})(q - t^{-1}))$, with $a_i \ne b_j$ and $b_j \ne q^{-s}$, we have

$$c_n = ((-1)^n q^{n(n-1)/2})^{s-r+1} \frac{(1 - a_1)^n_q \cdots (1 - a_r)^n_q}{(1 - b_1)^n_q \cdots (1 - b_s)^n_q (1 - q)^n_q}.$$

For $x > 0$, we have (analogous to p-adic integration, with q corresponding to $1/p$) Jackson integral

$$\int_0^x f(x)d_q(x) = (1 - q) \sum_{j=0}^{\infty} q^j f(q^j x)$$

subtracting such we can define integral from a to b, and then $\int_a^b D_q f(x)d_q x = f(b) - f(a)$. Sometimes $E^x_q := e^x_{1/q} = \sum q^{j(j-1)/2}x^j/[j]!$

is also looked upon as analog of exponential, so that for $t > 0$, we have integral formula

$$\Gamma_q(t) = \int_0^\infty x^{t-1} E_q^{-qx} d_q x = (1-q)_q^\infty / ((1-q)^{t-1}(1-q^t)_q^\infty),$$

for q-gamma function satisfying $\Gamma_q(t+1) = [t]\Gamma_q(t)$ and interpolation $\Gamma_q(n+1) = [n]!$.

For the finite difference calculus, we have $D_h(f(x)g(x)) = f(x)D_h g(x) + g(x+h)D_h f(x)$. We have

$$(x-a)_h^n := (x-a)(x-a-h)\cdots(x-a-(n-1)h), \quad D_h(x-a)_h^n = n(x-a)_h^{n-1}$$

and h-Taylor formula $f(x) = \sum (D_h^j f)(c)(x-c)_h^j/j!$ for polynomials f. Applied to the power function, we get $(x+b)^N = \sum \binom{N}{j} b_h^{N-j} x_h^j$. Exponential analog is $e_h^x := (1+h)^{x/h}$, which is its own D_h derivative. The h-definite integral is essentially equal width ($=h$) partition Riemann sum and we get the fundamental theorem of calculus in this case by telescoping.

One version of Euler-Maclaurin sum formula, valid for many functions and used often to accelerate convergence, is

$$\sum_{i=a}^{b-1} f'(i) = \sum_{n=0}^\infty \frac{B_n f^{(n)}(t)}{n!} \Big|_a^b.$$

It follows from the fact that the sum operator $\sum$ and the finite difference operator Δ, which is $e^{d/dx} - 1$ by the Taylor's theorem, are inverses of each other (by telescoping).

Here are creation-annihilation operators analogs (see [Koc96] for this, other versions, including q-analog case and references) mentioned in Section 15:

In Schrödinger representation, the creation-annihilation operators are unbounded operators on $L^2(\mathbb{R})$, given by $a^\pm := 2^{-1/2}(t + d/dt)$. They satisfy $[a^-, a^+] = 1$ and for Hermite functions ψ_n, we have

$$a^-\psi_n = \sqrt{n}\psi_{n-1}, \quad a^+\psi_n = \sqrt{n+1}\psi_{n+1}, \quad a^+a^-\psi_n = n\psi_n.$$

In Bargmann-Fock representation, the creation-annihilation operators are given on Hilbert space of entire functions of z by $\tilde{a}^+$ being multiplication by z and $\tilde{a}^-$ being derivative with respect to z. Hence $[\tilde{a}^-, \tilde{a}^+] = 1$ and for the basis $\psi_n = z^n$, we have

$$\tilde{a}^-\psi_n = n\psi_{n-1}, \quad \tilde{a}^+\psi_n = \psi_{n+1}, \quad \tilde{a}^+\tilde{a}^-\psi_n = n\psi_n.$$

In the p-adic case, on the continuous functions from $\mathbb{Z}_p$ to $\mathbb{Q}_p$, we have bounded operators $(a^+ f)(x) = x f(x-1)$ and $(a^- f)(x) = f(x+1) - f(x)$ with the same commutation relation and such that on Mahler basis $\psi_n = \binom{x}{n}$, we have

$$a^- \psi_n = \psi_{n-1}, \quad (a^- \psi_0 = 0), \quad a^+ \psi_n = (n+1)\psi_{n+1}, \quad a^+ a^- \psi_n = n\psi_n.$$

Greg Anderson (unpublished, but see [And94]) has developed better symmetric function theory, Schur functions etc. in this setting parallel to Macdonald's book treatment.

For more sophisticated dictionary in line with 7.7-7.8, see [Mum78].

Chapter 5

Zeta functions

The Riemann zeta function and its generalizations in various contexts are some of the most important functions in number theory.

In various contexts, the zeta functions have simple definitions (valid at least in some region) in terms of basic building blocks, such as integers or primes in the Riemann zeta function case, of the arithmetic structures. Nonetheless, they contain very interesting information about the deeper aspects of the structure in their special or leading values, congruences they satisfy, their analytic nature, functional equations, locations and multiplicities of their zeros and poles etc.

For example, divisibility information of Bernoulli numbers which essentially occur as special values of the Riemann zeta function gives information on class groups of cyclotomic fields. Its pole at $s = 1$ implies infinitude of primes, its non-vanishing on line $Re(s) = 1$ implies the prime number theorem and the Riemann hypothesis has immense number of well-known consequences. The analytic class number formula has orders of the class group, roots of unity, rank and finer information on unit group entering into it. The conjectured analytic properties and functional equations or Artin's entireness conjecture for Artin L-functions would imply Langlands conjectures for GL_2. Multiplicities at certain critical points give arithmetic information about ranks or cycles via Birch and Swinnerton-Dyer or Tate conjectures. There is a huge conjectural framework of Deligne, Beilinson, Bloch-Kato, Perrin Riou-Fontaine conjectures covering and extending special cases understood over more than hundred years, starting in fact, with Euler's determination of $\zeta(2n)$ and Dirichlet's class number formula.

In the function field case, the story of special zeta values in finite characteristics, studied initially by Carlitz and Goss, is still in the initial stages. We will see examples of very interesting special value results in some cases with some hints of rich underlying structures. We will see some results on analytic properties and zero distributions, but we do not know how to draw interesting consequences from those yet. We do not see if and how the Gauss sums and gamma functions should be a part of zeta story as in the classical case. We do not see any functional equations type results, hence we have separate special values results at positive and negative integers. Some of these special values results hint at some kind of relations between the two sets. This is somewhat similar to another non-archimedean situation of p-adic L-functions, where we do not have interesting functional equations and there are, for example, separate Stark type conjectures [Tat84, Chapter6] at $s = 0$ and at $s = 1$.

The story of multiplicities of trivial zeros (which classically is just a consequence of functional equations) is quite interesting implying that structural framework would be quite different.

In short, there are many interesting results, some analogous and some apparently quite different in the view of established analogies. The general situation is not even conjecturally understood, making the exploration even more exciting.

Rather than introducing new notation each time, we will use the similar notation for Riemann-Dedekind, for Artin-Weil or for Goss zeta functions. The context will make the use clear.

5.1   Zeta values at integers: Definitions

Let us start with the simplest zeta function: the Riemann zeta function $\zeta(s) = \sum n^{-s}$, where the sum is over positive integers and s is a complex variable with real part greater than one. For function field analogy, Artin considered n as the norm (which is the number of residue classes) of the ideal $n\mathbb{Z}$, as we consider also in the case of the Dedekind Zeta function, and defined his zeta function as $\sum \mathrm{Norm}(I)^{-s}$, where the sum is over nonzero ideals and s is again complex variable, with $\mathrm{Re}(s) > 1$. The sum is a complex number as in the Riemann zeta case. We have recalled the well-developed theory of these types of zeta functions in the first chapter. They satisfy many analogies. But they are simple rational functions of $t = q^{-s}$-variable.

The norm depends only on the degree and we thus ignore all the information except the degree. We can take the actual ideal or polynomial or a relative norm to the base instead: Carlitz used another function field analogy and considered special zeta values $\zeta(s) := \sum n^{-s}$, where now n runs through monic polynomials in $\mathbb{F}_q[t]$, i.e., monic generators of non-zero ideals. (With this basic idea, we can then consider various zeta and L-values, for example, those attached to finite characteristic valued representations by using product of characteristic polynomials of 'Frobenius at $\wp$ multiplied by $\wp^{-s}$'. We refer to [Gos96] and references therein for various such definitions.)

'Monic' is a sign condition similar to 'positive'. But note that positivity is closed under addition or multiplication, whereas monicity while closed under multiplication is not closed under addition in general.

Now polynomials can be raised to integral powers, and in particular, if s is a natural number, then the sum converges (as the terms tend to zero) to a Laurent series in K_∞.

Goss showed that if s is non-positive integer, then just grouping the terms for the same degree together, the sum reduces to a finite sum giving $\zeta(s) \in A$.

Example 5.1.1   For $A = \mathbb{F}_3[t]$, we have $\zeta(0) = 1 + 3 + 3^2 + \cdots = 1 + 0 + 0 + \cdots = 1$ and $\zeta(-3^n) = \zeta(-1)^{3^n} = 1$ because

$$\zeta(-1) = 1 + (t + (t+1) + (t-1)) + \cdots = 1 + 0 + \cdots = 1.$$

For $A = \mathbb{F}_2[t]$ on the other hand, $\zeta(-2^n) = 0$ because $\zeta(-1) = 1 + (t + (t + 1)) + 0 + \cdots = 1 + 1 = 0$. We leave it to the reader to verify $\zeta(-5) = 1 + t - t^3$ for $A = \mathbb{F}_3[t]$ using either the bound in the lemma below or the formula below.

In fact, the weaker version consisting of the first 3 lines of the proof of the following Theorem, which is essentially due to Lee, implies that grouped terms vanish for large degree. (See [Tha90; Tha95b] for the relevant history.)

For a non-negative integer $k = \sum k_i q^i$, with $0 \le k_i < q$, we let $\ell(k) := \sum k_i$, i.e., $\ell(k)$ is the sum of base q digits of k.

Theorem 5.1.2   *Let W be a $\mathbb{F}_q$-vector space of dimension d inside a field (or ring) $\mathcal{F}$ over $\mathbb{F}_q$. Let $f \in \mathcal{F} - W$. If $d > \ell(k)/(q-1)$, then $\sum_{w \in W}(f + w)^k = 0$.*

Proof.   Let $w_1, \cdots, w_d$ be a $\mathbb{F}_q$-basis of W. Then $(f+w)^k = (f + \theta_1 w_1 + \cdots + \theta_d w_d)^k$, $\theta_i \in \mathbb{F}_q$. When you multiply out the k brackets, terms involve at most k of θ_i's, hence if $d > k$, the sum in the theorem is zero, since we are summing over some θ_i, a term not involving it, and $q = 0$ in characteristic *p*. The next observation is that in characteristic p, $(a+b)^k = \prod (a^{q^i} + b^{q^i})^{k_i}$, hence the sum is zero, if $d > \ell(k)$, by the argument above. Finally, note that $\sum_{\theta \in \mathbb{F}_q} \theta^j = 0$ unless $q - 1$ divides j. Expanding the sum above by the multinomial theorem, we are summing multiples of products $\theta_1^{j_1} \cdots \theta_d^{j_d}$ and hence the sum is zero; because the sum of the exponents being $\ell(k) < (q-1)d$, all the exponents can not be multiples of $q - 1$. $\square$

For the applications, note that $A_{<i} = \{a \in A : \deg(a) < i\}$ form such $\mathbb{F}_q$ vector spaces whose dimensions are given by the Riemann-Roch theorem. Similarly, $A_{i+} = \{a \in A : \deg(a) = i, a \text{ monic}\}$ are made up of affine spaces as in the theorem.

In Section 5, we will interpolate Zeta on much bigger space, but if we just focus on the special values at integers, we can ignore the convergence questions etc. and just use the Theorem above. We can then work in the following simpler set-up.

If d_∞ is not necessarily one, we have to be careful about the signs, as they are no longer in the field of constants of K. Fix a sign function $\text{sgn} : K_\infty^* \to \mathbb{F}_\infty^*$.

(In examples, we usually take sgn such that x, y, t have sgn 1.) Now let S be a set of representatives of $\mathbb{F}_\infty^*/\mathbb{F}_q^*$.

We generalize our earlier definitions from 2.5: Let $A_+ := \{a \in A : \text{sgn}(a) \in S\}$. Define A_{i+} similarly. Elements of $\text{sgn} = 1$ are called 'positive' or 'monic'. By deg we will denote the residue degree over $\mathbb{F}_q$. By convention $\deg(0) = -\infty$.

Let H be the Hilbert class field of A. Recall that $K = H$ if $hd_\infty = 1$. Let L be a finite separable extension of K and let $\mathcal{O}_L = \mathcal{O}$ denote the integral closure of A in L.

Now we define the relevant zeta functions.

Definition 5.1.3   For $s \in \mathbb{Z}$, define the 'absolute zeta function':

$$\zeta(s, X) := \zeta_A(s, X) := \sum_{i=0}^{\infty} X^i \sum_{a \in A_{i+}} \frac{1}{a^s} \in K[[X]]$$

$$\zeta(s) := \zeta_A(s) := \zeta(s, 1) \in K_\infty.$$

By Theorem 5.1.2, $\zeta(s) \in A$ for integer $s \leq 0$. The results on the special values and the connection with Drinfeld modules later on, justifies the use of elements rather than ideals even when the class number is more than one. Also, we can look at a variant where a runs through elements of some ideal of A or we can sum over all ideals by letting s to be a multiple of class number and letting a^s to be the generator of I^s with appropriate sign. In the latter case, we have Euler product for such s, if $d_\infty = 1$.

If L contains H (note that this is no restriction if $hd_\infty = 1$, eg., for $A = \mathbb{F}_q[t]$), then it is known that the norm of an ideal $\mathcal{I}$ of $\mathcal{O}$ is principal. Let Norm$\mathcal{I}$ denote the generator whose sign is in S. Let us now define the relative zeta functions in this situation.

Definition 5.1.4   Put, for $s \in \mathbb{Z}$,

$$\zeta_\mathcal{O}(s, X) := \zeta_{\mathcal{O}/A}(s, X) := \sum_{i=0}^{\infty} X^i \sum_{\deg(\text{Norm } \mathcal{I})=i} \frac{1}{\text{Norm } \mathcal{I}^s} \in K[[X]]$$

$$\zeta_\mathcal{O}(s) := \zeta_{\mathcal{O}/A}(s) := \zeta_\mathcal{O}(s, 1) \in K_\infty.$$

The same remark as above applies to the second definition.

Finally, we define the vector valued zeta function, which generalizes both definitions above and works without assuming that L contains H. We leave the simple task of relating these definitions to the reader.

Definition 5.1.5   Let C_n ($1 \leq n \leq hd_\infty$) be the ideal classes of A and choose an ideal I_n in C_n^{-1}. For $s \in \mathbb{Z}$, define the vector $Z_\mathcal{O}(s, X)$ by defining its n-th component $Z_\mathcal{O}(s, X)_n$ via

$$Z_\mathcal{O}(s, X)_n := Z_{\mathcal{O}/A}(s, X)_n := \sum_{i=0}^{\infty} X^i \sum \frac{1}{(I_n \text{Norm } \mathcal{I})^s} \in K[[X]]$$

where I_nNorm$\mathcal{I}$ stands for the generator of this ideal with sign in S and the second sum is over $\mathcal{I}$ whose norm is in C_n and is of degree i.

$$Z_\mathcal{O}(s)_n := Z_{\mathcal{O}/A}(s)_n := Z_\mathcal{O}(s, 1)_n \in K_\infty.$$

Again the same remark as above applies to the second definition.

Also note that Z depends very simply on the choice of I_n's, with components of Z, for different choices of I_n's, being non-zero rational multiples of each other. In particular, the questions of when it (i.e., all the components) vanishes, when it is rational or algebraic etc. are independent of such choice.

We use X as a deformation parameter for otherwise discretely defined zeta values and hence we define the order of vanishing ord$_s$ of ζ, $\zeta_\mathcal{O}$ or $Z_\mathcal{O}$ at s to be the corresponding order of vanishing of $\zeta(s, X)$, $\zeta_\mathcal{O}(s, X)$

or $Z_O(s, X)$ at $X = 1$. This procedure is justified by the results described below. The order of vanishing is the same for s and ps, as the characteristic is p.

5.2 Values at positive integers

Clearly $\zeta(k)$ can be given by Euler product. Observe that, in contrast to the classical case where we have a pole at $s = 1$, here $\zeta(1)$, which can now be considered as an analog of Euler's constant γ, makes sense.

Euler's famous evaluation, $\zeta(m) = -B_m(2\pi i)^m/2(m!)$ for even m, has following analog:

Theorem 5.2.1 *(Carlitz [Car35]): Let $A = \mathbb{F}_q[t]$. Then for 'even' m (i.e., a multiple of $q - 1$), $\zeta(m) = -B_m\tilde{\pi}^m/(q - 1)\Pi(m)$.*

Proof. First note that $q - 1 = -1$ in the formula. Multiplying the logarithmic derivative of the product formula for $e(z)$ by z, we get

$$\frac{z}{e(z)} = 1 - \sum_{\lambda \in \Lambda - \{0\}} \frac{z/\lambda}{1 - z/\lambda} = 1 - \sum_{n=1}^{\infty}\sum_{\lambda}(\frac{z}{\lambda})^n = 1 + \sum_{n \text{ even}} \frac{\zeta(n)}{\tilde{\pi}^n}z^n$$

since $\sum_{c \in \mathbb{F}_q^\times} c^n = -1$ or 0 according as n is 'even' or not. But $z/e(z) = \sum B_n z^n/\Pi(n)$. □

Next we give the connection of these Bernoulli numbers with the class groups of cyclotomic fields, giving some analogs of Herbrand-Ribet theorems. The story will continue in the next section.

Let $\wp$ be a monic prime of A of degree d. Recall analogies

$$\Lambda_\wp =: e(\tilde{\pi}/\wp) \leftrightarrow \zeta_p =: e^{2\pi i/p}, \quad K(\Lambda_\wp) \leftrightarrow \mathbb{Q}(\zeta_p).$$

One also has 'maximal totally real' subfield

$$K(\Lambda_\wp)^+ =: K\left(\prod_{\theta \in \mathbb{F}_q^*} e\left(\frac{\theta\tilde{\pi}}{\wp}\right)\right) \leftrightarrow \mathbb{Q}(\zeta_p)^+ =: \mathbb{Q}\left(\sum_{\theta \in \mathbb{Z}^*} e^{\theta 2\pi i/p}\right).$$

Definition 5.2.2 Let C (C^+, $\tilde{C}$, $\tilde{C}^+$ resp.) denote the p-primary component of the class group of $K(\Lambda_\wp)$ ($K(\Lambda_\wp)^+$, ring of integers of $K(\Lambda_\wp)$, of $K(\Lambda_\wp)^+$ respectively).

Let W be the ring of Witt vectors of $A/\wp$. Then we've decomposition into isotypical components $C \otimes_{\mathbb{Z}_p} W = \oplus_{0 \leq k < q^d - 1} C(w^k)$ according to characters of $(A/\wp)^*$, where w is the Teichmüller character. (Similarly for $C^+, \tilde{C}, \tilde{C}^+$.)

Remark 5.2.3 Let $0 < k < q^d - 1$. It is easy to see that (a) if k is 'odd', then $c(w^k) \cong \tilde{c}(w^k)$, and $c^+(w^k)$ and $\tilde{c}^+(w^k)$ are trivial, (b) if k is 'even', then $c(w^k) \cong c^+(w^k)$ and $\tilde{c}(w^k) \cong \tilde{c}^+(w^k)$.

Theorem 5.2.4 *(Okada, Goss [Oka91; Gos96]): Let $A = \mathbb{F}_q[T]$. Then for $0 < k < q^d - 1$, k 'even', if $\tilde{C}(w^k) \neq 0$, then $\wp$ divides B_k.*

Proof. (Sketch) We define analogs of Kummer homomorphisms $\psi_i \colon \mathcal{O}_F^* \to A/\wp$ ($0 < i < q^d - 1$) by $\psi_i(u) = u_{i-1}$, where u_i is defined as follows. Let $u(t) \in A[[t]]$ be such that $u = u(\lambda)$ and define u_i to be $\Pi(i)$ times the coefficient of z^i in the logarithmic derivative of $u(e(z))$. Using the definition of the Bernoulli numbers, we calculate that the i-th Kummer homomorphism takes the basic cyclotomic unit $\lambda^{\sigma_a - 1}$ to $(a^i - 1)B_i/\Pi(i)$. If $\tilde{C}(w^k) \neq 0$, then by the component-wise version of Theorem 3.11.1 ('Gras conjecture': see remark after 5.3.7), $\psi_k(\lambda^{\sum w^{-k}(\sigma)\sigma^{-1}}) = 0$ and hence the calculation above implies that $\wp$ divides B_k. $\square$

Using her generalization of Theorem 3.11.1, Shu [Shu94b; Shu94a] has announced a generalization of this Theorem to any A. The converse is false and Gekeler has suggested the following modification (see also [Ang01]) using the Frobenius action restriction:

Conjecture (Gekeler [Gek90]) *If $\wp|B_{k'}$ for all k' such that $k' \equiv p^m k(q^d - 1)$ with $0 < k$, $k' < q^d - 1$, $k \equiv 0(q - 1)$ then $\tilde{C}(w^k) \neq 0$.*

We now look at the values of the zeta functions at the positive integers. The results obtained are analogs of the classical results with the base being $\mathbb{Q}$ (with $2\pi i$ as the relevant period) or an imaginary quadratic field (with the relevant period being the period of elliptic curve with complex multiplication by the field. In this second case, similar results for the absolute and relative zeta values may be known, but we do not know a suitable reference).

Let $\tilde{\pi}$ be a fundamental period of a sgn-normalized rank one Drinfeld A module ρ with the lattice corresponding to ρ being $\tilde{\pi}A$. This determines $\tilde{\pi}$ up to multiplication by an element of $\mathbb{F}_q^*$. This is an analog of $2\pi i$.

Theorem 5.2.5 *(Goss [Gos]) Let ρ be a sgn-normalized rank one Drinfeld module with corresponding lattice $\tilde{\pi}_\rho I_\rho$, where I_ρ is an ideal of A. Let J be an ideal of A, $\Lambda_{\rho,J}$ be the J-torsion of ρ and let $\alpha \in J^{-1}I_\rho$. Then for* *s a positive integer, we have*

$$M_s := \tilde{\pi}_\rho^{-s} \sum_{\substack{i \in I_\rho \\ \alpha+i \neq 0}} (\alpha + i)^{-s} \in H_1(\Lambda_{\rho,J}).$$

Proof. We will drop the subscripts ρ in the proof. Let $e(z)$ be the exponential function corresponding to ρ. It has coefficients in H_1. Also, $e(\alpha\tilde{\pi})$ is a J-torsion point of ρ and hence belongs to $H_1(\Lambda_J)$. Hence, if we write $1/e(z + \alpha\tilde{\pi}) = \sum c_n z^n$, then $c_n \in H_1(\Lambda_J)$. Now $e(z) = z \prod(1 - z/(\tilde{\pi}i))$, where the product runs over $i \in I - \{0\}$. Taking the logarithmic derivative, if $\alpha \notin I$, so that $\alpha + i \neq 0$, we get

$$\frac{1}{e(z+\alpha\tilde{\pi})} = \frac{1}{e(z)+e(\alpha\tilde{\pi})} = \sum_{i\in I} \frac{1}{z+\tilde{\pi}(\alpha+i)} = -\sum_{n=0}^{\infty}\sum_{i\in I} \frac{z^n}{(\tilde{\pi}(\alpha+i))^{n+1}}.$$

We get a similar equality if we multiply both sides by z to get rid of the pole at 0, if $\alpha \in I$. Comparison of the coefficients proves the theorem. $\square$

As an analog of Euler's theorem, generalizing in the straight fashion the much more precise theorem, Theorem 5.2.1 in the case $A = \mathbb{F}_q[t]$, we have,

Theorem 5.2.6 *Let s be a positive 'even' (i.e., a multiple of $q-1$) integer, then $\zeta_A(s)/\tilde{\pi}^s \in H_1$.*

Proof. The special case $\alpha = 0$ and $J = I = A$ of the previous theorem shows that $\tilde{\pi}^{-s} \sum a^{-s} \in H_1$, where a runs through the nonzero elements of A. By looking at the signs, we see that the sum is zero, if s is 'odd' and is $-\zeta(s)$, if s is 'even'. $\square$

Note that if we use a period of a Drinfeld module defined over H instead, the ratio belongs to H.

Example 5.2.7 By comparing the coefficients of z^{q-1} in the equation above, we get $\zeta(q-1)/\tilde{\pi}^{q-1} = -1/d_1$.

Now we turn to the relative zeta functions.

Theorem 5.2.8 *Let L be an abelian totally real (i.e., split completely at ∞) extension of degree d of K containing H and let s be a positive even (i.e., a multiple of $q-1$) integer. Then $R_s := \zeta_\mathcal{O}(s)/\tilde{\pi}^{ds}$ is algebraic. Furthermore, $R_s \in K(\Lambda_J)$ in the notation introduced below, $R_s^2 \in H_1$ and $R_s^{2n} \in H$ if $q^{d\infty} - 1$ divides $2dsn$.*

Proof. Let $G = \{\sigma_i : 1 \leq i \leq d\}$ be the Galois group of L over K. By the explicit class field theory, L is contained in the maximal totally real subfield $K(\Lambda_J)^+$ of $K(\Lambda_J)$, where J is the conductor of L and Λ_J denotes the J-torsion of ρ.

We prove the theorem only in the case $hd_\infty = 1$ (e.g., $A = \mathbb{F}_q[t]$), referring to [Tha95b] for the general case:

As G is abelian, the relative zeta function factors into L functions: $\zeta_O(s) = \prod_{\chi \in \hat{G}} L(\chi, s)$, where $L(\chi, s) := \prod(1 - \chi(P)P^{-s})^{-1} = \sum \chi(I)I^{-s}$, where I (P respectively) runs through (irreducible respectively) elements of A_+. Note that since s is even, I^s is independent of the choice of signs. (here the subtlety, important when p divides d, (see also the remarks below or [Gos, pa. 345]) is that we consider χ as coming from reductions to characteristic p of $\overline{\mathbb{Q}_p}$ valued characters, so that eg., if $G = \mathbb{Z}/p$ we still consider p characters (all trivial), rather than just one). Since this decomposition is through the Euler factors matching and since each Euler factor of ζ_O lies in K, it is enough to show the corresponding statements for $R'_s := \prod L(\chi, s)_J / \tilde{\pi}^{ds}$, where the subscript J indicates that Euler factors for J have been removed.

For $\sigma \in G$, let $\zeta_\sigma := \sum I^{-s}$, where the sum is over $I \in A_+$, prime to J and with the Artin symbol $\sigma_I = \sigma$.

Then we have $L(\chi, s)_J = \sum \chi(\sigma)\zeta_\sigma$. Furthermore, the ideal group corresponding to $K(\Lambda_J)^+$ consists of principal ideals generated by elements congruent to one modulo J. Hence ζ_σ is K-linear combination of sums in Theorem 5.2.5, which then implies that $R_s \in K(\Lambda_J, \mu)$, where μ is the root of unity of the order equal to the exponent of G, so that $\chi \in \mathbb{F}_q(\mu)$. Hence R_s is algebraic. To get a better control on the field which contains it, we use the Dedekind determinant formula: $\prod L(\chi, s)_J = \det(\zeta_{\sigma_i^{-1}\sigma_j})_{d \times d}$.

Hence $R'_s = \det(\zeta_{\sigma_i^{-1}\sigma_j}/\tilde{\pi}^s)$, where the entries lie in $K(\Lambda_J)$ by the above. The action of $\text{Gal}(K(\Lambda_J)/K)$ permutes the rows of the determinant, multiplying it by ± 1, hence $R_s^2 \in K$. $\quad\square$

Remarks 5.2.9 (1) We can not [Tha95b] replace H in the theorem, when $d_\infty = 1$, by K, in general. Classically, for arbitrary (not necessarily abelian over $\mathbb{Q}$) totally real number field F, it is known that $(\zeta_F(s)/(2\pi i)^{r_1 s})^2 \in \mathbb{Q}$.

(2) For the history and references on this general as well as the abelian case using L-functions, as we have done here, see references in [Tha95b] and see [Gos] for ideas about carrying over the proof in the general case. On the other hand, such an algebraicity result for the ratio of the relative

zeta value with an appropriate power of the period $2\pi i$ is not expected for number fields which are not totally real. But in our case, we can have such a result even if L is not totally real, as was noted in [Gos87]: Let L be a Galois extension of degree p^k of K, then since all the characters of the Galois group are trivial, the L-series factorization shows that $\zeta_{\mathcal{O}}(s) = \zeta(s)^{p^k}$ and the result follows then from Theorem 5.2.6. More elementary way to see this, when the degree is p and $\mathcal{O}$ is of class number one is to note that for $\alpha \in \mathcal{O} - A$, there are p conjugates with the same norms which then add up to zero, whereas for $\alpha \in A$, the norm is α^p.

5.3 Values at non-positive integers

We have seen how just grouping together the terms of the same degree gives $\zeta(-k) = \sum_{i=0}^{\infty}(\sum_{n \in A_{i+}} n^k) \in A$, for $k > 0$. Hence we have stronger integrality rather than rationality in the number field case, reflecting absence of pole at $s = 1$ in our case.

We also have the following vanishing result, giving the 'trivial zeros':

Theorem 5.3.1 *For a negative integer s, $\zeta((q-1)s) = 0$.*

Proof. If $k = -(q-1)s$, then

$$\zeta(-k) = \sum_{i=0}^{\infty} \sum_{a \in A_{i+}} a^k = -\sum_{i=0}^{\infty} \sum_{a \in A_i} a^k = 0$$

where the second equality holds since $\sum_{\theta \in \mathbb{F}_q} \theta^{q-1} = -1$ and the third equality is seen by using that the sum is finite and applying Theorem 5.1.2 with $W = A_{<m}$ for some large m. □

For $A = \mathbb{F}_q[t]$, another proof giving a formula as well as non-vanishing result parallel to the case of the Riemann zeta function can be given: For $k \in \mathbb{Z}_+$, $\zeta(-k) = 0$ if and only if $k \equiv 0 \mod (q-1)$. Also $\zeta(0) = 1$:

The proof [Gos79; Tha90] follows by writing a monic polynomial n of degree i as $th + b$ with h of degree $i - 1$ and $b \in \mathbb{F}_q$ and using the binomial theorem to get the induction formula

$$\zeta(0) = 1, \quad \zeta(-k) = 1 - \sum_{\substack{f=0,(q-1)|(k-f)}}^{k-1} \binom{k}{f} t^f \zeta(-f)$$

which shows $\zeta(-k) \in A$ since $\zeta(0) = 1$. If $(q-1)$ divides k, then induction shows $\zeta(-k) = 1 - 1 + 0 = 0$. If $(q-1)$ does not divide k, then there being no term in the summation corresponding to $f = 0$, $\zeta(-k) = 1 - tp(t) \neq 0$ where $p(t) \in A$.

We do not know for general A whether the values at odd integers are non-zero. But we have

Theorem 5.3.2 *Let $d_\infty = 1$ and s be a negative odd integer. Then $\zeta(s)$ is non-zero if A has a degree one rational prime $\wp$ (e.g. if $g = 0$ (this reproves $\mathbb{F}_q[t]$ case) or if q is large compared to the genus of K) or if $h = 1$. (Together they take care of the genus 1 case.) In fact, in the first case, $\zeta(s)$ is congruent to 1 mod $\wp$.*

Proof. The first part follows by looking at the zeta sum modulo $\wp$ and noticing that $\sum_{\theta \in \mathbb{F}_q} \theta^{-s} = 0$, hence the only contribution is 1 from the $i = 0$ term. Apart from $g = 0$, there are only 4 other A's with $hd_\infty = 1$ (these have no degree 1 primes) listed in 3.1.2. For $q = 2$, there are no odd integers. That leaves only two (ii) and (iii) of 3.1.2. In both cases, we look at contribution to the zeta sum for each i modulo x : $i = 0$ gives 1, $i = 1, 2, 3$ give 0 and for higher i's, with a modulo x we also get θa, $\theta \in \mathbb{F}_q$ and hence the contribution is zero as before. $\qquad\square$

Remark 5.3.3 In many particular examples of A, we can settle the non-vanishing by this method. If we ask, instead, for odd negative s's for which $\zeta(s)$ is always (i.e., for any A) non-zero, Theorem 5.1.2, says that $\zeta(s) = 1$, if $l(-s) < q - 1$. Again, this can be pushed farther: For example, if $l(-s) < 2(q-1)$ (and s odd negative as before), then by Theorem 5.1.2 and the $\mathbb{F}_q[t]$ case of the previous Theorem (by taking t to be an element of A of the lowest positive degree), we see that $\zeta(s)$ is nonzero.

Let us now turn to the relative zeta functions.

Theorem 5.3.4 *(Goss [Gos]) For a negative integer s, $\zeta_\mathcal{O}(s) \in A$. In fact, $Z_\mathcal{O}(s)_n \in A$ for every n.*

Proof. The idea is to decompose the sum in $\mathbb{F}_q$-vector spaces and use Theorem 5.1.2 to show that for large k, the k-th term of the zeta sum is zero. It is enough to show that for each ideal class $\mathcal{C}$ of $\mathcal{O}$, the contribution from $\mathcal{I} \in \mathcal{C}$ is zero. Let $\mathcal{J} \in \mathcal{C}$. Then $\mathcal{I}\mathcal{J}^{-1} = (i) \subset \mathcal{J}^{-1}$. Hence it is sufficient to show that

$$S_k := \sum_{\substack{i \in \mathcal{J}^{-1} - \{0\}/\text{units} \\ \deg \text{Norm}(i) = k}} \text{Norm}(i)^{-s} = 0.$$

Let $\infty_1, \cdots, \infty_n$ be the infinite places of L with relative residue degrees $f_1, \cdots, f_n$ and the corresponding embeddings $\sigma_1, \cdots, \sigma_n$ respectively. Then, writing the polar divisor as $-(i)_\infty = \sum m_l(\infty_l)$, we have $k = \deg \operatorname{Norm}(i) = d_\infty \sum m_l f_l$. We will show that parts of S_k corresponding to a fixed choice of $\{m_l\}$ also vanish. Namely, let $\overline{D} := \sum m_l(\infty_l)$, then we will show that if the $\deg(\overline{D})$ is large, then $\sum \operatorname{Norm}(i)^s = 0$, where the sum is over $S := \{i \in \mathcal{J}^{-1} - \{0\}/\text{units}, -(i)_\infty = \overline{D}\}$.

Once we fix a polar divisor, the ambiguity in i is only through a finite field. Hence, given $i \in S$, fix a representative $r(i)$ by demanding that the sign of $\sigma_1(r(i))$ is 1, for some choice of the sign in L_{∞_1}. $\mathcal{L} := \{j \in \mathcal{J}^{-1} : (j) + \sum (m_l - 1)(\infty_l) \geq 0\}$ acts on S by translation: $j \circ r(i) = r(i) + j$. This is an action because as k is assumed large, this translation does not change the top degree and hence polar divisor or a sign. Further, if $(j_1 + r(i))/(j_2 + r(i))$ is a unit, then since its polar divisor is zero, it has to be in a finite field, but then the choice of the sign makes it 1 and hence $j_1 = j_2$. In other words, there is no isotropy.

Now we show that the contribution of each orbit of S under this action is zero. In other words, we have to show that $\sum \operatorname{Norm}(j + r(i))^{-s} = 0$, where now the sum is over j running though $\mathcal{L}$, which is a $\mathbb{F}_q$-vector space. Now $\operatorname{Norm}(j + r(i))^{-s} = \mu \prod \sigma_l(j + r(i))^{-s}$, where μ is the root of unity to take care of the sign, and can be taken out of the sum, as it is the same for all j. Hence by a slight modification of Theorem 5.1.2, with $\prod(f_l + w)$ in the place of $f + w$ (see [Gos]), the sum is zero, when k is large.   □

Theorem 5.3.5   *For a negative integer s, $\zeta_{\mathcal{O}}((q-1)s) = 0$. In fact, $Z_{\mathcal{O}}((q-1)s)_n = 0$ for every n.*

Proof.   By the previous Theorem, the sum over i defining the zeta value is finite, so the proof follows exactly as in Theorem 5.3.1.   □

Remarks 5.3.6   The values of the Dedekind zeta function at negative integers are all zero, if the number field is not totally real. By Remarks 5.2.9, similar result does not hold in our case. In fact, we do not even need degree to be a power of the characteristic, as will be seen from examples 5.4.5. The ramification possibilities for the infinite places are much more varied in the function field case.

For $A = \mathbb{F}_q[t]$, Goss defined modification $\beta(k) \in A$ of $\zeta(-k)$, for $k \in \mathbb{Z}_{\geq 0}$, as follows:

$$\beta(k) := \zeta(-k) \text{ if } k \text{ is odd}, \quad \beta(k) := \sum_{i=0}^{\infty} (-i) \sum_{n \in A_i+} n^k \text{ if } k \text{ is even}.$$

In other words, deformation $\zeta(-k, X)$ of $Z(-k, 1) = \zeta(-k)$ has a simple zero at $X = 1$ if k is even and hence one considers $\frac{d\zeta}{dx}(-k, X)\big|_{X=1} = \beta(k)$ instead. For $k \in \mathbb{Z}_+, \beta(k) \neq 0$ and in fact

$$\beta(0) = 0, \quad \beta(k) = 1 - \sum_{f=0,(q-1)|(k-f)}^{k-1} \binom{k}{f} t^f \, \beta(k).$$

For general A, Goss similarly defines $\beta(k)$ by removing the 'trivial zero' at k of order d_k (see the next section for the definition and formulas) by $\beta(k) := (1 - X)^{-d_k} \zeta_A(-k, X)|_{X=1}$. One gets Kummer congruences and $\wp$-adic interpolations for $\beta(k)$'s, as for zeta values.

Recall the definitions in 5.2.2.

Theorem 5.3.7 (*Goss- Sinnott [GS85]*): *For $0 < k < q^d - 1$, $C(w^{-k}) \neq 0$ if and only if p divides $L(w^k, 1)$.*

Proof. (Sketch) The duality between the Jacobian and the p-adic Tate module T_p transforms the connection between the Jacobian and the class group in the Tate's proof of Stickelberger Theorem in Chapter 1 to $T_p(w^{-k})/(1 - F)T_p(w^{-k}) \cong C(w^{-k})$. On the other hand, we have a Weil type result: $\det(1 - F : T_p(w^{-k})) = L_u(w^k, 1)$. Here L_u is the unit root part of the L-function and hence has the same p-power divisibility as the complete L-function. Hence $\mathrm{ord}_p(L(w^k, 1))$ is the length of $C(w^{-k})$ as a $\mathbb{Z}_p[G]$-module. $\qquad\square$

The interpretation of the L-function as the characteristic polynomial in the above is precisely the result on which Iwasawa's main conjecture is based. Since this is already known, the Gras conjecture giving the component-wise version of Theorem 1.3.2, which follows classically from the main conjecture, is known here. This was recognized in [GS85].

Comparison with the corresponding classical result shows that we are looking at divisibility by p, the characteristic, rather than the prime $\wp$ relevant to the cyclotomic field. To bring $\wp$ in, we need to look at the finite characteristic zeta function of this chapter.

Theorem 5.3.8 (*Goss-Sinnott [GS85]*) $C(w^{-k}) \neq 0$ *if and only if* $\wp|\beta(k), 0 < k < q^d - 1$.

Proof. The identification $W/pW \cong A/\wp$ provides us with the Teichmüller character $w: (A/\wp)^* \to W^*$ satisfying $w^k(n \mod \wp) = (n^k \mod \wp) \mod p$. Hence the reduction of L value (since it also has 'trivial zero' factors missing) in the Theorem above modulo p is $\beta(k) \mod \wp$. $\qquad\square$

Remarks 5.3.9   (1) Recall that for 'odd' k, $\beta(k) = \zeta(-k)$, so the result is in analogy with Herbrand-Ribet theorem, but for 'even k' we get a new phenomenon. This is connected with the failure of Spiegelgungsatz for Carlitz-Drinfeld cyclotomic theory. Classically the leading terms at even k are conjectured to be transcendental, here they are rational, even integral. For values at positive integers on the other hand, the situation seems to be as expected with naive analogies.

(2) For simplicity, we will now restrict to the case $A = \mathbb{F}_q[T]$. Classically, Bernoulli numbers occur in the special values of the Riemann zeta function at both positive and negative integers and these values are connected by the functional equation for the Riemann zeta function. In our case, there is no simple functional equation known and in fact we get two distinct analogs of Bernoulli numbers B_k (or rather the more fundamental B_k/k) both connecting to class groups of cyclotomic extensions of K: those coming from the positive values relate to class groups of rings of integers, as in Theorem 5.2.4 in contrast to class groups of fields, as in Theorem 5.3.8.

Since $\zeta(-k)$ turns out to be a finite sum of n^k's, by Fermat's little theorem, the $\zeta(-k)$'s satisfy Kummer congruences enabling us to define a $\wp$-adic interpolation $\zeta_\wp$. On the other hand, the B_m satisfy analogs of the von-Staudt congruences and the Sylvester-Lipschitz theorem. We have now two distinct analogs of B_k/k: $-\zeta(-k+1)$ for $k-1$ 'odd' on one hand and $\Pi(k-1)\zeta(k)/\tilde{\pi}^k$, with k 'even' on the other. But the shift by one does not transform 'odd' to 'even' unless $q = 3$, and we do not know any reasonable functional equation linking the two.

5.4   Multiplicities of trivial zeros

Now we turn to the question of the order of vanishing. Classically, the answer is simple: The Euler product representation shows there are no zeros in the region where it is valid and hence the orders of vanishing of the trivial zeros (namely those at the negative even integer s) are easily proved by looking at poles of the gamma function factors in the functional equation. (For the number field situation in general, these are predicted by motives and these orders of vanishing are connected to K-theory and extensions of motives, so many structural issues and clues are at stake in this simple question.) We do not have functional equations. As described below, Goss gave a lower bound for orders of vanishing, when L contains H_1 and mentioned as an open question whether they are exact. The lower bounds match the naive analogies. But we will see that they are not exact orders and indeed that the patterns of extra vanishing are quite surprising in terms of the established analogies. The full situation is still not understood, even conjecturally.

The main idea of Goss is to turn the similarity in the definition of our zeta function with the classical one into a double congruence formula, using the Teichmüller character, and to use the knowledge of the classical L-function Euler factors to understand the order of vanishing. This is done as follows:

Let $\wp$ be a prime of A and let W be the Witt ring of $A/\wp$. The identification $W/pW \cong A/\wp$ provides us with the Teichmüller character $w : (A/\wp)^* \to W^*$ satisfying $w^k(a \mod \wp) = (a^k \mod \wp) \mod p$.

Now let $\Lambda_\wp$ denote the $\wp$-torsion of rank one, sgn-normalized Drinfeld module ρ of generic characteristic. Let L contain H_1 and let G be the Galois group of $L(\Lambda_\wp)$ over L. Then G can be thought of as a subgroup of $(A/\wp)^*$ and hence w can be thought of as a W-valued character of G. Let $L(w^{-s}, u) \in W(u)$ be the classical L-series of Artin and Weil in $u := q^{-sm}$, where m is the extension degree of the field of constants of L over $\mathbb{F}_q$.

Let $S_\infty := \{\infty_j\}$ denote the set of the infinite places of L and let G_j denote the Galois group of $L_{\infty_j}(\Lambda_\wp)$ over L_{∞_j}. Then $G_j \subset \mathbb{F}_q^*$. Given s, let $S_s \subset S_\infty$ be the subset of the infinite places at which w^{-s} is an unramified character of G. Then S_s does not depend on $\wp$. Put

$$\tilde{\zeta}_O(s, X) := \zeta_O(s, X) \prod_{\infty_j \in S_s} (1 - w^{-s}(\infty_j) X^{\deg(\infty_j)})^{-1}.$$

Theorem 5.4.1 *(Goss [Gos]) Let L contain H_1 and let s be a negative integer. Then $\tilde{\zeta}_O(s, X) \in A[X]$.*

Proof. (Sketch) Tracing through the definitions, the property of w mentioned above gives the double congruence formula $L(w^{-s}, X^m) \mod p = \tilde{\zeta}_O(s, X) \mod \wp$ for infinitely many $\wp$. But as the L function is known to be a polynomial, the result follows. $\qquad\square$

This gives the following lower bound for the order of vanishing:

Theorem 5.4.2 *(Goss [Gos]) Let L contain H_1 and let s be a negative integer, then the order of vanishing of $\zeta_O(s)$ is at least*

$$V_s := \mathrm{ord}_{X=1} \prod_{\infty_j \in S_s} (1 - w^{-s}(\infty_j) X^{\deg(\infty_j)}).$$

Remarks 5.4.3   Note that V_s depends on s only through its value modulo $q - 1$ and $V_s \leq [L : K]$. This is analogous to the properties of the exact orders of vanishing in the classical case, but not in our case as we will see.

Examples 5.4.4   (i) $L = K = H_1$: Then $V_s = 0$ or 1, according as s is 'odd' or 'even'. We have seen in 5.3 that for $A = \mathcal{O} = \mathbb{F}_q[t]$, the order of vanishing is V_s.

(ii) L is totally real extension of degree d of K containing H_1, i.e., ∞ splits completely in L: V_s is 0 or d according as whether s is 'odd' or 'even'. (Note that the bounds V_s in (i) and (ii) are in analogy with the orders of vanishing in the classical case.)

(iii) $L = K(\Lambda_{\wp})$: $V_s = (q^{\deg(\wp)} - 1)/(q - 1)$ for all s.

(iv) $L = \mathbb{F}_{q^n}(t)$ and $K = \mathbb{F}_q(t)$: $V_s = 0$ when s is 'odd', and for s 'even', $V_s = p^k$, if $n = p^k l$, $(l, p) = 1$.

(v) $L = \mathbb{F}_5(\sqrt{-t})$ and $K = \mathbb{F}_5(t)$: V_s is 1 or 0 according as whether 2 (not 4) does or does not divide s.

Let O_s denote the order of vanishing of the relevant zeta function at s. Then simple exact calculations using the Lee's vanishing bounds in 5.1.2 show that $O_s = V_s$ in the following examples:

Examples 5.4.5   (i) $A = \mathbb{F}_3[x, y]/y^2 = x^3 - x - 1$. O_s for $0 < -s < 28$ is 1 if s is even and 0 otherwise.

(ii) $A = \mathbb{F}_3[t]$ with $\mathcal{O} = \mathbb{F}_3[t, y]/y^2 = t^3 - t - 1$ and $\mathcal{O} = \mathbb{F}_3[\sqrt{t}]$ respectively, for $0 < -s < 31$, O_s is 1.

(iii) $A = \mathbb{F}_3[t]$ with $\mathcal{O} = \mathbb{F}_9[t]$, for $0 < -s < 179$, O_s is 1 if s is even and 0 otherwise.

The following simple calculation shows that the $V_s \neq O_s$ in general.

Examples 5.4.6   Let (a) $A = \mathbb{F}_2[x, y]/y^2 + y = x^3 + x + 1$, (b) $A = \mathbb{F}_2[x, y]/y^2 + y = x^5 + x^3 + 1$. In both these situations, $hd_\infty = 1$ and hence $K = H_1$. For small s we compute $\zeta(s, X)$ as follows: The genus of K is 1 and 2 for (a) and (b) respectively. Hence by the bounds obtained from the Theorem 5.1.2 and the Riemann-Roch theorem shows that for (a) and (b), $\zeta(-1, X) = 1 + 0 * X + (x + (x + 1)) * X^2 = 1 + X^2 = (1 + X)^2$. Hence the order of vanishing is 2 for $s = -2^n$. Similar simple calculation shows that the order of vanishing is 1 if $2^n \neq -s \leq 9$ for (a) and is 1 for $s = -7$ and 2 for other s with $-s \leq 9$ for (b).

In fact, more generally, we have

Theorem 5.4.7 *If $d_\infty = 1$, $q = 2$ and K is hyper-elliptic, then the order of vanishing of $\zeta(s)$ at negative integer s is 2 if $l(-s) \leq g$, where g is the genus of K.*

Proof. Let $x \in A$ be an element of degree 2 and let n be the first odd non-gap for A at ∞. Then the genus g of K is seen to be $(n - 1)/2$. Let $S_A(i)$ denote the coefficient of X^i in $\zeta_A(s, X)$. A simple application of Theorem 5.1.2 and the Riemann-Roch theorem shows that for $i > l(-s) + g$, $S_A(i) = 0$. Hence

$$\zeta_A(s, X) = \sum_{i=0}^{(n-1)/2} S_A(2i) X^{2i}$$

since $l(-s) \leq g = (n - 1)/2$. On the other hand, as x has degree 1 in $\mathbb{F}_q[x]$, we have $S_A(2i) = S_{\mathbb{F}_q[x]}(i)$, for $0 \leq i \leq (n - 1)/2$. Further, by Theorem 5.1.2, $S_{\mathbb{F}_q[x]}(i) = 0$, if $i > (n - 1)/2$, as $(n - 1)/2 \geq l(-s)$. Hence, $\zeta_A(s, X) = \zeta_{\mathbb{F}_q[x]}(s, X^2)$. Hence by Example 5.4.4 (i), the order of vanishing is $2 = 2 * 1$ as required. $\qquad\square$

Remarks 5.4.8 (i) We do not need a restriction on the class number of K, so apart from (a) and (b) this falls outside the scope of the theorem giving the lower bounds.

(ii) There are other examples of this phenomenon, with $q > 2$. See [Tha95b] and University of Arizona thesis (1996) of Javier Diaz Vargas.

(iii) Analogies between the number fields and the function fields are usually the strongest for $A = \mathbb{F}_q[y]$. But even in that case, here is an example of extra vanishing, when $A = \mathbb{F}_4[y]$ and $\mathcal{O} = \mathbb{F}_4[x, y]/y^2 + y = x^3 + \zeta_3$: Order of vanishing of $\zeta_{\mathcal{O}}$ at $s = -1$ is 2, as can be verified by direct computation, whereas the lower bound of Theorem 5.4.2 is 1.

We do not know yet whether in these cases, the order of vanishing is V_s, except for the extra vanishing exceptional situation mentioned in the theorem. This is true in the computational range and we have the following partial result (which can again be pushed further by the proofs along the same line), which shows that situation in general can only be worse as far as the pattern of zeros is concerned.

Theorem 5.4.9 *For the example (a) above, the order of vanishing of $\zeta(s)$ is 1 ($= V_s$), if $l(-s) = 2$ or if $-s \equiv 0, 3, 5$ or 6 mod 7.*

Proof. We know that $\zeta(s) = 0$ by Theorem 5.3.1. Hence we have

$$(\zeta(s,X)/(X-1))|_{X=1} = \frac{d}{dX}(\zeta(s,X))|_{X=1} = \sum_{i=0}^{\infty}\sum_{a\in A_{2i+1}} a^{-s} = \sum_{i=0}^{\infty}\sum_{a\in A_{2i}} a^{-s}.$$

Since the genus is 1, by the Riemann-Roch theorem and Theorem 5.1.2, we see that in the first case, the terms of the last sum vanish if $i > 1$, and hence the sum is $1 + x^{-s} + (x+1)^{-s} \neq 0$ as claimed. For $i > 1$, the terms of the second sum are all zero modulo y (as they can be decomposed as sums over pairs b and $b + y$), hence the sum is congruent to $1 + x^{-s} + (x+1)^{-s}$ modulo y, which takes care of the second case. $\square$

Remark 5.4.10 These results imply class group components vanishing by 5.3.8 and raise question about the meaning of real leading terms.

5.5   Zeta function interpolation on character space

Using the same idea of grouping the terms of the same degree together, Goss made a zeta function interpolation from these zeta values, by defining n^{-s}, allowing s to vary in a much bigger continuous space.

(a)   ∞-adic interpolation

Once we define n^s (or more generally, I^s for an ideal I, if we want to use non-principal base and arbitrary extensions) taking values in C_∞ say, then imitating the classical definitions we can define Riemann, Dedekind type Zeta (or partial zeta) functions, and more generally L functions attached to finite characteristic Galois representations that occur in Drinfeld modules theory. For simplicity, we stick to Riemann and Dedekind type zeta functions and refer to [Gos94; Gos; Gos96] for the general definitions.

The zeta function associated to the Carlitz module is $\zeta_{\mathbb{F}_q[t]}(s+1)$, because the arithmetic Frobenius acting as $\wp$ via the cyclotomic character makes geometric Frobenius $\wp^{-1}$ exactly as in the classical case of Tate motive.

Absolute convergence, meaning without grouping by degrees, and hence Euler product is valid on a smaller part of the exponent space, but the convergence after the grouping is on full exponent space (at infinity or at finite places).

In place of the complex plane $\mathbb{C}$ of the exponents s, Goss suggests the exponent space $S_\infty := C_\infty^* \times \mathbb{Z}_p$:

Fix a sign function and a positive (monic i.e., of sign 1 when $d_\infty = 1$)

uniformizer $\pi_\infty \in K_\infty$, say $1/t$ in $\mathbb{F}_q[t]$ case. For monic $\alpha \in K_\infty^*$, define its one-unit part

$$\langle \alpha \rangle := \alpha/\pi_\infty^{v_\infty(\alpha)}.$$

Since the p-th power spreads the power series expansion in characteristic p, the p^i-th power of a one unit tends to 1 as $i \to \infty$. Hence we can define a p-adic power of a one unit. (In fact, the similar construction is also usual in non-archimedean situation of p-adic integers.)

Definition 5.5.1

$$\alpha^s := x^{\deg(\alpha)} \langle \alpha \rangle^y = x^{-d_\infty v_\infty(\alpha)} \langle \alpha \rangle^y \in C_\infty, \quad \text{for } s = (x, y) \in S_\infty.$$

The motivation is as follows: If we want to define 's-th power homomorphism' from positive elements of K_∞^* to C_∞^*, then writing a positive α as $\pi_\infty^j \langle \alpha \rangle$, we need to specify x, the image of π_∞, and a homomorphism from one unit group (isomorphic to product of countably many copies of $\mathbb{Z}_p$) to C_∞^*. We choose only the simplest and natural p-adic power homomorphisms mentioned above from this huge group of possible homomorphisms. Another justification is that for a complex number $s = x + iy$, $n^x = |n^s|$ is the absolute value part, whereas $|n^{iy}| = 1$ in analogy with the degree and one unit parts above.

The powers, as we defined them, depend on the choice of uniformizer. The integral powers do not depend on the uniformizer : $\mathbb{Z}$ sits in S_∞ by the identification $j \in \mathbb{Z} \leftrightarrow (\pi_*^{-j}, j) \in S_\infty$, where π_* is some fixed d_∞-th root of π_∞.

The exponent space S_∞ is a topological group with respect to natural operations and topology and we write the group operation additively. Then, for positive α, β we have

$$\alpha^{s_0+s_1} = \alpha^{s_0} \alpha^{s_1}, \quad (\alpha\beta)^s = \alpha^s \beta^s.$$

We do not have $(n^{s_0})^{s_1} = n^{s_0 s_1}$, and we do not usually need it anyway in this context.

We will see that by grouping together the terms with n's of the same degree, $\zeta(s) := \sum n^{-s}$ makes sense on the whole of S_∞.

So we can restrict the first variable $s_0 = 1$ and consider the zeta function $\sum \langle n \rangle^{-s} \in K_\infty$ now with $s \in \mathbb{Z}_p$, analogous to the gamma function domain and range (see 4.5). Now the exponent space $\mathbb{Z}_p$ being a ring, we recover the lost property above, and can talk about values at fractions and now gamma and zeta have compatible domains. But we do not know how to interpret derivatives, values do not actually interpolate (as in gamma) and lose $n \to n$ character!

We do not know how to connect zeta on S_∞ to Goss' two-variable gamma function and our special values results are limited to integers which can be handled by easier approach. But we come back to S_∞ in 5.8.

With this definition, we can define analog of Riemann and Dedekind Zeta functions, when the base is $\mathbb{F}_q[t]$ or A with $h(A) = 1$ or for Dedekind zeta functions, more generally for the relative extensions containing H_A or H_1, so that the relative norms are ideals generated by positive elements.

When $\deg(x) > 0$, we have convergence and hence Euler product representation as usual.

Grouping the terms of the same degree gives a convergent series for any $s \in S_\infty$, by the following theorem [Gos79; Gos96], when we put $\langle n \rangle = 1 + w$, for n's of a given fixed degree d and note that the valuation estimate below grows quadratically in d, by the Riemann-Roch.

Theorem 5.5.2   *(Goss) Let W be a finite dimensional $\mathbb{F}_q$-vector subspace of F, a field over $\mathbb{F}_q$ with additive valuation v. Assume $v(w) > 0$ for all $w \in W$. Let $W_j := \{w \in W : v(w) \geq j\}$ and let $D = \sum \dim W_j$. Then,*

$$v\left(\sum_{w \in W} (1 + w)^i \right) \geq (q - 1)D, \quad \text{for all } i \in \mathbb{Z}_p.$$

Proof.   We can reduce to integral powers by noting that $(1 + w)^i = \sum_{j=0}^{\infty} w^j \binom{i}{j}$ and that the formula to be shown is independent of the power. Next, pick $\{w_1, \cdots, w_n\}$ so that $\{w_1, \cdots, w_{\dim W_j}\}$ is a basis for W_j, for all j. Let j_0 be so that $W_{j_0+1} = \{0\}$, but $W_{j_0} \neq \{0\}$. By using the binomial theorem and 5.1.2 applied to W_{j_0}, we find by induction that our sum has valuation at least $(q - 1) \sum j \dim(W_j/W_{j+1}) = (q - 1)D$. $\square$

If we try to extend to more general base A, we encounter two problems: The ideals may not be principal (i.e., $h(A) = h(K)d_\infty > 1$) and there may be more signs than constants in A (i.e., $d_\infty > 1$), even if we decide to just deal with elements rather than ideals.

Let us just indicate how to handle this, leaving the detailed exposition to [Gos96].

As the group of one units in C_∞^*, which contains the image of homomorphism $\langle \alpha \rangle^y$, is uniquely divisible and the quotient group of fractional ideals modulo those generated by positive elements is finite, there is unique extension $\langle I \rangle^y$ defined for fractional ideals I. Then we put $I^s := x^{\deg(I)} \langle I \rangle^y$, for $s = (x, y) \in S_\infty$ as before.

To handle signs, we proceed as in the gamma function case, by generalizing 'monic' elements to mean those with a sign in a fixed set S of representatives of $\mathbb{F}_\infty^*/\mathbb{F}_q^*$.

(b) $\wp$ adic interpolation

If p does not divide n, the well-known Fermat-Euler-Lagrange congruence says $n^{(p-1)p^j} \equiv 1 \mod p^{j+1}$, implying that if $s_1 \equiv s_2 \mod (p-1)p^j$, then $n^{s_1} \equiv n^{s_2} \mod p^{j+1}$. Now the Riemann zeta values at negative integers, which are heuristically sums of such terms, are rational and hence in $\mathbb{Q}_p$. These congruences thus suggest that once we remove the Euler factors at p, then for p-adically close exponents which are multiple of $p-1$, the corresponding zeta values should be p-adically close. This is in fact true and is equivalent to the Kummer congruences. This is justified by Mazur's integration approach to p-adic interpolation. See 5.12.

In our case, at negative integers, the corresponding sums are finite, so that the interpolation is immediate from the interpolations of the terms, since to check corresponding congruence of fixed zeta values, we need to look at only finitely many terms.

Let $\wp$ be a prime of A of degree d. If $\wp$ does not divide n and if $i_1 \equiv i_2$ mod $(q^d - 1)p^j$, then $n^{i_1} \equiv n^{i_2} \mod \wp^j$.

So for n relatively prime to $\wp$, n^i as a function of i interpolates to a continuous function on $S = \lim \mathbb{Z}/(q^d - 1)p^j \mathbb{Z}$.

Now by throwing out the Euler factor at $\wp$, i.e., using

$$\sum_{(n,\wp)=1} n^i = (1 - \wp^i) \sum n^i,$$

we get a continuous function $\zeta_\wp : S \to A_\wp$ such that $\zeta_\wp(-i) = (1 - \wp^i)\zeta(-i)$ if $i \in \mathbb{Z}_+$.

For any $k \in \mathbb{Z}$, we are just grouping with degree:

$$\zeta_\wp(k) = \sum_{d=0}^{\infty} \sum_{(\wp,n)=1, n \in A_d+} n^{-k} \in K_\wp.$$

This is because for large d the term is congruent to zero modulo high power of $\wp$, because of the Fermat congruences and vanishing of the corresponding terms for negative exponent by Theorem 5.1.2. The same reasoning for the valuation at infinity gives another proof of interpolation at infinity.

Note that since the values at negative even integers are all zero, we get identically zero function, for $q = 2$ and some trivial vanishing special values for general q.

We can again introduce another variable and still interpolate, as for negative integers the sums are finite. So we get $\zeta_\wp(s, X) \in A_\wp[[X]]$ and we can also let $X \in A_\wp$ (or $K_\wp$ or $\mathbb{C}_\wp$ say) to land in $A_\wp$ (or corresponding fields resp).

Define the exponent space $S_\wp := \mathbb{C}_\wp^* \times (\lim \mathbb{Z}/(p-1)p^j\mathbb{Z})$. For $s = (s_0, s_1) \in S_\wp$, define $n^s = s_0^{\deg(n)} n^{s_1}$, ideal exponentiation and zeta function analogously. We identify $k \in \mathbb{Z}$ with $(1, k)$.

Theorem 5.3.1 implies that its values at positive 'even' integers are zero. Some non-vanishing results at 'odd' integers are known. We will study special values in later chapters.

For analytic properties and more general setup, see [Gos96]. The experience with all zeta functions points to the importance of analytic properties (eg., converse theorems and implications to Langlands program or implications of analytic properties for Sato-Tate distribution or Ramanujan conjecture), and function fields should not be exceptions, but we do not see yet how to extract useful arithmetic information.

We just mention and refer to [Kap95; Gos96] for zeta functions of this type in local situation in higher dimensions.

5.6   Power sums

In this section, $A = \mathbb{F}_q[t]$.

Let us collect some simple results on power sums, due to Carlitz, Lee and Gekeler. For history, variants and references, see [Tha90].

Let $S_d(k) := \sum_{n \in A_{d+}} n^k$ ($k \geq 0$, unless stated otherwise). As we have already seen in Theorem 5.1.2, if $d > k$, $S_d(k) = 0$ and hence $\zeta(-k) = \sum_{i=0}^{\infty} S_d(k) \in A$.

One has much better results on vanishing of $S_d(k)$ and closed expressions for $S_d(k)$, which we now recall.

Let $\ell(k)$ be the sum of digits of k base q.

Theorem 5.6.1   *(Carlitz) Let $y \geq 1$. $S_d(y) \neq 0$ if and only if $y = m_0 + m_1, + \cdots + m_d$ where $q - 1 \mid m_j$, $m_j > 0$ for $1 \leq j \leq d$ and the sum is such that there is no carry over of digits base p.*

Corollary 5.6.2   *(Carlitz-Lee) (1) If $d > \ell(k)/(q-1)$, then $S_d(k) = 0$.*
(2) If $q = p$, then $S_d(k) = 0$ if and only if $d > \ell(k)/(q-1)$.

We postpone the discussion of the proof of this theorem, stated with incomplete proof by Carlitz, to Section 5.8. We leave the proof of the corollary to the reader. Part (1) of the Corollary also follows from Theorem 5.1.2.

Now let us see how, following Carlitz, the terms $S_d(k)$ (k any integer) of zeta values at positive or negative integers can be combined in a generating function which is essentially the logarithmic derivative of binomial coefficient function. (In fact, we have already used this in 4.15.) The idea is essentially the same as expressing the power sums in terms of elementary symmetric functions as in Newton's identities.

Theorem 5.6.3 *(Carlitz) If $\mathcal{B}_d(x)$ denotes the binomial coefficients $\left\{ \begin{smallmatrix} x \\ q^d \end{smallmatrix} \right\} = \binom{x}{q^d}$ for $\mathbb{F}_q[t]$ of Sections 4.15 and 4.16, then the quantity $-\frac{\mathcal{B}_d'(0)}{1-\mathcal{B}_d(x)}$ has Laurent series expansion $\sum_{n=0}^{\infty} S_d(n)x^{-n+1}$ at $x = \infty$ and $\sum_{n=0}^{\infty} S_d(-n-1)x^n$ at $x = 0$.*

Proof. Since $1-\mathcal{B}_d(x)$ has all monic polynomials of degree d as its simple zeros, by taking the logarithmic derivative, we see that

$$\sum_{a \in A_{d+}} \frac{1}{x-a} = -\frac{\mathcal{B}_d'(0)}{1-\mathcal{B}_d(x)}.$$

Developing the left hand side as geometric series in two different ways, we see that the Laurent series expansion is $\sum_{n=0}^{\infty}(\sum_{a \in A_{d+}} a^n)x^{-n+1}$ at $x = \infty$ and $\sum_{n=0}^{\infty}(\sum_{a \in A_{d+}} a^{-n-1})x^n$ at $x = 0$. $\square$

We note $\exp_1 x \log_1 = \sum \mathcal{B}_d(x)\tau^d$, and $\mathcal{B}_d'(0) = 1/l_d$.

We look at applications to zeta values in 8.1.

Using the explicit formulas for binomial coefficients in 4.15 and careful combinatorial arguments, we can deduce nice formulas in particular cases:

Corollary 5.6.4 *(1) (Lee) If $s < q$, then*

$$S_d(q^{\ell_1} + \cdots + q^{\ell_s} - 1) = (-1)^d[\ell_1]_d \cdots [\ell_s]_d/L_d$$

where $[h]_m := D_h/D_{h-m}^{q^m}$ if $m \le h$ and 0 otherwise.

In particular, $S_d(q^{k+d} - 1) = (-1)^d D_{d+k}/L_d D_k^{q^d}$.

In particular, $S_d(q^d - 1) = (-1)^d \Pi(q^d - 1)$.

(2) (Gekeler) Let $k < q^{d+1} - 1$ have base q expansion $k = \sum k_j q^j$. Then $S_d(k) = (-1)^\gamma \cdot M \cdot \prod_{j \le d} L_{d-j}^{q^j(k_j-(q-1))} g_k$ where $\gamma := d + \sum_{j < d}(d-j+1)k_j$ and M is the multinomial coefficient $\binom{k_d}{k_0', \ldots, k_d'}$. $k_j' := q - 1 - k_j (j < d)$ and $k_d' := \ell(k) - d(q-1)$.

5.7   Zeta measure

Since the zeta values at negative integers are really simple finite sums of powers, they satisfy Kummer congruences, because it just reduces to Fermat's little theorem on powers. Also, they can be expressed as integrals much more simply than in the classical case. Consider the basic compact opens $a + \wp^m A_\wp$, where without loss of generality, $a \in A$ has degree less than $m \deg(\wp)$. Let μ be the 'zeta' measure [Gos79] giving the measure 2 or 1 respectively to this set depending on whether a is monic or not. Then the k-th moment of μ is $\zeta(-k)$.

See 4.14 and 5.12 for some background on interpolations. For a quick and nice introduction to p-adic measures see [Kat75].

A $\mathbb{Z}_p$-valued p-adic measure on $\mathbb{Z}_p$ is just a $\mathbb{Z}_p$-linear map μ: Conti $(\mathbb{Z}_p, \mathbb{Z}_p) \to \mathbb{Z}_p$. One writes $\mu(f)$ symbolically as $\int_{\mathbb{Z}_p} f(x)\, d\mu(x)$.

If we use the binomial coefficient basis in 4.14.1, in the notation there, a $\mathbb{Z}_p$-valued measure μ on $\mathbb{Z}_p$ is uniquely determined by the sequence $\mu_k =: \int_{\mathbb{Z}_p} \binom{x}{k} d\mu(x)$ of elements of $\mathbb{Z}_p$. Any sequence μ_k determines a $\mathbb{Z}_p$-valued measure μ by the formula $\int_{\mathbb{Z}_p} f d\mu = \sum_{k \geq 0} f_k \mu_k = \sum_{k \geq 0} (\Delta^k f)(0) \mu_k$.

Convolution $*$ of measures μ, μ' is defined by $\int f(x)\, d(\mu * \mu')(x) = \int \int f(x+t)\, d\mu(x) d\mu'(t)$.

By comparing coefficients in $(1+m)^{x+t} = (1+m)^x (1+m)^t$ one sees

$$\binom{x+t}{k} = \sum \binom{x}{i}\binom{t}{k-i}.$$

Hence

$$\sum f_k (\mu * \mu')_k = \int \int \sum_{k \geq 0} f_k \sum_{i=0}^{k} \binom{x}{i}\binom{t}{k-i} d\mu(x) d\mu'(t) = \sum f_k \sum_{i=0}^{k} \mu_i \mu'_{k-i}.$$

In other words, if one identifies measure μ with $\sum \mu_k X^k$ then convolution $*$ on measures is just multiplication on the corresponding power series. This identification of $\mathbb{Z}_p$-valued measures on $\mathbb{Z}_p$ with power series is also called [Was97] Iwasawa isomorphism.

In the $\mathbb{F}_q[t]$ case, if we use the analog 4.14.2, with function field binomial coefficient analogs, an $A_\wp$-valued measure μ on $A_\wp$ (called just measure μ for short) is uniquely determined by the sequence $\mu_k = \int_{A_\wp} \binom{x}{k} d\mu$ of elements of $A_\wp$. Any sequence μ_k determines measure μ by $\int_{A_\wp} f d\mu = \sum_{k \geq 0} f_k \mu_k$.

Carlitz [Car40b] showed further that

$$\binom{x+t}{k} = \sum \binom{k}{i}\binom{x}{i}\binom{t}{k-i},$$

where the first binomial coefficient on the right is the classical and all others are function field binomial coefficients. So, if we identify measure μ with the divided power series $\sum \mu_k(X^k/k!)$ (recall the formal nature of $X^k/k!$ in the context of divided power series), then by a proof similar to that in the classical case above, convolution $*$ on measures is multiplication on the corresponding divided power series.

Classically [Kat75], the measure μ whose moments $\int_{\mathbb{Z}_p} x^k d\mu$ are given by $(1-a^{k+1})\zeta(-k)$ for some $a \geq 2$, $(a,p)=1$ has the associated power series $(1+X)/(1-(1+X)) - a(1+X)^a/(1-(1+X)^a)$. We need a twisting factor in front of the zeta values to compensate for the fact that the zeta values are rational rather than integral, in contrast to our case. Comparison of this result with the following result [Tha90], which answers a question of Anderson and Goss, is not well-understood.

Theorem 5.7.1   *For $A = \mathbb{F}_q[t]$, the divided power series corresponding to the zeta measure μ, namely the measure whose i-th moment $\int_{A_\mathfrak{p}} x^i d\mu$ is $\zeta(-i)$, is given by $\sum \mu_k(X^k/k!)$ with*

$$\mu_k = (-1)^m \quad if \quad k = cq^m + (q^m-1), \ 0 < c < q-1, \ and \ \mu_k = 0 \ otherwise.$$

Proof.   Let $k = \sum_{j=0}^w k_j q^j$, $0 \leq k_j < q$, $k_w > 0$. Write $\binom{x}{k} = \sum b_i x^i$. Then we want to compute $\mu_k = \sum b_i \zeta(-i)$.

First consider the case $k \neq q^{w+1} - 1$, so that by (1) of 5.6.2, we have

$$\mu_k = \sum b_i \sum_{n \in A_{\leq w}+} n^i = \sum \binom{n}{k} = \sum_{n \in A_w+} \binom{n}{k - k_w q^w},$$

since the binomial coefficients $\binom{n}{q^w}$ are zero or one according as whether the degree of n is less than w or equal to it. The binomial coefficient being summed being a polynomial in n of degree $k - k_w q^w$, the sum is zero by (1) of Theorem 5.6.2, if $k - k_w q^w < q^w - 1$. On the other hand, using the last statement of (1) of Corollary 5.6.4 when $k - k_w q^w = q^w - 1$, we see that the sum is $(-1)^w$. This finishes the proof in this case.

When $k = q^{w+1} - 1$, we do the same calculations, but now for degrees w and $w + 1$. We leave it to the reader. $\square$

Remarks 5.7.2   (1) This result can be viewed as giving many interesting identities for $\zeta(-k)$'s. This result is trivially true for $q = 2$. For a non-trivial variant, with 'beta-measure', as well as for other proofs, see [Tha90].

(2) In fact, $k = cq^m + (q^m - 1)$ iff deg $\zeta(-k) = $ deg $\Pi(k)$ for $k > 0$.

(3) The k's for which $\mu_k \neq 0$ can be characterized as those 'odd' k's for which any smaller positive integer has each base q digit no larger than the corresponding base q digit of k. For such k's, all the binomial coefficients $\binom{k}{i}$ are nonzero modulo q. When q is a prime, this last property characterizes such k's among the 'odd' numbers. Some other properties of k's are described in [Tha90; Gos]. Recall that we have also seen interesting influence of base q digits on orders of vanishing for the zeta function.

(4) Yang [Yan01] calculates the zeta measure for k_∞, with Carlitz basis with t replaced by $1/t$.

5.8   Zero distribution

The most famous open question about the zeta function is whether the Riemann hypothesis saying that 'the non-trivial zeros s of the Riemann zeta function $\zeta(s)$ all lie on the line $Re(s) = 1/2$' is true or false. In addition to its intrinsic interest about the tight structural restrictions on the zero location, it has many interesting arithmetic consequences.

In the function field situation, for the zeta function for $A = \mathbb{F}_p[t]$, Daqing Wan [Wan96b] showed by calculating the Newton polygon a remarkable result that for a fixed $y \in \mathbb{Z}_p$, if $\zeta(x, y) = 0$, then $x \in K_\infty$, analog of the real line, and further that all the zeros are simple. Note that since C_∞, where the zeros can lie a priori, has infinite degree over K_∞, in contrast to $[\mathbb{C}, \mathbb{R}] = 2$, this is in some sense more remarkable. But we do not yet have many interesting arithmetic applications of this zero distribution restrictions and also the situation for general A and more general zeta functions seems different, as we will see.

In fact, the Newton polygon calculation can also be done by using Carlitz assertions 5.6.1, which were justified by Javier Diaz-Vargas for $\mathbb{F}_p[t]$ in his University of Arizona thesis (see [DV96]) and (after some preliminary results by Bjorn Poonen) for $\mathbb{F}_q[t]$ by Jeff Sheats [She98], combinatorist at University of Arizona then. Surprisingly, while for most of the things we have seen so far, it does not matter whether q is prime or not, in this case the assertion for $\mathbb{F}_q[t]$ needed much more extensive combinatorial effort. We will only discuss $q = p$ case fully and then briefly discuss the general A situation, referring to [Gos00; Gos03; Gos96] for more extensive discussion and evidence.

We start with the proof [DV96] of Theorem 5.6.1 in case $q = p$.

Proof. Given y a positive integer, we say that integers $\{m_j\}$ satisfy $(\mathcal{H})$ if and only if they satisfy the conditions of Theorem 5.6.1. From now on we assume $q = p$, but we keep both q and p in the notation.

If $f = a_d + \cdots + a_1 t^{d-1} + t^d$ then

$$f^y = \sum_{y=m_0+\cdots+m_d} \frac{y!}{m_0! m_1! \cdots m_d!} (t^d)^{m_0} (a_1 t^{d-1})^{m_1} \cdots a_d^{m_d}.$$

Now, by the theorem of Lucas, $\dfrac{y!}{m_0! m_1! \cdots m_d!}$ is prime to p, the characteristic of K, if and only if the sum $y = m_0 + \cdots + m_d$ is such that there is no carry over of digits base p. If this is impossible then clearly $S_d(y) = 0$. On the other hand,

$$S_d(y) = \sum_{a_1,\ldots,a_d \in \mathbb{F}_q} \sum_{y=m_0+\cdots+m_d} \frac{y! \, a_1^{m_1} \cdots a_d^{m_d}}{m_0! m_1! \cdots m_d!} t^{dm_0+(d-1)m_1+\cdots+m_{d-1}}.$$

Changing the order of the summation and using the relation,

$$\sum_{a \in \mathbb{F}_q} a^h = 0 \text{ if and only if } (q-1) \nmid h \text{ or } h = 0,$$

we see that a necessary condition for the non-vanishing of $S_d(y)$ is that $(q-1) \mid m_j$, and $m_j > 0$ for $1 \le j \le d$. Finally if $y = m_0 + m_1 + \cdots + m_d$ where the $\{m_j\}$ satisfy $(\mathcal{H})$, we may select such m_j in order to maximize $dm_0 + \cdots + m_{d-1}$. To finish the proof that $S_d(y) \ne 0$, we now show that this maximum is unique. Notice that max $dm_0 + \cdots + m_{d-1} = dy - \min m_1 + \cdots + dm_d$, so in order to find the maximum, we must minimize "the weight" $m_1 + \cdots + dm_d$ such that $y = m_0 + m_1 + \cdots + m_d$ subject to $(\mathcal{H})$. Of all such decompositions of y, choose those such that

$$m_d \le m_{d-1} \le \cdots \le m_1.$$

Now of these decompositions of y choose those with minimum m_d, then the ones with minimum m_{d-1}, and so on. We call this solution $\{m_i\}$ the **greedy solution**. It is enough to show that given any non-greedy solution $\{m_i'\}$ of $(\mathcal{H})$, we can modify it to get another solution with smaller weight. Without loss of generality, $m_d' \le m_{d-1}' \le \cdots \le m_1'$. Now each m_i' is a sum of $k(p-1)$ p-powers. We can assume that $k = 1$ (throwing $(k-1)(p-1)$

of them in m_0' decreases the weight). If j is the largest index such that $m_j' \neq m_j$, then $m_j' > m_j$. If

$$m_j = \sum_{k=1}^{p-1} p^{i_k}, \ i_k \le i_{k+1}, \text{ and } m_j' = \sum_{k=1}^{p-1} p^{i_k'}, \ i_k' \le i_{k+1}',$$

then choose the least k such that $r := i_k < i := i_k'$. This implies that for some $t < j$, $p^r \in m_t'$. Define $\{\tilde{m}_i\}$ by

$$\tilde{m}_j = m_j' - p^i + p^r, \quad \tilde{m}_t = m_t' + p^i - p^r, \quad \tilde{m}_i = m_i' \text{ for } i \neq j, t.$$

Since $q = p$, $\tilde{m}_j$ is still divisible by $q - 1$ regardless of r and i. But the weight of the solution $\{\tilde{m}_i\}$ is $w' - (j - t)(p^i - p^r)$ less than w', the weight of the non-greedy solution.   □

Remarks 5.8.1   We have in fact proved that

$$\deg S_d(y) = dm_0 + (d - 1)m_1 + \cdots + m_{d-1},$$

where $\{m_i\}$ is the greedy solution.

Theorem 5.8.2   *Let* $A = \mathbb{F}_q[t]$. *If* $\zeta_A(x, y) = 0$, *then* $x \in K_\infty$. *Further, all the zeros* x *of* ζ_A *are simple.*

Proof.   The idea is that the theorem and the remark above allows us to calculate the Newton polygon which gives information on zeros. We will prove it only for $q = p$ a prime, see [She98] for the general case. Now

$$\zeta(s) = \sum_{d=0}^{\infty} (1/x)^d \sum_{f \in A_d+} <f>^{-y}.$$

For each fixed $y \in \mathbb{Z}_p$, define $\zeta_y(x) := \zeta(x, -y)$. Let $v_d(y)$ be the ∞-adic valuation of the coefficient of $(1/x)^d$ in $\zeta_y(x)$. The **Newton polygon** of $\zeta_y(x)$ is then the convex hull in the plane $\mathbb{R}^2$ of the lattice points $(d, v_d(y))$, $d \ge 0$. The ∞-adic valuation of the zeros of $\zeta_y(x)$ are completely described by its Newton polygon. Namely, if the Newton polygon has a side of slope λ whose horizontal projection is of length l, then $\zeta_y(x)$ has exactly l zeros with valuation λ. First we deal with the case where y is a positive integer and then generalize to any p-adic integer. For y a positive integer, we have $v_d(y) = dy - \deg S_d(y)$. By the remark above,

$$v_d(y) = dy - (dm_0 + \cdots + m_{d-1}) = m_1 + 2m_2 + \cdots + dm_d,$$

where $\{m_i\}$ is the greedy solution. If $y = y_0 + y_1 p + \cdots + y_k p^k$ is the decomposition of y with base p, then clearly the greedy solution is obtained as follows:

$$m_d = y_0 + y_1 p + \cdots + y_{j_d}^* p^{j_d}, \ y_{j_d}^* \le y_{j_d},$$

where $y_0 + y_1 + \cdots + y_{j_d}^* = p - 1$. Next, m_{d-1} is obtained in the same way by considering the decomposition of $y - m_d$ base p and so on. Finally $m_0(d) = y - (m_1(d) + \cdots + m_d(d))$. If such construction is impossible then $S_d(y) = 0$ by Theorem 5.6.1, and we set $v_d(y) = \infty$. If not, then

$$v_d(y) = m_1(d) + 2m_2(d) + \cdots + dm_d(d).$$

Notice that by construction $m_{d-r}(d) = m_{d-r-1}(d-1)$ for $r = 0, \ldots, d - 2$. Using this we calculate the slope $\lambda(d)$ of the Newton polygon from $(d - 1, v_{d-1}(y))$ to $(d, v_d(y))$:

$$\lambda(d) = v_d(y) - v_{d-1}(y) = m_1(d) + m_1(d-1) + m_2(d-1) + \cdots + m_{d-1}(d-1).$$

Finally, $\lambda(d)$ is a strict increasing function of d since

$$\lambda(d) - \lambda(d - 1) = m_1(d) > 0,$$

for $d \ge 1$, if we put $\lambda(0) = 0$. This implies that each slope of the Newton polygon has horizontal projection one and so the zeros of $\zeta_y(x)$ are simple and lie in K_∞.

Assume now that $y = y_0 + y_1 p + \cdots + y_i p^i + \cdots \in \mathbb{Z}_p - \mathbb{N}$ such that the number of nonzero digits is infinite. Then choose enough digits $y_0, y_1, \ldots, y_k$ such that for $\bar{y} = y_0 + y_1 p + \cdots + y_k p^k$ we can define allowable nonzero $m_0(d), m_1(d), \ldots, m_d(d)$. The number of digits that you choose for that to happen is unimportant for our purposes since they only change $m_0(d)$ but not $m_1(d), m_2(d), \ldots, m_d(d)$, and so the valuation $v_d(\bar{y})$ does not change. Hence by approximating $y \in \mathbb{Z}_p$ p-adically as a limit of $n \in \mathbb{N}$, we see

$$v_d(y) = v(\lim_{n \to y} t^{-dn} \sum_{f \in A_d+} f^n) = \lim_{n \to y} v(t^{-dn} \sum_{f \in A_d+} f^n) = v_d(\bar{y}).$$

Hence all the results about the slope $\lambda(d)$ of the Newton polygon hold for any p-adic integer. $\qquad \Box$

Whereas the 'simplicity' and 'reality' of these zeros is just quite difficult to prove for $\mathbb{F}_q[t]$, it does not even hold for general A:

Example 5.8.3   Example 5.4.6 (b) of class number one and genus two which showed higher order vanishing at trivial zeros also shows (inseparable) zeros of multiplicity two and lying outside K_∞ for $y = 2^k + 1$, because of the relation between ζ_A and Carlitz zeta found in the proof of Theorem 5.4.7.

5.9   Low values and multi-logarithms

We computed elementary symmetric functions of A_i and A_{i+}, so we can compute corresponding power sums, as in Newton's formulas. In fact, logarithmic derivative is much simpler for a $\mathbb{F}_q$-linear function. We can express low zeta values in terms of multi-logs.

The harmonic series $\sum 1/n$, which can be thought of as the 'value' of the Riemann zeta function at $s = 1$ and also as the 'value' of logarithm $-\log(1 - x) = \sum x^n/n$ at $x = 1$, diverges. (But this connection between zeta and logarithm is quite useful for Dirichlet L-values at 1 and logarithms of cyclotomic units.) But in our case, $\zeta(1)$ makes good sense.

For $A = \mathbb{F}_q[t]$ and the Carlitz module (as we have normalized it, see [AT90, Sec. 3] for historical remarks), we have $\log(z) = \sum z^{q^d}/l_d$ and $\zeta(1) = \log(1)$ (see below), and if we define the k-th multilogarithm by $\log_k(z) = \sum z^{q^d}/l_d^k$ by naive analogy, we get

Theorem 5.9.1   *Let $A = \mathbb{F}_q[t]$, and $\log$ and $\log_k$ be the logarithm and the naive multi-logarithm for the Carlitz module. Let $0 < r < q$. Then*

$$S_d(-rp^n) = 1/l_d^{rp^n}, \quad \zeta(rp^n) = \log_{rp^n}(1).$$

Proof.   The first statement follows from straight calculation from Theorem 5.6.3, and it implies the second statement.   □

We discuss v-adic counterparts in 8.1. But note that the Theorem implies that the sum of reciprocals of monic polynomials prime to t and of degree $d > 0$ is $1/l_d - 1/(tl_{d-1}) = t^{q^d}/(tl_d)$, so that $\zeta_t(1) = \log_t(t)/t$, where the t-adic logarithm, defined by the same series, converges at t. (Note that for Carlitz logarithm $\log(t)/t = (1 - 1/t)\log(1)$, by its functional equation, but t-adic log does not converge at 1.)

In fact, we will see in Chapters 7 and 8, that there is a good notion of the logarithm for the n-th tensor power of the Carlitz module which is a good deformation of $\log_n$ and $\zeta(n)$ can be expressed as the value of (the canonical co-ordinate of) $\log_n$ at some algebraic point. For $n > 1$, the n-th tensor power is no longer a Drinfeld module, so in this section, we stick to $n = 1$.

As we have seen for many quantities such as factorials, binomial coefficients etc., the view-points which coincide for $A = \mathbb{F}_q[t]$ diverge in general, but sometimes not by too much. Here also, as we will see in one simplest case of $g = h = 1$, there is a connection between zeta and log, though not as simple as before. For a general situation, as well as more examples, we refer to [Tha92b; And94; And96b], and 8.9 where we discuss Anderson's theorem in a slightly different context.

We now apply the results of Section 4.15 to special zeta values. Here is a class number 1 example (iii) of 3.1.2.

Theorem 5.9.2 *Let $A = \mathbb{F}_3[x, y]/y^2 = x^3 - x - 1$. We have*

$$\sum_{a \in A_{i+}} \frac{1}{a} = \frac{1}{L_i} = \frac{y(\frac{1}{y})^{3^i}}{l_i} + \frac{(\frac{1}{y})^{3^{i-1}}}{l_{i-1}} + \frac{(-\frac{1}{y^6} - \frac{1}{y^8})^{3^{i-2}}}{l_{i-2}} + \frac{(-\frac{1}{y^{27}})^{3^{i-3}}}{l_{i-3}}$$

$$\zeta(1) = yl(\frac{1}{y}) + l(\frac{1}{y} - \frac{1}{y^6} - \frac{1}{y^8} - \frac{1}{y^{27}}) = log(y - 1)$$

$$e(\zeta(1)) = y - 1.$$

Proof. From the ρ_x given in 3.1.2 (iii) and from 2.4.3 (1), we see that $x_1 = y(x^3 - x)$, $y_1 = y(y^3 - y)$, $y_2 = y^9 + y^3 + y$. So Theorem 4.15.5 gives

$$f_i = \frac{[i]_y - y[i]_x}{[i]_x - 1}, \quad g_i = \frac{[i]_y - y^{3^i}[i]_x}{[i+1]_x + 1}.$$

Using the values of g_i, g_{i-1} and g_{i-2} thus obtained we express the right-hand side of the first claim as $1/\nu_i l_i$ where ν_i is a rational function with coefficients in $\mathbb{F}_3$ of x, y, $x^{3^{i-2}}$ and $y^{3^{i-2}}$. On the other hand, from Theorems 4.15.4 and 4.15.5, we see that μ_i is also such a function. The direct comparison, the details of which we omit, of the complicated expressions thus obtained for these rational functions μ_i, ν_i shows that, in fact, $\mu_i = \nu_i$, thus proving the first claim. From the formula for g_i given above, we see that deg $g_i = -3^i$. By induction, we then deduce that deg $l_i = -(3/2)(3^i - 1)$. Hence the logarithm series converges, when deg $z < -3/2$. Summing the first equality over i running from 0 to ∞, we get the first equality in the second claim. The second equality there follows by using functional equations $a \log(z) = \log(\rho_a(z))$, with $a = y$ and $z = 1/y$. The final formula follows by exponentiating. $\qquad\square$

Remarks 5.9.3 The relation above is not the 'simplest' form of the relation between L_i's and l_i's. For example, using the recursion relations for l_i's obtained from the functional equation of logarithm, with $a = y$, $a = x$ and by evaluation of g_i above respectively, we get the following three equalities

$$\frac{1}{L_i} = \frac{1}{l_i} + \frac{y^{3^{i-1}}}{l_{i-1}} + \frac{1}{l_{i-2}} = \frac{1 - [i]_x}{l_i} - \frac{y^{3^i}}{l_{i-1}} = \frac{y^{3^i+1} - 1 - [i]_x^2 - [i]_x}{([i+1]_x + 1)l_i}.$$

For applications to transcendence, see Chapter 10. We will return to this topic with much stronger results once we develop the tools for them.

Remarks 5.9.4 In the number field case, Zagier and others have looked at connections between the values at the Dedekind zeta functions $\zeta_K(s)$, which are relative zeta functions with base $\mathbb{Q}$ and the multilogarithms, where $\zeta_{\mathbb{Q}}(n) = \log_n(1)$ by comparing the definitions, but for other number fields we get more complicated identities with interesting connections to K-theory and motives. While we may expect such type of results here also, the results above and in Chapter 8 are of different nature in that we look at the absolute (rather than relative) zeta functions and change the base (rather than top).

Finally, we briefly discuss some **open questions** raised by these special value results. The main questions are how to predict the exact orders of vanishing in general and whether there is any reasonable functional equation relating the values at positive integers to those at the negative integers and how to predict algebraic incarnations of the special values.

The results of Goss and Sinnott, mentioned above, come via congruence with the classical L- function and hence relate the arithmetic of the class group to the V_s-th term of the expansion of the zeta value at $X = 1$, rather than the leading term. It is not clear what information the leading term provides in the cases of the extra vanishing.

There are also simpler open questions such as whether $\zeta(s)$ can be zero at odd negative s or whether there are any non-totally real extensions (which are not Galois of p-power degree) for which the values at positive even integers are related to the periods as in Theorem 5.2.8 or whether anything special happens at the positive integers in the cases of extra vanishing. Another open question, already alluded to is whether Theorem 5.2.8 generalizes to arbitrary totally-real extension perhaps via Eisenstein series.

5.10 Multizeta values

In addition to his evaluation of the zeta function at positive even integers and negative integers, and his guessing (together with partial verification) of the functional equation for the zeta, Euler also studied 'multizeta' values

$$\zeta(s_1, \cdots, s_k) := \sum_{n_1 > \cdots n_k > 0} \frac{1}{n_1^{s_1} \cdots n_k^{s_k}}$$

of 'depth' k and 'weight' $\sum s_i$, for positive integers s_i, with $s_1 > 1$, giving some nice relations between them, such as $\zeta(2, 1) = \zeta(3)$, for whose proof he employed the technique of divergent series, which was not well-understood then.

These multizeta values are iterated sums as well as iterated integrals: Recall that, given functions f_i of integers, we define iterated sums as

$$\text{It} \sum_1^N f_1, \cdots f_k := \sum_{N > n_1 \cdots > n_k > 0} f_1(n_1) \cdots f_k(n_k)$$

$$= \sum_{N > n_1 > 0} f_1(n_1) \text{It} \sum_1^{n_1} f_2, \cdots, f_k.$$

Given one-forms $w_i = f_i(x)dx$, we define iterated integrals as

$$\text{It} \int_0^t w_1, \cdots, w_k := \int_{t > x_1 > \cdots x_k > 0} f_1(x_1) \cdots f_k(x_k) dx_1 \cdots dx_k$$

$$= \int_0^t (w_1 \text{It} \int_0^x w_2, \cdots w_2, \cdots, w_k).$$

Then

$$\zeta(s_1, \cdots, s_k) = \text{It} \sum_1^{\infty} \frac{1}{n^{s_1}}, \cdots, \frac{1}{n_k^s}$$

$$= \text{It} \int_0^1 \underbrace{\frac{dt}{t} \cdots \frac{dt}{t} \frac{dt}{1-t}}_{s_1 \text{ factors}} \underbrace{\frac{dt}{t} \cdots \frac{dt}{1-t}}_{s_2 \text{ factors}} \cdots \cdots$$

In recent years, the interest in the arithmetic of and relations between these values has been rekindled, partially through important works of Drinfeld, Deligne, Zagier and others: They occur naturally as coefficients of Drinfeld associator, with connections to mathematical physics and knot invariants, as well as to the studies of iterated extensions of Tate motives over $\mathbb{Z}$, of fundamental group of projective line minus three points, and of Feynman path integral renormalization. See MathSci net for references.

The iterated sums and integral expressions above show that products of multizeta values are integral linear combinations of multizeta values, as we can see from considering different possible orderings of n_i's and x_i's. These give so-called shuffle identities.

Here we begin the study of their function field variants. As usual, we will see some interesting analogies as well as interesting differences. We will focus on the case $A = \mathbb{F}_q[t]$. The definitions and some easy results immediately generalize to general A.

(a)   Complex valued multizeta

It is straightforward to define complex valued multizeta by generalizing Artin-Weil approach in depth one case: n^s gets replaced by $|n|^s$, where $|n|$ is the norm q^d of n of degree d and $n_i > n_{i+1}$ gets replaced by corresponding inequality of norms and you sum over all non-zero ideals (or what is the same, over their monic generators) leading to definition

$$\zeta(s_1, \cdots, s_k) := \sum_{\substack{n_i \in A+ \\ |n_1| > \cdots > |n_k|}} \frac{1}{|n_1|^{s_1} \cdots |n_k|^{s_k}}.$$

We then have, after summing the resulting geometric series and simple algebraic manipulations, for example,

$$\zeta(s_1, s_2) = \sum_{d_1=1}^{\infty} \frac{q^{d_1}}{q^{d_1 s_1}} \sum_{d_2=0}^{d_1-1} \frac{q^{d_2}}{q^{d_2 s_2}} = \frac{q^{1-s_1}}{(1 - q^{1-s_1})(1 - q^{2-s_1-s_2})}$$

and more generally

$$\zeta(s_1, \cdots, s_k) = \frac{q^{(1-s_1)(k-1) + (1-s_2)(k-2) + \cdots + (1-s_{k-1})}}{(1 - q^{(1-s_1)})(1 - q^{(1-s_1)+(1-s_2)}) \cdots (1 - q^{(1-s_1)+\cdots+(1-s_k)})}$$

so that not only we can evaluate the values at positive integers, but we get a simple expression for its continuation for complex s_i. For general A, we will have more complicated expressions involving the gap data.

When you multiply multizetas, we do get sum-shuffle identities, but which are slightly different than their classical counterpart, because now there are q^d monic polynomials of size $|n| = q^d$, rather than unique positive integer of size $|n|$ making a difference in the counting: When the degrees are equal, on the left hand side of identities below we get $q^d q^d$ terms, whereas the last term on the right hand side gives q^d terms. Hence, instead of classical shuffle identity $\zeta(s_1)\zeta(s_2) = \zeta(s_1, s_2) + \zeta(s_2, s_1) + \zeta(s_1 + s_2)$ we now get

$$\zeta(s_1)\zeta(s_2) = \zeta(s_1, s_2) + \zeta(s_2, s_1) + \zeta(s_1 + s_2 - 1).$$

Instead of combinatorial shuffle argument we gave, this can also be verified using the explicit formula given above and the usual $\zeta(s) = \sum q^d / q^{ds} = 1/(1 - q^{1-s})$.

Since we have explicit formulas in general, these types of identities may not be that interesting except for the understanding of analogies.

(b) Finite characteristic variants

For the rest of 5.10, we will reserve letters n, n_i etc. for monic polynomials and let $d = \deg(n)$ and $d_i = \deg(n_i)$. In this section, s and s_i will be positive integers. So multizeta values will refer to values at such s_i's.

Let us discuss several variants that arise when we try, as in the passage from the Artin zeta to the Carlitz zeta, to replace $|n|^s$, which is a complex number for complex s, by n^s, which is a rational function of t for integer s.

Just doing this, leads to the following definition (where subscript d refers to the ordering by degree and should not be confused with degree indices).

Definition 5.10.1

$$\zeta_d(s_1, \cdots, s_k) := \sum_{\substack{n_i \in A+ \\ |n_1| > \cdots > |n_k|}} \frac{1}{n_1^{s_1} \cdots n_k^{s_k}}.$$

Other variants arise when we try to order n's rather than $|n|$'s: We may use some total order, say the dictionary or lexicographic (use arbitrary order on $\mathbb{F}_q$ and extend in natural way to polynomials) order (which is also the simultaneous $\wp$-ordering in sense of 4.14).

For the rest of 5.10, we write $n_1 > n_2$ to mean $n_1 - n_2$ is monic.

Definition 5.10.2 We write ζ_l for the multizeta we get by using lexicographic order.

We write ζ (there should be no confusion with previous complex valued ones) for what we get with $>$, where $n_1 > \cdots > n_k$ is taken to mean monic n_i's with $n_i > n_{i+j}$ for $j > 0$.

We write $\bar{\zeta}$ when we take it to mean monic n_i's with just $n_i > n_{i+1}$.

In depth ≤ 2, $\zeta = \bar{\zeta}$, and all zetas of course agree with usual zeta in depth one. There are other possible variants depending on whether we allow only distinct n_i's (the difference with $\overline{\zeta}$ will arise only in depth more than q) or we only take n_k monic, by interpretation $n_1 > n_2 \cdots > n_k > 0$, but this gives variant with bad properties for special values and shuffle identities.

In contrast to the classical case, clearly

Theorem 5.10.3 *All these variants converge when $s_i > 0$.*
All satisfy identity of form $\zeta(s_1, \cdots, s_k)^p = \zeta(ps_1, \cdots, ps_k)$.

Since $d_i := \deg(n_i)$, $d_1 > d_2$ implies $n_1 > n_2$, but the converse is not true. But $n_1 > n_2$ implies $d_1 \geq d_2$, unless $p = 2$.

For $q = 2$, $n_1 > n_2$ is equivalent to $n_1 \neq n_2$, and for $q = 3$, when there are two signs as in the classical case, we have mutually exclusive cases $n_1 > n_2$ or $n_1 = n_2$ or $n_2 > n_1$ as in the classical case. But none of the possibilities may hold for $q > 3$, and when $p = 2$, cases at both end either both hold or don't! Hence,

Theorem 5.10.4 *Let $p = 2$. Then $\zeta(s_1, \cdots, s_k)$ is independent of the ordering of s_i's. If two of the s_i's are equal, then it is zero. Further, $\overline{\zeta}(s_1, \cdots, s_k)$ is the same as $\overline{\zeta}(s_k, s_{k-1}, \cdots, s_1)$.*

Proof. The first statement follows because $n_i > n_j$ is equivalent to $n_j > n_i$. Then we can assume $s_1 = s_2 = s$ without loss of generality. Hence the second statement follows because we are summing over terms $1/((n_1 n_2)^s x) + 1/((n_2 n_1)^s x) = 0$. The last statement follows from the fact that $n_i > n_{i+1}$ implies $n_{i+1} > n_i$. ☐

In the classical case, the multizeta values (at positive arguments) are never zero, since the terms are positive numbers.

Theorem 5.10.5 *Let $q = 3$. Then $\zeta(s, s, s) = \overline{\zeta}(s, s, s)$.*

Proof. The difference of the two zeta values is the sum of $1/(n_1 n_2 n_3)^s$, where $n_1 > n_2 > n_3 > n_1$. But such tuples occur 3 times adding to zero in characteristic 3. ☐

Theorem 5.10.6 *(1) ζ_l satisfies classical sum-shuffle identities.*
(2) When $q = 3$, we have classical sum-shuffle identity of depth 2:
$\zeta(s_1)\zeta(s_2) = \zeta(s_1, s_2) + \zeta(s_2, s_1) + \zeta(s_1 + s_2)$.
(3) $\zeta_d(s_1, \cdots, s_k)$ satisfy classical sum-shuffle identity, if $q \geq \sum s_i$ (i.e., when weight is not more than q).

Proof. For (1) and (2), the same proof as in the classical case works. As for part (3), with $S_d(-s_i) = \sum n^{-s_i}$, where the sum is over monics of degree d, we can write

$$\zeta_d(s_1, \cdots, s_k) = \sum_{d_1 > \cdots > d_k \geq 0} S_{d_1}(-s_1) \cdots S_{d_k}(-s_k)$$

and apply the shuffle argument to d_i's using the fact 5.9.1 that $S_d(-s) = 1/l_d^s$, for $0 \leq s \leq q$, where the logarithm of Carlitz module is $\sum z^{q^i}/l_i$. (This allows us to calculate some ζ_d values, such as $\zeta_d(s,s) = (\zeta(2s) - \zeta(s)^2)/2$, when $p \neq 2$ and $2s \leq q$.) $\qquad\square$

The sum shuffle identities don't carry the same information: For example, $2\zeta_l(s,s) = \zeta(2s) - \zeta(s)^2$ does not give any information on $\zeta_l(s,s)$, in contrast to the classical case, when $p = 2$, since then both sides are trivially zero.

Theorem 5.10.7   *When $q = 2$, $\zeta_l(s,s) = 0$, if and only if $s = 2^r$, $r \geq 0$.*

Proof.   By taking out the power of 2 using Theorem 5.10.3, we assume, without loss of generality, that s is odd. The terms with $d_i \leq 1$ contribute degree $-s - 1$, whereas each of the other terms is of degree $\leq -2s$. So if $s > 1$, there is no cancellation giving non-vanishing.

When $s = 1$, there is cancellation when you truncate to a fixed degree: We can see this by induction on degrees so that in the induction step n_1 can be of fixed degree. Then the total of $1/(n_1 n_2)$'s with both n_1 and n_2 of the same degree cancells out the sum of $1/(n_1 n_2)$ with n_2 being of lower degree, because $1/(n_1 n_2) = (1/n_1 - 1/n_2)/(n_2 - n_1)$ and the mapping from pairs of distinct n_1, n_2's of same degree to pairs $n_2 - n_1, n_1$ and $n_2 - n_1, n_2$ (each decides the other pair) is bijective. $\qquad\square$

Even when $q = 3$, when you consider depth more than 2, with the interpretation '$n_i > n_{i+1}$' mentioned above, any tuple of distinct integers can be ordered (as we can verify by induction, using the three cases above), but not necessarily uniquely, as in $n > t + 1 > t > t - 1 > t + 1$ for any monic polynomial of degree more than one. On the other hand, there is no ordering possible beween 3-tuple t, $t + 1$, $t - 1$ with the interpretation '$n_i > n_{i+j}$'. Because of this, in both interpretations, in depths more than two, the sum shuffle identities do not follow as in the classical case, even when $q = 3$.

Now let us begin evaluating these multizeta values when we can.

Definition 5.10.8   We call multizeta value of weight w 'eulerian', if it is an algebraic multiple of $\tilde{\pi}^w$.

By the Carlitz result, in depth 1, values of 'even' weight are eulerian, eg., when $q = 2$, all depth one values (as are the values of geometric gamma function in this case) are eulerian. Here is a generalization to any depth.

Theorem 5.10.9   *When $q = 2$, $\zeta(s_1, \cdots, s_k)$ are eulerian.*
   In fact, they are given explicitly in the proof, in terms of lower depths or in terms of the depth 1 (i.e., the usual zeta).

Proof.   We are summing over mutually distinct non-zero polynomials n_i. So first of all, $\zeta(s_1, \cdots, s_k)$ is independent of the order of s_i's. Next, when we multiply out $\zeta(s_1) \cdots \zeta(s_k)$, apart from $\zeta(s_1, \cdots, s_k)$ that we get in all unequal case, we have to look at a few n_i's being equal and we get lower depth zetas as follows: We have

$$\zeta(s_1, s_2) = \zeta(s_1)\zeta(s_2) + \zeta(s_1 + s_2)$$

and more generally $\zeta(s_1, \cdots, s_k)$ is the sum of $\zeta(s_1) \cdots \zeta(s_k)$ and all the depth $j < k$ zeta values with the first entry being a sum of $k - (j - 1)$ of the s_i's and the other $j - 1$ entries being the rest of s_i's. This is a symmetric form of the recursion.
   Alternately, it is easy to see that $\zeta(s_1, \cdots, s_k)\zeta(s_{k+1})$ is

$$\zeta(s_1, \cdots, s_{k+1}) + \zeta(s_1 + s_{k+1}, s_2, \cdots, s_k) + \cdots + \zeta(s_1, \cdots, s_{k-1}, s_k + s_{k+1}).$$

You can lower depths inductively, from either formula: eg., $\zeta(s_1, s_2, s_3)$ is

$$\zeta(s_1)\zeta(s_2)\zeta(s_3) + \zeta(s_1 + s_2, s_3) + \zeta(s_1 + s_3, s_2) + \zeta(s_2 + s_3, s_1) + \zeta(s_1 + s_2 + s_3$$
$$= \zeta(s_1)\zeta(s_2)\zeta(s_3) + \zeta(s_1 + s_2)\zeta(s_3) + \zeta(s_1 + s_3)\zeta(s_2) + \zeta(s_2 + s_3)\zeta(s_1$$

But all depth one values are eulerian and weigths are preserved in the recursion, proving the theorem.   □

Theorem 5.10.10   *When $q = 2$, $\overline{\zeta}(s_1, \cdots, s_k)$ are eulerian.*
   In fact, they are given explicitly in the proof, in terms of lower depths or in terms of depth 1 (i.e., the usual zeta).

Proof.   For depth not more than two, $\zeta = \overline{\zeta}$. In depth three, for example, we have cases: $n_1 \neq n_2, n_2 \neq n_3$ or $n_1 = n_2, n_2 \neq n_3$ or $n_1 \neq n_2, n_2 = n_3$ or $n_1 = n_2 = n_3$ and the corresponding relation

$$\zeta(s_1)\zeta(s_2)\zeta(s_3) = \overline{\zeta}(s_1, s_2, s_3) + \zeta(s_1 + s_2, s_3) + \zeta(s_1, s_2 + s_3) + \zeta(s_1 + s_2 + s_3).$$

Hence, $\overline{\zeta}(s_1, s_2, s_3) = \zeta(s_1, s_2, s_3) + \zeta(s_1 + s_3, s_2)$.

Non-symmetric recursion is easier, looking at cases $n_k \neq n_{k+1}$ or $n_k = n_{k+1}$, we get

$$\overline{\zeta}(s_1, \cdots, s_k)\zeta(s_{k+1}) = \overline{\zeta}(s_1, \cdots, s_{k+1}) + \overline{\zeta}(s_1, \cdots, s_{k-1}, s_k + s_{k+1}).$$

Hence straight inductive argument shows

$$\overline{\zeta}(s_1, \cdots, s_k) = \sum_{r=1}^{k} \sum_{0=j_0<j_1<j_2<\cdots<j_r=k} \prod_{a=0}^{r-1} \zeta(s_{j_a+1} + \cdots + s_{j_{a+1}}).$$
□

Theorem 5.10.11 *If $p \neq 2$ and s is 'even', $\zeta_l(s, s)$ is eulerian. More generally, if $p > k$, and s is 'even', then depth k zeta value $\zeta_l(s, \cdots, s)$ is eulerian.*

If $q = 3$ and s is even, then $\zeta(s, s)$ is eulerian.

Proof. Both depth two statements follow from the sum-shuffle identities and the Carlitz result. In the case of depth k and ζ_l, note that given any distinct monic n_i's, the term $1/(n_1 \cdots n_k)^s$ occurs only once in ζ_l, whereas it appears $k!$ times in $\mu(s, \cdots, s)$ where $\mu(s_1, \cdots, s_k) := \sum 1/(n_1 \cdots n_k)^s$, where the sum is over all distinct monic n_i's. But when s_i's are 'even', exactly as in $q = 2$ case, we can see that μ values are eulerian (by induction on k). Dividing by $k!$ (which is allowed since $p > k$) gives the result. □

Theorem 5.10.12 *When $q = 2$, $\zeta_d(2, 1)$ or $\zeta_d(1, 2)$ is not eulerian.*
When $q = 3$, $\zeta_d(2, 2)$ is not eulerian.

We postpone the proof of this theorem to Chapter 10, as it is an application of methods of 7.6, 8.1 and Jing Yu's theorems on transcendence referred in Chapter 10.

Theorem 5.10.13 *We have $2\zeta_d(1, 1) = 2\zeta(1, 1) = \zeta(1)^2 - \zeta(2)$.*
In particular, $\zeta_d(1, 1)$ and $\zeta(1, 1)$ are the same, if $p \neq 2$, but unequal for $p = 2$. In fact, for $p = 2$, $\zeta(1, 1) = 0$ and for $q = 2$, $\zeta_d(1, 1) = \zeta(2)/(t^2 + t)$.
On the other hand, $\zeta_d(2, 2)$ and $\zeta(2, 2)$ are the same for $q > 3$, but unequal for $q = 3$.
For $q > 3$, $2\zeta_d(2, 2) = \zeta(2)^2 - \zeta(4)$.

Proof. For $p = 2$, each quantity is zero, so we can assume that $p \neq 2$. Hence

$$\zeta(1,1)-\zeta_d(1,1) = \sum_{d=0}^{\infty} \sum_{\substack{n_1>n_2 \\ d_1=d_2=d}} \frac{1}{n_1 n_2} = \sum_{d} \sum_{n_2,d_2<d} \frac{1}{n_2} \sum_{n_1,d_1=d} \left(\frac{1}{n_1-n_2} - \frac{1}{n_1}\right) = 0,$$

where the last equality follows because n_1 and $n_1 - n_2$ run through the same set, for a fixed n_2 of smaller degree. This proves the first equality.

On the other hand, by 5.9.1,

$$\zeta_d(1,1) = \sum_{d=0}^{\infty} \frac{1}{l_d} \sum_{k=0}^{d-1} \frac{1}{l_k} = \sum_{d>k} \frac{1}{l_d l_k}$$

so that $2\zeta_d(1,1) + \sum 1/l_k^2 = (\sum 1/l_k)^2$, by usual shuffle argument. So the proof is complete by appealing to 5.9.1 (note that the second equality is just shuffle equality when $q = 3$). When $p = 2$, we have seen in 5.10.4 that $\zeta(1,1) = 0$, but $\zeta_d(1,1) = \sum_{d \geq 0, j>0} 1/(l_d l_{d+j})$ has valuation that of $1/l_1$ and hence is non-zero. For $q = 2$, the claimed evaluation of $\zeta_d(1,1)$ follows from $\sum_{d=0}^{k} 1/l_d^2 = (t^2 + t) \sum_{d=0,j>0}^{k+1} 1/(l_d l_{d-j})$, which follows by induction (and subtraction of consecutive values) from $1/l_k^2 = (t^2 + t)/l_{k+1} \sum_0^k 1/l_d$, which again follows from induction subtraction of consecutive values.

When $q = 3$, we also have shuffle identity $2\zeta(2,2) + \zeta(4) = \zeta(2)^2$. On the other hand, by argument as above, for any q, we have $2\zeta_d(2,2) + \sum 1/l_k^4 = (\sum 1/l_k^2)^2 = \zeta(2)^2$. But by 5.9.1, for $q > 3$, $\sum 1/l_k^4 = \zeta(4)$, and not for $q = 3$. ∎

The computer calculations, not done extensively suggest that the second part of the theorem generalizes considerably to $\zeta_d(s, \cdots, s) = \overline{\zeta}(s, \cdots, s)$, if $p \neq 2$ and $s_0(q - k) < q - 1$, $1 < k \leq q$. It is easy to see that $d = 1$ contribution is the same under these conditions.

We have seen that, when $q = 2$, $\zeta(2,1) = \zeta(1)\zeta(2) + \zeta(3) = \zeta(1)^3 + \zeta(3) = \bar{\pi}^3/((t^4 + t)(t^4 + t^2)) \neq 0$.

Theorem 5.10.14 *Let $p > 2$. Then $\zeta(2,1) = 0$ and more generally,*

$$\zeta(k,1) = \sum_{j=2}^{k/2} (\zeta(k+1) - \zeta(j)\zeta(k+1-j)) \text{ for even } k < q$$

$$= \frac{1}{2}(\zeta(k+1) - \zeta(\frac{k+1}{2})^2)$$

$$+ \sum_{j=2}^{(k-1)/2} (\zeta(k+1) - \zeta(j)\zeta(k+1-j)) \text{ for odd } k < q.$$

The same formula does not hold for $k = q$.

Proof. Note that $1/n_2 + 1/(n_1 - n_2) = n_1/(n_2(n_1 - n_2))$. Hence,

$$\zeta(2,1) = \sum_{n_1 > n_2} \frac{1}{n_1^2 n_2}$$

$$= \sum_{d=0}^{\infty} [(\sum_{n_1, d_1 = d} \frac{1}{n_1^2})(\sum_{n_2, d_2 < d} \frac{1}{n_2}) + \sum_{n_1, n_2, d_2 < d_1 = d} \frac{1}{n_1^2(n_1 - n_2)}]$$

$$= \sum_{d=0}^{\infty} \sum_{n_1, n_2, d_2 < d_1 = d} \frac{1}{n_1 n_2(n_1 - n_2)}$$

$$= \sum_{d=0}^{\infty} \sum_{n_2, d_2 = d} \frac{1}{n_2^2} \sum_{n_1, d_1 > d} (\frac{1}{n_1 - n_2} - \frac{1}{n_1})$$

$$= 0$$

where the last equality follows because, for a fixed n_2, the values through which n_1 and $n_1 - n_2$ run are the same and thus cancel.

Fix $k < q$. Write $S_j := \sum_{d, j \geq 0} 1/(l_d^j l_{d_j}^{k+1-j})$. Then

$$\zeta(k,1) = \sum_{n_1 > n_2} \frac{1}{n_1^k n_2}$$

$$= \sum_{d \geq 0} \sum_{n_1, n_2, d_2 < d_1 = d} \frac{1}{n_1^{k-1} n_2(n_1 - n_2)}$$

$$= \sum_{d \geq 0} \sum_{n_2, d_2 = d} \frac{1}{n_2^2} \sum_{n_1, d_1 > d} \frac{1}{n_1^{k-2}} (\frac{1}{n_1 - n_2} - \frac{1}{n_1})$$

$$= \sum_{d \geq 0} \sum_{n_2, d_2 = d} \frac{1}{n_2^2} \sum_{n_1, d_1 > d} \frac{1}{(n_1 - n_2)n_1^{k-2}} - \sum_{d \geq 0} \sum_{n_2, d_2 = d} \frac{1}{n_2^2} \sum_{n_1, d_1 > d} \frac{1}{n_1^{k-1}}$$

$$= \sum_{d \geq 0} \sum_{n_2, d_2 = d} \frac{1}{n_2^3} \sum_{n_1, d_1 > d} \frac{1}{n_1^{k-3}} (\frac{1}{n_1 - n_2} - \frac{1}{n_1}) - S_2$$

$$= \cdots$$

$$= -(S_{k-1} + \cdots + S_3 + S_2)$$

since eventually as the power of n_1 in the first term becomes zero, we get zero by reducing to the end of calculation of $\zeta(2,1)$ above. To finish the proof, note that by 5.9.1, $S_j + S_{k+1-j} + \sum 1/l_d^{k+1} = (\sum 1/l_d^j)(\sum 1/l_d^{k+1-j})$, and that $\sum 1/l_d^i = \zeta(i)$, when $i \leq q$, but not for $i = q + 1$. $\square$

Similar calculation gives

$$\zeta(1,k) = \zeta(1)\zeta(k) - \zeta(k+1) - \zeta_d(k-1,2) + \zeta_d(k-2,3) + \cdots + (-1)^k \zeta_d(2, k-1)$$

for $k < q$. Using the sum-shuffle for ζ_d, this can be expressed into just ζ's of depth one, when k is 2 or not divisible by 2 (so that alternating signs work the right way): eg. $\zeta(1,2) = \zeta_d(1,2) + \zeta_d(2,1) = \zeta(1)\zeta(2) - \zeta(3)$, for $p \neq 2$, which is consistent (since $\zeta(2,1) = 0$) with the classical sum-shuffle relation, even when $q > 3$. On the other hand, $\zeta(1,3) + \zeta(3,1) = \zeta(1)\zeta(3) - \zeta(2)^2$, for $q > 3$, which is different from classical counterpart. In general, for $p \neq 2$, and for $k < q$ not divisible by 2, $\zeta(1,k)$ is

$$\frac{(-1)^{(k-1)/2}}{2}(\zeta(\frac{k+1}{2})^2 - \zeta(k+1)) + \sum_{j=1}^{(k-1)/2} (-1)^{j-1}(\zeta(j)\zeta(k+1-j) - \zeta(k+1)).$$

Classically, $\zeta(2,1) = \zeta(3)$. In our case, none of the multizetas of depth more than one can be depth one zeta value, for the simple reason that all the terms in the multizeta sum are of degree less than zero, once the depth is more than one, whereas the depth one zeta values are 1-units at ∞.

Theorem 5.10.15   *When $q = 2$, $\zeta(k,l) = 0$ if and only if $k = l$.*

Proof.   We have already proved the 'if' part. By Theorems 5.10.3 and 5.10.4, we can assume that k and l are not both even and that $k > l$.

First case is when l is odd: The contribution by $n_1 = 1$ and $d_2 = 1$ is $1/t^l + 1/(t+1)^l = 1/t^{l+1} + \cdots$ and the other contributions are of still lower degree, thus $\zeta(k,l)$ is of degree $-(l+1)$ and hence non-zero.

It remains to consider $l = 2^r m$, $r \geq 1$, m is odd and k is odd: Now the $n_1 = 1$, $d_2 = 1$ contribution is of degree $-2^r(m+1)$, and if $k > 2^r(m+1)$, there is no cancellation as above, proving the result. Otherwise $2^r m < k < 2^r(m+1)$, and there is no cancellation to higher degree $-(k+1)$ you get with $n_2 = 1$, $d_1 = 1$, unless $k = 2^r(m+1) - 1$, when those contributions add and cancel leaving the degree to be $-(k+2)$, because the binomial coefficient 'k choose 2' is odd. Other contributions are again of lower degree giving you non-zero value. We omit the straightforward details.   □

Remarks 5.10.16   (1) This together with Theorems 5.10.4 and 5.10.10 show that when $q = 2$, $\overline{\zeta}(s, s, s') = \zeta(s, s+s') + \zeta(s, s)\zeta(s') = \zeta(s, s+s') \neq 0$, but the value is eulerian.

(2) Let us compare the properties of the variants: While ζ_d satisfies congruence relations with the complex multizeta, and ζ_l satisfies all sum-shuffle identities, their relations with periods do not seem as good as those of ζ and $\overline{\zeta}$. Further, ζ_l (and lexographic ordering) is not even invariant with respect to the automorphisms $t \mapsto t + \theta$, $\theta \in F_q$ of $F_q[t]$ together with

the sgn. The other variants of multizeta are invariants with these because, if $n_1(t) - n_2(t)$ is monic, then $n_1(t + \theta) - n_2(t + \theta)$ is monic.

(3) For $q = 2$: calculation shows that $\zeta_l(2,1)/\zeta(1)^3$ and $\zeta_l(2,1)/\zeta(3)$ and $\zeta_l(3,1)/\zeta(1)^4$ have very complicated continued fractions with many small beginning partial quotients, so that it appears that the corresponding ζ_l values are at least not rational multiples of $\tilde{\pi}^w$ with w being the weight and $\zeta_l(2,1)$ is not simply related to $\zeta(3)$.

(c) Interpolations

Definition 5.10.17 For Z standing for any of the variants defined above, keeping the summation conventions on n_i's the same, we define corresponding formal power series by

$$ Z(s_1,\cdots,s_k,X_1,\cdots,X_k) := \sum n_1^{-s_1}\cdots n_k^{-s_k} X_1^{d_1}\cdots X_k^{d_k}. $$

Hence $Z(s_1,\cdots,s_k,X_1,1,\cdots 1) = \sum n_1^{-s_1}\cdots n_k^{-s_k} X_1^{d_1}$ and $Z(s_1,\cdots,s_k,X,\cdots,X) = \sum n_1^{-s_1}\cdots n_k^{-s_k} X^{d_1+\cdots+d_k}$.

Theorem 5.10.18 *When s_i's are non-positive integers, these formal power series are polynomials, for Z being ζ, ζ_d or $\overline{\zeta}$.*

Proof. Coefficient of $X_1^{d_1}\cdots X_k^{d_k}$ is zero when some d_i is large in ζ_d case, and also when further, some large d_i is unequal to other degrees in ζ or $\overline{\zeta}$ case, because it has $S_{d_i}(-s_i) = 0$, as a factor then. (When $p \neq 2$, $d_1 \geq \cdots \geq d_k$, but we won't need this.) Hence assume, without loss of generality, that $d_1 = \cdots = d_r \neq d_j$ for $j > r$, then the coefficient has as a factor the sum of the form $\sum n_1^{-s_1}(n_1 - m_2)^{-s_2}\cdots(n_1 - m_r)^{-s_r}$, where n_1 runs throgh all (monic) polynomials of degree d_1 and m_i run through some of degree $< d_1$. (This is because once we have such a term for some choice of m_i's and n_1, we can substitute for n_1 any other degree d_1 monic polynomial, because the relative differences and hence the inequalities all remain unchanged.) Hence binomial expansions and vanishing of $S_{d_1}(-s)$ (here $-s \leq -(s_1 + \cdots + s_r)$) implies vanishing when d_1 is large. $\square$

The computer calculations (done only for $q = 2$ so far) suggest that the same is true for ζ_l: Indeed, when d_i are distinct, coefficients of $X_1^{d_1}\cdots X_k^{d_k}$ are the same for all variants and vanish, when one of the exponents is large. Consider $k = 2$, $s_1 = s_2 = -1$ and degree d contribution: For $i \leq 2d$, and $a + b = i$, suppose we count in how many ways we get t^a term from n_1 and t^b from n_2, then the number is multiple of p. In fact, each t^a and t^b comes power of q number of ways. It seems that if the number is μ, then total contribution is $1 + 2 + \cdots + \mu$ or $1 + 2 + \cdots + \mu - 1$ and in both cases it is zero in characteristic p.

Equating variables to one, we get interpolation of multizeta values at non-positive exponents leading to an element in $\mathbb{F}_q[t]$ as in the depth one case. Since n^s satisfy Kummer congruences, these values interpolate, giving a continuous $\wp$ adic function defined on $\mathbb{Z}_p^k$ taking values in $\mathbb{F}_q[t]_\wp$ by defining $Z_\wp(s_1, \cdots, s_k)$ to be the $\wp$-adic limit of $Z(s_{1_i}, \cdots, s_{k_i})$ as for sequences $-s_{r_i}$ tending to infinity and satisfying $s_{r_i} \equiv s_r$ modulo $\mathrm{Norm}(\wp)^i(\mathrm{Norm}(\wp) - 1)$ (and thus p-adically tending to s_r through a correct congruence class). As $n^{-s_{r_i}}$ tends $\wp$-adically to 0 or n^{-s_r} respectively according as $\wp$ divides n or not, the $\wp$ adic interpolation has a similar sum expression (convergent in $\wp$-adic topology), but where all n_i's are prime to $\wp$.

Remarks 5.10.19 (1) If the n_i's in the sum were 'independent', then we would have had $\prod(1 - \wp^{-s_i})$ as Euler factor. But because of complicated constraints, we do not have a relation with multizeta values via a simple Euler factor as in depth one case now.

(2) When $q = 2$, multizeta values at negative integers are zero and hence the interpolations are identically zero: Since we are dealing with finite sums involving $S_d(k)$'s, we see by straight manipulations that, for example, $\zeta(-a, -b, X_1, X_2) = \zeta(-a, X_1)\zeta(-b, X_2) - \zeta(-(a+b), X_1X_2)$, hence equating variables to one, we get the result for depth two from the known result in depth one, and so on. In the depth one and $q = 2$ situation of 5.5 (b), when the interpolation is identically zero, since the zero is simple, we can interpolate $\beta(k)$'s of 5.3 by the same techniques. In higher depth, there are several choices available of 'leading terms' according to whether we look at 'total degree' or in X_1 etc.

Examples 5.10.20 Let $Z = \zeta$, and m be a negative integer. For $p \neq 2$, $Z(0, m) = \zeta(m) - 1$ and $\zeta(1, m, X_1, X_2) = \sum_{d=1}^{\infty}(S_d(-m+1) + S_d(-m))(X_1X_2)^d$, so that $\zeta(1, m) = \zeta(m-1) + \zeta(m) - 2$. Note that this is zero, for $-(m - 1) < q - 1$, as both the zeta values are then one. For $q = 3$, on the other hand, it is never zero, as the zeta value with the even argument is then zero and the other value is congruent to $1 \bmod t$, by 5.3.1 and the discussion following it.

5.11 Analytic properties of zeta and Fredholm determinant

For special values results, we focused on simple zeta functions, but following classical analogies, Gekeler and Goss have defined and studied zeta and L-functions in more general setting and strong analytic properties are conjectured and partially proved for them by Goss, Wan and Taguchi, Pink and Böckle. We now discuss some of these definitions and results. We restrict to Drinfeld modules setting, and refer to [TW96; TW97; Gos96; Böc02] for more general settings of higher dimensional A-modules, ϕ-sheaves and shtukas. We start with L-function of Drinfeld modules in finite characteristic.

Definition 5.11.1 Let $\wp$ be a prime of A and $\mathbb{F}_\wp := A/\wp$. Let X be a scheme of finite type over $\mathbb{F}_\wp$. Let ρ be a Drinfeld A-module of rank r over X. Thus for a closed point $x \in X$, we get a rank r Drinfeld module $\rho^{(x)}$ over its residue field, say $r(x)$ of degree d_x over $\mathbb{F}_\wp$. Let F_x be the Frobenius morphism of $\rho^{(x)}$ over $r(x)$. Fix a prime v of A, other than the characteristic $\wp$. Let $H_v^1 := \mathrm{Hom}_{A_v}(T_v, A_v)$, where T_v is the v-adic Tate module of ρ.

Then we define the L function of ρ to be

$$L(\rho, T) := \prod_{x \text{ closed point of } X} \mathrm{Det}(I - T^{d_x} F_x | H_v^1)^{-1}.$$

Example 5.11.2 Note that without F_x, the product on the right is (the reduction modulo p of) the classical zeta function $\zeta(X, T)$ of X and hence is rational function of T by Dwork's result. Let's now take $A = \mathbb{F}_q[t] = X$, consider the constant family $\rho_t = \iota(t) + \tau$ of rank one. Then since $\rho_\wp = \tau^{\deg \wp}$ for monic $\wp$, we see that $L(\rho, T) = \zeta(X, \wp T)$. More generally, if ('sign-normalized') ρ has height and rank equal to r, then $L(\rho, T) = \zeta(X, \wp T^r)$ is again rational.

Theorem 5.11.3 *(Taguchi-Wan [TW96]) When $A = \mathbb{F}_q[t]$, the L function defined above is a rational function of T and is independent of the choices made in the definition.*

Gekeler also considered full zeta functions by looking at alternate products of characteristic polynomials on $H_v^i := \wedge^i H_v^1$, with analogy with the classical abelian variety case. See [Gek91] for some calculations.

Now in analogy with Artin-Hasse-Weil L-functions, we define L functions for Drinfeld modules in generic characteristic:

Definition 5.11.4   Let ρ be a Drinfeld A-module of generic characteristic over X, a scheme of finite type over A. For (non-zero) prime $\wp$ of A, the fiber $\rho_\wp$ of ρ over $\wp$ is a Drinfeld A-module over $X_\wp/\mathbb{F}_\wp$. We then define the L function of ρ to be

$$L(\rho, s) := \prod_\wp L(\rho_\wp, \wp^{-s}),$$

where the interpretation of $\wp^{-s}$ is explained in 5.5 for general A and $s = (x, y) \in S_\infty$ in Goss' character space.

Examples 5.11.5   Let ρ be the Carlitz module for $A = \mathbb{F}_q[t]$, then $L(s) = \zeta(s + 1)$.

Theorem 5.11.6   *(Taguchi-Wan [TW96]) Let $A = \mathbb{F}_q[t]$ and let ρ and X be as in definition. Then (i) for each fixed $y \in \mathbb{Z}_p$, the $L(\rho, s)$ is meromorphic as a function of x $(1/x?)$, (ii) It is continuous in y-variable, (iii) At negative integer y, it is a rational function in x.*

Let us discuss the ideas behind the proofs. Goss proved special cases by using Theorem 5.5.2 and its generalizations.

Daqing Wan realized that Dwork's p-adic (and christalline) technology to handle rationality and meromorphicity questions can be well-adapted in function field situation, and proved (together with Taguchi in more general case) conjectures of Goss in case of $A = \mathbb{F}_q[t]$. We refer to [Wan96a; TW96] for general discussion, (and discussion of entireness or v-adic situation, for finite v) but we sketch the ideas by focusing on the simplest case of the Carlitz zeta function and then indicating how the generalizations go.

For the Carlitz zeta function, with extra degree-counting variable X as in 5.1, we have $\zeta(s, X) = \prod(1 - X^{\deg(\wp)}\wp^{-s})^{-1}$. Factoring monic $\wp$ of degree d as $(t - x)(t - x^q) \cdots (t - x^{q^{d-1}})$, we see that, if we let $n = r = 1$ and $B(z) = B_s(z) = (t - z)^{-s}$ in 2.11.5, then

$$\zeta(s, X)(1 - Xt^{-s}) = L(B/\mathbb{G}_m, X) = \mathrm{Det}(I - XF_B)$$

because we are avoiding prime $\wp = t$ by excluding $x = 0$, whereas all others are accounted for. We can use affine space formula of 2.11.5 to get the full zeta function:

$$\zeta(s, X) = \mathrm{Det}(I - XF_B^{\{1\}}).$$

In this case, in fact, we do not need to use this general formula and what we need follows immediately from the first formula above by the observation that F_B and $F_B^{\{1\}}$ only differ in that the first has zeroth row and column extra and that $b_0 = t^{-s}$ the $(0,0)$-th entry of the first determinant is $1 - Xt^{-s}$, whereas all the other entries in the zeroth row are zero

Example 5.11.7 (1) When s is negative $B_s(z)$ is just a polynomial, leading to only finitely many non-zero b_i's For example, when $q = 3$, $s = -5$, a straight calculation shows that the infinite determinant giving $\zeta(-5, X)$ reduces to the two-by-two determinant corresponding to $1 \leq i, j \leq 2$, because the diagonal entries are one and below diagonal entries are zero after the second row, thus giving $(1 - t^3 X)(1 + tX) - (X)(-t^4 X) = 1 + (t - t^3)X$ as it should.

(2) When $s = 1$, we have $b_i = t^{-i-1}$ and we get a truly infinite determinant. The n-th diagonal entry is $1 - X/t^{nq-(n-1)}$ and thus it easily follows that $\zeta(1, X) = 1 - X(1/t^q + 1/t^{2q-1} + \cdots) + X^2 \quad + \cdot$, and the coefficient of X is $-1/(t^q - t)$ as it should from Section 5 6 formulas

Though they look complicated, it might be possible to get interesting special value results from this determinantal formula, which are quite successful in giving good analytic information as we will see

This thus reduces the study of analytic properties of ζ to those of F_B and B The technique applies to quite general zeta and L-functions of Drinfeld $\mathbb{F}_q[t]$-modules of arbitrary rank over general schemes For example, simple linear algebra calculation of Frobenius with standard basis shows that for the zeta function for Drinfeld module $\rho_t = \sum_{i=0}^{r} t_i \tau^i$, with $t_r = 1$ we can take corresponding B to be r cross r matrix which is essentially the companion matrix of the τ-polynomial above and $B(x)$ is just its value at x, i e. reduction for the corresponding maximl ideal

The general A-case looks complicated In addition, there are difficulties with ideal exponentiation etc. as we have seen for zeta function.

5.12   Note on classical interpolations

We now recall interpolations (or more generally p-adic counterparts) of some basic functions in the simplest case of field of rational numbers For full account of some of the things here, [Kob80, Kob77] are good references. Interpolation is still an art and not fully developed science yet, in that we do not yet have a good unified view of all interpolation processes, but we (hopefully) recognize a good interpolation when we see it

Below we will assume $p \neq 2$ for simplicity, otherwise some formulas need slight modification, such as replacing p by $2p$, as usual.

For functions like exponential and logarithm, we just use the same Taylor series, but now p-adically in whatever domain it converges, to get $\exp_p$ and $\log_p$. We can enlarge the domain by using Dwork's trick or Iwasawa choice etc., so that we can evaluate at say roots of unity and get p-adic expression for additive character.

If we try to take straight p-adic limits of gamma values $\Gamma(n) = \prod_{j=1}^{n-1} j$, as n approaches x, p-adically, then letting n large, we see that the limit would be zero, because of many p's in the product. Overholtzer [Ove52] dropped terms in the product and interpolated $\prod_0^{n-1}(1 + ip)$ p-adically, by Wilson's theorem $((p-1)! \equiv -1 \mod p)$ generalization to uniformly continuous function on $\mathbb{Z}_p$, when $p \neq 2$. A very successful interpolation was obtained by Morita [Mor75] by interpolating $(-1)^n \prod_1^{n-1} \bar{j}$, where $\bar{j}$ is j, if p does not divide j, and it is 1 otherwise. This drops less number of terms (only those divisible by p) while alternating signs, and again interpolates by generalization of Wilson's theorem.

Why should zeta function have a p-adic interpolation?: Since n^{k_1} is p-adically close to n^{k_2}, if k_1 and k_2 are the same modulo $p-1$ and close p-adically, we may expect the same from $\zeta(-k_1)$ and $\zeta(-k_2)$ under the same conditions, because $\zeta(-k)$ is morally $\sum n^k$, and the Fermat congruences imply $n^{k_1} \equiv n^{k_2} \mod p^k$, when $k_1 \equiv k_2$ modulo $(p-1)p^k$ and either $(n, p) = 1$ or $k_1 \geq k_2 > k$. When we approach k p-adically through high powers, the terms divisible by p should contribute zero in limit and we should get Euler factor at p out. The problem is that when it is actually this sum (i.e. $-k > 1$), it is infinite sum which is transcendental (known at even k and expected in general), and for positive k's when it is rational (so that p-adic), it is not such a sum. On the other hand, there is Euler's famous formal calculation which gives the right answer. It is essentially

$$\zeta(1-k) = \sum n^{k-1} = (\frac{d}{dt})^{k-1} \sum e^{nt}|_{t=0} = (\frac{d}{dt})^{k-1}(\frac{1}{1-e^t}-1)|_{t=0} = -\frac{B_k}{k}.$$

Let us look at several approaches to zeta function interpolation in detail, focusing only on Riemann zeta function, though most of them generalize. We look at only those approaches close to function field counterparts we have discussed. For more, and for different treatments, see books by Hida, Borevich-Shafarevich, Ireland-Rosen and [Kat75; Was97; Iwa72; Kob80; Kob77].

(i) **Bernoulli congruences approach**: This approach uses these correct values $\zeta(1 - k) = -B_k/k$. First we generalize the congruence modulo

p proof in Borevich-Shafarevich to modulo p^n: Assume first that $p \neq 2$ and let g be a primitive root modulo p^n. From

$$\sum \frac{B_m(g^m - 1)t^{m-1}}{m!} = \frac{g}{(e^t - 1 + 1)^g - 1} - \frac{g}{e^t - 1} = \sum c_i(e^t - 1)^i$$

we see that c_i are p-integral. Since $(e^t - 1)^i$ is an integral linear combination of terms $e^{rt} = \sum r^n t^n / n!$ and we have $r^{m+w(p-1)p^{n-1}} \equiv r^m \mod p^n$, whenever $m \geq n$, $w > 0$ and $g^m - 1$ are also periodic modulo $(p-1)p^{n-1}$, we see that when m is not divisible by $p - 1$, B_m/m are p-integral and satisfy $B_{m+w(p-1)p^{n-1}} \equiv B_m \mod p^n$ when $m > n$ and $w > 0$. This congruence is equivalent to what Kummer proved via Mahler transform.

Thus if we define $\zeta_p(-s) := \lim \zeta(-s_i)$, where $s_i \to \infty$ (this is forced by $m > n$ condition above) and p-adically approach s and $s_i \equiv s \mod p - 1$, we see that $\zeta_p(s)$ extends to a continuous function from $\mathbb{Z}_p$ to $\mathbb{Z}_p$, because such a congruence class is dense in $\mathbb{Z}_p$.

Let us now show that for $s \in \mathbb{Z}_{>0}$, $\zeta_p(-s) = (1 - p^s)\zeta(-s)$. We have to show that as $m \to \infty$, $B_{n+(p-1)p^m}/(n + (p-1)p^m)$ p-adically approaches $(1 - p^{n-1})B_n/n$:

Recall that $B_n = (\lim_{k\to\infty} \sum_1^{p^k} a^n)/p^k$, considering the same expression for $B_{n+(p-1)p^m}$, we see that as $m \to \infty$, as the contribution of a's divisible by p tends to zero, we can restrict to a's not divisible by p. Hence we are looking at limit as $k \to \infty$ of $\sum_1^{p^k} a^n/p^k - \sum_1^{p^{k-1}} p^{n-1}a^n/p^{k-1}$ which is $(1 - p^{n-1})B_n$ as claimed.

In fact, we can avoid the generating function argument above completely and can use this expression of B_n to give direct proof of p-adic interpolation of values once the Euler factor is removed by proving full Kummer congruences in the same way, because as we saw, taking away Euler factor corresponds to summing over a prime to p, so that Fermat's congruence applies.

(ii) **Measure and integration approach**: Mazur's measure approach: [Kat75; Kob77; Kob80] or Cassels' 'Local fields' book. We move from sums to integrals on $\mathbb{Z}_p$:

$$\int_{\mathbb{Z}_p} f d\mu := \lim \sum_0^{p^n - 1} f(a)\mu(a + p^n \mathbb{Z}_p),$$

for suitable measures μ. Using Bernoulli-zeta identity (or distribution property of Bernoulli polynomials, mentioned in 4.16) we manipulate a suitable measure such that the moments $\int x^k d\mu$ are (essentially, there is a factor to take care of integrality) $\zeta(-k)$, so that integral on $\mathbb{Z}_p^* = \mathbb{Z}_p - p\mathbb{Z}_p$ is $(1 - p^k)\zeta(-k)$, and since Fermat congruence says that if k_i and k_2 are in the same congruence class modulo $p - 1$ and close p-adically, then x^{k_1} is close to x^{k_2}, so are their integrals (on compact $\mathbb{Z}_p^*$), giving us Kummer congruences on Zeta values and interpolation. (The measure in our case, see 5.7, is much simpler.)

(iii) **Kubota–Leopaldt's approach** is based on the p-adic limit formula in 4.16 for Bernoulli numbers, and the observation that if f is locally analytic on $\mathbb{Z}_p$, then $\lim p^{-n} \sum f(j)$ exists, where the sum is over $0 \le j < p^n$ and the limit is as $n \to \infty$. Hence if we restrict to $k \equiv k_0$ mod $p - 1$ and write $j = \langle j \rangle w(j)$, with Teichmüller character w, then

$$(1 - p^{k-1})\zeta(1 - k) = (1 - p^{k-1})(-B_k/k) = -k^{-1} \lim p^{-n} \sum \langle j \rangle^k w(j)^{k_0}$$

interpolates, where the sum is over $0 \le j < p^n$ such that p does not divide j. Similarly, we can define p-adic L functions and partial zeta functions. For example, taking limit as N tends to ∞, and as N tends p-adically to 0,

$$\overline{\zeta}_p(s, x) := \lim \sum_{0 \le j < N, (p, x+j)=1} \langle x + j \rangle^{1-s}/(s - 1).$$

(iv) **Eisenstein series approach**: Serre's modular approach looks at Fourier expansion of Eisenstein series. The fact that the constant term is zeta value is important and occurs in great generality of various L-functions and is basis of Langlands approach to analytically continue L-functions via continuing Eisenstein series. Serre achieved a p-adic continuation by p-adically continuing the Eisenstein series noting that its coefficients for non-constant terms are just divisor functions, thus finite sums of powers and hence satisfy Fermat-Kummer congruences. We will see in 6.2 in our function field case that Eisenstein series defined by Goss have constant terms zeta values, but have complicated Fourier expansions, so that such approach is not clear, especially because of lack of understanding about functional equation.

(v) **Grouping the sums approach**: Stark's beautiful approach [Sta87] towards p-adic zeta function is via partial p-adic zeta functions which are just defined by grouping the sums correctly (very close to our function field approach!), by taking limit as N tends to ∞, and as N tends

p-adically to $-x$,

$$\zeta_p(s, x) := \lim \sum_{0 \le j < N, (p, x+j)=1} \langle x + j \rangle^{-s}, \quad x, s \in \mathbb{Z}_p.$$

A quick connection between Stark and Kubota-Leopaldt approaches can be seen as follows. Consider following two simple calculations (we drop the precise hypotheses for simplicity).

(a) Let $g(x) := \lim(\sum f(x + j))/N$, where the sum is over $0 \le j < N$ and the limit is as N becomes large and tends p-adically to 0. Then $g(x + 1) - g(x) = \lim(f(x + N) - f(x))/N = f'(x)$.

(b) Let $h(x) := \lim \sum f'(x + j)$, where the sum is over $0 \le j < N$ and the limit is as N becomes large and tends p-adically to $-x$. Then $h(x + 1) - h(x) = \lim f'(x + N) - f'(x) = f'(0) - f'(x)$.

In particular, if $f'(0) = 0$, then $-g(x)$ and $h(x)$ differ by a constant. Let $f(x+j)$ be $\langle x+j \rangle^{1-s}/(1-s)$, if p does not divide $x+j$, and let it be 0 otherwise. Then we get Stark and Kubota-Leopaldt zeta functions $\zeta_p(s, x)$ and $\overline{\zeta}_p(s, x)$. Since they differ in a constant function of x, when we take character sums, we of course get the same L-functions. So we can easily prove one from the other.

The trade-off is that in Stark's approach the sum is closer to classical definition, but the values are more complicated, involving $B_{k+1}(0)$, and for forming L-or zeta functions from the partial ones, we need to vary the p-adic limits of N, by varying x.

Though the definition is not as nice looking, Kubota-Leopaldt special values are classical (up to Euler factor at p) and forming L or zeta functions from the partial ones we always take p-adic limit of N as zero. Stark's and diamond's log-gamma are also connected via observations above.

Since $\log \Gamma(x + 1) - \log \Gamma(x) = \log(x)$, Diamond made p-adic analog of the log gamma function as $\lim p^{-n} \sum (x+j) \log_p(x+j) - (x+j)$, where the sum is over $0 \le j < n$ and the limit is as $n \to \infty$. This has nice property, but it is defined on complement of $\mathbb{Z}_p$ in $\mathbb{C}_p$, whereas $\log_p \Gamma_p$ gives a nice analog on $\mathbb{Z}_p$. See [Kob80] for more on this.

Again, Stark's approach to p-adic gamma is connected by the two observations above.

Chapter 6

Higher rank theory

So far we have essentially looked only at rank one theory (and complex multiplication theory rank one over new base) in detail. Now we look at the higher rank objects; for example, rank two objects, such as analogs of elliptic curves, modular forms, hypergeometric functions, and also higher rank objects such as higher rank modular forms which have no direct classical analogs.

To keep analogies visible, we use classical notations for analogous objects such as function field Goss modular forms j, Δ, E_k etc. or function field hypergeometric functions.

6.1 Elliptic modules

In Chapter 2, we have already described how in the rank two situations, the analogies are closer with the elliptic curves situation. Let us quickly review: The lattice rank is then two, the endomorphism ring structures are similar, torsion possibilities are similar. For example, in finite characteristic, for the rank two, we have only two possibilities as in the elliptic curves case, namely ordinary or supersingular, rather than intermediate possibilities in the higher ranks.

Consider an elliptic curve E over $\mathbb{Q}$. If it has complex multiplications, Deuring proved that it has ordinary reduction for half (in terms of density) the primes and supersingular for the rest of the places of good reduction. On the other hand, Elkies proved the existence of infinitely many supersingular primes in general. Brown proved the existence of infinitely many supersingular primes for 'most' of the rank two, generic characteristic Drinfeld $\mathbb{F}_q[t]$-modules. Poonen (see [Poo98] also for references to papers of Elkies, Brown and Chantal David) corrected the list of possible exceptions in Brown's paper and provided examples (eg., $\rho_t = t(1 - \tau)^2$, when $p \neq 2$) which have no **supersingular prime** and explained representation theoretic obstructions (which do not exist for number fields) for infinitute. The general situation is an interesting open question.

The comparison of Weierstrass equations and the Drinfeld modules reveals interesting analogies and gives some modular forms [Gos80c]: For a fixed a, the coefficient of τ^i in ρ_a, considered as a function of the lattice L corresponding to ρ, is a **modular form** (see below for more analytic details) of weight $q^i - 1$. This follows from the commutation relations $\tau^i c = c^{q^i} \tau^i$. For example, if a rank 2 Drinfeld module ρ for $\mathbb{F}_q[t]$ is given by $\rho_t = t + G\tau + \Delta\tau^2$, then as $L \to \lambda L$, $(G, \Delta) \to (\lambda^{1-q}G, \lambda^{1-q^2}\Delta)$. This should be compared with $y^2 = x^3 - g_2 x - g_3$, with modular forms g_2, g_3. In fact, $j := G^{q+1}/\Delta$ is a weight 0 modular function parametrizing the isomorphism classes of these Drinfeld modules, again parallel to the elliptic curve situation. Algebraicity and transcendence properties (see Chapter 10) of this j-function, as well as factorizations of singular moduli [Dor91; HY98] are in parallel with the elliptic curves case. Also note that if Δ vanishes, we get a degeneration (bad reduction) of the Drinfeld module to rank one. This corresponds to the fact that Δ, which plays the role of the discriminant, is a cusp form. (Note that for general A, there are several Δ's, essentially one for each a.) On the other hand, we have seen a strong contrast in the study of rational points, as now Mordell-Weil rank is no longer finite. But we can still look at the rational torsion questions.

In a major breakthrough study, Mazur had shown boundedness of rational torsion for any elliptic curve over a given number field by studying the rational points on modular curves $X_1(n)$ (whose non-cuspidal points classify elliptic curves with a point of order n) and in particular proved a complete list of possibilities over $\mathbb{Q}$, thus settling Levi-Ogg conjecture. Building on this and work of Kamienny, Merel proved uniform boundedness where we just fix a degree of a number field rather than the number field.

In the Drinfeld modules setting, Poonen [Poo97] studied the corresponding questions and proved **uniform boundedness** for rank one as well as analog of the result of Manin bounding $\wp$-primary part of rational torsion for the rank two $\mathbb{F}_q[t]$-modules over a given function field F: Having such a $\wp^n$-torsion, gives a F-rational point on Drinfeld modular curve $X_1(\wp^n)$, which being non-isotrivial of genus at least two for large n has only finitely many F-rational points by analog of Mordell conjecture over function fields proved by Samuel and for each corresponding j-invariant, the rational torsion over the Drinfeld modules having that j is uniformly bounded.

The reason that we can not imitate the classical proof of boundedness of rational torsion of the elliptic curve over (characteristic zero) function field using the genus computation of $X_1(n)$ is that Drinfeld modular curve $X_1(I)$ being a curve over $\mathrm{Spec}(A)$, it only implies boundedness of L-torsion for rank 2 ('non-iso-trivial') Drinfeld A-modules, where L is now a function field over the 'constants' K, rather than a finite extension of K suggested by the usual analogy.

The stronger conjectures are still open.

Now over number fields, the Tate isogeny conjecture and **Shafarevich finiteness conjectures** follow from each other for abelian varieties. On the other hand, in our case, we see immediately that the family $\rho_t = t + G\tau + \tau^2$ depending on G contains infinitely many non-isomorphic rank 2 Drinfeld modules, with good reduction everywhere (so not only the support of the discriminant is bounded, but the discriminant is one).

This is in contrast to the classical situation, where (by the Faltings theorem, which was known as the Shafarevich conjecture) there are only finitely many isomorphism classes of abelian varieties over a given number field K and of given dimension, with a good reduction outside a fixed finite set of places of K. The fact that the discriminant (for any non-constant a, the corresponding discriminant, i.e., the top coefficient for ρ_a, is enough for our purposes, since ρ_a determines ρ) is quite unrestricted here in contrast to the classical case where the bound on the discriminant also bounds g_2 gives this different behavior in the reduction theory. In fact, we do not have any good definition for a **'conductor'** of a Drinfeld module: The usual way to get the exponent at p from l-adic representations fails as for all $\wp$'s we are still in characteristic p. A crude candidate like the product of primes in the support of the minimal discriminant is an isogeny invariant, but fails to give any refined information and it does not satisfy any Szpiro conjecture type bounds: The naive analogies suggest exponent $q+1$ in place of 6 in Szpiro conjecture. But even in semistable case, where classically the conductor is exactly this product, the discriminant exponent can be arbitrary (as we can see in the example $\rho_t = t + \tau + \wp^n\tau^2$) and Szpiro type inequality does not hold for any exponent. More generally, we see that the local exponents in the conductors can not stay bounded, in contrast to $p > 3$ case over number fields, if such an inequality is to hold.

One of the highlights of the study of arithmetic of elliptic curves is the **rank of the group of rational points** and its connection with the Birch and Swinnerton-Dyer conjecture. We saw in 2.9 that the finite generation theorem such as Mordell-Weil theorem does not hold for Drinfeld modules. Another possibility of getting hold of Mordell-Weil type rank r in our case might be to use points over finite fields and get rank out of distributions of traces of Frobenius at p. Recall that the first version of Birch-Swinnerton-Dyer conjecture, expected (but not known) to be equivalent to their L-function version, says that $\prod_{p \leq x} N_p/p$ is asymptotic to $c(\log x)^r$. Here $N_p = |E(\mathbb{F}_p)| = p + 1 - \text{Trace}(\text{Frob}_p)$. Hence we will look at (see also [HY00; Dav01]) distribution of $\text{Frob}_\wp$'s. First note that if we make a similar definition then the product converges as $x \to \infty$, as terms tend to one! So similar use, if any, is unknown. We will focus on rank two, though similar calculations can be done in general.

Example 6.1.1   Let ρ be $\mathbb{F}_q[t]$-module over $\mathbb{F}_q[t]$ given by $\rho_t = t + e\tau + \tau^2$. Let $\wp$ be a prime of degree d. We will use the same notation for reduction mod $\wp$, except we use θ for image of t to avoid confusion. Then we have $\wp = \prod_0^{d-1}(t - \theta^{q^i})$.

Let

$$M_k := \begin{pmatrix} 0 & t - \theta^{q^k} \\ 1 & e^{q^k} \end{pmatrix}.$$

Then M_0 is just the companion matrix for τ, and since we are in τ-linear rather than linear setting, the matrix for τ^k is $F_k := M_0 M_1 \cdots M_{k-1}$ and $\text{Frob}_\wp$ is M_d.

First, for simplicity, let $e = 0$. In this case, we have complex multiplications by $\mathbb{F}_{q^2}[t]$. If d is odd, the diagonal entries and thus trace of F_d, which is the trace of $\text{Frob}_\wp$ for ρ, is zero. Eigenvalues are just $\pm\sqrt{\wp}$. If d is even, then we get diagonal F_d with the diagonal entries (i.e., eigenvalues) $\prod(t - \theta^{q^i})$ and $\prod(t - \theta^{q^j})$, where i and j run respectively through even and odd values $\leq d - 1$. So, in particular, the trace of Frobenius in this case is zero, if d is odd, and it is of degree $d/2$, if d is even and $p \neq 2$ (the degree drops when $p = 2$).

Note that this 'half ordinary, half supersingular' behavior is as in the classical complex multiplication case, but not quite, as the natural densities do not exist (as in the 'prime number theorem' situation in 1.3) because of the wild jumps in norms. Also, $d/2$ is the Riemann hypothesis bound reached here.

Now let us look at the general case. From definitions it follows that

$$\text{If} \quad F_k = \begin{pmatrix} \alpha_k & \beta_k \\ \gamma_k & \delta_k \end{pmatrix} \quad \text{then}$$

$$F_{k+2} = \begin{pmatrix} \alpha_k(t - \theta^{q^k}) + \beta_k e^{q^k} & \beta_k(t - \theta^{q^{k+1}}) + e^{q^{k+1}}(\alpha_k(t - \theta^{q^k}) + \beta_k e^{q^k}) \\ \gamma_k(t - \theta^{q^k}) + \delta_k e^{q^k} & \delta_k(t - \theta^{q^{k+1}}) + e^{q^{k+1}}(\gamma_k(t - \theta^{q^k}) + \delta_k e^{q^k}) \end{pmatrix}.$$

First we claim that if d is even and $p \neq 2$, then again degree of the trace of Frobenius at $\wp$ is $d/2$. The proof follows by straight induction on d by applying the matrix entry calculation above to induction hypothesis that α_d and δ_d are monic of degree $d/2$ and that β_d (γ_d respectively) has degree at most (less than respectively) $d/2$.

Next we claim that if d is odd, then the degree is $(d-1)/2$, unless the trace from $A/\wp$ to $\mathbb{F}_q$ of e is zero. (For example, if $e \in \mathbb{F}_q - \{0\}$, then this is exactly when p divides d.) Again the proof follows as above with the induction hypothesis that β_d and γ_d are monic of degrees $(d+1)/2$ and $(d-1)/2$ respectively and $\alpha_d = (\sum_0^{d-2} e^{q^j})t^{(d-1)/2} + \text{lower}$ and $\delta_d = (\sum_0^{d-1} e^{q^i})t^{(d-1)/2} + \text{lower}$, where i and j run through even and odd values respectively.

6.2 Modular forms

We have already mentioned the coefficients of the Drinfeld modules arising as (C_∞-valued rather than $\mathbb{C}$-valued) modular forms. The **automorphic forms** considered by Weil, Jacquet, Langlands, Drinfeld are basically $\mathbb{C}$-valued (or F-valued for any characteristic zero field F, since in the absence of archimedean places no growth conditions needed and all arise from those over $\mathbb{Q}$ by tensoring) functions ϕ on $G(K)\backslash G(\mathcal{A})/KZ(K_\infty)$, where $G = GL_2$ say. See [DH87; GR96; Tei] for the discussion in Drinfeld modules settings as well as comparison.

Following analogies with the classical upper half-plane approach, as well as sections of multi-differentials over modular curves approach, Goss [Gos80c; Gos80d; Gos80a] considered C_∞-valued **modular forms** on **Drinfeld upper half-plane** $\Omega := C_\infty - K_\infty$ (compare $\mathcal{H}^\pm = \mathbb{C} - \mathbb{R}$) in the rank 2 situation which we will focus on.

We replace Ω by Drinfeld symmetric space $\Omega^{r-1} := \mathbb{P}^{r-1}(C_\infty)$ minus all K_∞-rational hyper-planes, for the general rank r situation. (Analogous p-adic symmetric spaces that Drinfeld then defined have been of great use in moduli theory, representation theory and work on p-adic Birch and Swinnerton-Dyer conjecture.)

The automorphic forms, on the other hand, live on a tree (=Bruhat-Tits building for $Gl_2(K_\infty)$= tree of norms) (and its quotients) constructed as a nerve of special covering of Ω. See [Dri74; DH87; Tei].

Our main references are papers mentioned above and works of Gekeler where he developed the theory extensively, for example, [Gek86; Gek99; Gek88; Gek83; Gek85a; GR96; Gek01] and various articles in [G+97].

Put $\text{Im}(z) := \text{Inf}_{x \in K_\infty} |z - x|$. Then $\text{Im}(\gamma z) = |\text{Det}(\gamma)||cz + d|^{-2}\text{Im}(z)$ for $\gamma \in GL_2(K_\infty)$. The sets $\Omega_c := \{z \in \Omega : \text{Im}(z) \geq c\}$ give open admissible neighborhoods of ∞ (not to be confused with the place ∞ of K) in the rigid analytic topology. Ω is connected but not simply connected.

Let e denote the exponential for the Carlitz module, i.e., corresponding to $\Lambda = \tilde{\pi} A$. Then $q_\infty(z) = 1/e(\tilde{\pi}z)$ is a uniformizer which takes a neighborhood of ∞ to the neighborhood of origin and since it is invariant with respect to translations from A, it can be used for q_∞-expansions (analogs of $q = e^{2\pi i z}$-expansions).

Definition 6.2.1   Modular form of weight k (nonnegative integer), type m (integer modulo $q-1$ (or rather the cardinality of $\text{Det}(\Gamma) \subset \mathbb{F}_q^*$)) for Γ is $f : \Omega \to C_\infty$ satisfying $f(\gamma z) = (\text{Det}(\gamma))^{-m}(cz + d)^{-k}f(z)$, for $\gamma \in \Gamma$ and which is rigid holomorphic and holomorphic at cusps.

Remark 6.2.2   Since $dq_\infty = -\tilde{\pi}q_\infty^2 dz$ (in contrast to $dq = (2\pi i)qdz$), the holomorphic differentials correspond to double-cuspidal forms.

As an analog of **Dedekind product formula** into cyclotomic factors:

$$\Delta(z) = (2\pi i)^{12}q \prod_{n \in \mathbb{Z}_{>0}} (1 - q^n)^{24} = (2\pi i)^{12}q \prod ((q^{-n} - 1)q^{\text{Norm}(n)})^{24},$$

for Δ as above, Ernst Gekeler [Gek85b] proved

$$\Delta = -\tilde{\pi}^{q^2-1}q_\infty^{q-1} \prod_{a \in A+} (C_a(q_\infty^{-1})q_\infty^{\text{Norm}(a)})^{(q^2-1)(q-1)}.$$

This has been generalized recently to higher ranks by Hamahata [Ham02].

For $A = \mathbb{F}_q[t]$ and $\Gamma = GL_2(A)$, the algebra of modular forms of type 0 is $C_\infty[g, \Delta]$ and the algebra for all types is $C_\infty[g, h]$ here h is a Poincaré series of type 1 and weight $q+1$ defined by Gekeler. We have $h^{q-1} = -\Delta$.

Eisenstein series $E^{(k)}(z) = \sum'_{a,b \in A}(az + b)^{-k}$ are of weight k, type 0.

The constant terms, by construction, are zeta values. But in general, the coefficients of q_∞-expansions of modular forms, which are very rich arithmetically in the classical case, are very poorly understood objects so far.

For the rest of this section, we will focus only on $A = \mathbb{F}_q[t]$ situation which is developed more than the general case.

Remarks 6.2.3 (1) Hecke operators can be defined as usual, but now they are totally multiplicative: we have $T_{\wp^n} T_\wp = T_{\wp^{n+1}} + q^d T_{\wp^{n-1}} T_\wp$ as usual, but the $q^d = 0$ now!

This makes associated Galois representations abelian. Böckle has recently given an analog of Deligne's construction to attach Galois representation to modular forms and proved an analog of Eichler-Shimura isomorphism.

(2) We have $T_\wp E^{(k)} = P^k E^{(k)}$ and $T_\wp \Delta = P^{q-1}\Delta$, where $\wp = (P)$ for monic generator P. So the eigenvalues do not determine the form (this happens even in weight two). Multiplicity one fails and Hecke action is not semi-simple.

(3) Gekeler [G+97, pa. 184] proved that there are no cusp forms of weight one for any arithmetic subgroup of $Gl_2(A)$, in contrast to the classical case mentioned at the end of Chapter 1.

Because of this total multiplicativity, we can associate $L_f(s) := \prod(1 - c_\wp \wp^{-s})^{-1}$, but then this Dirichlet series is indexed by $a \in A$ whereas the q_∞-expansions are indexed by $n \in \mathbb{Z}$ and the usual connection $\sum c_n q^n \leftrightarrow \sum c_n n^{-s}$ does not make sense. (The arithmetic meaning of the q_∞-expansion coefficients is not understood even for the Eisenstein series. For Eisenstein series associated to totally real fields, the q_∞-expansion has not been understood : This is one reason why we have not yet been able to imitate Siegel's proof mentioned in 5.2 and remove the abelian hypothesis there.)

There are other ways to attach L-functions to f due to the work of Drinfeld, Schneider, Teitelbaum, Gekeler and Goss which we now describe:

Let z be co-ordinate on Ω. Let $U := \{P \in \Omega : q^{-1} < |z(P)| < q, |z(P) - \lambda| > q^{-1}, \text{for} \lambda \in \mathbb{F}_q\}$. Then translates $U(\gamma)$ of U by $\gamma \in GL_2(K_\infty/GL_2(\mathcal{O}_\infty)$ give a special rigid covering of Ω. Associated to it we can define an infinite homogeneous tree $\mathcal{T}$ with $q+1$ edges leaving every vertex, where the opens $U(\gamma)$'s correspond to its vertices and the overlapping annuli correspond to the edges. Drinfeld constructed this as tree of norms and it is also the usual Bruhat-Tits building for $PGL_2(K_\infty)$.

Modular form of weight k and type m for Γ gives rise to a Γ-invariant harmonic cochain (i.e., function c on (oriented) edges e of $\mathcal{T}$ such that

$\sum_{e \to v} c(e) = 0$ and $c(e) = -c(e^-))$ c_f of weight k and type m (i.e., with values in $V(1-k, 1-m)$). Here V is the standard two-dimensional representation of $GL_2(C_\infty)$ and $V(n,i) := (\mathrm{Det})^i \otimes \mathrm{Sym}^{n-1}(V^*)$ (essentially space of homogeneous forms in two variables X and Y of degree $n-1$) and $V(-n, -i) := \mathrm{Hom}(V(n,i), C_\infty)$.

In fact, $c_f(e)$ is given on the basis by $\mathrm{Res}_e(f)(X^i Y^{k-2-i}) = \mathrm{Res}_e z^i f(z) dz$, where the residue is in the annulus corresponding to e. The Eisenstein series have zero residues, but for $k \geq 2$, the residue map is an isomorphism between the space of cusp forms of weight k and type m for a group Γ and the space of harmonic cochains of weight k and type m for Γ. In fact, the inverse process is integration: A harmonic cochain c gives rise to a 'measure' (we will not go into the technicalities of this integration theory) μ_c on $\mathbb{P}^1_{K_\infty}$ (which can be identified naturally with the set of ends of T) and $f(z) = \int_{\mathbb{P}^1} d\mu_{c_f}(x)/(z-x)$ for the cusp form f.

Once we have this integration theory, we define the L-functions as usual by Mellin transforms: $\int t^{s-1} d\mu_{c_f}$. Goss defines a two-variable ($s \in S_\infty$) L-function as before by exponentiating positives and also a one variable ($s \in \mathbb{Z}_p$) by exponentiating one units. These take values in the representation space above. So for weight 2, we have C_∞-valued L-function.

For $A = \mathbb{F}_q[t]$ and the full modular group, we have a functional equation $L_f(s) = (-1)^{1-m} L_f(k-s)$. We also have the formula

$$a_j = \int_{K_\infty / \tilde{\pi} A} e(x)^{j-1} d\mu_f(x)$$

for the coefficients of q_∞-expansion $f(z) = \sum a_j q_\infty^j$.

Special values and links to the arithmetic of q_∞-expansion need to be investigated further.

The existence of this finite characteristic valued L-function and such L-function defined by Goss for Grossencharacters (which can be thought of as GL_1-automorphic forms with finite characteristic values) suggests that there might be C_∞-valued automorphic (or modular) L functions attached to C_∞-valued representations. Such representations are not well-understood so that it is not known whether for some good class of such automorphic adelic representations, we can imitate Langlands type local component definition of L-functions.

Relations with characteristic 0-valued theory

We already saw in 5.4.1, at the usual L-functions level, a congruence relation between classical and finite characteristic versions.

Both the double coset space used in the definition of the automorphic forms (zero characteristic valued) and Ω used in the definition of modular forms (finite characteristic valued) are linked with T. In fact, Drinfeld set up a natural bijection between harmonic cochains on T of weight 2 with values in F and F-valued automorphic forms which transform like a special representation at component at ∞. Analyzing this correspondence together with Teitelbaum's correspondence mentioned above, Gekeler and Reversat [GR96] showed (at least for $A = \mathbb{F}_q[t]$, there seem to be some technical difficulties in general) that double cuspidal modular forms of weight two, type one, and with $\mathbb{F}_p$-residues (such forms generate over C_∞ those with C_∞- residues, i.e., the usual C_∞-space of such modular forms) are the reductions mod p of the automorphic cusp forms special at ∞.

For higher weights and ranks the connection between the modular forms versus the automorphic forms is not well-understood.

6.3 Galois representations

In this section, which will not be needed elsewhere in this book, we will quickly review (in an over-simplified manner, which is hopefully still helpful to some readers) great developments that have taken place in the study of Galois representations for function fields without going into technical details (which I do not know anyway), which can be found in [Laf02; Lau02] and references therein and below.

Drinfeld's main motivation in introducing Drinfeld modules of rank r and studying their moduli was to try to prove function field analog of Langland's conjectural correspondence between the Galois representations and automorphic representations for Gl_r, generalizing the class field theory situation corresponding to $r = 1$.

To modular cusp eigen-forms of weight $k \geq 2$, Deligne (Shimura for weight 2, using Eichler-Shimura relation) associated (compatible system of) l-adic Galois representations of dimension 2 by choosing appropriate piece cut out by Hecke action of suitable l-adic etale cohomology of elliptic modular curves so that traces of Frobenius elements at p are p-th Fourier coefficients of the modular forms etc. (Given a weight one new form, looking at Galois representations associated to higher weight congruent forms, Deligne-Serre then succeeded in associating corresponding 2 dimensional complex representation, as in Langlands original conjectures. The more interesting converse is still open, though a lot of progress is made using class field theory and base change to do various solvable cases.)

To carry over such project over function fields, taking his clues from elliptic curves-lattice connections, and Lubin-Tate local class field theory (see Serre's article in [CF67] or original paper of Lubin and Tate referred there) which uses formal A-modules, Drinfeld introduced (what are now called) Drinfeld modules. Drinfeld used Drinfeld modular curves and associated to automorphic representations with special component at ∞ (analog of weight 2 cusp forms), corresponding Galois representations by finding in Tate modules of Jacobians of these modular curves. Because of non-archimedean nature of infinite primes, instead of complex cohomology and differential forms connection to modular forms, Drinfeld connected to automorphic forms using rigid cohomology, Bruhat-Tits trees and representations as harmonic cochains.

Deligne in his letter to Drinfeld in 1975 [DH87] showed how to deduce **local Langlands conjecture for** Gl_2 from Drinfeld's global recipe to get Galois representation from an automorphic one, together with earlier results of Weil and Jaquet-Langlands etc. which from properties of Artin L-functions for function fields, such as analog of Artin's entireness conjecture and converse results proved by Weil applied to the results on functional equations, local constants proved by Jacquet, Langlands and Deligne, Gelbart etc. allow to go in the other direction establishing the correspondence in this local case.

In the same letter, Deligne also pointed out that function field analog of **modularity conjecture** (of Shimura-Taniyama-Weil) can be deduced from combining Drinfeld's work with earlier works:

Let us start by recalling original version: Elliptic curve over $\mathbb{Q}$ of conductor N is modular of level N, in the sense that there is a non-constant morphism $X_0(N) \to E$ defined over $\mathbb{Q}$. This is now a well-known theorem of Wiles (and collaborators), which proved Fermat's last theorem by reducing it to Ribet's earlier work.

The function field version is that if E is elliptic curve over a function field K of conductor $I\infty$ and having a split multiplicative reduction at place, say ∞ of K, then there is a non-constant morphism from Drinfeld modular curve $X_0(I)$ (parametrizing isomorphism classes of Drinfeld A-modules with Γ_0-level structures, for A corresponding to K and ∞ as usual) to E defined over K.

Note that if elliptic curve over F is not isotrivial, its j-invariant has some pole, where it has a multiplicative reduction, which may not be of split type over F, but will be split over a quadratic extension.

The detailed arguments for this unpublished result were given eg. in [GR96]. Here is a brief sketch of the main ingredients:

By works of Weil, Grothendieck, Langlands, Deligne on L-functions in function field situations, it follows that L function of E satisfies the hypothesis of Weil's converse theorems giving an automorphism form f (of level I, special at ∞, Hecke eigenform with integer eigenvalues etc.) which has same L function as E (together with finite character twists). Drinfeld's work now leads to a factor A_f of the Jacobian of $X_0(I)$, which has the same L-functions and is thus isogenous to E by Zahrin's proof of Tate's isogeny conjectures for abelian varieties over function field.

Drinfeld matched galois and automorphic representations by comparison of Grothendieck-Lefschetz trace formula on etale cohomology side and (function field version of) Arthur-Selberg trace formula on representation theory side for various test functions to match representations on both sides.

This way, Drinfeld succeeded in establishing **Langlands correspondence in the function field case for** Gl_2, but, at first, only partially, since there were restrictions at ∞ on the representations, coming from the choice of ∞ coming up in Drinfeld modules set up.

Since Drinfeld modules set-up works for any r, this gave approach to Langlands conjectures for Gl_r in function fields.

Since the moduli varieties are now of relative dimension $r-1$, the complications of good compactifications arise. Deligne conjectured (and Pink later proved) that Lefschetz trace formula works even in such non-compact cases, if we 'twist by large enough power of Frobenius'.

Modulo such complications in trace formula, Kazdhan and Flicker generalized Drinfeld's work on Langlands correspondence to Gl_n from Gl_2, again with restrictions at ∞, coming from their use of Drinfeld modules.

Meanwhile Drinfeld used more general concept of **Shtukas, or Frobenius-Hecke sheaves** (these are more suitable to handle all ∞'s at once, see Chapter 7 for a little more details) to get rid of the conditions at infinity, thus establishing full Langlands correspondence for Gl_2 over function fields, by analyzing l-adic cohomology of the algebraic stack of rank r Shtukas.

Laumon, Rappoport and Stuhler (1993) established **local Langlands correspondence for** Gl_r (matching l and ϵ factors etc.) using Deligne's approach and as well as the progress on the general rank case, and Shtukas and their division algebra (forms of Gl_n, but now leading to compact modular varieties easier to deal with) variants due to Stuhler.

In 2002, by this line of attack using Shtukas, but getting over formidable technical obstacles, Lafforgue proved **full Langlands correspondence for Gl_n over function fields**:

Theorem 6.3.1   *Let K be a function field of characteristic p. Fix prime $l \neq p$ and integer $r \geq 1$.*

Then there is a bijection $\pi \to \sigma(\pi)$ between the set of irreducible cuspidal automorphic representations π of $Gl_r(A_K)$ which have central character of finite order and the set of of irreducible representations $\sigma : G_K \to Gl_r(\overline{\mathbb{Q}_l})$, unramified outside a finite set of primes and with determinant of finite order. This bijection has the property that Frobenius and Hecke eigenvalues (or equivalently the local L factors) of π and $\sigma(\pi)$ match.

Remarks 6.3.2   (1) In the number field case, since the lattice dimension is bounded by 2, we do not have analogs of Drinfeld modular varieties for $r > 2$. We do not have Shimura varieties for Gl_r in the number field case. We can look at Shimura varieties for other reductive groups, but because of existence of non-stable conjugacy classes now, the trace formula techniques are much more difficult, in addition to difficulties with compactifications, archimedean places etc. Moduli of abelian varieties lead to symplectic groups rather than general linear groups. Only in $r = 2$ case, these are close as $CSp_2 = PSl_2$.

(2) In 1983 paper, Drinfeld also gave direct geometrical construction of automorphic representations from a single l-adic galois representations, bypassing converse theorems, which need a compatible system of l-adic representations. So in fact, (up to a character twist) we can always fit a l-adic representation in this case in a compatible system. This is in contrast with the number field case, where in contrast to the function field discrete situation, we have higher dimensional deformation spaces, so that we can have transcendental eigenvalues for a single representation and so no hope to expect it from algebraic geometry and compatible systems.

(3) We can ask about p-adic, $\wp$-adic (occurring from torsion say) or finite field valued representations. The situation seems not fully clear even conjecturally. Modifying Drinfeld's argument, de Jong has associated eigenform with usual relations to a two dimensional continuous, everywhere unramified $\mathbb{F}_l((t))$ valued absolutely irreducible representation of G_K with trivial determinant. Böckle and Khare have used this together with lifting techniques and Lafforgue result to give partial results related to analog of Serre's conjectures, in r-dimensions, for function fields: For example, they prove that a representation $\overline{\rho} : G_K \to Gl_2(\mathbb{F})$, where F is a finite field of

characteristic $l \neq p$, which is unramified outside a finite set of primes and is irreducible when base changed to $K\overline{\mathbb{F}_q}$ is the reduction mod l of an l-adic representation associated to a cusp eigenform.

For Drinfeld modules over finite fields, the analog of **Tate isogeny theorem** was proved [Dri77b] by Drinfeld. For Drinfeld modules of generic characteristic (in fact, in much more general setting), the analog of **Tate conjecture/Faltings theorem** was established by Tamagawa and Taguchi [Tam94; Tag95; Tag96], by a method, quite different from Faltings in the classical case and inspired by previous work of Anderson, of approximating solutions of '$\wp$-adic linear Frobenius equations'. Taguchi [Tag93; Tag91] also proved the **semisimplicity** of the Galois representation on the Tate module, for both finite and generic characteristic Drinfeld modules. Taguchi [Tag99] proved that a given L-isogeny class of Drinfeld A-module contains only finitely many L-isomorphism classes, for L a finite extension of K. Oliver Watson has recently proved the analog of Tate conjecture in the equi-characteristic case, in his 2003 University of Pennsylvania thesis.

Classically, there is a well-known theorem of Serre on **the image of Galois representation** obtained from torsion of elliptic curves. Pink [Pin97] showed that if ρ has no more endomorphisms than A, then for a finite set S of places $v \neq \infty$, the image of $\mathrm{Gal}(K^{sep}/K)$ in $\prod_{v \in S} GL_n(A_v)$ for the corresponding representation for rank n Drinfeld modules is open. Note that this is weaker than Serre type adelic version, but much stronger (unlike the classical case) than the case of one prime v, because we are dealing with all huge pro-p groups here, even though the primes v change. So the simple classical argument combining p-adic and l-adic information to go from the result for one place to the result for finitely many places does not work). This has been generalized and improved in recent works of Pink, Traulsen and Gardeyn.

The finite characteristic valued L-series can be attached to these finite characteristic valued representations in a way analogous to the classical case, using the exponent space of David Goss. We saw that interesting special values, zero distribution results have been established for analogs of Riemann and Dedekind zeta functions, but not yet in rank two or more general situation. See [Gos96] for this general theory including its analytic aspects. The **cohomological aspects** have been developed by Taguchi, Wan [TW96] and more recently by Pink and Böckle [Böc02].

6.4 DeRham Cohomology

Drinfeld modules are not varieties, but are analogs of motives in that they do give rise to compatible cohomology realizations. While this view point has not been developed fully satisfactorily from the categorical point of view in the literature, mainly because many of the interesting applications and explorations have been done without a need of a full development, hopefully this will be done soon. We only present some of the ingredients used later.

The lattices and the Tate modules corresponding to Drinfeld modules immediately lead to analogs of Betti and $\wp$-adic homology, cohomology realizations for them in the usual fashion: For example, if ρ is a Drinfeld A-module over C_∞ with corresponding lattice Λ, then the Betti realizations are given by

$$H_*(\rho) = \Lambda, \quad H^*(\rho, C_\infty) = \operatorname{Hom}_A(\Lambda, C_\infty),$$

where lower or upper positioning of the asterisque denotes co- or contravariance as usual. For abelian varieties analogy, these are the first cohomology. Since these are the only ones we are going to consider, and since weights are $1/r$ in contrast to the abelian variety case, following Anderson we drop the numbering. Note also that Λ is an A-module.

Deligne, realizing that Grothendieck's interpretation of DeRham cohomology for abelian varieties in terms of Lie algebra of the universal additive extension can be carried out in this context also, had defined, in the Drinfeld modules context, the DeRham cohomology, Gauss-Manin connection and considered quasi-periodic functions. He communicated this to us in 1987 during the special program on Drinfeld modules at the Institute for Advanced Study. The equivalent formulations and further developments, such as interpretations with bi-derivations and proof of DeRham isomorphism, were then achieved by Anderson, Gekeler and Yu.

For completeness, we give an account here (pointing out explicitly how things work, but generally omitting the verifications) based on their work, following closely Anderson's lecture notes in addition to [Gek89a; Gek89b; Gos94; Yu90] which can be consulted for additional discussion. For additional details and history in the classical case going back to Rosentlich, Severi, Weil, Barsotti etc. see [MM74; Ser88] and references therein.

(a) Elliptic curves case: Motivation

We motivate by drawing parallel with the elliptic curves case, which we now recall. The elliptic curves being abelian varieties in dimension one are closer to Drinfeld modules, especially to those of rank two. Many parts of the following formulation work for any abelian variety and any Drinfeld module and even generalize to higher dimensions discussed in later chapter.

Consider an elliptic curve over $\mathbb{C}$ given in Weierstrass model $y^2 = 4x^3 - g_2 x - g_3$, given the usual group structure with the neutral element e being the point at infinity. Let c_1, c_2 be an oriented basis for its first singular cohomology $H_1(E)$ with $\mathbb{Z}$ coefficients. The DeRham cohomology $H^1_{DR}(E)$ is obtained by modding out the space of differentials of the second kind (i.e., meromorphic differentials with zero local residues, i.e., no logarithmic differential terms) by the exact ones. For our model, we can use the usual basis consisting of holomorphic differential (also called meromorphic of first kind) dx/y together with $x\,dx/y$ with double pole at e.

Then we have fundamental periods $\omega_j := \int_{c_j} dx/y$ spanning the period lattice Λ, giving uniformization $z \to (\wp(z), \wp'(z)) : \mathbb{C}/\Lambda \to E$ via Λ-periodic Weierstrass $\wp$ function, $\wp(z) := 1/z^2 + \sum(1/(z-\lambda)^2 - 1/\lambda^2)$, with λ running over non-zero periods. We also have basic quasi-periods $\eta_j := \int_{c_j} x\,dx/y$ and corresponding quasi-periodic function $\zeta(z) := 1/z + \sum(1/(z-\lambda) + 1/\lambda + z/\lambda^2)$ of Weierstrass: $\zeta(z+\omega_j) + \eta_j = \zeta(z)$. In fact, $-\zeta'(z) = \wp(z)$.

Let P represent an oriented fundamental parallelogram enclosing the origin: $z_0 \to z_0 + \omega_1 \to z_0 + \omega_1 + \omega_2 \to z_0 + \omega_2 \to z_0$. Then by Cauchy's residue theorem, we have $\int_P \zeta(z)dz = 2\pi i$, whereas by the quasi-periodicity evaluation on the pair of parallel sides gives the value $\omega_1 \eta_2 - \omega_2 \eta_1$. Thus we have Legendre's determinantal relation $\omega_1 \eta_2 - \omega_2 \eta_1 = 2\pi i$.

In co-ordinate-free language, the second exterior power of the first cohomology of the elliptic curve is identified with the second cohomology via the cup product, which in turn, is identified with the second cohomology of the projective line, via say x co-ordinate. Following these maps for DeRham as well as Betti cohomology and connecting via the DeRham isomorphism gives the Legendre relation, since the period in dimension two of the projective line is $2\pi i$ and the period in dimension one of the elliptic curve is the two-by-two matrix with rows (ω_i, η_i).

Since the determinant does not vanish, in particular, we see that periods and quasi-periods span lattice $\tilde{\Lambda}$ of four dimensions in $\mathbb{C}^2$. This is, in co-ordinate free language, the reflection of the DeRham isomorphism

$$\eta \to (c \to \int_c \eta) : H^1_{DR}(E) \xrightarrow{\sim} \mathrm{Hom}(H_1(E), \mathbb{C}).$$

Let us now see (omitting many details) how the natural interpretation of $\mathbb{C}^2/\tilde{\Lambda}$ as an algebraic group (generalized Jacobian) representing the universal additive extension $E^\sharp$ of E and the interpretation of $\mathrm{Lie}(E^\sharp)^v$ as $H^1_{DR}(E)$ works:

Let $E^\sharp$ be the group of divisors of E of degree zero supported away from identity e modulo the divisors of functions f such that df/f vanishes at e; and let V be its subgroup generated by principal divisors. Then by the theory of generalized Jacobians, $E^\sharp$ is an algebraic group with subgroup V being isomorphic to $\mathbb{G}_a$ and we have an exact sequence $0 \to V \to E^\sharp \to E \to 0$, by interpretation of E as its own Jacobian. In fact, this is an universal extension of E by $\mathbb{G}_a$ in that any other extension is obtained by pushing this.

In fact, the generalized Abel-Jacobi map sending a generalized divisor class of D to integration over its boundary of dx/y and xdx/y is an isomorphism from $E^\sharp$ to $\mathbb{C}^2/\tilde{\Lambda}$. Avoiding co-ordinates, this says that map sending D to the map taking η to the integration above is isomorphism from $E^\sharp$ to $H^0(E, \Omega_E(2[e]))^v/H_1(E)$. Passing to the tangent space at the neutral element, we see that $\mathrm{Lie}(E^\sharp)^v$ is isomorphic to $H^0(E, \Omega(2[e]))$, which is as we have seen above, isomorphic to $H^1_{DR}(E)$, via the natural map sending η to its class. Hence we have canonical isomorphism

$$\mathrm{Lie}(E^\sharp)^v = H^1_{DR}(E).$$

This is the description which generalizes to Drinfeld modules well, as we will see. If you have an additive extension together with a Lie splitting:

$$0 \to \mathbb{G}_a \xrightarrow{i} E' \xrightarrow{\pi} E \to 0, \quad 0 \to \mathrm{Lie}(\mathbb{G}_a) \underset{s}{\overset{\mathrm{Lie}(i)}{\underset{\longleftarrow}{\longrightarrow}}} \mathrm{Lie}(E') \xrightarrow{\mathrm{Lie}(\pi)} \mathrm{Lie}(E) \to 0,$$

then using $E^\sharp \xrightarrow{f} E'$ given by the universal property, we get an element $s \circ \mathrm{Lie}(f) \in \mathrm{Lie}(E^\sharp)^v$. In fact, the elements of $\mathrm{Lie}(E^\sharp)^v$ classify such extensions with Lie splitting giving equivalent interpretation of the DeRham cohomology in terms of extensions data.

(b)   Drinfeld modules case

We will give alternate description of Anderson, Gekeler, Yu which is more useful for our immediate purposes and then will just point out the correspondence explicitly, leaving out the details and functorial properties etc. necessary for deeper study to the references.

Let us now translate from the language of extensions into the terminology of derivations and bi-derivations more in line with usual thinking in terms of the differential forms.

Let A be a ring and M be a left A-module, then a derivation of A to M is just an additive map $a \to f_a : A \to M$ such that $f_{ab} = af_b + bf_a$. With $\Delta := \ker(a \otimes a' \to aa' : A \otimes A \to A)$ and $\delta := a \to a \otimes 1 - 1 \otimes a : A \to \Delta$, we get universal derivation $d : A \to \Omega := \Delta/\Delta^2$ in that composition with d gives isomorphism from $\mathrm{Hom}_A(\Omega, M)$ to $\mathrm{Der}(A, M)$ for all M.

Now let N be an (A, A)-bimodule, i.e., equipped with both left and right A-module structures, compatible in the sense that $(an)a' = a(na')$ (but not necessarily with $an = na$). Then bi-derivation of A into N is an additive map $a \to \eta_a : A \to N$ such that

$$\eta_{ab} = a\eta_b + \eta_a b.$$

The map δ defined above is then universal bi-derivation in that the composition with δ gives isomorphism from $\mathrm{Hom}_{A,A}(\Delta, N)$ to $\mathrm{BiDer}(A, N)$.

For $n \in N$, $a \to an - na$ are bi-derivations called inner. We have

$$\mathrm{BiDer}(A, N)/\mathrm{InnerBiDer}(A, N) = \mathrm{Hom}_A(\Omega, N/\Delta N),$$

provided that N is $A \otimes A$-flat, where ΔN is the submodule of N generated by $an - na$'s.

Definition 6.4.1   Let ρ be a rank r Drinfeld A-module over F. Let $M = \mathrm{Hom}(\rho, \mathbb{G}_a) = F\{\tau\}$ be (A, A)-bimodule with usual left action via multiplication by elements of A and right action of A via ρ, so that $\eta_{ab} = a\eta_b + \eta_a\rho_b$. (Note these are just composition actions with natural actions on the two sides.) Let $N := M\tau = F\{\tau\}\tau$ be the submodule of those morphisms with trivial Lie action. Let

$$H^*_{DR}(\rho) := \mathrm{BiDer}(A, N)/\mathrm{InnerBiDer}(A, N).$$

Remarks 6.4.2   (1) The only algebraic group homomorphism from E to $\mathbb{G}_a$ being the trivial one, the analogy breaks down at this point.

(2) It is easy to verify that $\delta^0 := \delta_\rho : a \to \rho_a - a$ is a bi-derivation.

(3) When $A = \mathbb{F}_q[t]$, we can define a bi-derivation η by prescribing it on a generator t and then using the definition of bi-derivations to calculate it on any element. In particular, for $1 \le i \le r$, we will use the bi-derivations $\delta_t^i := \tau^i$

(4) For general A, a bi-derivation η is fixed by η_a for a single non-constant a: We can in fact solve for coefficients of η_b inductively by using consistency condition implied by the definition of bi-derivation and $\eta_{ab} = \eta_{ba}$. Enough supply of bi-derivations for general A and ρ is given below in Gekeler's proof of DeRham isomorphism.

(5) Given a morphism between Drinfeld modules, just composition gives morphism between bi-derivations inducing morphism of DeRham cohomologies.

(6) The definition we have adopted agrees with Deligne's $\mathrm{Lie}(E^\sharp)^v = H^1_{DR}(E)$ of the last section, with the elliptic curve replaced by a Drinfeld module. In fact, a bi-derivation $\eta \in H^*_{DR}(\rho)$ gives rise to a Lie-split additive extension ρ^η as follows:

Let ρ^η be $\mathbb{G}_a \times \mathbb{G}_a$ as an algebraic group (considered as column vectors) with action of A via the matrix

$$\rho_a^\eta = \begin{pmatrix} a & -\eta_a \\ 0 & \rho_a \end{pmatrix}.$$

The property being bi-derivation translates to ring homomorphism property $\phi_a^\eta \phi_b^\eta = \phi_{ab}^\eta$. By the definition of N, the matrix reduces to the scalar matrix a at the Lie level and thus gives the splitting. See more discussion in 7.4.

Let us now take $F = C_\infty$. In analogy with the classical case, we think of bi-derivations as differentials of second kind and inner ones as exact, while C_∞-multiples of $\delta^0 := a \to \rho_a - a$ as differentials of the first kind. The quasi-periodic functions and integration over cycles $\lambda \in \Lambda$ work as follows:

Definition 6.4.3   For $\eta \in \mathrm{BiDer}(A, N)$, let $F_\eta : C_\infty \to C_\infty$ be the unique entire function such that

$$\eta_a e_\rho(z) = a F_\eta(z) - F_\eta(az), \quad F_\eta(z) = O(z^q).$$

Then for $\lambda \in \Lambda$, put

$$\int_\lambda \eta = F_\eta(\lambda).$$

Remarks 6.4.4   (1) In fact, F_η's are readily found by using the equation in the definition for any non-constant $a \in A$. Another way is to use the equivalent formulation below. The definition also implies the quasi-periodicity of F_η with quasi-periods $F_\eta(\lambda)$: $F_\eta(z+\lambda) = F_\eta(z) + F_\eta(\lambda)$ (this is just additivity) and $F_\eta(a\lambda) = aF_\eta(\lambda)$. Note that $z - e(z)$ is quasi-periodic with quasi-periods being the periods λ's.

(2) Given bi-derivation η, $f\eta$ and $\eta\rho_a$ are also bi-derivations with corresponding functions $fF_\eta(z)$, $F_\eta(az)$ and quasi-periods $fF_\eta(\lambda)$ and $aF_\eta(\lambda)$ respectively.

(3) In notation of 4.14 (b), the equation in the definition implies $-d_\tau F_{\delta^1} = e_\rho$. This is analog of $-\zeta'(z) = \wp(z)$ of Weierstrass theory.

The definition immediately gives equivalent formulas for cycle integration without explicit use of the quasi-periodic functions F_η as

$$\int_\lambda \eta = -\lim_{\deg(a)\to\infty} \eta_a e_\rho(\lambda/a) = -\sum_{n=0}^\infty a^n \eta_a e_\rho(\lambda/a^{n+1}),$$

with $a \in A$ of $\deg(a) > 0$.

Since $F_{\delta^0}(z) = z - e_\rho(z)$, we recover the periods $\int_\lambda \delta^0 = \lambda$ from the periods of the differentials of the first kind, whereas exact η leads to $\int_\lambda \eta = 0$, because $\eta_a = af - f\rho_a$ gives $F_\eta(z) = fe_\rho(z)$.

Tracing through the definitions, it can be shown that the exponential map $e^\eta : \text{Lie}(\rho^\eta) \to \rho^\eta$ can be described by $e^\eta(z_1, z_2) = (z_1 + F_\eta(z_2), e_\rho(z_2))$, and that the notion of the integral defined above is consistent with the notion implicit in Deligne's definition.

Gekeler [Gek89a] (and Anderson by different method, in unpublished work) proved that DeRham map induced by cycle integration above is isomorphism:

Theorem 6.4.5 *The DeRham morphism*

$$DR : H^*_{DR}(\rho, C_\infty) \to H^*(\rho, C_\infty), \quad DR(\eta) := \chi_\eta := F_\eta|_\Lambda$$

obtained from the integral pairing is onto isomorphism.

Proof. We just give Gekeler's recipe (see [Gek89a] for details of proof) for the bijectivity:

Given A-character $\chi : \Lambda \to C_\infty$, its preimage under DR is given by

$$\eta_a^\chi := al_\chi - l_\chi\rho_a, \quad l_\chi := \sum_{i\geq 1}\Big(\sum_{0\neq\lambda\in\Lambda} \chi(\lambda)/\lambda^{q^i}\Big)\tau^i,$$

which despite its a priori infinite series appearance can be shown to be a reduced representative in a sense that τ-degree of η_a is not more than

$r \deg a$, as can always be achieved by using the remainder obtained by dividing by δ^0. Further the corresponding quasi-periodic function is

$$F_\chi(z) = \sum_{\lambda \in \Lambda} \chi(\lambda)(e_\rho(z)/(z-\lambda))^q.$$

(Note that we need additive function restricting to χ on Λ and $e(z)/(z-\lambda)$ is essentially additive characteristic function of Λ and we need q-th power for convergence.)   □

Let us now see how the Legendre relation version proof of DeRham isomorphism works in our case, following Anderson and Gekeler and restricted to rank 2 for simplicity:

We work with $A = \mathbb{F}_q[t]$ and rank two Drinfeld module $\rho_t = t + G\tau + \Delta\tau^2$ over C_∞ with period lattice Λ and exponential e, to make connections with the elliptic curve case described above and to identify the modular form that enters the picture now:

For $\lambda \in \Lambda$ and $i \geq 0$, consider

$$f_\lambda^{(i)}(u) := -\sum_{n=0}^{\infty} e(\lambda/t^{n+1})^{q^i} u^n \in C_\infty[[u]].$$

Then $\lambda = \mathrm{Res}_{u=t} f_\lambda^{(0)}(u)du$ and $\int_\lambda \delta^1 = f_\lambda^{(1)}(t)$. For $\lambda_1, \lambda_2 \in \Lambda$ consider 'Wronskian' $W(u)$ defined to be the determinant of the two-by-two matrix with (i,j)-th entry $f_{\lambda_i}^{(j-1)}(u)$. Then $\mathrm{Res}_{u=t} W(t)dt$ is the determinant of $\int_{\lambda_i} \delta^{j-1}$. If we let λ_i to be the generators of Λ, it is then analog of the Legendre determinant above.

Here is the way the **Legendre relation** comes about in our case. See [Gek89b] for details of a different argument: We see that $\rho_t(f_\lambda(u)) = u f_\lambda(u)$, where τ-acts trivially on u. Using these relations for $\lambda_i \in \Lambda$ and eliminating the middle τ term (i.e., $f^{(1)}$ terms) shows that $\rho_t'(W(u)) = uW(u)$ where $\rho_t' = t - \Delta\tau$. (When we develop the tensor formalism in the next chapter, we will see that ρ' is the determinant of ρ.) This equation is also satisfied by $f_{\lambda'}'(u)$, where f' and λ' are defined for ρ' the same way as f and λ for ρ. It can be shown that there is unique $\lambda' \in \Lambda'$, the lattice corresponding to ρ' such that $W(u) = f_{\lambda'}'(u)$ and further that it is a generator for Λ' if λ_1 and λ_2 generate Λ. Hence taking residue at $u = t$ shows that determinant is λ'. This is the Legendre relation. Notice that ρ' is isomorphic to the Carlitz module C: $\rho' = c^{-1}Cc$, where $c^{q-1} = -\Delta$. Hence $\lambda' = \tilde\pi/(-\Delta)^{1/(q-1)}$, which is $\tilde\pi$ if we take normalized ρ with $\Delta = -1$. There is sign ambiguity as in the classical case for the determinant and $\tilde\pi$, so may be the better way to state the relation is by taking $q-1$-th powers of both sides.

In fact, $\lambda' = -\tilde{\pi}^q/h$ where h is the modular form discussed in Section 6.2. Summarizing, we have

Theorem 6.4.6 *Let $A = \mathbb{F}_q[t]$ and suppose $\rho_t = t + G\tau + \Delta\tau^2$ is a rank two Drinfeld module with corresponding A-lattice Λ. Suppose λ_1 and λ_2 generate Λ and that η_1 and η_2 are quasi-periods corresponding to δ^1. Then $\lambda_1\eta_2 - \lambda_2\eta_1 = \tilde{\pi}/(-\Delta)^{1/(q-1)}$. In particular, the left side is $\tilde{\pi}$, if we take isomorphic model with $\Delta = -1$.*

In 4.7 (a) and 4.11, we saw two complex multiplication cases where the periods were expressed as gamma values. As an application of Legendre relation, let us see how the quasi-periods yield other gamma values. This has been generalized fully in higher dimensions by Brownawell and Papanikolas [BP02] using work of Anderson and Sinha. We will see more about this in Chapter 8.

In the setting of Theorem 4.7.3, if we specialize to rank two in the Chowla-Selberg for constant field extensions, then with $A = \mathbb{F}_q[t]$, we have rank two period lattice $\tilde{\pi}_2\mathbb{F}_{q^2}[t] = \tilde{\pi}_2 A + f\tilde{\pi}_2 A$, where f is a primitive $q^2 - 1$-th root of unity. So $\lambda_1 = \tilde{\pi}_2$ and $\lambda_2 = f\tilde{\pi}_2$ generate the period lattice and if we denote the corresponding quasi-periods for δ^1 by η_1 and η_2, then Legendre relation shows that quasi-period (note that multiplication rings acts on quasi-periods) $\eta_2 - f\eta_1$ is $\tilde{\pi}/\tilde{\pi}_2$ up to an algebraic quantity and thus, this quasi-period is $(1/(1-q^2))!^q/(q/(1-q^2))!$ up to an algebraic quantity. On the other hand, we saw in 4.7.3, that the period $\tilde{\pi}_2$ is $(q/(1-q^2))!^q/(1/(1-q^2))!$. This should generalize to any rank.

Exactly similarly, in situation of 4.11 we get rank two situation if we specialize to $q = 3$, then the reflection formula together with Legendre relation shows that similar quasi-period is $\Gamma(-1/t)$, while the period is $\Gamma(1/t)$, up to an algebraic quantity. Summarizing,

Theorem 6.4.7 *(1) Let $A = \mathbb{F}_q[t]$. Then $(q/(1-q^2))!^q/(1/(1-q^2))!$ and $(1/(1-q^2))!^q/(q/(1-q^2))!$ are periods and quasi-periods respectively (up to simple algebraic quantities) of the Carlitz module over $\mathbb{F}_{q^2}[t]$, considered as rank 2 A-module with complex multiplication by $\mathbb{F}_{q^2}[t]$.*

(2) Let $A = \mathbb{F}_3[t]$. Then $\Gamma(1/t)$ and $\Gamma(-1/t)$ are periods and quasi-periods respectively (up to simple algebraic quantities) of the Carlitz module ρ over $\mathbb{F}_3[\rho[t]] = \mathbb{F}_3[\sqrt{-t}]$, considered as rank 2 A-module with complex multiplication by $\mathbb{F}_3[\rho[t]]$.

Thiery [Thi] proved that in complex multiplication case, we have additional relation between the periods and quasi-periods analogous to the classical result of Eisenstein:

Theorem 6.4.8   *Let ρ be a rank two Drinfeld $A = \mathbb{F}_q[t]$-module with period lattice Λ generated by λ_1 and λ_2 and having complex multiplication, say $\gamma \in \overline{K} - K$ with $\gamma\Lambda \subset \Lambda$. Let $\eta_i = F_\eta(\lambda_i)$ be quasi-periods. Then there is $\beta \in \overline{K}$ such that $(\eta_2 - \beta\eta_1)/\lambda_1 \in \overline{K}$:*

Proof.   We have natural action of γ on bi-derivations and their one dimensional quotient by the inner ones, so that there is $\kappa \in \overline{K}$ such that $F_\eta(\gamma z) - \kappa F_\eta(z) = n'z - ne_\rho(z)$, where the right side is just the quasi-periodic function corresponding to inner $an - n\rho_a$. With $z = \lambda_1$, noting that $\gamma\lambda_1 = a_1\lambda_1 + a_2\lambda_2$, with $a_i \in A$ and $a_2 \neq 0$, implies the claim.   □

We will see application in Chapter 10.

Remarks 6.4.9   (1) The DeRham cohomology is a rank r A-module, with rank one part of first kind given by multiples of δ^0. Note that in this Hodge decomposition, we have $h^{1,0} = 1$ and $h^{0,1} = r - 1$, the two being equal for rank two, the case closest to the elliptic curves.

(2) For $A = \mathbb{F}_q[t]$, we have a nice reduced basis δ^i, where $0 \leq i \leq r$ for the DeRham cohomology.

(3) Drinfeld modules one dimensional, so the universal vectorial is same as the universal additive extension.

6.5   Hypergeometric functions

We now look at analogs of hypergeometric functions [Tha95a; Tha00] in the setting of function fields over finite fields. We focus on $A = \mathbb{F}_q[t]$ case, and only remark on the general case, which is not much explored beyond possible definitions. We will see analogs of the differential equations, integral representations, transformation formulas, continued fractions and show how analogs of various special functions and orthogonal polynomials occur as their specializations. There are two analogs: one with the parameters a, b, c etc. in characteristic zero domain and one with characteristic p domain. We will see that analogs of several crucial properties are shared between them, giving us some confidence that these are good analogs. The appearance of two distinct analogs is as in the case of the gamma functions for the function fields, but we do not understand their place in the general scheme as well, because, for example, we do not know their role, if any, in the two cyclotomic theories, as in the gamma case.

The theory and literature on hypergeometric and q-hypergeometric functions is vast, so there are indeed a huge number of open questions. But we have focused on only the simplest properties and it seems that the main open question is that of geometric interpretation. Since historically hypergeometric functions and especially their q-analogs also got backed by basic mathematical structures (such as Kac-Moody algebras, quantum groups) much later, we are hopeful that a good geometric understanding of the hypergeometric functions will emerge with good integral formulas, for example, coming from DeRham set-up in section 6.4. Is there an interesting p-adic (or $\wp$-adic) theory analogous to Dwork's theory [Dwo90]?

For the classical material, we refer to and use the notation of [Sla66]. We use the same notation for the analogs defined below, but the context will make clear what we are referring to.

Let $A = \mathbb{F}_q[t]$. We write $d_i = D_i$ of Sections 2.5, 4.5(a), 4.15 and $l_i = (-1)^i L_i$ in notation of 2.5 (but note that it was denoted by L_i in 4.15). Then we have $e(z) = \sum z^{q^i}/d_i$ and $l(z) = \sum z^{q^i}/l_i$, for the Carlitz exponential and logarithm.

(a) The first analog

The first analog arises as a solution of Gauss differential equation analog. But let us just first define and discuss it a little and then look at the differential equations.

Definition 6.5.1 For $n \in \mathbb{Z}_{\geq 0}$, $a \in \mathbb{Z}$, define $(a)_n$ (which should really be $(a)_{q^n}$):

$$(a)_n := \begin{cases} d_{n+a-1}^{q^{-(a-1)}} & \text{if } a \geq 1 \\ 1/l_{-a-n}^{q^n} & \text{if } n \leq -a \geq 0 \\ 0 & \text{if } n > -a \geq 0. \end{cases}$$

Then we have

$$(1)_n = d_n, \qquad (a)_{n+1} = [n+a]^{q^{-(a-1)}}(a)_n^q,$$

and for $a \neq 0$, we have

$$(a)_n = (a+1)_{n-1}^q, \qquad (a+1)_n = [n+a]^{q^{-a}}(a)_n.$$

Definition 6.5.2 For $a_i, b_i \in \mathbb{Z}$, for which it makes sense (see below), we define

$$_rF_s := {}_rF_s(a_1, \cdots, a_r; b_1, \cdots, b_s; z) := \sum_{n=0}^{\infty} \frac{(a_1)_n \cdots (a_r)_n}{(b_1)_n \cdots (b_s)_n d_n} z^{q^n}.$$

Remarks 6.5.3 (1) When $b_j > 0$, the terms are well-defined and when $a_i \leq 0$ the series terminates, just as in the classical case. When $a_i, b_j > 0$, the series then converges for $z = 0$ only, for all z or for z with $\deg z < \sum(b_j - 1) - \sum(a_i - 1)$ according as whether $r > s + 1$, $r < s + 1$ or (the balanced case) $r = s + 1$ respectively. This division into the three cases is parallel to the classical case. For example, Gauss series $_2F_1$ is balanced, and if eg. $a + b = c + 1$, then it converges for $\deg z < 0$. Below, when we state some formulas, we will leave it to the reader to figure out the domain of validity.

(2) Now we explain some motivation, analogies and the splicing at $a = 0$ of $(a)_n$ we have used by making use of both (see 2.10, 4.14 also) d_i and l_i. Classically, $(a)_n = a(a + 1) \cdots (a + n - 1) = (a + n - 1)!/(a - 1)!$. With the Carlitz' analog of the factorial (Chapter 4), the right-hand side should be replaced by d_{n+a-1}/d_{a-1} if we think of $a + n - 1$ and $a - 1$ taking the place of q^{a+n-1} and q^{a-1} respectively. Here, some extra twistings of q-powers come up in shifting the factorial and also we do not divide to get linear term normalized to be z. This is done for simplicity and to get good specializations. If we do normalize the linear term to be z, we get the radius of convergence to be one in the balanced case, parallel to the classical case. This is related to the fact that in the differential equation analog given below, d_r (see 4.14 (b)) is linear only with scalars in $\mathbb{F}_q$, unlike d/dz which is linear.

For our purposes, it seems that an appropriate analogy is the following: Think of d_n, which is the Carlitz factorial at q^n, as (see 4.13.1 (4)) factorial at n and try to use the recursion equation to extend the definition for negative n. We immediately run into problems of dividing by zero, which we interpret to be poles. Then, we do have poles at negative integers, analogous to the classical situation. The same problem is encountered in extending $(a)_n$ to non-positive a's. We then renormalize or reinterpret $(a)_n$ by picking 'residue' or in other words specifying $(0)_0 = 1$ and then continue using the same recursion relations from that point on-wards. The process is justified, a posteriori, by the results below proving analogs of various classical properties, and it also fits with the classical normalizations

$(1)_0 = (0)_0 = 1$ coming from $(a)_{-n} = (-1)^n/(1-a)_n$.

Differential operator formalism:

For $a \in \mathbb{Z}$, consider the operators (Δ and $\Delta^{(w)}$ appear in [C5] and were introduced in 4.15 (b))

$$\Delta_a g(z) := g(tz) - t^{q^{-a}} g(z), \Delta := \Delta_0, \Delta^{(w)} := \Delta_0 \cdots \Delta_{1-w}, d_\tau := \Delta^{1/q}.$$

Then $d_\tau^w = (\Delta^{(w)})^{1/q^w}$. Analogies are as follows: We consider series of the form $\sum a_i z^{q^i}$ instead of $\sum a_i z^i$ in the classical case. Hence we consider the 'q-th power' operator, which takes z^{q^i} to $z^{q^{i+1}}$, as analogous to the 'multiplication by z' operator in the classical case. We consider Δ_a to be an analog of $zd/dz + a$, d_τ as an analog of d/dz and d_τ^w as an analog of d^w/dz^w, some of which we have already seen in 4.15 (b). Also,

$$(z\frac{d}{dz} + a)z^n = (n+a)z^n \iff \Delta_a z^{q^n} = [n+a]^{q^{-a}} z^{q^n}$$

$$(z\frac{d}{dz}) \cdots (z\frac{d}{dz} - w + 1) = z^w \frac{d^w}{dz^w} \iff \Delta^{(w)} = (d_\tau^{(w)})^{q^w}.$$

Also, $\Delta_a g(z) = (\Delta g(z)^{q^a})^{q^{-a}}$ is an analog of classical $(zd/dz + a)g(z) = 1/z^a(zd/dz)(z^a g(z))$. Note $\Delta_a = \Delta - [-a]$ (for $a \in \mathbb{Z}$). The operators Δ_a's commute among themselves and with the constants. We have $\Delta_a \tau = \tau \Delta_{a+1}$ and so $d_\tau^w = \tau^{-w} \Delta_0 \cdots \Delta_{1-w}$.

Properties of the first analog

Gauss Differential equation: We want to show that $_rF_s$ satisfies an analog of the Gauss differential equation. The convenient form for the differential equation turns out to be the product form. The analog of the Gauss differential equation is

$$\prod \Delta_{a_i} {}_rF_s = d_\tau \prod \Delta_{b_j-1} {}_rF_s \qquad (a_i, b_j, b_j - 1 \neq 0).$$

Let us verify that $_rF_s$ satisfies it: If $a_i, b_j \neq 0$, we have

$$\Delta_r F_s = \sum_{n=0}^{\infty} \frac{(a_1)_n \cdots (a_r)_n}{(b_1)_n \cdots (b_s)_n} \frac{t^{q^n} - t}{d_n} z^{q^n}$$

$$= \sum_{n=1}^{\infty} \frac{(a_1+1)_{n-1}^q \cdots (a_r+1)_{n-1}^q}{(b_1+1)_{n-1}^q \cdots (b_s+1)_{n-1}^q} (\frac{z^{q^{n-1}}}{d_{n-1}})^q$$

$$= {}_rF_s(a_1+1, \cdots, a_r+1; b_1+1, \cdots, b_s+1; z)^q.$$

Hence we have

$$d_{\tau r}F_s(\{a_i\};\{b_j\};z) = {}_rF_s(\{a_i+1\};\{b_j+1\};z).$$

On the other hand, if $a_i \neq 0$, we have

$$\Delta_{a_i\,r}F_s = {}_rF_s(a_1,\cdots,a_i+1,\cdots,a_r;\{b_j\};z)$$

and if $b_j \neq 1$, we have

$$\Delta_{b_j-1\,r}F_s = {}_rF_s(\{a_i\};b_1,\cdots,b_j-1,\cdots,b_s;z).$$

Hence ${}_rF_s$ satisfies analog of the Gauss differential equation above.

We will concentrate on $F := {}_2F_1$ leave it to the reader to see what generalizes to ${}_rF_s$ in a similar fashion.

It is straightforward to verify another solution $F(a+1-c,b+1-c;2-c;z)q^{1-c}$ parallel to the classical $z^{1-c}F(a+1-c,b+1-c;2-c;z)$ in addition to $F(a,b;c;z)$ in the case of Gauss ${}_2F_1$.

Now we study the solutions more systematically. First we seek the solutions of the form $y = \sum c_n z^{q^{n+g}}$. Equating the coefficients of $z^{q^{g-1}}$, we get the indicial equation $[g+c-1]^{q^{-c}}[g]^{q^{-1}} = 0$ analogous to [Sla66, 1.3.3], which shows that $g = 0$ or $g = 1-c$. Further,

$$c_{n+1} = \frac{[n+g+a]^{q^{-a+1}}[n+g+b]^{q^{-b+1}}}{[n+g+c]^{q^{-c+1}}[n+g+1]}c_n^q.$$

The two solutions $F(a,b;c;z)$ and $F(1+a-c,1+b-c;2-c;z)q^{1-c}$ correspond to two g's and to particular choices of c_0's. Now d_τ is right linear, but not left linear, so together with solution $y(z)$ we get solutions $y(\theta z)$ depending on choice of c_0, in contrast to the classical case where we get $\theta y(z)$. (Of course, for special cases such as $y(z) = z^{q^n}$, eg. with $a = -n$, $c = 1-n$ both are equivalent.) In particular, if we think of these solutions as functions (rather than non-commutative formal power series in τ) the radius of convergence is affected by the choice of c_0. For $c_0 = 1$, radius is one at ∞ and at $\wp$ of large degree. For more see [Tha00].

Kummer solutions: Kummer gave six different series solutions [Sla66, 1.3] to Gauss equation, two each at 0, 1 and ∞. We now look for analogs of those. In 2.1, we have mentioned the solution at $z = 0$. Solutions around $z = 1$ or $z = \infty$ involve typically power series in $(z-1)$ or $1/z$. We note that $(z-1)^{q^n}$ or z^{-q^n} are not linear functions. We do not know any analog of Kummer solution at 1, but to look at 'solutions around infinity'

we rather look at τ^{-n} i.e we seek a solution of Gauss equation of the form
$y = \sum c_n z^{q^{-n+g}}$.

Definition 6.5.4 For $n \geq 0, a \in \mathbb{Z}$, put

$$(a)_{-n} := \begin{cases} l^{q^{-n}}_{n+a-1} & \text{if } a \geq 1 \\ 1/d^{q^a}_{-a-n} & \text{if } n \leq -a \geq 0 \\ 0 & \text{if } n > -a \geq 0. \end{cases}$$

Then we have

$$d_{-n} := (1)_{-n} = l_n^{q^{-n}}, \quad (a)_{-n-1} = [-n-a]^{q^{a-1}}(a)_{-n}^{q^{-1}},$$

and for $a \neq 0$ we have

$$(a)_{-n} = (a+1)^{q^{-1}}_{-n+1}, \quad (a+1)_{-n} := [-n-a]^{q^a}(a)_{-n}.$$

It is easy to verify that parallel to Kummer's solutions $(-z)^{-a}F(a, 1 + a - c; 1 + a - b; 1/z)$ and $(-z)^{-b}F(b, 1 + b - c; 1 + b - a; 1/z)$ we have the solutions $\overline{F}(a, 1 + a - c; 1 + a - b, z)^{q^{-a}}$ and $\overline{F}(b, 1 + b - c; 1 + b - a; z)^{q^{-b}}$ with

$$\overline{F}(a, b; c; z) := \sum_{n=0}^{\infty} \frac{(a)_{-n}(b)_{-n}}{(c)_{-n}d_{-n}} z^{q^{-n}}.$$

Secondly, if $a, b, c > 0$, these are only formal solutions, since straight valuation calculation shows that the series converges only for $z = 0$, both at ∞ or $\wp$. On the other hand, we get honest solutions in the terminating cases. But note that in that case, we do not get any new solution than ones listed in 2.1.

Remarks 6.5.5 (1) The notation $(a)_{-n}$ clashes with the earlier notation $(a)_n$ when $n = 0$ and that means if we want to consider a bilateral series, we need to choose c_0 appropriately to get the same term for $n = 0$.

(2) Another important issue is what happens when the two solutions in series around zero coincide (i.e., when $c = 1$). Classically one gets a logarithm function from the bigger space of functions than series around zero. What should be a good space to look at, when there is only one $\mathbb{F}_q$-linear series solution? Are Kummer solutions at other points to be found in such spaces? May be the formal solutions we found are related to actual solutions in bigger function spaces, analogous to the classical Stokes line phenomenon.

Ratios of terms: The ratios c_{n+1}/c_n of consecutive terms of $_rF_s$ are rational functions of n, q^n respectively in the classical and q-analog cases respectively. In our case, c_{n+1}/c_n^q is a q-power power of a rational function of t^{q^n}. The q-analogs of the classical hypergeometric series are obtained when n is replaced by q^n in a certain fashion. At this naive level, the $A := \mathbb{F}_q[t]$ analog seems to be $[n] := t^{q^n} - t$ or just t^{q^n}.

Linear relations between contiguous functions: There are fifteen linear relations [Sla66, pa. 13-14] due to Gauss between the contiguous (i.e., those with all parameters the same, except for one pair, which differs by one) hypergeometric functions $_2F_1$. For example: $(b-a)F = bF(b+1) - aF(a+1)$ and $(1-z)F + (c-b)c^{-1}zF(c+1) - F(a-1) = 0$, which is linear in z. An analog, easy to verify term-wise, of the first one is $[b-a]^{q^{-b}}F = F(b+1) - F(a+1)$ and an analog, easy to verify term-wise, of the second one is $F - F^q + [c-b]^{q^{-c+1}}F(c+1)^q + [-a+1]F(a-1) = 0$.

Recalling that multiplication by z corresponds to raising to the q-th power, we can write this more suggestively to show that it is 'linear in Z' (symbol Z denoting the raising to the q-th power operator) as follows $(1 - Z)F + [c-b]^{q^{-c+1}}ZF(c+1) + [-a+1]F(a-1) = 0$.

Integral representation: We do not have analog of Euler's integral formula for hypergeometric functions, but we do have analogs [Tha95a; Tha00] which are formal reinterpretation of Barnes contour integral formula (as residues) [Sla66, pa. 24] or Hadamard convolution integral formulas, (as corresponding to term by term product of power series). But good interpretation with relationship to periods is missing so far.

Specializations: Analogs of many classical special functions, such as Carlitz exponential, binomial coefficients, Bessel functions, Legendre and Jacobi polynomials, are specializations of the hypergeometric functions defined above. As they have not been put to use yet, we do not give here the formulas, but refer to [Tha95a; Tha00] for details.

Continued fraction: By Lemma 9.3.5 and the recursion relation of $(a)_n$'s we obtain [Tha95a] continued fractions (similar to those for $e(z)$ recalled in 9.2) for the hypergeometric series with a nice (but quite different from the classical analog) pattern, with the partial quotients involving the products of (q-power powers of) $[n]$'s compared to the products of n's appearing in the classical result of Gauss [Sla66, pa.15].

For evaluations of some interesting simple continued fractions in terms of these hypergeometric functions, in spirit of Lehmer's results in classical case, see [Tha96b, 6.1].

Summation formula: The usual trick [Sla66, pa. 27] of putting $z = 1$ to cancel terms in the linear relations does not work. Instead, we proceed as follows: Let us write $\sum_{j=0}^{\mu} f_j(t)T^j := (T - t^{q^{-(k+\mu-1)}}) \cdots (T - t^{q^{-k}})$. Then we see that for $m, k, \mu > 0$ we have

$$F(-m, k + \mu; k; 1) = \sum_{r=0}^{m} \frac{d^{q^{-(k+\mu-1)}}_{r+k+\mu-1}}{d^{q^{-(k-1)}}_{r+k-1} l^{q^r}_{m-r} d_r}$$

$$= \sum_{r=0}^{m} \frac{[r + k + \mu - 1]^{q^{-(k+\mu-1)}} \cdots [r + k]^{q^{-k}}}{l^{q^r}_{m-r} d_r}$$

$$= \sum_{r=0}^{m} \frac{(t^{q^r} - t^{q^{-(k+\mu-1)}}) \cdots (t^{q^r} - t^{q^{-k}})}{l^{q^r}_{m-r} d_r}$$

$$= \sum_{j=0}^{\mu} f_j(t) \{ \frac{t^j}{q^m} \}.$$

Hence the terms are zero for $j < m$. In particular, this vanishes when $\mu < m$, parallel to implication of Vandermonde's evaluation [Sla66, pa. 2] $(-\mu)_m/(k)_m$ of the left hand side in the classical case.

Connection with tensor products: We will see in the next chapter that the exponential $\exp_m$ of the m-th tensor power of the Carlitz module evaluated at the column vector (which has canonical meaning) with the top entry z and the other entries zero is the column vector whose j-th entry is $_0F_{m-1}(-; 1, \cdots, 1; z)$ if $j = 1$ and $_0F_{m-1}(-; 2, \cdots, 2, 1, \cdots, 1; z)^q$ otherwise, where there are $j - 1$ 2's. For example, writing columns as rows for convenience, we have $\exp_4(z, 0, 0, 0)$ given by

$$(_0F_3(-; 1, 1, 1; z), _0F_3(-; 2, 1, 1; z)^q, _0F_3(-; 2, 2, 1; z)^q, _0F_3(-; 2, 2, 2; z)^q).$$

(b) The second analog

Definition 6.5.6 Let n be a nonnegative integer and let $a \in C_\infty$. We define $(a)_n$ to be $e_n(a)$.

Now, $(t^n)_n = d_n$. (Note that a and n now belong to the rings of different characteristics.)

Definition 6.5.7 For $a_i, b_i \in C_\infty$ for which it makes sense (see below) we define

$$_rF_s := {}_rF_s(a_1, \cdots, a_r; b_1, \cdots, b_s; z) := \sum_{n=0}^{\infty} \frac{(a_1)_n \cdots (a_r)_n}{(b_1)_n \cdots (b_s)_n d_n} z^{q^n}$$

Remarks 6.5.8 (1) If $b_j \notin A$, then the terms are well-defined and $a_i \in A$ gives the terminating case. Notice that we do not need any sign condition in contrast to the classical case and the first analog. If $a_i, b_j \notin A$, it is easy to calculate the radius of convergence from the degrees of $e_i(z)$. We then see that the degree of the coefficient of z^{q^n} is $nq^n(r-s-1) - (r-s)q(q^n - 1)/(q-1)$. So the series then converges for $z = 0$ only, for all z or for z with $\deg z < q/(q-1)$ according as whether $r > s+1$, $r < s+1$ or (the balanced case) $r = s+1$ respectively. This division into three cases is parallel to the classical case. In particular, we see that $_2\mathcal{F}_1$ converges if $\deg z < q/(q-1)$, which is exactly the radius of convergence of the logarithm and in that sense similar to the classical case.

(2) Classically, $(a)_n = (-1)^n n! \binom{-a}{n}$, hence analogies with binomial coefficients in 2.5 and 4.14 motivate our definition.

Properties of the second analog

Specializations: An analog of the binomial series $(1 + z)^n = \exp(n(\log(1+z)))$ is $e(al(z))$. In particular, for $a \in A$, we get $C_a(z)$. These are specializations of $_2\mathcal{F}_1$ in a manner parallel to the classical case. We have $_2\mathcal{F}_1(a, b; b; z) = e(al(z))$. This provides a formula for the exponential similar to the classical case:

$$\lim_{\deg a \to \infty} {_2\mathcal{F}_1}(a, b; b; \frac{z}{a}) = e(z).$$

We have, of course, the ready-made specialization $_0\mathcal{F}_0(-;-;z) = e(z)$. Similarly, we have expression for the 0-th Bessel function as

$$\lim_{\deg a,b \to \infty} {_3\mathcal{F}_1}(a, b, c; c; \frac{z}{ab}) = J_0(z).$$

The following expression for the logarithm seems different from the usual classical expression: $(-z)_2F_1(1,1;2;z) = \log(1 - z)$ (which corresponds rather to $\sum z^{q^n}/[n]$):

$$\lim_{a \to 0} \frac{_2\mathcal{F}_1(a, b; b; z)}{a} = \lim \sum \frac{e_i(a)}{ad_i} z^{q^i} = \sum \frac{e_i'(0)}{d_i} z^{q^i} = l(z).$$

Euler and Kummer transformations: Keeping in mind the analog of the binomial series explained above and the fact that we are dealing with additive rather than multiplicative situation, an analog of the Euler's transformation formula $_2F_1(c - a, c - b; c; z) = (1 - z)^{c-a-b} {_2F_1}(a, b; c; z)$ in our case is $_2\mathcal{F}_1(c - a, c - b; c; z) = e((c - a - b)l(z)) + {_2\mathcal{F}_1}(a, b; c; z).$

Similarly, closely related analog of Kummer's theorem for the confluent hypergeometric function $e^{-z} {}_1F_1(a;b;z) = {}_1F_1(b-a;b;-z)$ in the classical case, we have analog $e(-z) + {}_1\mathcal{F}_1(a;b;z) = {}_1\mathcal{F}_1(b-a;b;z)$.

Sum formula: The fact that we get sum instead of the product as in the classical case of Euler transformation implies that we get the following sum formulas as analogs of the product formulas [S1] pa. 13: If $u - x - y = 1 + n + v + w - m$, then ${}_2\mathcal{F}_1(x,y;u;z) + {}_2\mathcal{F}_1(1-n-v,1-n-w;1-n-m;z)$ $= {}_2\mathcal{F}_1(u-x,u-y;u;z) + {}_2\mathcal{F}_1(v-m,w-m;1-n-m;z)$.

Ratios of terms: In analogy with the classical, q-analogs and first analog, if c_n denotes the n-th term in the definition of ${}_r\mathcal{F}_s$ and if $a_i, b_j \in K - A$, then c_{n+1}/c_n^q is an algebraic function of t^{q^n} as we will see in Chapter 8. See 8.5.3 and [Tha95a, pa. 230] for some explicit examples.

Remarks 6.5.9 (1) We do not have differential interpretation for the second analog yet.

(2) In the $\wp$-adic setting Kochubei (see [Koc00; Koc99; Koc98] and recent preprints) has proved several basic theorems such as Cauchy problem, convergence vs. formal solutions etc. on these difference-differential-Frobenius equations and has studied analogs of several classical equations such as those for power functions, logarithms. In regular case, unique solution converges in some disk, whereas in singular case, existence of formal solution implies convergence in some disk, unlike the classical case.

(3) How about generalization to A other than $\mathbb{F}_q[t]$? As we have seen, there are good generalizations of Carlitz' $e(z)$, $l(z)$, d_i and l_i and binomial coefficients arising from the sign-normalized Drinfeld modules of Hayes. We can hence proceed to define hypergeometric functions in analogous fashion using these generalized ingredients. In fact, there are a couple of ways to generalize as we have seen in 4.15 (see also Chapter 8). But more pressing problem seems to be to get a better understanding of this simplest case.

Chapter 7

Higher dimensions and geometric tools

Higher dimensional generalizations of Drinfeld modules were first considered by Gross, Drinfeld, Stuhler and Anderson [And86]. It was quickly understood that naive viewpoint of generalizing simple actions to matrix setting encounters many difficulties and that geometric viewpoint of Shtukas developed by Drinfeld [Dri77a; Dri87; Mum78] provides good objects in higher dimensions, if at the expense of obscuring a little the connections with the familiar language. Some of the resulting objects were systematically developed and exploited for arithmetical purposes of our interest by Anderson, who pointed out some subtleties, solved a basic correspondence problem connecting the geometric and analytic formulations posed by Drinfeld, proved a conjecture of Gross and understood them in a formalism parallel to motives: Though Drinfeld modules are objects different from varieties, their Betti (lattice), l-adic (Tate modules) incarnations occurring in Drinfeld's work and DeRham incarnation pointed out by Deligne gives hints in this direction in the one dimensional case. Various cohomology realizations are obtained by Anderson by simple operations of algebra.

The general motivic formalism has not been worked out in detail and several crucial differences with the classical case, especially in the connection with L and Γ functions are already apparent. Understanding of the special values of Zeta functions in the right formalism is an interesting open problem, while in the number field case the situation seems to be better understood: at least, then we have very general conjectures with good evidence in their favor.

Rather than using this analogy with motives, Anderson developed direct efficient tools, borrowing inspiration from Krichever-Drinfeld dictionary and related soliton theory to prove strong results on special values of zeta and gamma functions and Gauss sums implying, by work of Jing Yu and others, extremely strong related transcendence results.

With the understanding thus gained of some concrete situations, may be general picture can be better seen by using geometric tools of Shtukas again, by developing the necessary tools and the language; just as it is fruitful to study elliptic curves first with concrete Weierstrass equations before moving on to abelian varieties with more general techniques.

Drinfeld A-modules of rank r over C_∞ correspond to A-lattices of rank r in C_∞. In d dimensions, we look at A-lattices in C_∞^d. As in the classical case, not all lattices give algebraic modules, and also in other direction, there is a not well-understood condition (geometric situation is better) for getting a lattice from a module. More generally, we now look at commutative subrings, such as the co-ordinate rings of curves minus some points, inside the endomorphism ring of $\mathbb{G}_a^d$, which is a ring of size d square matrices with entries being polynomials in Frobenius. The additional freedom allows more places at infinity. So, for example, we can try to get objects such as analogs of Fermat motives with multiplications by full cyclotomic field, whereas Drinfeld modules had restricted us to level t or constant field extensions only. This has applications (see 8.4-8.8) to gamma values at general fractions, generalizing the results of Chapter 4.

Instead of taking up the development of whole 'complex multiplication by A' machinery at the start, following Anderson, we will first restrict to the t-modules and corresponding t-motives: namely the simplest case of $\mathbb{F}_q[t]$-modules, which simplifies notation and already offers enough interesting twists to begin with.

In fact, the general development is not carried out as thoroughly in the literature, but see [Tam94; Pot99] and van der Heiden's University of Groningen thesis of 2003, for some work along this line. Things quickly get complicated and geometric formulation may be better suited for applications in general.

For the applications in the next chapter, Drinfeld's geometric Shtuka description given below becomes even more useful in higher dimensions, but we first look at equivalent algebraic and analytic descriptions, without stressing the underlying geometry. We will first work in the affine setting of rings (say co-ordinate ring of the projective line minus ∞) and modules and data at ∞. This has the advantage of simplicity and immediate connection to applications in the next chapter. Then we move to the setting of projective line and corresponding schemes and sheaves and dictionary between the two languages.

We only outline some proofs referring to the original sources for details.

7.1 *t*-modules and *t*-motives

Let $A = \mathbb{F}_q[t]$. Let F be an A-field ($\iota : A \to F$) which is perfect, i.e.,
$F^p = F$. This requirement is put to make structure theory of $F\{\tau\}$-modules
and both left and right division algorithms and resulting elementary divisor
theory work well. Further, if α is not a p-th power in a field of characteristic
p, then, for example, the algebraic group defined as the kernel of $(x, y) \to$
$x + x^p - \alpha y^p : \mathbb{G}_a^2 \to \mathbb{G}_a$ is not isomorphic to $\mathbb{G}_a$ over it, but becomes
isomorphic to $\mathbb{G}_a$ over the extension obtained by adjoining the p-th root of
α to it.

When convenient, we will even assume F to be algebraically closed, as
we will not pay attention to the smallest possible fields of definition.

Let $\theta := \iota(t) \in F$. As usual in the complex multiplication theory, we
should keep the roles of the field of multiplication and the field of definition
separate, even when the two fields are the same. In the previous chapters,
we have ignored this when it did not matter. In this chapter, when we
speak of C_∞ as the field of definition, it should be thought of as a copy of
C_∞ obtained by θ replacing t.

The non-commutative ring $\mathrm{End}_F(\mathbb{G}_a^d)$, can be identified with the ring
of d cross d matrices with entries in $F\{\tau\}$. (We will work over $\mathbb{F}_q$ and use
τ rather than τ_p.) Hence an endomorphism t of $\mathbb{G}_a^d$ can be described as
$t(X) = \sum M_i \tau^i(X)$.

In the one dimensional case, the restriction to the $i = 0$ term (i.e.
tangent or Lie algebra or linear level) gives a homomorphism from A to
F, which we identified with the structure morphism ι. If we want to keep
such connection, two natural generalizations would be to ask for M_0 to
be the scalar matrix θ or to ask for all its eigenvalues to be θ. We will
see shortly that simple requirement of taking tensor products of Carlitz
modules shows that the first condition is too restrictive and so we impose
the second condition instead. Hence we allow the differential to be θ plus
a nilpotent matrix:

Definition 7.1.1 A *t*-module G (of dimension d) over F is an algebraic
group isomorphic to $\mathbb{G}_a^d$, and equipped with an endomorphism t of G, such
that $(t-\theta)^d \mathrm{Lie}(G) = 0$. A morphism of *t*-modules is *t*-equivariant morphism
of underlying algebraic groups.

Hence the one dimensional *t*-modules are Drinfeld A-modules, except
we also allow trivial linear action.

To define tensor product of *t*-modules, we look at dual notion of functions, namely the set $M(G) := \mathrm{Hom}(G, \mathbb{G}_a)$ of F-algebraic group homomorphisms. It inherits group structure $(m_1 + m_2)(g) := m_1(g) + m_2(g)$ from $\mathbb{G}_a$ and composing with the endomorphisms of G in $F[t]$ on left or endomorphisms $F\{\tau\}$ of $\mathbb{G}_a$ on right:

$$G \xrightarrow{t} G \xrightarrow{m} \mathbb{G}_a \xrightarrow{f,\tau} \mathbb{G}_a.$$

It is thus a $F[t]$-$F\{\tau\}$ bimodule [DH87, 2.7]. We will follow the formulation of [And86].

Definition 7.1.2 Let $F[t, \tau]$ denote the non-commutative ring generated by t, τ and all $f \in F$ subject to

$$t\tau = \tau t, \quad ft = tf, \quad \tau f = f^q \tau.$$

We sometimes write $F[\tau]$ for the non-commutative subring $F\{\tau\}$ of the earlier chapters. Note that the commutation relations are those dictated by the interpretation above.

Then $M(G)$ is a left $F[t, \tau]$ module and the morphisms correspond to $F[t, \tau]$-linear maps. Clearly, $M(G)$ is free over $F[\tau]$ on the basis $m_1, \cdots, m_d : G \to \mathbb{G}_a$ defining an isomorphism $G \xrightarrow{\sim} \mathbb{G}_a^d$.

Tensor product (defined below) corresponds to the tensor product over $F[t]$ with τ (coming from $\mathbb{G}_a$-action) acting diagonally.

Notation for matrices: We will use I for identity matrix of size d and E_{ij} for size d (elementary) matrix with the (i, j)-th entry one and other entries zero. For typographical reasons, we may sometimes write column vector $X_{d \times 1}$ as a row vector $X = (x_1, \cdots, x_d)$ instead. For a matrix M, we write $M^{(i)}$ for the matrix obtained by raising each entry of M to the q^i-th power.

Examples 7.1.3 (1) G_{linear} (also denoted just as $\mathbb{G}_a$ and called a trivial t-module) has the underlying group $\mathbb{G}_a$ and $t(x) = \theta x$. Then $M(\mathbb{G}_a)$ is free over $F[\tau]$ on basis $m_1 = id_{\mathbb{G}_a} : \mathbb{G}_a \to \mathbb{G}_a$ and

$$tm_1 = (x \to m_1(\theta x)) = (x \to \theta m_1(x)) = \theta m_1.$$

(2) $G_{Drinfeld}$ has the underlying group $\mathbb{G}_a$ and $t(x) = \theta x + f_1 x^q + \cdots + f_r x^{q^r}$, with $f_i \in F$, $f_r \neq 0$, $r > 0$. Then $M(G_{Drinfeld})$ is free rank one over $F[\tau]$ on basis $m_1 = id_{\mathbb{G}_a}$ and $tm_1 = (\theta + f_1 \tau + \cdots + f_r \tau^r)m_1$. Note that it is free of rank r over $F[t]$ with basis $m_1, \tau m_1, \cdots, \tau^{r-1} m_1$ for example.

(3) If ϕ and ψ are Drinfeld $\mathbb{F}_q[t]$-modules of rank r and s with corresponding τ-basis m_1 and n_1, as in the example above, then the tensor product of

their t-motives has τ-basis $p_i = \tau^{i-1}m_1 \otimes n_1$, $(1 \leq i \leq r)$, $p_{r+i} = m_1 \otimes \tau^i n_1$, $(1 \leq i \leq s)$ and t-basis $\tau^i m_1 \otimes \tau^j n_1$ $(0 \leq i < r, 0 \leq j < s)$, and it is straightforward to write down the matrix of corresponding t-action. As a simple example, very useful for our applications, of a tensor product, let us calculate the n-th tensor power $C(n) := C^{\otimes n}$ of the Carlitz module C. (It is an analog of the Tate twist motive $\mathbb{Z}(n) = \mathbb{Z}(1)^{\otimes n}$, as we will see): By definition, it is free of rank one over $F[t]$ with basis $m_1 := m^{\otimes n}$, where $\tau m = (t - \theta)m$, for basis m for the Carlitz module as above. As τ acts diagonally, we then have $\tau(m^{\otimes n}) = (t - \theta)^n m^{\otimes n}$. Then $m_i := (t - \theta)^{i-1} m_1$, $1 \leq i \leq n$ is its basis as $F[\tau]$-module. Hence the dimension d is n and we see that $tm_i = m_{i+1} + \theta m_i$, for $i < n$ and $tm_n = \tau m_1 + \theta m_n$. Identification with standard projections $m_i := (x_1, \cdots, x_n) \to x_i : \mathbb{G}_a^n \to \mathbb{G}_a$ gives the t-action in these co-ordinates as

$$t = \begin{pmatrix} \theta & 1 & & \\ & \ddots & \ddots & \\ & & \ddots & 1 \\ \tau & & & \theta \end{pmatrix} = (\theta I + N) + E_{n1}\tau$$

where $N = (n_{i,j})$ is the nilpotent matrix with $n_{i,i+1} = 1$ for $1 \leq i < n$ and 0 otherwise.

(4) Since $M(G_{Drinfeld}) = \oplus_{i=1}^r F[t]\tau^{i-1}m_1$, we have $\wedge^r M(G_{Drinfeld}) = F[t]m$ with $m := m_1 \wedge \tau m_1 \wedge \cdots \wedge \tau^{r-1}m_1$. Hence

$$\tau m = \tau m_1 \wedge \cdots \wedge \tau^r m_1 = (-1)^{r-1}\tau^r m_1 \wedge \tau m_1 \cdots \tau^{r-1}m_1 = (-1)^{r-1}f_r^{-1}(t-\theta)m$$

by using the action of (2). Hence the top exterior power of the rank r Drinfeld t-module is isomorphic to the Carlitz module, eg. if $f_r = (-1)^{r-1}$ or more generally over $F(\mu)$ where $\mu := (f_r(-1)^{r-1})^{1/(q-1)}$, as we can see by changing the basis from m to μm.

Remarks 7.1.4 (1) Example 3 shows that if we want to have tensor products of Drinfeld modules, we need to generalize to higher dimensions and allow addition of nilpotent matrix to the scalar matrix θI at the Lie algebra level and also that in interesting examples as above, the top degree coefficient matrix need not be invertible.

(2) In one dimension, just the non-triviality condition gives nice objects, namely the Drinfeld modules. But since, for example, matrices do not commute in general and non-zero does not mean invertible, many nice things that we are used to in one dimension, such as the correspondence between

the top degree and rank, defined with lattices or cardinalities of torsion say, do not work in higher dimensions, unless we put extra conditions.

(3) For various explicit calculations with tensor products, such as decomposition into symmetric and exterior parts, see [Ham93].

Example 7.1.5 (i) If the top non-zero coefficient matrix, say M_r, is nilpotent, then for the endomorphism t^d the rd-th coefficient is zero.

(ii) Consider the t-module with t-action given by θI plus a strictly upper (or lower) triangular matrix, eg., the two dimensional t-action $\theta I + E_{12}\tau$. First, note that t-torsion (in fact, a-torsion for any non-zero a, for the same reason) just consists of the zero vector. Next, if we define the corresponding exponential e by the resulting functional equation as usual, then its kernel is just zero: Let X belong to the kernel and let $(e_1, e_2) := e(\theta^{-1}X)$. Then $(0,0) = e(X) = (\theta e_1 + e_2^p, \theta e_2)$. So $(e_1, e_2) = 0$. Hence $\theta^{-n}X$ belongs to the kernel by induction on n. But by the inverse function theorem, the kernel of the exponential is discrete and hence it has to be zero.

Our examples suggest that to get a good notion of rank, we might add the requirement that $M(G)$ is free of finite rank, r say, as $F[t]$-module:

Definition 7.1.6 A t-motive M (over F) is a left $F[t, \tau]$-module, free of finite rank as a module over $F[t]$ as well as over $F[\tau]$ and satisfying $(t - \theta)^n(M/\tau M) = 0$ for some n.

A morphism of t-motives is just a $F[t, \tau]$-linear map.

The dimension $d(M)$ is defined to be the rank of M as $F[\tau]$-module ('τ-rank') and the rank $r(M)$ is defined to be the rank as $F[t]$-module ('t-rank').

An abelian t-module is a t-module G for which $M(G)$ is finitely generated as $F[t]$-module.

Tensor product of two t-motives is defined to be tensor product over $F[t]$ with τ acting diagonally.

Remarks 7.1.7 (1) If H is a sub-t-module of a t module G, then G is abelian if and only if H and G/H are abelian.

(2) That 'the tensor product of t-motives is a t-motive' is a non-trivial and unpublished result of Anderson. See 7.3.

(3) We have canonical isomorphism $\mathrm{Hom}_F(\mathrm{Lie}(G), F) \xrightarrow{\sim} M(G)/\tau M(G)$.

(4) A t-motive can then be described by matrices $M_t \in M_{d \times d}(F[\tau])$ or $M_\tau \in M_{r \times r}(F[t])$ giving the corresponding actions on some basis. Changing the basis by matrix C replaces M_τ by $C^{-\tau}M_\tau C$, as τ is only a semi-linear transformation.

(5) If M is t-motive over F of rank one and dimension d, then $M_\tau = f(t - \theta)^d$, for some $f \in F$, as can be immediately seen by definition of t-motive and dimensions. In particular, over the algebraic closure of F, it is isomorphic to $C^{\otimes d}$ of 7.1.3 (3).

Lemma 7.1.8 (*[And86, Lemma 1.4.5]) Any left $F[t, \tau]$-module finitely generated as a $F[t]$ module and as a $F[\tau]$-module is free of finite rank over $F[t]$ if and only if free of finite rank over $F[\tau]$.*

Theorem 7.1.9 (*[And86, Theorem 1]) The functor $G \to M(G)$ is anti-equivalence of categories of abelian t-modules and t-motives.*

Examples (2) and (3) of 7.1.3 are t-motives, while (1) is not.

Remark 7.1.10 We follow the terminology of [And86]. Note that terminology of [Gos96] differs in the definitions of t-modules and t-motives, in that what is called t-motive here is 'abelian' there, so that the theorem above has versions with both or none of the t-modules and t-motives 'abelian'. But as we saw that without finite generation as $F[t]$-module, we get badly behaved examples and tensor products are wild, so that motives terminology is not that good. In any case, there should be no confusion, as we usually deal with 'abelian' examples anyway.

In light of dualities in Section 2.10, Anderson found it convenient to define related additional notions:

Definition 7.1.11 A dual t-motive M^* is a left $F[t, \tau^{-1}]$-module, free of finite rank over $F[t]$ as well as $F[\tau^{-1}]$ and satisfying $(t-\theta)^n (M^*/\tau^{-1}M^*) = 0$ for some n.

A morphism of dual t-motives is a left $F[t, \tau^{-1}]$-module homomorphism. Given a t-module G, let $M^*(G) := \mathrm{Hom}(\mathbb{G}_a, G)$.

Remarks 7.1.12 (1) $M^*(G)$ can be identified with $M_{d\times 1}(F[\tau^{-1}])$, where d is the dimension of G. It is a left $F[t, \tau^{-1}]$-module, under natural composition actions on left and right: i.e. natural structure as $F[\tau^{-1}]$-module and with $t \circ \sum \tau^{-j} m_j := M_t^* \sum \tau^{-j} m_j$, where the adjoint of M_t is obtained via taking adjoints of all entries of the transpose matrix. (Note that $(AB)^* = B^* A^*$ then). The abelian t-modules give rise to dual t-motives.

(2) $M^*(G)$ is free of rank d as a left $F[\tau^{-1}]$-module, and if G is an abelian t-module, $M^*(G)$ is a dual t-motive. The functor M^* is covariant in contrast to M which is contravariant, and its advantage comes from simple descriptions that $M^*(G)/(\tau^{-1} - 1)M^*(G)$ 'is' the algebraic group G and $M^*(G)/\tau^{-1}M^*(G)$ 'is' its Lie algebra. Further, the differentials and

residues duality, occurring in uniformization and period lattice construction of [And86] that we describe later, is replaced by adjoint duality, giving a smoother construction. We will say no more about these notions, but refer the reader to [ABP] instead to look at the technical advantages at work.

7.2 Torsion

For abelian t-modules, pathology of the example (ii) of 7.1.4 disappears and we do get the expected correspondence between the rank and the cardinality of torsion:

Let F be algebraically closed. Let $a \in A \subset F[t, \tau]$ be prime to the characteristic (i.e., prime to $t - \theta$ i.e., $a(\theta) \neq 0$). Then the set of a-division points $G_a(F) := \{g \in G(F) : ag = 0\}$ is an A-module.

To express torsion in terms of M, we have the following theorem [And86], [Mum78, Pa. 147]:

Theorem 7.2.1 *As an A-module, $G_a(F)$ is canonically isomorphic to $\mathrm{Hom}_A((M/aM)^\tau, (1/a)Adt/Adt)$. In particular, it is $A/(a)$-module of rank $r(M)$.*

Proof. We have canonical isomorphisms

$$G_a(F) = \mathrm{Hom}_{F[t,\tau]}(M/aM, \mathrm{Hom}(A/(a), F)) = \mathrm{Hom}_A((M/aM)^\tau, \mathrm{Hom}(A/(a), \mathbb{F}_q))$$

The first one is just $g \to (m \to (a' \to m(a'g)))$ and the second uses surjectivity of Lang's $\tau - 1$ isogeny for $Gl_n(F)$ to get a good basis. The claim follows from the perfect pairing $(a', w) \to \mathrm{Res}_{t=\infty}(a'w) : A \times Adt \to \mathbb{F}_q$. $\square$

A t-module G, with $G_a(F)$ being a $A/(a)$-module of rank (some fixed) r, for all a as above, has been called **regular** by Jing Yu [Yu97]. Hence abelian implies regular.

The usual connection of $\wp$-adic cohomology obtained from inverse limits of torsion sequences and Betti cohomology (lattice) with torsion thus shows that they need to be tensored with differentials to make them commute with tensor products. This was pointed out in [And86].

7.3 Purity

Since we can now take our t-action to be, for example, a direct product of Drinfeld modules of different ranks, it need not be 'pure'.

Unlike in the one dimensional case or for geometric notion of Shtukas studied below, in this naive viewpoint, the purity or Riemann hypothesis has to be enforced by definition:

Unlike the varieties over finite fields case, where some Frobenius power is linear and we can talk about eigenvalues, here τ is only semi-linear and its eigenvalues are not well-defined. Though over the reductions, as in the classical case, we can follow similar path.

By 7.1.7 (5), looking at $\wedge^r M$, we see that the determinant of the size r matrix M_τ is $f(t - \theta)^d$, for $f \in F^*$. If M is 'pure', in the sense that all 'eigenvalues' have the same absolute values, then all eigenvalues should have valuation ('weight') d/r. This heuristics leads to correct answer with proper definitions below. We have already seen this weight $1/r$ in the one dimensional case. Note that unlike the classical case, where the weights are half-integral, even for Drinfeld modules the weights can have arbitrary denominators.

Remark 7.3.1 Even if we allow change of basis with extended scalars, the absolute values of 'eigenvalues' remain the same in one dimension, since if $\tau m = \lambda m$, then $\tau(p(t)m) = p(t)^{\tau-1}\lambda(p(t)m)$ and $p(t)^{\tau-1}\lambda$ has the same absolute value as that of λ. But in higher dimensions, we can change the basis by higher dimensional matrices rather than scalar $f(t)$ as above and the following simple example shows that even the absolute values of the eigenvalues are not well-defined. For example, the identity matrix I of size two has 1 (with t-valuation zero) as a double eigenvalue, but $C^{\tau-1} = C^\tau I C^{-1}$ has characteristic polynomial $X^2 - (t+1)X + 1$ and thus has two eigenvalues with t-valuation 1 and -1 respectively, for the change of basis matrix C given by its two rows (t, γ^{-1}) and $(\gamma t - \gamma, 1)$, with $\gamma^{q-1} + \gamma^{1-q} = 1$.

This τ-linear situation is somewhat analogous to F-crystal / Dieudonne modules context as is apparent in [Dri77c; And86; TW96]. To enforce purity, it is convenient to proceed as follows [And86] by killing most of the information other than the absolute value and using a lattice:

Definition 7.3.2 Call M pure, if it is free, finitely generated as $F[t]$-module and there is a $F[[1/t]]$-lattice L in $M((1/t)) := M \otimes_{F[t]} F((1/t))$ such that $t^u L = \tau^v L$, with $u, v > 0$.

See 7.8 for how this condition generalizes naturally the condition in one dimension coming about from the Shtuka description, and how the proof of the following theorem is related to a tower related to a Shtuka. This purity condition enforces Riemann hypothesis at the finite characteristic reductions.

Theorem 7.3.3 *If M and L satisfy the conditions of the definition, then M is a pure t-motive and $u/v = d(M)/r(M)$.*

Proof. Consider the increasing filtration $M_j := M \cap t^{(j+N)u} L$, where N is large enough so that $M + t^N L = M((1/t))$. Then we have isomorphisms

$$M_{j+1}/M_j \to t^{(j+N+1)u} L/t^{(j+N)u} L = \tau^{(j+N+1)v} L/\tau^{(j+n)v} L.$$

Thus $t^u M_j + M_j = M_{j+1} = \tau^v M_j + M_j$, implying that M is finitely generated over $F[\tau]$ and is thus a t-motive and that $\dim_F(M_{j+1}/M_j) = ur(M) = vd(M)$ implying the claim. □

Definition 7.3.4 For a pure t-motive M, we define its weight to be $w(M) := d(M)/r(M)$.

Remarks 7.3.5 (1) By definition of the tensor product and rank, we see that $r(M \otimes M') = r(M)r(M')$. If M and M' are pure, with $t^u L = \tau^v L$ and $t^{u'} L' = \tau^{v'} L'$ as in the definition of purity, then we have $t^{uv'+u'v}(L \otimes L') = \tau^{vv'}(L \otimes L')$, and so the tensor product of pure motives is pure and $w(M \otimes M') = w(M) + w(M')$.

(2) If M is pure, its (non-zero) $F[t, \tau]$-submodule M' is also pure of same weight (use $L' := M'((1/t)) \cap L$), and its (nonzero) $F[t, \tau]$-quotient M'' without $F[t]$-torsion is also pure of same weight (use L'' to be the induced image of L in $M''((1/t))$).

Theorem 7.3.6 *Let G be a t-module of dimension d given by $(\theta + N) + \sum_{i=1}^s M_i \tau^i$, with M_s invertible. Then it is abelian t-module of rank sd, pure of weight $1/s$.*

In particular, Drinfeld $\mathbb{F}_q[t]$-modules of rank r are pure of weight $1/r$.

Proof. We leave it to the reader to verify that $M(G)$ has $F[\tau]$-basis m_i ($i = 1$ to d) and $F[t]$-basis $\tau^j m_i$ ($j = 0$ to $s - 1$) and the purity criterion works with $L := L_k := \oplus \tau^j m_i \otimes F[[1/t]]$, where j runs through 0 to a sufficiently large k, and in that case, we have $tL = \tau^s L$.

The proof is similar to the $d = 1$ (and so $r = s$) case in which we indicate more details: We have $tL_k = tL_k + L_k$, because $a \otimes b = t(a \otimes (1/t)b)$ implies $L_k \subset tL_k$. Now $tL_k + L_k = L_{k+r}$: The right side clearly contains the left, hence it is enough to show that the left side contains $\tau^j m_1$, for $j \leq k + r$. For $j \leq k$, it is clear. Otherwise, if $tm_1 = (\theta + g_1 \tau + \cdots + g_r \tau^r)m_1$, then $\tau^j m_1 - t(\tau^{j-r} g_r^{1/q^{j-r}} m_1)$ reduces the power and we can proceed by induction. Finally, $L_{k+r} = \tau^r L_k + L_k = \tau^r L_k$: The first equality is clear, hence it is enough to show that $\tau^j m_1 \in \tau^r L_k$, for $j \leq r$, i.e., $m_1 \in \tau^b L_k$ for $b \leq r$. Note that $1/(t - \theta) \in F[[1/t]]$ and write $(t - \theta)m_1 = \tau f(\tau)m_1$, with polynomial f of degree at most $r - 1$. Choose $k > r(r - 1)$. Then $l := f(\tau)^b m_1 \otimes 1/(t - \theta)^b \in L_k$ and $\tau^b l = m_1 \otimes 1$ as desired. □

Hence we have purity in dimension one, and 7.1.7 (5) shows that rank one is also automatically pure (of weight d).

In the local case, parallel to Dieudonne-Manin work classifications using slopes of crystals/Dieudonne modules, in the mixed characteristic case, Anderson proved (unpublished) similar theorem in equal characteristic case: the matrix can be normalized to a diagonal matrix with entries positive rational powers of t. We define these exponents to be the weights of the t-motive and then pure just means only one weight. With this as a definition of weight, over a finite field, pure motives give pure Frobenius eigenvalues of these weights. Using these slope filtration results, Anderson has proved that tensor product of t-motives is a t-motive generalizing 7.3.5 (1).

7.4 Exponential, period lattice and uniformizability

Let us look at the analytic theory over $F = C_\infty$. As mentioned before, C_∞ is now the completion of an algebraic closure of $\mathbb{F}_q((1/\theta))$, with $\iota : A \to C_\infty$ and $\iota(t) = \theta$ and θ is now transcendental over $\mathbb{F}_q$.

A t-module G given by $(\theta+N)+\sum_{i=1}^k M_i\tau^i$ of dimension d over C_∞ has associated **exponential** $\exp_G(X) = X + \sum_{i=1}^\infty e_i X^{(i)}$, which is a $\mathbb{F}_q$-linear power series with coefficients e_i being square matrices of size d, which is entire in the sense that it gives a function $\exp : C_\infty^d \to C_\infty^d$ and satisfies

$$\exp((\theta + N)X) = t\exp(X) = (\theta + N)\exp(X) + \sum_{i=1}^d M_i \exp(X)^{(i)}.$$

(For trivial t-module, the exponential is just $e(z) = z$. We will usually avoid this case.)

The entireness follows [And86; Gos96] as in the one dimensional case, by estimating the sizes of coefficients obtained by solving the recurrence relations given by the functional equation.

Iterating, we have

$$\exp([a]X) = d[a] \exp(X), \quad \text{for } a \in A,$$

where we write $[a]$ for the corresponding action and $d[a]$ for its differential action at the Lie level.

Example 7.4.1 Let ρ be Drinfeld $\mathbb{F}_q[t]$-module of rank r. Given bi-derivation basis, for example δ^i, $0 \leq i < r$ of Remark 6.4.2 (3), we can consider 'quasi-periodic' extension of ρ by $\mathbb{G}_a^{r-1}$: This r dimensional t-motive is given by r cross r matrix M_t whose lower $r-1$ cross $r-1$ block is just θ times the identity matrix, and (i, i) entries δ_t^{i-1} and other entries in first row zero. This motive is not abelian, as we get just θ action on all except the first co-ordinates. But comparison with exponential equation and equations for quasi-periodic functions corresponding to the bi-derivations in 6.4 shows that the corresponding exponential function is just

$$\exp(x_1, \cdots, x_r) = (e_\rho(x_1), x_2 + F_{\delta^1}(x_1), \cdots, x_r + F_{\delta^{r-1}}(x_1))$$

with period vectors consisting of period and quasi-periods of ρ. For more discussion, see [BP02].

In the classical case, when the dimension is more than one, not all the lattices correspond to abelian varieties. We have a similar situation: Given a lattice and action on Lie, we may not get corresponding uniformizable, pure t-motive. In fact, [And86, Cor. 3.3.6] shows that for pure uniformizable t-motive, $C_\infty[t]$-span of period lattice is the Lie algebra C_∞^d.

Example 7.4.2 Suppose Λ is rank one discrete lattice $A(0, \lambda)$ in $\mathrm{Lie}(E) = C_\infty^2$, where the tangent level action of t is just by a scalar matrix θ. This does not satisfy the condition. But we will see that in fact this is period lattice of $C^{\otimes 2}$ in characteristic 2, for $\lambda = \tilde{\pi}^2$. The Lie action as described in 7.1.3 (3) is not scalar.

Good conditions (analog of Riemann form conditions) for when the lattices correspond to some abelian t-modules (or just t-modules, or better classes discussed below) are not known yet and are currently under investigation.

In contrast to the one dimensional case, the surjectivity of the exponential is not automatic for higher dimensions. In other words, it fails to provide uniformization in general, even for abelian, pure t-motives:

Example 7.4.3 Here is a two-dimensional example of Coleman [And86]:

$$tX = \theta X + \begin{pmatrix} 0 & 1 - c^{q+1} \\ 1 - c^{p+1} & 0 \end{pmatrix} X^{(1)} + \begin{pmatrix} c^{1+q+q^2} & 0 \\ 0 & c^q \end{pmatrix} X^{(2)},$$

where $c^{-1} + c = \theta$ and $|c| < 1$. It is easy to verify that

$$\lambda(X) := cX + \begin{pmatrix} 0 & -c^{q+1} \\ -c^{q+1} & 0 \end{pmatrix} X^{(1)} + \begin{pmatrix} c^{1+q+q^2} & 0 \\ 0 & 0 \end{pmatrix} X^{(2)}$$

is an automorphism of $\mathbb{G}_a^2$ satisfying $tX = \lambda(X) + \lambda^{-1}(X)$ and $\lambda(\exp(X)) = \exp(cX)$ ('functoriality' of the exponential) for the exponential exp. Thus the kernel of exp is stable under the multiplication by c, so as in Example 7.1.5, the kernel is zero. But invertibility of the top coefficient implies by Theorem 7.3.5, our t-module is abelian and so by Theorem 7.2.1, it has non-trivial t-torsion say. By the functional equation of exp, it can not be in the image, since exp has zero kernel. Hence exp is not surjective.

Now we turn to a criterion for uniformizability:

Definition 7.4.4 Let $C_\infty\{t\}$ denote the ring of power series $\sum c_i t^i$, with coefficients $c_i \in C_\infty$ and tending to zero. For a t-motive M, put $M\{t\} := M \otimes_{C_\infty[t]} C_\infty\{t\}$, with the usual action of $C_\infty[t]$ and with the τ-action $\tau(m \otimes (\sum c_i t^i)) = (\tau m) \otimes (\sum c_i^q t^i)$.

Definition 7.4.5 For an abelian t-module G, with the corresponding exponential $\exp_G : \mathrm{Lie}(G) \to G(C_\infty)$, put $H_*(G) := \mathrm{Kernel}(\exp_G)$.

For a t-motive M, put $H^*(M) := M\{t\}^\tau$. Say that M is rigid analytically trivial, if the evident map $H^*(M) \otimes_{\mathbb{F}_q[t]} C_\infty\{t\} \to M\{t\}$ is an isomorphism.

In [And86; Gos96], ring of power series of coefficients in a finite extension of K_∞ (in 2.1.1 and 2.3.3 of [And86] this condition is meant instead of what is stated) and tending to zero is used in place of $C_\infty\{t\}$. But [ABP, prop.3.1.3] shows that when M is defined over a finite extension of K_∞, as in all our application below, it does not matter. If the t-motive is defined over C_∞ instead, it is not clear whether the conditions of the theorem are still equivalent, the issue being Weierstrass preparation theorem tool that we have for finite extensions of K_∞.

Anderson [And86, Theorem 4, 2.12.1] proved

Theorem 7.4.6 *The following properties of an abelian t-module G defined over a finite extension of K_∞ are equivalent: (i) The $\mathbb{F}_q[t]$-rank of $H_*(G)$ is $r(G)$, (ii) The exponential $\exp_G$ is surjective, (iii) $M(G)$ is rigid analytically trivial.*

If these properties hold, $H_*(G)$ is canonically isomorphic to $\mathrm{Hom}_A(H^*(M(G)), A\mathrm{d}t)$.

If these properties hold, G is called **uniformizable**.

We omit the proof (giving some related ideas, recipes and examples instead), except for the easy implication (ii) implies (i): If (ii) holds, then we have uniformization $G/H_*(G) \xrightarrow{\sim} G$ provided by the exponential. Then, on one hand, the dimension of the t-torsion over $\mathbb{F}_q$ is the rank by 7.2.1 and on the other hand, since the t-torsion is $\exp(H_*(G)/t)$, it is $\mathbb{F}_q[t]$-rank of the lattice $H_*(G)$, and hence (i) follows.

By Remark 7.1.7 (4), and the definition of rigid analyticity, M being rigid analytically trivial is equivalent to the existence of matrix $C \in M_{r \times r}(C_\infty\{t\})$ solving the '$\tau - 1$' equation $C^\tau C^{-1} = M_\tau$. In particular, when $r = 1$, this simplifies to the condition that the formal solution $C = (\prod_{i=0}^\infty M_\tau^{(i)})^{-1}$ (note that any $\mathbb{F}_q[t]$ multiple would also be a solution as τ acts trivially on $\mathbb{F}_q[t]$) belongs to $C_\infty\{t\}$.

Anderson's theorem gives residue formula for recovering the period lattice from the motive description or in other words from solution C above. The connection of this solution with period is apparent, when we look at rigid analytically trivial condition as comparison isomorphism of appropriate cohomologies.

In the Calitz module case, $M_\tau = t - \theta$, so that $C = (-\theta)^{1/(q-1)} \prod_{i=0}^\infty (1 - t/\theta^{q^i})^{-1}$. Note that its residue at $t = \theta$ of the product solution above is seen to be (identifying θ with t at the end for easy comparison) $(-t)^{1/(q-1)} t \prod_{i=1}^\infty (1 - 1/t^{q^i-1})^{-1}$, i.e. the expression of the Carlitz fundamental period we saw in 2.5.

Let us take another look at the calculation in 2.5 recovering period lattice from 'division sequences' (see also 7.2.1) which make up the Tate module which is essentially the completion of the lattice:

Consider a Drinfeld $\mathbb{F}_q[t]$-module ρ. A t-division sequence $\beta = (\beta_0, \beta_1, \cdots)$ is any sequence with $\rho_t(\beta_{j+1}) = \beta_j$, for $j \leq 0$ and $\rho_t(\beta_0) = 0$. Let M be a motive corresponding to ρ as in 7.1.3. (2), with identification $M/\tau M$ with C_∞ sending m_1 to 1. Then we see that the evaluation at m_1 sends $\mathrm{Hom}_{C_\infty[t,\tau]}(M, C_\infty\{t/\theta\}dt/(t - \theta))$ injectively inside $C_\infty\{t/\theta\}dt/(t-\theta)$ with image consisting of exactly $f_\beta(t) = \sum \beta_j t^j \in C_\infty[[t]]$ satisfying $t f_\beta(t) = \rho_t(f_\beta(t))$ (which just means that β is a division sequence as above) converging in disc $|t|_\infty \leq 1$, or equivalently $(t - \theta)f_\beta(t)$ converging in $|t|_\infty \leq |\theta|_\infty$. As in 2.5, since there is a period λ such that $\beta_j = e_\rho(\lambda/\theta^{j+1})$, $f_\beta(t) = -\lambda/(t - \theta)$ plus a power series converging in $|t|_\infty \leq |\theta|_\infty$. The upshot then is that the period lattice is just the residue at $t = \theta$ of the image above (both considered as A-submodules of $\mathrm{Hom}_{C_\infty}(M/\tau M, C_\infty)$).

For Drinfeld modules for general A, the formula is similar, except we have to replace the Tate disc by appropriate affinoid subspace of $\overline{X}$ and take residues at point ξ corresponding to structure morphism in the notation of the last section of this chapter.

Now we describe the recipe how to recover the period lattice as the image in $\mathrm{Lie}(M)$ of $W := CM_{r \times 1}(\mathbb{F}_q[t])$ via 'residue' for a general t-motive M in terms of basis and co-ordinates:

As we have seen in 7.1, we can identify M with d cross 1 matrices with entries in $C_\infty[\tau]$ as well as r cross 1 matrices with entries in $C_\infty[t]$ after choice of basis. Suppose we are working with such bases, and let B be the d cross r matrix whose rows are just inverse images of co-ordinate vectors e_i ($i = 1$ to d) under these identifications. Then the period lattice in C_∞^d is the residue at $t = \theta$ of BW.

Let us work it out in detail for the motive $C^{\otimes n}$ of 7.1.3 (3) in the basis chosen there: We have $n = d$, $r = 1$ and $M_\tau = (t - \theta)^n$, so that $C = c_n(t)$ here is the n-th power of that in the Carlitz module case above and the i-th entry of B is $(t - \theta)^{i-1}$. Hence writing elements of W as $c_n(t)f(t) = \sum f_i(t - \theta)^i$ the corresponding element in the period lattice is the d cross 1 vector $(f_{-1}, \cdots, f_{-n})$. In particular, the fundamental period, λ say, has the last co-ordinate $\tilde{\pi}^n$, where $\tilde{\pi}$ is the period of the Carlitz module and the period lattice can be described as $\{d[a]_n \lambda : a \in \mathbb{F}_q[t]\}$, where $d[a]_n$ is action (see 7.6 also) of a at Lie level. The proofs of residue formula [AT90, pa. 178] are essentially the same as the one we saw in 2.5.

Remarks 7.4.7 (1) It follows from the Theorem that the tensor product of uniformizable t-modules is uniformizable.

(2) An abelian sub t-module of an abelian, uniformizable t-module is [Yu89, 5.3] also uniformizable.

(3) It follows [BP02, 2.1.2] from the Theorem that if G is uniformizable with corresponding period lattice Λ and $\Lambda' \subset \mathrm{Lie}(G)$ is A-lattice in which Λ has finite index, then Λ' is the period lattice corresponding to some uniformizable, abelian G', with natural isogeny $G \to G'$: In fact, if T is the image in $G(C_\infty)$ of Λ' under $\exp_G$, then $G' = G/T$.

(4) By the theorem we know that dimension one is uniformizable, and by 7.1.7 (5), we also see that the rank one is uniformizable. We have in fact, calculated the C above.

Theorem 7.4.8 [And86, 3.3.6, 2.12.2] *If G is pure, uniformizable, then the corresponding period lattice $H_*(G)$ generates $\mathrm{Lie}(G)$ as $C_\infty[t]$-module.*
Uniformizable, abelian t-module is determined by its period lattice.

Some more equivalent conditions for uniformizability, non-examples and discussion of issues can be found in recent work by Pink, Gardeyn, Hartl.

Let us now define   [And86; BP02] **Hilbert-Blumenthal-Drinfeld (HBD) modules and modules of complex multiplication (CM) type:**

Let $A = \mathbb{F}_q[t]$, $K = \mathbb{F}_q(t)$, L_+ be a finite separable extension of K of degree d in which ∞ splits completely, and $\mathcal{O}_+$ be the integral closure of A in L_+. Let $\sigma_i : L_+ \to C_\infty$, $1 \leq i \leq d$, be the embeddings extending the embedding of K via the structure morphism ι.

Definition 7.4.9   A HBD module with multiplications by $\mathcal{O}_+$ is a pure, uniformizable, abelian t-module G of dimension d such that A-action extends to a map of A-algebras $\Psi : \mathcal{O}_+ \to \mathrm{End}(G)$ in such a way that the induced representation σ on $\mathrm{Lie}(G)$ is isomorphic to the direct sum of σ_i's.

Generalizing Drinfeld's correspondence in 2.4 between Drinfeld modules and lattices in one dimension via kernel of exponential, Anderson proved [And86, Theorem 7]:

Theorem 7.4.10   *The following categories are equivalent:*

(a) Category of HBD modules with multiplications by $\mathcal{O}_+$, with morphisms the t-module homomorphisms which are $\mathcal{O}_+$-equivariant.

(b) Category of lattices in C_∞^d invariant under $\sigma(\mathcal{O}_+)$, with morphisms the C_∞-linear maps on C_∞^d which carry one lattice into the other and commute with $\sigma(\mathcal{O}_+)$.

Now let L be a finite separable extension of L_+ in which each infinite place of L_+ is totally ramified. Choose extensions of σ_i's to embeddings $L \to C_\infty$, which we again denote by σ_i and denote the set of σ_i's by S. Let Λ be a discrete A-submodule of C_∞^d of rank $[L : K]$ which is invariant under the action via σ_S of some order of L. Let $\mathcal{O}$ be the maximal such order. We then think of Λ as having real multiplications by $\sigma(\mathcal{O}_+ \cap \mathcal{O})$ and complex multiplications by $\sigma_S(\mathcal{O})$. Then Λ is [BP02] the period lattice of a uniformizable abelian t-module G of dimension d and the A-action extends to an injection of A-algebras $\Psi : \mathcal{O} \to \mathrm{End}(G)$ in such a way that the induced representation on $\mathrm{Lie}(G)$ is isomorphic to the direct sum of σ_i's in S. Then G is called **CM module** of CM type (L, S) and L is called CM field of G. See [BP02] for many basic results on these concepts as well as on development of quasi-periods in higher dimensions.

Remarks 7.4.11   Since abelian varieties are uniformizable and their cohomology (say H^1) is pure, may be, uniformizability and purity should have been added for a good notion of abelian. As pointed out in [BP02], there are abelian t-modules over algebraically closed fields which are not semi-simple up to isogeny. The truth of the corresponding statement is unknown for 'pure, abelian' or for 'pure, abelian, uniformizable'.

7.5 Cohomology realizations

Anderson gives simple constructions for the realizations of a t-motive M:
   The Betti realization is (see 7.4)

$$M_B := (M \otimes_{F[t]} F\{t\})^\tau,$$

where $F\{t\}$ is the convergent power series ring on unit disc.
   For M not of characteristic $\wp$, the $\wp$-adic realization is (see 7.2)

$$M_\wp := (M \otimes_{F[t]} \varprojlim F[t]/\wp^n)^\tau.$$

For M of characteristic $\wp$, the crystalline realization is

$$M_{\mathrm{cris}} := (M \mod \wp) \otimes F[t]_\wp.$$

The de Rham realization is

$$M_{DR} := \tau M/(t - \theta)\tau M.$$

Let us see why for Drinfeld modules, it is the same as Deligne's definition of 6.4 of de Rham cohomology: See end of 6.4 (a) for the description of the de Rham cohomology in terms of Lie split extensions. In terms of t-motives, we are thus looking at extensions of $F[t, \tau]/(t - \theta)F[t, \tau]$ by M as $F[t, \tau]$ modules, such that the corresponding extension of F by $M/\tau M$ (see 7.1.7 (3)) as $F[t]$-modules, splits. This is equivalent to extension of $F[t, \tau]/(t-\theta)F[t, \tau]$ by τM and thus using the resolution $F[t, \tau] \overset{(t-\theta)}{\to} F[t, \tau]$, the de Rham cohomology is computed to be M_{DR} as defined above.
   The Hodge filtration on the deRham is defined by its i-th piece F^i being the image of $(t - \theta)^i M \cap \tau M$ in it.

Remarks 7.5.1   (1) Since $(t - \theta)^d M \subset \tau M$, $F^{d+1} = 0$.
   (2) Hence for Drinfeld modules of rank r, we have $F^0 = H_{DR}$ of rank r, F^1 of rank 1 and $F^i = 0$ for $i > 0$, consistent with description in 6.4.
   (3) For $C^{\otimes n}$, we have $F^i = H_{DR}$ of rank one for $0 \le i \le n$ and $F^i = 0$ for $i > n$.

Hence M is a good object amenable to good linear algebra constructs giving various realizations which are concrete objects involving matrices with entries being polynomials or power series in Frobenius with coefficients in various fields. From these basic $F[t]$-modules, we can then get various cohomology and homology realizations and appropriate comparison isomorphisms (by using duals and tensoring with differentials, as seen in 7.2 and 7.4) to fit in the motivic framework. Unfortunately, thorough analysis of the resulting formalism has not been carried out yet, but see [And86; Gos94] and recent work of Richard Pink on the Hodge structures in function field situation. For some calculations of 'extensions' and connections with bi-derivations of 6.4, see [PR03]. For Gekeler's description of crystalline cohomology in terms of language in 6.4, see [Ang97].

7.6   Example: Carlitz-Tate twist $C^{\otimes n}$

We now look at $C^{\otimes n}$ of 7.1.3 in more detail, with an eye to application to Zeta values in the next chapter. We leave the straightforward verifications involving matrix manipulations and estimates to the reader referring to [AT90] for more details.

By definition $C^{\otimes n}$ is of rank one. Hence it is simple and without complex multiplication, i.e., $\text{End}(C^{\otimes n})$ is $\mathbb{F}_q[t]$: See [Yu91, 1.5, 1.6] for more details. It is of dimension n, by 7.3.4 and 7.3.5 (1), uniformizable by 7.4.6 (1) and pure t-motive, by 7.3.5 (1). By abuse of notation, we will identify $\theta = t$, as it will not matter: We will write $[a]_n \in \text{End}(\mathbb{G}_a^n)$ for the image of a, so that $[t]_n(x_1, \cdots, x_n) = (tx_1 + x_2, \cdots, tx_{n-1} + x_n, tx_n + x_1^q)$. If we write $d[a]_n$ for the coefficient of τ^0, then it is the matrix representing the endomorphism of $\text{Lie}(\mathbb{G}_a^n)$ induced by $[a]_n$ and in particular, we have $d[a]_n(*, \cdots, *, x) = (*, \cdots, *, ax)$: The largest quotient of $\text{Lie}(\mathbb{G}_a^n)$ on which the derivative and multiplication actions of A coincide is isomorphic to $\mathbb{G}_a$ with the isomorphism $\ell_n : \text{Lie}(\mathbb{G}_a^n) \to \mathbb{G}_a$ given by the last co-ordinate. We will thus play a special attention to this canonical co-ordinate, as it will play an important role in the applications to periods (see 7.4 and the Theorem below) and zeta values (Theorem 8.1.1).

Let us collect these facts together with the period lattice description seen in 7.4:

Theorem 7.6.1   *The t-module $C^{\otimes n}$ is abelian, uniformizable, simple t-module without complex multiplication. It is of rank one and dimension n and pure of weight n.*

There is $\lambda \in C_\infty^n$ with the last entry $\tilde{\pi}^n$ such that the period lattice of $C^{\otimes n}$ is $\{d[a]_n\lambda : a \in A\}$.

Write $\exp_n$ and $\log_n$ for the corresponding exponential function and its inverse function: the logarithm normalized with $\log_n(0) = 0$. Thus $\log_n[a]_n = d[a]_n \log_n$ and $\exp_n d[a]_n = [a]_n \exp_n$. Let us fix n and write E for the elementary matrix E_{n1} and $\exp_n = \sum E(i)\tau^i$, $\log_n = \sum L(i)\tau^i$. Then $[t]_n = t + N + E\tau$ implies

$$\exp_n(t + N) = (t + N + E\tau), \quad \log_n(t + N + E\tau) = (t + N)\log_n .$$

This gives recursions

$$(t + N)E(i + 1) + EE(i)^{(1)} = E(i + 1)(t^{q^{i+1}} + N),$$

$$(t + N)L(i + 1) = L(i + 1)(t^{q^{i+1}} + N) + L(i)E.$$

We have $E(0) = L(0) = 1$ and since N is nilpotent, the recursions can be solved by geometric series to give

$$E(i + 1) = \sum_{j=0}^{2n-2} \frac{\mathrm{ad}(N)^j(EE(i)^{(1)})}{(t^{q^{i+1}} - t)^{j+1}}, \quad L(i + 1) = -\sum_{j=0}^{2n-2} \frac{\mathrm{ad}(N)^j(L(i)E)}{(t^{q^{i+1}} - t)^{j+1}},$$

where we have, as usual, $\mathrm{ad}(X)^0(Y) = Y$ and $\mathrm{ad}(X)^{j+1}(Y)$ is the commutator $[X, \mathrm{ad}(X)^j(Y)]$.

Remarks 7.6.2 (1) Thus the canonical last co-ordinate ℓ_n of $\log_n(X)$ is given by the formula $\sum_{i=0}^{n-1}(-1)^i\Delta^i\mathcal{L}_n(x_n - i)$, where $\mathcal{L}_n(x) = \sum_{i=0}^{\infty}(-1)^{in}x^{q^i}/L_i^n$ is naive poly-logarithm corresponding to the Carlitz logarithm and Δ is our analog of zd/dz operator. In particular, the last co-ordinate seems to be a good deformation of the naive poly-log $\ell_n(\log_n(0, \cdots, 0, x)) = \mathcal{L}_n(x)$.

(2) Let us work out the exponential in case $n = 2$: With $k_i := 1/[i] + 1/[i - 1]^q + \cdots + 1/[1]^{q^{i}-1}$, we have

$$E(i) = \begin{pmatrix} 1/D_i^2 & -2k_i/D_i^2 \\ [i]/D_i^2 & (-2[i]k_{i-1}^q - 1)/D_i^2 \end{pmatrix}.$$

In particular, in characteristic 2, we have

$$\exp_2(x_1, x_2) = (\sum_{i=0}^{\infty} x_1^{q^i}/D_i^2, \sum_{i=0}^{\infty}([i]x_1^{q^i} + x_2^{q^i})/D_i^2).$$

So, in this case, if $p(t) \in \mathbb{F}_q[t]$, $(0, \tilde{\pi}^2p(t))$ is a period of $\exp_2$.

(3) More generally, comparison with the exponential and logarithm of Carlitz module in 2.5 shows that the first and the last co-ordinates are multi-exponential (Bessel) and multi-logarithms respectively. When p divides n, the formulas are especially simpler. The connections with hypergeometric functions is mentioned in 6.5.

We now record ∞-adic and v-adic convergence properties of these formal series in $X = (x_1, \cdots, x_n)$ proved by straight estimates [AT90] on the recursions:

Theorem 7.6.3

(i) The series $\exp_n(X)$ converges for all $X \in C_\infty^n$.

(ii) The series $\log_n(X)$ converges when $|x_i| < |t|^{i-n+(nq/(q-1))}$ and for such X, we have $\exp_n(\log_n(X)) = X$.

(iii) The series $\exp_n(X)$ converges v-adically when $v(x_i) > 1/((2n-1)(q^{\deg v} - 1))$.

(iv) The series $\log_n(X)$ converges v-adically, when $v(x_i) > 0$. (We write the value as $\log_{n,v}(X)$.)

Finally, we record the Galois action on the torsion for $C^{\otimes n}$:

Theorem 7.6.4   *[AT90] Let $a \in A$ be non-zero.*

The a-torsion in C_∞ of $C^{\otimes n}$ is isomorphic to A/a as an A-module.

If v does not divide a and λ denotes an a-torsion point, then the arithmetic Frobenius σ_v acts via $\sigma_v \lambda = [v^n]_n \lambda$.

Hence the image of the Galois representation $Gal(K^{sep}/K) \rightarrow Aut_A(Kernel[a]_n) = (A/a)^*$ consists of n-th powers in the group, and so there is no K-rational non-trivial a-torsion, unless $q - 1$ divides n.

7.7   Drinfeld dictionary in the simplest case

We will first describe the Drinfeld's geometric approach for the simplest case of rank and dimension one, namely for Drinfeld modules of rank one. For convenience, as well as with the belief that it always helps to see things in many different perspectives, we ignore efficiency and use a slightly different equivalent language than the one used in the general case in the next section. We use the notation of 2.2 and follow Anderson's formulation [Tha93b]:

Let F be an algebraically closed field containing $\mathbb{F}_q$ and of infinite transcendence degree. Let $\overline{X}$ denote the fiber product of X with $Spec(F)$ over $\mathbb{F}_q$. We identify closed points of $\overline{X}$ with F-valued points of X in the obvious way. For $\xi \in X(F)$, let $\xi^{(i)}$ denote the point obtained by raising the coordinates of ξ to the q^i-th power. We extend the notation to the divisors on $\overline{X}$ in the obvious fashion. For a meromorphic function f on $\overline{X}$, let $f|_\xi$ denote the value (possibly infinite) of f at ξ and let $f^{(i)}$ denote the pull-back of f under the map $id_X \times \mathrm{Spec}(\tau^i) : \overline{X} \to \overline{X}$, where $\tau := x \to x^q : F \to F$. For a meromorphic differential $\omega = fdg$, let $\omega^{(i)} := f^{(i)}d(g^{(i)})$. Fix a $\overline{\mathbb{F}_q}$-point $\overline{\infty}$ above ∞.

Fix a local parameter $t^{-1} \in \mathcal{O}_{\overline{X},\overline{\infty}}$ at $\overline{\infty}$. For a nonzero $x \in \mathcal{O}_{\overline{X},\overline{\infty}}$, define $\deg(x) \in \mathbb{Z}$ and $\mathrm{sgn}(x) \in F$ to be d_∞ times the exponent in the highest power of t and the coefficient of the highest power respectively, in the expansion of x as Laurent series in t^{-1}, with coefficients in F. Note that $\mathrm{sgn}(x) \in \mathbb{F}_\infty^*$ for nonzero $x \in K$.

Shtukas and associated functions (See [Dri77a; Dri87; Mum78] for details):

Theorem 7.7.1 *Let ξ, $\eta \in X(F)$, a divisor V of $\overline{X}$ of degree g and a meromorphic function f on $\overline{X}$ be given such that*

$$V^{(1)} - V + (\xi) - (\eta) = (f).$$

If $\xi \neq \eta^{(s-1)}$ $(|s| < g)$, then

$$H^i(\overline{X}, \mathcal{O}_{\overline{X}}(V - (\eta^{(-1)}))) = 0 \quad (i = 0, 1).$$

Proof. There is nothing to prove if $g = 0$, so assume $g > 0$. Define divisors V_i for $i \in \mathbb{Z}$ by the rules

$$V_0 := V - (\eta^{(-1)}), \quad V_{i+1} := (\eta^{(i-1)}) + V_i$$

and set $\mathcal{L}_i := \mathcal{O}_{\overline{X}}(V_i)$. By the Riemann-Roch theorem, it will be enough to prove that $h^0(\mathcal{L}_0) = 0$. Set

$$S := \{1 - g \le s \le g : h^0(\mathcal{L}_s) = h^0(\mathcal{L}_{s-1}) + 1\}.$$

Then the cardinality of S is g, because $h^0(\mathcal{L}_{-g}) = 0$ and $h^0(\mathcal{L}_g) = g$, by the Riemann-Roch theorem. Let $s \in S$ be such that $s < g$. Then there exists a global section e of $\mathcal{L}_s$ not vanishing modulo $\mathcal{L}_{s-1}$ and consequently, $fe^{(1)}$ is a global section of $\mathcal{L}_{s+1}$ not vanishing modulo $\mathcal{L}_s$, since $\xi \neq \eta^{(s-1)} = \mathrm{supp}(\mathcal{L}_{s+1}/\mathcal{L}_s)$ by the hypothesis. So $s + 1 \in S$, $S = \{1, 2, \cdots g\}$. $\square$

Remark 7.7.2 In the original version of this theorem of Drinfeld, η is $\overline{\infty}^{(-1)}$ and $\xi \neq \overline{\infty}^{(s)}$ for all s and hence the condition such as $\xi \neq \eta^{(s)}$ for $|s| < g$ is not made explicit. This refined version is due to Greg Anderson, who has made use in his work on solitons of the flexibility thus obtained. In what follows though, we only need the original version.

Corollary 7.7.3 *Let ξ, $\eta \in X(F)$. Let V be an effective divisor of degree g on $\overline{X}$ such that $V^{(1)} - V + (\xi) - (\eta)$ is principal. Then $\eta^{(-1)}$, ξ (resp. μ, where μ is a $\mathbb{F}_q$-rational point of X) can not belong to the support of V, provided that for $|s| < g$, $\xi \neq \eta^{(s-1)}$ (resp. $\xi \neq \eta^{(s)}$). If W is an effective divisor of degree $d < g$ and $\xi \neq \eta^{(g-d+s-1)}$ for $|s| < g$, then $W^{(1)} - W + (\xi) - (\eta)$ can not be principal.*

Proof. The fact that $\eta^{(-1)}$ is not in the support of V, follows immediately for the lemma. If ξ (resp. μ) belongs to the support of V, we apply this fact to $W = V - (\xi) + (\eta)$, $\xi^{(1)}$, $\eta^{(1)}$ (resp. $W = V - (\mu) + (\eta)$, ξ, $\eta^{(1)}$) in place of V, ξ and η to get a contradiction. To prove the last claim, apply the Theorem with $V := W + (\eta) + (\eta^{(1)}) + \cdots + (\eta^{(g-d-2)}) + (\xi^{(-1)})$ if $d < g-1$ (resp. $V := W + \eta$ otherwise), and with $\xi^{(-1)}$ and $\eta^{(g-d-1)}$ (resp. ξ and $\eta^{(1)}$) in place of ξ and η in the Theorem. $\square$

We review the language of Drinfeld modules again in this notation.

Let $\iota : A \to F$ be an embedding of A in F. By a Drinfeld A-module ρ relative to ι we will mean such with normalized with respect to sgn, of rank one and generic characteristic, but we will drop these words. Let H and H_1 be as in Chapter 3. Write its exponential as $e(z) = \sum e_i z^{q^i}$ and logarithm as $l(z) = \sum l_i z^{q^i}$.

Fix a transcendental point $\xi \in X(F)$. Then evaluation at ξ induces an embedding of A into F. In our earlier notation, $\xi = \iota$. By solving the corresponding equation on the Jacobian of X, we see that for some divisor V, $V^{(1)} - V + (\xi) - (\overline{\infty})$ is principal. A Drinfeld divisor V relative to ξ is defined to be an effective divisor of degree g such that $V^{(1)} - V + (\xi) - (\overline{\infty})$ is principal. From 7.7.1 and Riemann-Roch, it follows that Drinfeld divisor is the unique effective divisor in its divisor class. (In particular, there are h such divisors.) Hence there exists a unique function $f = f(V)$ with $\mathrm{sgn}(f) = 1$ and such that $(f) = V^{(1)} - V + (\xi) - (\overline{\infty})$. By abuse of terminology, we call f a Shtuka. (In fact, in our context, Shtuka is a line bundle $\mathcal{L}$ on $\overline{X}$ with $\mathcal{L}^{(1)}$ being isomorphic to $\mathcal{L}(-\xi + \overline{\infty})$ and in our case, with $\mathcal{L} = \mathcal{O}_{\overline{X}}(V)$, f realizes this isomorphism.)

Drinfeld bijection: The set of Drinfeld divisors V (relative to ξ) is in natural bijection with the set of Drinfeld A-modules ρ (relative to ξ) as follows. (See [Mum78] for details of the proof.) Let $f = f(V)$ as in 0.3.4.

Then

$$1, f^{(0)}, f^{(0)} f^{(1)}, f^{(0)} f^{(1)} f^{(2)}, \ldots$$

is an F-basis of the space of sections of $\mathcal{O}_{\overline{X}}(V)$ over $\overline{X} - (\infty \times_{\mathbb{F}_q} F)$. Define $\rho_{a,j} \in F$ by the rule

$$a := \sum_j \rho_{a,j} f^{(0)} \cdots f^{(j-1)}.$$

Then the ρ corresponding to V is given by $\rho_a := \sum \rho_{a,j} \tau^j$. (Reader should verify the relations).

Remark 7.7.4 If we take ξ to be a $\overline{\mathbb{F}_q}$-point of $\overline{X}$ not above ∞, instead of a transcendental point, as we have done, there is still a Drinfeld bijection with Drinfeld modules of finite characteristics. See the next section.

Theorem 7.7.5 *We have*

$$e(z) = \sum_{n=0}^{\infty} \frac{z^{q^n}}{(f^{(0)} \cdots f^{(n-1)})|_{\xi^{(n)}}}.$$

Proof. Note that by 7.7.3 the coefficients on the right hand side are never infinite. To see that the right hand side satisfies the correct functional equations for $e(z)$, divide both sides of the equation above defining $\rho_{a,j}$ by $f^{(0)} \cdots f^{(n-1)}$ and evaluate at $\xi^{(n)}$. $\square$

By Riemann-Roch theorem, there exists a unique global section $\omega = \omega(V)$ of $\Omega_{\overline{X}/F}(-V + 2(\overline{\infty}))$ with the leading term in the Laurent expansion in t at $\overline{\infty}$ being dt times an element of sgn one.

Theorem 7.7.6 *We have*

$$l(z) = \sum_{n=0}^{\infty} (\text{Res}_\xi \frac{\omega^{(n+1)}}{f^{(0)} \cdots f^{(n)}}) z^{q^n}.$$

Proof. The necessary relations to show that $l(z)$ is the formal inverse of $e(z)$ come from the fact that the sums of the residues of differentials figuring in the expression vanish. $\square$

Examples 7.7.7 (1) For Carlitz module for $\mathbb{F}_q[t]$, the genus is zero and thus the Drinfeld divisor is empty, so that $f = t - t|_\xi$ and $\omega = dt$. The reader should verify the formulas for coefficients of the exponential and the logarithm.

(2) For hyperelliptic X given by $y^2 = P(x)$, with P being a polynomial of degree $2g + 1$, writing $\overline{V}$ for the divisor obtained by applying

the hyperelliptic involution to the Drinfeld divisor $V = \sum \xi_i$, we have $\omega = c \prod (x - x_{\xi_i}) dx/y$, with appropriate c to fix the sign condition. Note that $(\omega) = V + \overline{V} - 2\overline{\infty}$.

(3) More examples of V and f are provided in Section 8.2.

7.8 Krichever/Drinfeld dictionary in more generality

Now we come to geometric sheaf theoretic description of elliptic and abelian modules due to Drinfeld. Inspired by Krichever's theory of algebraic-geometric solutions of KdV equations, Drinfeld introduced a geometric concept of a Shtuka (the Russian word means 'a piece of something') and established a natural bijection between Drinfeld modules and (special) Shtukas.

The motivation in Mumford's words [Mum78] is 'A remarkable dictionary discovered by Krichever based on suggestions in the work of Zakharov and Shabat, where they attempted to find a common formalism for the inverse scattering method of integrating certain non-linear partial differential equations'.

The main references, which should be consulted for the details, as well as differential operator case and its connections with other parts of mathematics, are [Mum78; Dri77a; And86; BS97; SW85; Tha01; Pot99; Gos96] and references therein.

Following Mumford, we give the dictionary first in the simplest case of 'rank' and 'dimension' and 'the degree of the infinite place' all being one, in both the differential operator case and a field operator case (we will only discuss the Drinfeld case):

Krichever dictionary: Let F be a field of characteristic 0. Then there is a natural bijection between the sets of data 'X a complete (possibly singular) curve over F, P a smooth F-rational point on it, a fixed isomorphism of the tangent space at P with F and a torsion free, rank one sheaf $\mathcal{F}$ on X with $h^0 = h^1 = 0$' and the data 'Commutative subring R (containing F) of $F[[t]][d/dt]$ having two operators of relatively prime orders, and with R_1 and R_2 identified if there is $u \in F[[t]]^\times$ with $R_1 = uR_2u^{-1}$'.

Now let F be a field, and σ an automorphism of F of infinite order. Let $\mathbb{F}$ be the fixed field. Let $F\{\sigma\}$ be the ring of maps $A : F \to F$ of the form $A(f) = \sum_{i=0}^{N} a_i \sigma^i(f)$. We start with a slightly generalized form of the dictionary of the last Section in a slightly different terminology:

Drinfeld dictionary: With the notation as above, there is a natural bijection between the sets of data

(A) X a complete, reduced, irreducible curve over $\mathbb{F}$, P a smooth $\mathbb{F}$-rational point on it, $\mathcal{F}$ a torsion free rank one coherent sheaf on $X_F := X \times_{\mathbb{F}} F$ with $h^0 = h^1 = 0$ and isomorphism $(1_X \times \sigma)^* \mathcal{F} \equiv \mathcal{F}(P - P')$ for some smooth point P' (distinct from P) on X_F

and the data

(B) Commutative subring R of $F\{\sigma\}$ containing $\mathbb{F}$, having two operators of relatively prime degree (in σ) and with R_1 and R_2 identified, if there is $u \in F^\times$ with $R_1 = uR_2u^{-1}$.

More general (still in dimension one) Drinfeld dictionary [Mum78] asserts a natural bijection between sets of data

(A) (a) X reduced, irreducible complete curve over $\mathbb{F}$, such that $X_F = X \times_{\mathbb{F}} F$ is irreducible (it is then automatically reduced [Mum78]).

(b) $P \in X$ a regular closed point. Let $P_F := P \times_{\mathbb{F}} F \subset X$.

(c) A torsion free, coherent sheaf $\mathcal{F}$ (of $\mathcal{O}_{X_F}$-modules) on X_F such that $h^0(\mathcal{F}) = h^1(\mathcal{F}) = 0$.

(d) A maximal flag of subsheaves $\mathcal{F} = \mathcal{F}_0 \supset \mathcal{F}_{-1} \supset \cdots \mathcal{F}_{-s} = \mathcal{F}(-P_F)$, where length of any successive quotient is one.

(e) A homomorphism of sheaves $\alpha : (1_X \times \sigma)^* \mathcal{F} \to \mathcal{F}$ on $X_F - P_F$, which is not surjective, such that, on X_F, α carries $\mathcal{F}_k$ to $\mathcal{F}_{k+1}$

and the data

(B) A commutative subring $R \subset F\{\sigma\}$, with R strictly containing $\mathbb{F}$, with $R \cap F = \mathbb{F}$, where we identify R with uRu^{-1}, for $a \in F$.

Here is the recipe to go back and forth. For details on the proof, see [Mum78; Dri77a].

From (A) to (B): For $n \in \mathbb{Z}$, define tower $\mathcal{F}_n$ via $\mathcal{F}_{n+s} := \mathcal{F}_n(P_F)$. Let $s_0 \in \Gamma(\mathcal{F}_1)$. Put $s_n = \alpha(1 \times \sigma)^* s_{n-1}$. Then the sections $s_n \in \Gamma(\mathcal{F}_{n+1}) - \Gamma(\mathcal{F}_n)$, for $n \geq 0$ give basis of $\Gamma(X_F - P_F, \mathcal{F})$. In this basis, if the action of $R = \Gamma(X - P, \mathcal{O}_X)$ on $\Gamma(X_F - P_F, \mathcal{F})$ is expressed by say $rs_0 = \sum r_n s_n$, then $r \to \sum r_n \sigma^n$ gives the required embedding of R into $F\{\sigma\}$.

From (B) to (A): Given R as in the data, σ-degree gives a gradation $R_n = r \in R : \deg(r) \leq n$. Then X is just the Proj of the graded ring $\mathcal{R} := \oplus R_n$. Consider the module $M = F\{\sigma\}$ over $R \otimes_{\mathbb{F}} F$, where R acts by right multiplication and F by left. Put $M_n := \{m \in M : \deg_\sigma(m) \leq n\}$ and $\mathcal{M} = \oplus M_n$ and $\mathcal{M}[n]$ be $\mathcal{M}$ with grading shifted by n. Then $\mathcal{F}_{n+1} := \tilde{\mathcal{M}}_n$. If $e \in R_1$ represents 1, the affine open $e \neq 0$ corresponds to $\text{Spec}(R)$, whereas the divisor $e = 0$ corresponds to the point P. Multiplication by e gives $\mathcal{F}_{n+1} \hookrightarrow \mathcal{F}_{n+2}$, which is isomorphism on $e \neq 0$.

Remarks 7.8.1 (1) If X_F is non-singular, then the torsion-free sheaf $\mathcal{F}$ is a vector bundle of rank r. Further, if $\sigma = \tau$ for perfect F of finite characteristic, we reduce to Drinfeld module case, with R being A, $\mathbb{F}$ being $\mathbb{F}_q \subset F$, the field of definition, P being ∞, P' being the characteristic $\iota : A \to F$.

(2) When X_F is singular, R is an order in A of the corresponding normalization. We saw this variant in Hayes approach to do explicit class field theory. In geometric class field theory, it naturally corresponds to generalized Jacobians.

(3) If F is perfect of finite characteristic, we can also use $\sigma = \tau^{-1}$, for example. These issues come up in study of adjoints, hypergeometric functions as well as Anderson's new dual t-motives point of view. Hellegouarch [Hel92] has investigated general σ variants and σ-transcendence questions.

(4) As before, $\chi = 0$ implies $h^0 = h^1 = 0$ for $\mathcal{F}$ which is as in the data except for this, and the proof of this in 7.7 also shows the basis property.

(5) Note that P_F consists of $\deg(P)$ regular points, and the sheaves $\mathcal{F}_n$ are thus locally free of rank $s/\deg(P)$ in a neighborhood of P_F. In the Drinfeld module case, this is its rank.

(6) The sheaf description of the torsion is as follows: If $\wp$ is a closed point of $X - P$, and $F \subset L$, with σ denoting extension to L, then $\wp$ torsion in L is given by $\mathrm{Hom}_{(F,\sigma)}(\mathcal{F}/\wp\mathcal{F}, L)$.

The dictionary above is dictionary between elliptic modules and elliptic sheaves. Drinfeld then introduced [Dri87; Laf02] more general concept of a **Shtuka** (called F-sheaf or FH-sheaf also).

Definition 7.8.2 Drinfeld's Shtuka is a pair $\mathcal{F}_0$, $\mathcal{F}_1$ on X_F, with

$$\mathcal{F}_0 \overset{\alpha}{\hookrightarrow} \mathcal{F}_1 \overset{\beta}{\hookleftarrow} (1 \times \sigma)^* \mathcal{F}_0,$$

with lengths of cokernels of α and β being one. The support of the cokernel of α is called the zero of the Shtuka, and that of β the pole.

Remarks 7.8.3 (1) The pole and the zero of the universal Shtuka give a morphism from the algebraic stack Sht^r of rank r Shtukas to $X \times X$ which is smooth of pure relative dimension $2r - 2$. (Compare Theorem 3.7.1.)

(2) In $r = 1$ case we considered, this algebraic stack Sht^1 is just the pullback of Lang isogeny $\mathcal{L} \to \mathcal{L}^{-1} \otimes_{\mathcal{O}_F \otimes X} \mathcal{L}^\tau$ on algebraic stack of line bundles over X to $X \times X$ via the Abel-Jacobi map $(\infty, 0) \to \mathcal{O}_{F \otimes X}(\infty - 0)$.

(3) In Chapter 1, we have seen Lang's geometric class field theory giving abelian extensions as pull-backs under isogenies of generalized Jacobians,

which parametrize line bundles. Using the other side of the dictionary, we can write down equations for these extensions, as in Chapter 3. For more detailed discussion, see [Mum78, pp. 151-152].

(4) The Shtukas arising from towers $\mathcal{F}_n$ (see [Mum78, p. 151]) are characterized by (i) the zero is disjoint from $P :=$ {the pole and its conjugates over $\mathbb{F}$}, (ii) restricted to P, the σ^{-1}-semi-linear map $\alpha^{-1}\beta$ of the $\mathbb{F}$-vector space $\mathcal{F}_0/m_P\mathcal{F}_0$ is nilpotent.

Let us see how Anderson's notion of 'purity' above is modeled on (ii): Here $\alpha^{-1}\beta$ stands for τ^{-1}, and in terms of modules rather than sheaves, when we look at the completion at P, the condition above says that $\tau^{-v}M((1/t)) \subset F[[1/t]]M((1/t))$, which after choosing a basis and clearing t powers t^u in denominators boils down to existence of $F[[1/t]]$ lattice L in $M((1/t))$ (spanned by this basis) satisfying $t^u L = \tau^v L$, as in the definition of purity.

Chapter 8

Applications to Gauss sums, Gamma and Zeta values

In this chapter, we apply the geometrical as well as higher dimensional techniques of the last chapter to the study of arithmetic objects and issues introduced and discussed in previous chapters.

In the first section, we will explore algebraic incarnation of Carlitz zeta values on tensor powers of Carlitz module. Applications to transcendence will be discussed in Chapter 10. We have seen similar phenomenon in 5.9 using just Drinfeld modules. The phenomenon seems quite general, but we do not have good specific conjectures predicting or explaining the meaning of the algebraic incarnations in a general setting.

In the next two sections, we connect Shtukas to Jacobi sums. This leads to better understanding of prime factorizations of Gauss sums. It also suggests studying a different gamma function (which is the same as that of 4.5 for $\mathbb{F}_q[t]$). We study its interpolations at finite and infinite primes and prove a new Gross-Koblitz analog at the finite primes.

In the next four sections, we study Anderson's solitons and their applications. These generalize many of the formulas we have looked at before. We also look at Gross-Koblitz situation for geometric gamma and in more general perspective.

The last two sections deal with a log-algebraicity theorem of Anderson, its implications to cyclotomic theory, relation with Kummer's theorems and conjectures. We present a new proof and formulas, which seem to indicate existence of interesting underlying structure waiting to be explored, for the resulting quantities.

8.1 $C^{\otimes n}$ and $\zeta(n)$

We give explicit algebraic incarnation of (transcendental) Carlitz zeta value $\zeta(n)$ (and its v-adic counterparts), for n a positive integer, on the Carlitz-Tate motives $C^{\otimes n}$. We will see the applications to transcendence of zeta values in Chapter 10.

Let $A = \mathbb{F}_q[t]$ and we will use notation of 7.6. Let us write, temporarily in this section, V_n for the vector $(0, \cdots, 0, 1)$ of length n.

Theorem 8.1.1   *Let $A = \mathbb{F}_q[t]$. There exists (constructed explicitly below) A-valued point Z_n of $C^{\otimes n}$ such that the canonical last co-ordinate of $\log_n(Z_n)$ is $\Gamma(n)\zeta(n)$.*

If we put $Z_{n,v} := [v^n - 1]_n Z_n$, for a monic irreducible $v \in A$, then the canonical last co-ordinate of $\log_{n,v}(Z_{n,v})$ is $v^n \Gamma(n) \zeta_v(n)$.

Further, for 'odd' n, the point Z_n is not a torsion point.

Proof.   We use the connection between the binomial coefficients and the power sums giving the terms of the zeta functions explained in 5.6 to construct the point Z_n as follows.

Let $G_n(y) := \prod_{i=1}^{n}(t^{q^n} - y^{q^i})$, so that $G_i(t^{q^k}) = (l_k/l_{k-i})^{q^i}$ and so for the binomial coefficient notation of 5.6, $\mathcal{B}_k(x) = \sum_{i=0}^{k} G_i(t^{q^k})/d_i(x/l_k)^{q^i}$. Hence if we define $H_n(y) \in K[y]$ by

$$\sum_{n=0}^{\infty} \frac{H_n(y)}{n!} x^n = (1 - \sum_{i=0}^{\infty} \frac{G_i(y)}{d_i} x^{q^i})^{-1},$$

then the fact that $n!m!$ divides $(n+m)!$ for the Carlitz factorial, implies first of all that with $H_{n-1}(y) = \sum h_{ni} y^i$, the coefficients h_{ni} belong to A. Further by 5.6.3, we see that $H_n(t^{q^k})/l_k^{n+1} = n! S_k(-(n+1))$.

We define $Z_n \in C^{\otimes n}(A) = A^n$ by

$$Z_n := \sum [h_{ni}]_n(t^i V_n).$$

Straight degree estimates [AT90] show that for $i \leq \deg H_{n-1}$, $t^i V_n$ is in the region of convergence for $\log_n$. Hence we have

$$\ell_n(\log_n(Z_n)) = \sum \ell_n(d[h_{ni} \log_n(t^i V_n)) = \sum h_{ni} \ell_n(\log_n(t^i V_n))$$
$$= \sum h_{ni} \frac{t^{iq^k}}{l_k^n} \qquad\qquad = \Gamma(n)\zeta(n).$$

Notice that $\log_n$ is multi-valued function. With $z_n := \sum d[h_{ni}]_n(t^i V_n) \in K_\infty^n$, we have $\exp_n(z_n) = Z_n$.

If Z_n were a torsion point, when n is 'odd', then since $\tilde{\pi}^n$ is the last co-ordinate logarithm of zero, $\zeta(n)$ would be a rational multiple of $\tilde{\pi}^n$. But we know (eg. from its fractional degree $q/(q-1)$ we calculated in 2.5) that $\zeta(n)/\tilde{\pi}^n$ is not in K_∞, when n is 'odd'.

Notice that the calculation proceeds degree by degree, so that instead of dealing with $\zeta(s) \in K_\infty$, we can also work with $\zeta(s, X) \in K[[X]]$, so multiplication by $1 - v^n$, which is just removing the Euler factor at $s = -n$ as in 5.5 (b), then gives the corresponding sum where the power sums are now over a prime to v, thus giving v-adic zeta value (see [AT90] for detailed treatment paying attention to convergence questions). □

Examples 8.1.2 : (1) Directly from the recipe above or from 5.9.1 and 7.6, we see that for $n = rp^k$, with $0 < r < q$, we have $Z_n = [\Gamma(n)]_n (0, \cdots, 0, 1)$.

(2) It follows by easy induction from the definition of t-action on $C^{\otimes n}$ that for $k < n$, $[t^k]_n V_n$ has its $n - j$-th co-ordinate $\binom{k}{j} t^{k-j}$ for $0 \le j \le k$ and 0 for $j > k$. In particular, $[t^q]_{q-1} V_{q-1} = [t]_{q-1} [t^{q-1}]_{q-1} V_{q-1} = (0, \cdots, 0, 1, t) = [t]_{q-1} V_{q-1}$. In other words, for $n < q$, V_n is a torsion point if (and only if, by the Theorem) $n = q - 1$ and then it is torsion point of order $t^q - t$. (We had already seen this for $q = 2$.)

(3) $Z_{q+1} = (1, 0, \cdots, 0) + [t^q - t]_{q+1} V_{q+1}$.

(4) If $p = 3$, Z_{14} is the zero vector.

Remarks 8.1.3 (1) We will see in 10.5 that for 'even' n, (i.e., all n if $q = 2$) Z_n is a torsion point.

(2) For small n, it can be directly checked that denominator of 'B_n/n' that we calculated in Chapter 4 kills Z_n, when n is 'even'. I do not know a general proof without using results of Chapter 10. If we develop canonical heights for $C^{\otimes n}$, it should be possible to verify the torsion, non-torsion results directly.

(3) We will see in Chapter 10 the applications to getting good multizeta values by connecting the candidates to points on tensor powers.

8.2 Shtuka and Jacobi sums

We now show how the analogs of function field Jacobi sums of 4.1 can be obtained from a Shtuka of 7.7. We apply this to obtain some results on the prime factorization of analogs of Gauss sums and to prove, in the next section, an analog of the Gross-Koblitz formula for general function field, generalizing the results in 4.8. In 4.3, we saw that for general A, Gauss sums made up from $\wp$-torsion need not lie above $\wp$, so that the Gross-Koblitz analog of 4.8 does not generalize immediately. For this purpose, we introduce and interpolate a new analog of gamma function, agreeing with that of 4.5 for $\mathbb{F}_q[t]$, but not in general.

We now use notation of 7.7. First we recall Gauss and Jacobi sums from Chapter 4: Let $\wp$ be a prime of A of degree d. Choose an A-module isomorphism $\psi : A/\wp \to \Lambda_\wp$ (an analog of additive character) and let χ_j ($j \mod d$) be $\mathbb{F}_q$-homomorphisms $A/\wp \to F$, indexed so that $\chi_j^q = \chi_{j+1}$ (special multiplicative characters which are q^j-powers of 'Teichmüller character'). We can identify $\chi_j(z)$ with $a|_{\theta(j)}$, for some geometric point θ above $\wp$, if $a \in A$ is such that $a \mod \wp$ is z. Then we define basic Gauss sums

$$ g_j := g(\chi_j) := - \sum_{z \in (A/\wp)^*} \chi_j(z^{-1})\psi(z). $$

The g_j are nonzero. We define Jacobi sums J_j by $J_j := g_{j-1}^q/g_j$. Then J_j is independent of the choice of ψ. We put $J := J_0$.

For the Carlitz module, it was shown in 4.2 that $J_j = -(t - \chi_j(t))$ and Stickelberger factorization of g_j was easily obtained from this. Note for example that $g_j^{q^d-1} = J_j J_{j-1}^q \cdots J_{j-d+1}^{q^{d-1}}$.

Now we show that the Jacobi sums made up from $\wp$ torsion of a Drinfeld module can be interpreted as specializations at geometric points above $\wp$ of a meromorphic function, obtained from the Shtuka corresponding to the Drinfeld module, on curve cross its Hilbert cover. Hence the strange factorizations of the Gauss sums get related to the divisor of this function.

Theorem 8.2.1 *Let V and f be the Drinfeld divisor and the Shtuka respectively, corresponding to a sgn-normalized Drinfeld module ρ (of rank one and generic characteristic) via the Drinfeld bijection in 7.7. Then with the Jacobi sums J_j defined using $\wp$-torsion of ρ and normalized as above, we have*

$$ f|_{\theta(j)} = J_j. $$

Proof. First note that a $\overline{\mathbb{F}}_q$-point θ of X over a prime $\wp$ of A can not belong to the support of V, otherwise the specialization of the defining equation of f to $\xi = \theta$ would contradict 7.7.3. So let h_0 be a $q^d - 1$-th root of $(f f^{(1)} \cdots f^{(d-1)})|_\theta$ and define h_i inductively by $h_i := h_{i-1}^q/f|_{\theta(i)}$. Then h_i depends only on i modulo d. Specializing the equation defining $\rho_{a,j}$ in

Drinfeld bijection recipe of 7.7 at $\theta^{(i)}$ gives

$$a|_{\theta^{(i)}} h_i = \sum \rho_{a,j} h_{i-j}^{q^j}.$$

Hence with $\phi(z) := \sum \chi_i(z) h_i$, we see that $\phi(az) = \sum \rho_{a,j} \phi(z)^{q^j}$ and that $\phi(z)$ is a $\wp$-torsion point. In other words, ϕ is an analog of additive character, as above, and so $\phi(z) = \psi(uz)$ for some $u \in (A/\wp)^*$. It then follows that $h_j = \chi_j(u)g_j$. Hence

$$f|_{\theta^{(j)}} = \frac{h_{j-1}^q}{h_j} = \frac{g_{j-1}^q}{g_j} = J_j$$

as claimed. $\square$

(a) Gauss sums and Theta divisor

Let us see how the factorization of Gauss sums in general depends on the geometry of the theta divisor.

The relation $\xi \leftrightarrow V$ defines a correspondence, say W on $X \times Y_1$, where Y_1 is the 'Hilbert cover' corresponding to H_1. Let U be the restriction of the transpose of W at $\theta \times Y_1$. Then the Theorem above and integrality of $\rho_{a,j}$ show that $(J) = qU^{(-1)} - U + (\theta) - I$, where I is supported on ∞. This determines the divisor of J completely, if $h = d_\infty = 1$. In general, I can be calculated from 4.3.11.

When $d_\infty = 1$, using 7.7.1 together with the moduli description of theta divisor in terms of cohomology jumps of line bundles mentioned in 1.2, Greg Anderson has given the following nice interpretation of W, in terms of the theta divisor:

Let $\alpha : X \to J$ be the embedding of X in the Jacobian J of X, via $x \to x - \infty$, i.e., the Abel map of 1.2. Define the Hilbert cover $\pi : Y \to X$ as a pullback $\beta : Y \to J$ of α under the Lang's isogeny $\tau - 1 : J \to J$, so that $\mathcal{L} \in Y(M)$ means $\mathcal{L}^{(1)} \equiv \mathcal{L}(\xi - \infty)$. (Examples of such $\mathcal{L}$'s are $-V + g\infty$.) Then W is the pullback of the theta divisor under $-\alpha \circ p_1 - \beta \circ p_2 : X \times Y \to J$. This is because, with these identifications, μ is in support of V corresponding to ξ if and only if $h^0(V - \mu) > 0$ if and only if $h^0(\mathcal{L}^{-1}(\infty - \mu + (g - 1)\infty)) > 0$ if and only if $\mathcal{L}^{-1}(\infty - \mu) \in \Theta(M)$.

(b) Examples and applications

We saw in 7.7 that for the Carlitz module for $\mathbb{F}_q[t]$, $f = t - t|_\xi$. This shows that the Theorem gives the correct value for the Jacobi sums recalled just before it.

Next let K be a rational function field with ∞ being any place of K. Let ρ be a Drinfeld A-module. We can write $H = \mathbb{F}_\infty(t)$, with $\text{sgn}(t) = 1$. Then it is easy to see that $f = t - t|_\xi$. Hence the prime factorization of Gauss sums in this case is quite analogous to the $\mathbb{F}_q[t]$ case above, except that place $\overline{\infty}$ corresponding to ρ replaces ∞.

Now let us look at the four A's of class number one listed in 3.1.2. In Theorem 4.3.2, we computed J and its divisor, but with different notation and in Theorem 4.16.5 we computed f. In practice usually, as in the case of these examples, to produce f explicitly so that $f|_{\xi^{(i)}} = e_{i-1}^q/e_i$ and $f|_{\theta^{(i)}} = g_{i-1}^q/g_i$, the computation consists of eliminations on the relations occurring in the proof of the Theorem 8.2.1 with $a = x, y$. Now we summarize the results of those computations:

We put $\overline{x} := x|_\xi$ and $\overline{y} := y|_\xi$. While giving the divisor of J, by abuse of notation, instead of the quantities corresponding to θ, we use those corresponding ξ.

(i) $A = \mathbb{F}_2[x, y]/y^2 + y = x^3 + x + 1$: We have

$$f = \frac{\overline{x}(x + \overline{x}) + y + \overline{y}}{x + \overline{x} + 1}.$$

If $\xi + 1$ is the point where x is $\overline{x} + 1$ and y is $\overline{x} + \overline{y} + 1$, then $V = (\xi + 1)$. (Note that this point corresponds to the automorphism σ of example 1 of Theorem 4.3.2.) By the recipe above it follows (and can be verified from 4.3.2) that

$$(J) = 2(\underline{\xi + 1})^{(-1)} - (\underline{\xi + 1}) + (\underline{\xi}) - 2(\infty)$$

where $\underline{\xi + 1}$ is the point where x is $\overline{x} + 1$ and y is $\overline{x} + \overline{y}$.

(ii) $A = \mathbb{F}_4[x, y]/y^2 + y = x^3 + \zeta_3$: We have

$$f = \frac{\overline{x}^2(x + \overline{x}) + y + \overline{y}}{x + \overline{x}}.$$

If $\underline{\xi}$ denotes the point where x is $\overline{x}$ and y is $\overline{y} + 1$, then $V = \underline{\xi}$ and

$$(J) = 4(\underline{\xi})^{(-1)} - (\underline{\xi}) + (\underline{\xi}) - 4(\infty)$$

(iii) $A = \mathbb{F}_3[x, y]/y^2 = x^3 - x - 1$: We have

$$f = \frac{-\overline{y}(x - \overline{x}) + y - \overline{y}}{x - \overline{x} - 1}.$$

If $\xi + 1$ (resp. $\xi - 1$) denotes the point where x is $\bar{x} + 1$ (resp. $\bar{x} - 1$) and y is $\bar{y}$, then $V = (\xi + 1)$ and

$$(J) = 3(\xi - 1)^{(-1)} - (\xi - 1) + (\xi) - 3(\infty).$$

(iv) $A = \mathbb{F}_2[x, y]/y^2 + y = x^5 + x^3 + 1$. We have

$$f = \frac{(\bar{x} + x)(\bar{x}^4 + \bar{x}^3 + (1 + x)\bar{x}^2) + \bar{y} + y}{\bar{x}^3 + x\bar{x}^2 + (1 + x)\bar{x} + x^2 + x}.$$

Let $\underline{\xi} + 1$, $\underline{\xi}$, $\underline{\xi} + 1$ be the points where x is $\bar{x} + 1$, $\bar{x}$, $\bar{x} + 1$ resp. and y is $\bar{y} + \bar{x}^2$, $\bar{y} + 1$, $\bar{y} + \bar{x}^2 + 1$ resp. Then $V = (\underline{\xi}) + (\underline{\xi} + 1)^{(1)}$ and

$$(J) = 2((\underline{\xi}) + 2(\underline{\xi} + 1)^{(-1)})^{(-1)} - ((\underline{\xi}) + 2(\underline{\xi} + 1)^{(-1)}) + (\underline{\xi}) - 4(\infty)$$

(c)   The case $g = d_\infty = 1$

After the easy genus zero case, the next easiest case is when $g = d_\infty = 1$ and the resulting elliptic curve is naturally mapped to itself considered as its Jacobian via $x \to (x) - (\infty)$.

To understand the factorization of Gauss and Jacobi sums, we need V, which is obtained by solving the defining equation of f on the Jacobian of X. In other words, V 'is' $-(\tau - 1)^{-1}(\xi)$. (The point ∞, when $d_\infty = 1$, is used in the embedding of the curve in the Jacobian.) Examples (a), (b) and (c) are of genus one. Hence the Jacobian is the elliptic curve, so in fact $V = (-(\tau - 1)^{-1}(\xi))$ can be verified directly with the results given above by computing it on the elliptic curve. (Note that the theta divisor is just a point in this case.)

Here is the example 3.1.2 (iii) above: $y^2 = x^3 - x - 1/\mathbb{F}_3$. Then

$$(\tau - 1)(x, y) = (x^3, y^3) \oplus (x, -y) = (x - 1, -y)$$

as follows by usual algorithms for addition and inverse on elliptic curve. So $\tau - 1 = \sigma^2 \mu$, where $\sigma(x, y) := (x + 1, y)$ is the order three automorphism of A, as in Theorem 4.3.2, and the order two automorphism $\mu(x, y) := (x, -y)$ is just the multiplication by -1 on the elliptic curve. So $-(\tau - 1)^{-1} = \mu(\tau - 1)^{-1} = \sigma$ as we saw.

We saw above how to calculate f (and thus V) from the analytic as well as algebraic descriptions. Drinfeld bijection recipe in 7.7 shows how to go back in the other direction to ρ or e.

We showed in 4.16.5 how in $g = d_\infty = 1$ case to write down f 'parametrically' in terms of the coefficients of sgn-normalized ρ. Let us show (see [And94], but there is a sign and notation difference) how to get it directly.

In fact, in this case of elliptic curve over $\mathbb{F}_q$, it is its own Jacobian as usual using ∞ to get the embedding $X \to J$ via $x \to (x) - (\infty)$. We get by geometric class field theory, the Hilbert cover of A as pull-back of A by the Lang isogeny $\tau - 1$ of the Jacobian:

Let us, for simplicity, assume $p \neq 2$ and write $A = \mathbb{F}_q[x, y]/(y^2 = P(x))$ as usual, with cubic P and x, y considered monic. Note that $-(a, b) = (a, -b)$ for these co-ordinates under the group law. The Lang isogeny being a surjective isogeny from the curve to itself, we get a point (or rather h points) $V := (\omega, \eta)$ such that

$$(\omega^q, \eta^q) - (\omega, \eta) = -(x, y) = (x, -y)$$

for a generic point $\xi = (x, y)$ and the Abel-Jacobi correspondence between the points of the curve and divisors shows that this equation means that V is a Drinfeld divisor in this case. It is defined over $H = H_A$ which is just $\mathbb{F}_q(\omega, \eta)/\omega^2 = P(\eta)$. The integral closure $B = \mathcal{O}_H$ is given by $B = \mathbb{F}_q[\omega, \eta, \beta]/(\omega^2 = P(\eta))$, with β^{-1} being the product of $\omega - \theta$'s as θ runs through elements of $\mathbb{F}_q$ such that $P(\theta)$ is a square. This is because these correspond to $\mathbb{F}_q$-rational points of the elliptic curve (Jacobian) and hence are in kernel of the Lang isogeny and hence correspond to inverse image of ∞, so we have poles there.

The f function is also visible in these co-ordinates: By the group law on the elliptic curve above we know that $(x, -y)$, $(\omega, -\eta)$ and (ω^q, η^q) are collinear points, so that

$$\mathcal{D} := \begin{vmatrix} 1 & 1 & 1 \\ x & \omega & \omega \\ y & -\eta & \eta^q \end{vmatrix} = 0.$$

Now considered as a function of x and y with $\omega, \eta \in F$ considered as constants, the determinant is a degree 3 function on our curve X with zeros $\xi = (x, y)$ (with x, y now being a copy in M), conjugate $V^c := (\omega, -\eta)$ of V and $V^{(1)}$ because of the equation above, since the determinant vanishes if the columns are the same. Since the degree 2 function $x - \omega$ has zeros V and V^c, we see that the divisor of $\mathcal{D}/(x - \omega)$ is $V^{(1)} - V + \xi - \infty$ and is thus f up to normalization by a constant multiple to get $\operatorname{sgn}(f) = 1$. A straight calculation shows that

$$f = \frac{\mathcal{D}}{(x - \omega)(\omega^q - \omega)}.$$

In other words, for genus one base, we can easily explicitly produce

Drinfeld-Hayes modules using this uniform recipe and combined with 4.16 this allows us to compute many arithmetic quantities of interest.

8.3   Another Gamma function

Let $F = C_\infty$. Let ρ be a Drinfeld A-module of rank one and generic characteristic. Let $e(z)$ and $l(z)$ be the corresponding exponential and logarithm, with the kernel of exponential being $\tilde{\pi}\mathcal{A}$ with $\mathcal{A}$ an ideal of A. We have seen in that $e(z)$ and $l(z)$ are $\mathbb{F}_q$-linear and hence the coefficients of z^k in their power series expansion vanish if k is not a power of q.

Theorem 8.3.1   *The coefficients e_n of z^{q^n} in $e(z)$ are never zero, in each of the following situations. (a) X has a closed point of degree one (for example, when $d_\infty = 1$ or q is large compared to g, or when $g < 2$). (b) $g < 6$ and K is not $\mathbb{F}_2(x, y)/y^4 + xy^3 + (x^2 + x)y^2 + (x^3 + 1)y + x^4 + x + 1 = 0$. (This K can be characterized [LMQ75] as the unique class number one field of genus three and having no prime of degree two.)*

Proof.   We will first establish two criteria for non-vanishing of e_n's. For simplicity, first consider the situation where $d_\infty = 1$. By the Riemann-Roch, in the contributions to the coefficient of z^{q^n} coming from the product formula for the exponential $e(z) = z \prod(1 - z/\tilde{\pi}a)$ there is a unique term of highest degree and hence the coefficient is nonzero. This argument also gives non-vanishing of various coefficients of $e(z)$, depending on the gaps in Riemann-Roch, when $d_\infty > 1$, but in general there are many terms of the same highest degree canceling the top degree, so the full result does not follow this way. But using Galois conjugation we can conclude non-vanishing of e_n for ρ corresponding to one ideal class from the corresponding statement for different ideal class. This leads to the first criterion:

(C1): If there is a $\mathbb{F}_q$-divisor D with $h^0(D) = n$, then $e_n \neq 0$.

The case (a) is settled by (C1) since we can twist the divisor by appropriate multiple of the degree one prime to get the desired dimension.

On the other hand, note that e_n is zero if and only if $\xi^{(n)}$ belongs to the support of V, so a priori there are at most g (so none if $g = 0$) exceptions n to the following hypothesis:

$$e_n \neq 0 \quad n \in \mathbb{N}. \tag{$\mathcal{H}$}$$

The second criterion is

(C2): $e_n \neq 0$, if A has a prime of degree dividing n.

Since otherwise the specialization of the defining equation of f to ξ at a $\overline{\mathbb{F}_q}$-point above the prime contradicts 7.7.3. This together with the $d_\infty = 1$ case gives a different proof for the case (a).

In particular, if we let n_0 to be the smallest (it is easy to calculate n_0) positive integer such that A has primes of all degrees dividing $n > n_0$, then $e_n \neq 0$ for $n > n_0$. Since there is always a $\mathbb{F}_q$-divisor of any prescribed degree, $h^0 = n$ can be achieved, by Riemann-Roch, if $n \geq g$. Hence by (C1), $e_n \neq 0$, if $n \geq g$. We can also see this by Weil bounds and (C2), unless $q = 2$ and $g \leq 6$, for then $n_0 \leq g$.

Let n be an exception to $\mathcal{H}$. By above, $1 \leq n < g$. Now A contains constants and hence $h^0 = 1$ is possible, so $n \neq 1$. For D of degree $2g - 2$, $h^0(D) = g - 1 + h^0(\mathcal{K} - D)$, where $\mathcal{K}$ is the canonical divisor. If $h > 1$, we can choose $\mathcal{K} - D$ nontrivial and get $h^0 = g - 1$ showing $n \neq g - 1$ in that case. But by [LMQ75], $g \geq 4$ implies $h > 1$ and there are only two class number one, genus 3 fields. Among them the one not listed in (b) has a prime of degree two and (C2) applies. We have shown $2 \leq n \leq g - 2$, except possibly for K listed in (b). For D of degree $2g - 3$, we have $h^0(D) = g - 2 + h^0(\mathcal{K} - D)$ and $\mathcal{K} - D$ being of degree one, if its $h^0 > 0$, then we have a closed point of degree one and there are no exceptions. So in any case, $2 \leq n \leq g - 3$. Consider now $g = 5$. Only possibility to be ruled out is $n = 2$. For D of degree 2, $h^0(D) = h^0(\mathcal{K} - D) - 2$. If it is zero, we apply (C1); otherwise we apply (C2) to see that $n \neq 2$.   □

We have proved that there are at most $g - 4$ exceptions (except possibly for the K of the theorem) and it is easy to extend the last part of the proof to show that there are at most $g - 5$. The Weil bounds together with (C2) give much better asymptotic bound of the order $2\log_q 2g$. We have more evidence for $\mathcal{H}$ than that listed in the theorem and it is possible that $\mathcal{H}$ is always satisfied. But as we can not settle it in full generality,

we will assume $\mathcal{H}$ below in this Section.

See remarks 8.3.12 for discussion of what can be done without the hypothesis.

We can then write $e(z) = \sum z^{q^i}/d_i$, with $d_i \in H_1$, since ρ is sgn-normalized.

Definition 8.3.2   For $n \in \mathbb{N}$, we define the factorial $\Pi(n)$ of n as follows: Write $n = \sum n_i q^i$, $0 \leq n_i < q$ and put $\Pi(n) := \prod d_i^{n_i}$.

Let $\wp$ be a prime of A of degree d and let θ be an $\overline{\mathbb{F}_q}$- point of $\overline{X}$ above $\wp$. If $w \in K_\wp$ is a local parameter at θ, we put

$$\tilde{d}_i := \tilde{d}_{i,w} := \frac{d_i}{d_{i-d}w^{l_i}}$$

where l_i is chosen so that $\tilde{d}_i$ is a unit at θ.

One motivation for this definition, at least when $n = q^i$, is that classically the factorials are given by the reciprocals of the Taylor coefficients of the exponential. A motivation for the definition for arbitrary n and for the $\wp$-adic interpolation is from gamma function of 4.5 with which this coincides when $A = \mathbb{F}_q[t]$.

Theorem 8.3.3 *If u is any local parameter at θ, there is a nonzero $c \in \mathbb{F}_{q^d}$ such that with $w = cu$, $\tilde{d}_i$ tends to one θ-adically as i tends to infinity.*

Proof. Let us write $f_i := f|_{\xi^{(i)}}$. Then by 7.7.5, we have $d_i/d_{i-1}^q = f_i$ and hence $y_i := \tilde{d}_i/\tilde{d}_{i-1}^q = f_i/f_{i-d}w^{e_i}$ where e_i is such that y_i is a unit at θ. Now as i tends to infinity, f_i tends $f|_{\theta^{(i)}} = f|_{\theta^{(i-d)}} \neq 0$. Hence y_i tends to one as i tends to infinity. Now if c is chosen, so that for some sufficiently large i_0, $\tilde{d}_{i_0}$ is a one-unit, then this shows that $\tilde{d}_i$ is a one unit for $i \geq i_0$. But since $\tilde{d}_i$ are one units and y_i tends to one, $\tilde{d}_i$ tends to one (because q power map spreads the series expansions in characteristic p) as claimed. $\square$

Let w be a local parameter at θ such that $\tilde{d}_i$ is a one unit for large i, then this implies that $\tilde{d}_i$ tends to one.

Definition 8.3.4 Define $\wp$-adic factorial $\Pi_\wp(z) := \Pi_w(z)$ for $z \in \mathbb{Z}_p$ as follows. Write $z = \sum z_i q^i$, $0 \leq z_i < q$ and put $\Pi_\wp(z) := \prod \tilde{d}_i^{z_i}$.

In the case of Carlitz module, this agrees with Goss interpolation of 4.5 and the following theorem gives another proof of Goss' result in that case.

Theorem 8.3.5 *When $h_A = 1$, if we choose w to be a monic prime $\wp$ of A of degree d, then $-\tilde{d}_i \to 1$ as i tends to infinity.*

Proof. Let us put $f_i := f|_{\xi^{(i)}}$. Then by 7.7.5, we have

$$d_i/d_{i-d} = (f_i f_{i-1}^q \cdots f_{i-d+1}^{q^{d-1}})(f_{i-d} \cdots f_1^{q^{i-d-1}})^{q^d-1}.$$

Hence the θ-adic sign of d_i/d_{i-d} is the same as that of the first bracket specialized at θ. By Theorem 8.2.1, it is then the θ-adic sign of $g_i^{q^d-1}$. Now $g_0^{q^d-1} \equiv -\wp \mod \wp^2$ can be seen, when $h_A = 1$ exactly as in Lemma 4.2.5 and Theorem 4.2.6. Since $(-1)^{q^i} = -1$, this finishes the proof. $\square$

(a)   Analog of Gross-Koblitz

For this subsection, let F be an algebraic closure of $K_\wp$, ξ be the tautological point, i.e., the F-valued point of X corresponding to $K \hookrightarrow K_\wp$ and let θ be the Teichmüller representative in the residue disc of ξ. Note that even though we used the ∞-adic completion earlier, the Taylor coefficients of $e(z)$, being in H_1, can be thought of as elements of F.

Theorem 8.3.6   *Let $0 \leq j < d$. If μ is the valuation of g_j at ξ, then we have $g_j = \zeta w^\mu / \Pi_\wp(q^j/(1-q^d))$, where ζ is a $q^d - 1$-th root of unity.*

Proof.   (Compare with 4.8.1): For the purposes of this proof, μ_i will be suitable integers. Let

$$M_j := \Pi_\wp(q^j/(1-q^d)) = \lim_{m\to\infty} \tilde{d}_j \tilde{d}_{j+d} \tilde{d}_{j+2d} \cdots \tilde{d}_{j+md}.$$

Call the quantity after the limit sign T_m, so that $M_j = \lim T_m$. Then

$$T_m = \frac{d_j \cdots d_{j+md}}{d_j \cdots d_{j+md-d}} w^{\mu_1} = d_{j+md} w^{\mu_1}.$$

Since $d_i = f_i d_{i-1}^q$, we have

$$T_m = f_{j+md} f_{j+md-1}^q \cdots f_{j+1+(m-1)d}^{q^{d-1}} T_{m-1}^{q^d} w^{\mu_2}.$$

But as m tends to infinity, by Theorem 8.2.1, $f_{l+md} \to f|_{\theta^{(l)}} = J_l$. Hence

$$M_j^{1-q^d} = J_j J_{j-1}^q \cdots J_{j-d+1}^{q^{d-1}} w^{\mu_3} = g_j^{q^d-1} w^{\mu_3}$$

as claimed.   □

Remark 8.3.7   Unlike the classical or the $\mathbb{F}_q[t]$ case 4.8.1, this theorem is weaker in that it does not tell us what the valuation μ or the root of unity is. When $h_A = 1$, both can be specified using the Theorem before the previous and the results of 8.2. But unlike the two cases above, μ can depend on $\wp$, not only through its degree. For example, by (iii) of Section 8.2, the valuation at $\wp$ of $(g_0 g_1 g_2)^2$ is 3 or 1 according as whether $\wp$ is y or $y+x$. In general, the dependence of μ on $\wp$ seems complicated and deserves further investigation.

(b)   Interpolation at ∞ for new Gamma

In this subsection, let $d_\infty = 1$, unless mentioned otherwise and let F be an algebraic closure of K_∞, ξ be the tautological point, i.e., the F-valued point of X corresponding to $K \hookrightarrow K_\infty$ and let $\overline{\infty}$ be the Teichmüller representative in the residue disc of ξ. For a nonzero $b \in \mathcal{O}_{\overline{X},\overline{\infty}}$, let $\overline{b} :=$ $b/t^{\mathrm{val}\ b}$ be the unit part of b with respect to t.

Theorem 8.3.8 *There exists a nonzero $c \in \mathbb{F}_\infty$ such that as i tends to infinity, $c^{q^i} \overline{d}_i$ tends to one, $\overline{\infty}$-adically*

Proof. We have $f_i := d_i/d_{i-1}^q = f|_{\xi^{(i)}}$. By 7.7.3 f has a pole at $\overline{\infty}$ of order one (see below for $d_\infty > 1$ case). This together with the fact that $\mathrm{sgn}(f) = 1$ implies that for given N, for sufficiently large i, $d_i/d_{i-1}^q = t^{q^i} +$ terms of degree less than $q^i - N$ in t. This in turn implies that $\overline{f}_i$ tends to one, as i tends to infinity. Now, if c is chosen so that for some sufficiently large i_0, the unit $c^{q^{i_0}} \overline{d}_{i_0}$ is in fact a one-unit, then the facts that the q-power spreads the series expansion in characteristic p, together with $\overline{f}_i \to 1$ prove the claim. $\qquad\square$

Theorem 8.3.9 *As i tends to infinity, the degree of d_i tends to zero, p-adically.*

Proof. This follows from considering the degrees of both sides in the expression for d_i/d_{i-1}^q in the Theorem above and induction on i. $\qquad\square$

We first interpolate the one-unit part as follows.

Definition 8.3.10 For $z \in \mathbb{Z}_p$, we define ∞-adic factorial by

$$\overline{\Pi}_\infty(z) := \overline{\Pi}_t(z) := \prod (c^{q^i} \overline{d}_i)^{z_i}$$

where as usual $z = \sum z_i q^i$, with $0 \le z_i < q$.

Using the Theorem above, we can define Π_∞ by 'putting back the degree part', just as in 4.5.

Remark 8.3.11 Main open problem here is to understand the arithmetic significance of $\Pi_\infty(-1)$, generalizing the $A = \mathbb{F}_q[t]$, in which it is the period of the Carlitz module, by Theorem 4.7.5. This result together with v-adic result Theorem 8.3.6 for general A suggest that it should have some arithmetic significance. Let us say a few more words about this question: Let us use the short-form $f_{ij} := f^{(i)}|_{\xi^{(j)}}$. In the Carlitz case of genus 0, we have switching-symmetry $f_{ij} = -f_{ji}$. In general, up to simple algebraic factors arising from residue calculations, $\tilde{\pi}^{-1}$ is $\prod_{i=1}^\infty f_{i0}$ and $\Pi_\infty(-1)^{-1}$ is $\prod_{j=1}^\infty f_{0j}$, as can be seen from formulas in 7.7 giving d_i and l_i in terms of specializations of f. So the switching-symmetry mentioned above connects the two in the Carlitz case.

How are these two quantities obtained by the switch of the two copies of X related? Are we getting some new transcendental invariant of A other than $\tilde{\pi}$ or are they simply related algebraically? At present, we have no evidence pointing either way.

If A, with $d_\infty = 1$, of genus $g > 0$ is hyperelliptic (for example, just to fix gap structure), we can calculate degree of $\tilde{\pi}$ (for principal ideal class, so that lattice of the sgn-normalized ρ written as $\tilde{\pi}A$) to be $(2q - q^{g+1})/(q - 1)$, using 4.7.5 and Riemann-Roch to calculate degrees of D_i's. Calculating degrees of d_i's by recursion formulas obtained by ρ_x-action, where x is of degree 2, and using $\zeta(q - 1) = -\tilde{\pi}^{q-1}/d_1$ to calculate degree of d_1, we can check that $\deg \Pi_\infty(-1) = \deg \tilde{\pi} + g q^{g+1}$.

We now give different analytic proofs for these two theorems, using the product formula for the exponential.

By the product formula, for large i, the degree and the sign of d_i is the same as that of $\tilde{\pi}^{q^i-1}$ times the product of all nonzero elements of A (there are $q^i - 1$ of these) of degree not more than $i + w$ with $w = g - 1 + \deg A$ by Riemann-Roch. By Theorem 4.7.5, the $\tilde{\pi}$ power contributes the sign $(-1)^{(q^i-1)/(q-1)} = (-1)^i$. And as the product of all nonzero elements of $\mathbb{F}_q$ is -1, the contribution from the product of the $q^i - 1$ elements is also $(-1)^{(q^i-1)/(q-1)} = (-1)^i$, making the total contribution 1. Hence the sign is 1. Now we give a sketch of another proof of the fact that the gap between the powers of two top degrees of t in the Laurent expansion of d_i tends to infinity, so that $\bar{d}_i \to 1$: Let us put D_i to be the product of all monic elements of A of degree i, then the term producing the top degree is $\tilde{\pi}^{-q^i+1}/(D_0 \cdots D_{i+w})^{q-1}$ up to a sign, hence by Theorem 4.7.5, the one-unit part of this tends to one. Hence it remains to show that the difference of degrees between those of $1/(D_0 \cdots D_{i+w})^{q-1}$ and sum of the reciprocals of all other possible products of $q^i - 1$ distinct elements of A tends to infinity with i. We group the terms in the sum by taking one element of the largest degree (say $i + w + r$, with $r > 0$) of fixed sign as variable and keeping the other $q^i - 2$ elements fixed. Then the gap is clearly at least $-\deg(\sum 1/a) + i + w + r$ where this sum is over monic elements of degree $i + w + r$. Now this tends to infinity as claimed, by the formula for $A_{i0} = \sum 1/a$ in 4.15 and Riemann-Roch. But in doing this, we have included some terms involving products with repetitions, this leads to consideration of $\sum 1/a^2$ etc. which are easily seen to have sufficiently small degrees. This shows that Theorem 8.3.8 holds with $c = 1$.

Now again by Theorem 4.7.5, when i tends to infinity, the degree of the product tends p-adically to the degree of $\tilde{\pi}$ and hence the total degree tends to zero p-adically. This gives another proof of Theorem 8.3.9.

For any d_∞, if q^i is the number of elements of $\mathcal{A}$ of degree less than or equal to m say, then, just as above, there being unique contribution of the top degree, we can show that $e_i \neq 0$ and calculate the degree and sign of d_i in a similar fashion. For example, let n be sufficiently large and $i = nd_\infty - w$. Then by the argument above $\deg(d_i)$ is, up to addition of a contribution independent of n, $(q^{d_\infty} - 1) \sum_{j=1}^n j d_\infty q^{j d_\infty - w - d_\infty} + \deg(\tilde{\pi}^{q^i - 1})$. Hence the degree of $d_i / d_{i-d_\infty}^{q^{d_\infty}}$ is up to addition of a contribution independent of n (in fact from what we show, this contribution is zero, but we do not need this for the argument) equal to $d_\infty q^i$. This is compatible with the formula for d_i / d_{i-1}^q in the proof of Theorem 8.3.8, exactly when $m = 1$. In other words, f has a simple pole at $\overline{\infty}$. Together with 7.7.3, this means that $\overline{\infty}$ is not in the support of V. This argument also shows how to find a formula for c, in terms of $\operatorname{sgn}(\tilde{\pi})$.

Remarks 8.3.12 In case, $\mathcal{H}$ is not satisfied, we can define the gamma function and interpolations on smaller subsets of $\mathbb{Z}_p$, so as to avoid the j-th digit in q-base expansion, whenever e_j is zero. We know that there are at most g such j's. With this (awkward) restriction, the situation is then similar, even the Gross-Koblitz formula holds, for large d and for j non-exceptional. Another (equally awkward) alternative is to define, say $d_i = 1$, if $e_i = 0$. Then Gamma function is everywhere defined, but Gross-Koblitz statement has to be corrected by a correction factor in H_1.

In case of logarithm, we have better theorem than Theorem 8.3.1:

Theorem 8.3.13 *The coefficients l_n of z^{q^n} are never zero.*

Proof. By Galois conjugation, it is sufficient to prove the theorem in the case when the lattice corresponding to ρ is $\tilde{\pi}A$, i.e., ρ corresponds to the principal ideal class. Write $z/e(z) = \sum b_j z^j$. Carrying out proof of 4.16.1 in this more general notation, we get $l_i = b_{q^i - 1}$. (As a side application, this shows $\deg(l_i) = (1 - q^i) \deg \tilde{\pi}$, if ρ corresponds to $\tilde{\pi}A$, because of the connection of $z/e(z)$ and even zeta values as in 5.2.1.) Taking the logarithmic derivative of the product formula for e, this is in turn equal to $-\tilde{\pi}^{1-q^i} \sum 1/a^{q^i - 1}$, where the sum is over nonzero elements of A. The sum does not vanish, because the top degree contribution to it is the sum over $a \in \mathbb{F}_q^*$ and equals $q - 1 = -1 \neq 0$. This proves that $l_i \neq 0$. $\qquad\square$

Remarks 8.3.14   If we let $(\omega) = \overline{V} + V - 2(\overline{\infty})$, with $\overline{V}$ being an effective divisor of degree g, then by 7.7.6, $l_n = 0$ if and only if $\xi \in V + \overline{V}^{(n+1)}$, which by 7.7.3 is equivalent to $\xi \in \overline{V}^{(n+1)}$. So there are at most g exceptional n's (none, if $g = 0$). The fact that ξ does not belong to the support of $\overline{V}^{(1)}$ follows exactly as in 7.7.1 and 7.7.3, by replacing V by $\overline{V}^{(1)}$ and q^i-th powers by q^i-th roots. Hence by copying the proof of $(C2)$ of 8.3.1, we obtain another proof of the fact that $l_n \neq 0$, under the hypothesis that X has a finite closed point of degree one. In fact, we thus see a stronger result that ξ does not belong to $\overline{V}^{(n)}$, for $n \in \mathbb{Z}$. This result is not true in general without the hypothesis as can be seen from example (ii) of Section 8.2 (b).

When ξ is a transcendental point, the fact that ξ does not belong to the supports of V and $\overline{V}^{(1)}$ also follows from 7.7.5 and 7.7.6 together with the analytic theory, which gives the existence of the exponential and the logarithm.

8.4   Fermat motives and Solitons

In this Section, we let $A = \mathbb{F}_q[t]$ be our base and $a \in A$, $f \in A_+$, such that $a/f \in K - A$.

We have looked at Drinfeld modules with complex multiplications by (integral closure of A in) cyclotomic fields, such as constant field extensions $K(\mu_n)$ of K or Carlitz-Drinfeld cyclotomic extensions $K(\Lambda_\wp)$, where we are restricted to $\wp$ of degree one, so that there is only one place at infinity. This had applications to Gamma values, Gauss sums etc. (at all places). To handle general $K(\Lambda_f)$, we need t-motives which can handle many infinite places. Applying deformation techniques from soliton theory which occur in Krichever theory to the arithmetic case of Drinfeld dictionary, Anderson achieved this [And; And94; ABP]. The applications to special values and transcendence go way beyond the classical counterparts this time, because the t-motives occurring can have arbitrary fractions as weights in contrast to the classical motives (Anderson's 'ulterior motives' are not classical motives, but formal gadgets to serve similar purpose).

Interpolating partial gamma products, as described below, by algebraic functions on the cyclotomic curve times itself, Anderson constructed these objects and showed that their periods relate to Gamma values, whereas they also give ideal class annihilators fitting coherently to give Hecke characters. (Hayes also produced these Hecke characters in a different way). This vastly generalizes very simple examples we looked at eg. in 4.11, 6.4.

We saw that part of gamma products for $\Gamma(a/f)$, restricted to degree d, are essentially binomial coefficients $\{a/f, q^d\}$, which are coefficients of t^{q^d} in $e(l(t)a/f)$, which can be thought of as deformation of torsion $e(l(0)a/f)$, since $\bar{\pi}$ is a value of $l(0)$. In function fields, we have the luxury of introducing more copies of variables (tensor products, products of curves, generic base change) with no direct analog for number fields. (In other words, we do not know how to push polynomials versus numbers analogy to multi-variable polynomials.) So we can treat this deformation and in fact realize these functions on products, so that coefficient of t^{q^d} is specialization of this function at graph of q^d-power partial Frobenius.

The tools used by Anderson for his first construction [And; And94] were using Drinfeld-Krichever dictionary to transport methods of solving soliton equations such as KdV. As this will take us too far, we refer to these papers and [Tha01] for motivations and some indications/references. Existence of interpolating functions called solitons, was not that explicit at first ([And, 6.1] and [Tha99a, p. 313]), but Anderson found much better explicit constructions later on and we essentially follow very nice account in [ABP]. The nice interpolation formulas turned out to be expressible better using notions of adjoints of 2.10.

First approach used Moore determinant and theta, tau function determinant approach, new approach used torsion and adjoint torsion (see Ore's connection between the two: Theorem 2.10.3).

Our notation is different from that of [ABP]. In particular, t and T are interchanged. Also [ABP] notation for adjoints is different. See 5.2.9 of [ABP]. We keep notation consistent with 2.10.

Let $A = \mathbb{F}_q[t]$. Let $\tilde{t}$ denote a fixed $q - 1$-st root of $-t$. Let e be the exponential for the Carlitz module and let $\bar{\pi}$ be the period of the Carlitz module for A normalized so that with $\mathbf{e}(z) := e(\bar{\pi}z)$, we have $\mathbf{e}(1/t) = \tilde{t}$.

For $f = \sum f_i T^i \in C_\infty[[T]]$, we define n-th twist $f^{(n)} := \sum f_i^{q^n} T^i$. Consider

$$\Omega(T) := \tilde{t}^{-q} \prod_{i=0}^{\infty} (1 - \frac{T}{t^{q^{i+1}}}).$$

In Section 2.5, we saw that $E(T) = 1/\Omega^{(-1)}(T)$ is the generating function of t-power torsion in that

$$\frac{1}{\Omega^{(-1)}(t)} = \sum_{i=0}^{\infty} \mathbf{e}(\frac{1}{T^{i+1}}) t^i.$$

This followed by matching the first coefficients and recursion on coefficients from the functional equation

$$(T - t)\Omega = \Omega^{(-1)}$$

with Carlitz module recursion on t-power torsion.

Now put $\Omega^{(-1)}(T) = \sum a_i T^i$, $a_i \in K_\infty(\tilde{t})$. Let Res $: K_\infty \to \mathbb{F}_q$ be the unique $\mathbb{F}_q$-linear functional with kernel $\mathbb{F}_q[t] + 1/t^2\mathbb{F}_q[[t]]$ and with $\text{Res}(1/t) = 1$, i.e. the usual residue function for parameter t. For $x \in K_\infty$, put

$$\mathbf{e}_*(x) := \sum_{i=0}^{\infty} \text{Res}(t^i x) a_i.$$

For example, $\mathbf{e}_*(1/t^{n+1}) = a_n$, for $n \geq 0$ and 0 otherwise. Hence

$$\Omega^{(-1)}(t) = \sum_{i=0}^{\infty} \mathbf{e}^*(\frac{1}{T^{i+1}}) t^i.$$

So $\mathbf{e}_*(1/t) = 1/\tilde{t}$. From the functional equation above, we get recursion on a_i's which implies that

$$t\mathbf{e}_*(x)^q + \mathbf{e}_*(x) = \mathbf{e}_*(tx)^q.$$

Comparison with adjoint $C_t^* = t + \tau^{-1}$ of Carlitz module shows that $\mathbf{e}_*(a/f)$ are q-th roots of f-torsion points of this adjoint.

Remark 8.4.1   (1) Note that $\mathbf{e}_*$ is not the adjoint of the exponential and is not given by a $\sum w_i \tau^i$ formula which is convergent (recursion leads only to formal series), though by definition we see that $\mathbf{e}_*$ is $\mathbb{F}_q$-linear and continuous in ∞-adic topology.

(2) Thus the duality is reflected in the generating functions of t-power tower Carlitz torsion and of its adjoint (up to q-power) being inverses of each other.

(3) By calculation as in the proof of 2.10.3, Anderson has showed

$$\mathbf{e}^*(a/f) = \prod_{b \in A_{<\deg f}, \text{Res}(ab/f)=1} \mathbf{e}(b/f)^{-1}.$$

We will omit the proof, as we will not need it.

What has this to do with the interpolation of partial gamma products or, what is the same, of binomial coefficients? The key is that the recursion scheme

$$\left(\frac{z}{q^{N-1}}\right)^q - \left(\frac{z}{q^{N-1}}\right) = (t^{q^N} - t)\left(\frac{z}{q^N}\right)$$

can be matched by telescoping:

$$\left(\prod_{i=0}^{N-1}(T-t^{q^i})^{-1}\right)^{(1)} - \prod_{i=0}^{N-1}(T-t^{q^i})^{-1} = (t^{q^N}-t)\prod_{i=0}^{N}(T-t^{q^i})^{-1}.$$

Now from the functional equation $\Omega^{(-1)} = \Omega(T-t)$,

$$\Omega^{(N)}/\Omega^{(-1)} = \prod_{i=0}^{N}(T-t^{q^i})^{-1}, \text{ if } N \geq 0 \text{ and } = \prod_{i=1}^{|N|-1}(T-t^{q^i}) \text{ otherwise.}$$

Combining with

$$(T-t)^{-1} = -\sum_{n=0}^{\infty} T^n/t^{n+1} = -\sum_{i=0}^{\infty}\binom{1/t^{n+1}}{q^0}T^n,$$

we see

$$\Omega^{(N)}/\Omega^{(-1)} = -\sum_{i=0}^{\infty}\binom{1/t^{n+1}}{q^N}T^n, \text{ if } N \geq 0.$$

Lemma 8.4.2 *For $x \in K_\infty$ and $N \in \mathbb{Z}$ such that $\deg x \leq \min(-1, N)$,*

$$\sum_{i=0}^{\infty} \mathbf{e}_*(t^{-i-1})q^{N+1}\mathbf{e}(t^i x) = -\left(\frac{x}{q^N}\right), \text{ if } N \geq 0, = \operatorname{Res}(t^{-N-1}x) \text{ otherwise.}$$

Proof. Since both sides depend $\mathbb{F}_q$-linearly and continuously on x, it is enough to prove this for $x = t^{-n-1}$, with $-n-1 \leq \min(-1, N)$. By the formula just before the lemma, and similar displayed formulas above with $\mathbf{e}$ and $\mathbf{e}_*$, this is just a calculation of coefficient of T^i in $\Omega^{(N)}/\Omega^{(-1)}$. $\square$

Definition 8.4.3 *For $f \in A_+$, we say that families $\{a_i\}$, $\{b_j\}$ ($i, j = 1$ to $\deg f$) of elements in A are f-dual if $\operatorname{Res}(a_i b_j/f) = \delta_{ij}$.*

Since the square matrix with (i, j)-th entry $\operatorname{Res}(t^{\deg f+i-j-1}/f)$ is lower triangular with 1's along the diagonal, the pairing $(a, b) \to \operatorname{Res}(ab/f)$: $A/f \times A/f \to \mathbb{F}_q$ is perfect and hence f-dual families exist and are easy to write down after straight calculation. For example, for $f = t^d$, $a_i = t^i$, $b_j = t^{d-1-j}$ works.

Theorem 8.4.4   *Fix $f \in A_+$ and f-dual families $\{a_i\}$, $\{b_j\}$ as in the definition above. Fix $a \in A$ with $\deg a < \deg f$. Then*

$$\sum_{i=1}^{\deg f} \mathbf{e}_*(a_i/f)^{q^{N+1}} \mathbf{e}(b_i a/f) = -\binom{a/f}{q^N}$$

for all $N \geq 0$. Also, if $a \in A_+$,

$$\sum_{i=1}^{\deg f} \mathbf{e}_*(a_i/f)\mathbf{e}(b_i a/f)^{q^{\deg f - \deg a - 1}} = 1.$$

Proof.   For $N \in \mathbb{Z}$, the left side of the first equation is

$$\sum_{i=1}^{\deg f} \sum_{n=0}^{\infty} \mathbf{e}_*(1/t^{n+1})^{q^{N+1}} \mathrm{Res}(t^n a_i/f)\mathbf{e}(b_i a/f) = \sum_{n=0}^{\infty} \mathbf{e}_*(1/t^{n+1})^{q^{N+1}}\mathbf{e}(t^n a/f),$$

by $\mathbb{F}_q$-linearity of Residue and $\mathbf{e}$ and definition of dual family. This in turn equals to $-\binom{a/f}{q^N}$ for $N \geq 0$ and to 1, if $a \in A_+$ and $N = \deg a - \deg f$, by Lemma above and definition of $\mathbf{e}_*$. $\qquad\qquad\square$

Definition 8.4.5   Let $U_f/\mathbb{F}_q$ be the irreducible, non-singular plane algebraic curve in (T, z)-plane defined by primitive f-th cyclotomic polynomial factor of $C_f(z)$ with T substituted for t. Let $X_f/\mathbb{F}_q$ be the nonsingular projective model of U_f. The closed points of X_f in the complement of U_f are said to be at infinity. For $a \in A$ prime to f, put $\xi_a := (t, \mathbf{e}(a/f))$. (Then $\{\xi_a\}$ is the collection of $\overline{K}$-valued points of U_f above the point $T = t$ of the affine T-line.) Put $\overline{U_f} := \overline{K} \otimes_{\mathbb{F}_q} U_f$ and $\overline{X_f} := \overline{K} \otimes_{\mathbb{F}_q} X_f$. For $x \in X_f(\overline{K})$, put $x^{(n)} \in X_f(\overline{K})$ to be the point obtained by applying the q^n-th power automorphism of $\overline{K}$ to the co-ordinates of K. Finally, let ∞_X be the formal sum of the $\overline{K}$-valued points of X_f at infinity, multiplied by $(q-1)$ and viewed as a divisor of $\overline{X_f}$.

Note that there are $|(A/f)^*|/(q-1)$ points at infinity for X_f, each ramified to order $q-1$ and they are $\mathbb{F}_q$-rational. So no new closed points appear in $\overline{X_f}$ and $\infty_{X_f}^{(1)} = \infty_{X_f}$.

Definition 8.4.6   For $x = a_0/f \in f^{-1}A - A$, put

$$g_x := 1 - \sum_{i=1}^{\deg f} \mathbf{e}_*(a_i/f)(C_{a_0 b_i}(z)|_{t=T}).$$

Remark 8.4.7 (1) It is independent of the choice of f-dual families in the expression.

(2) This meromorphic function on $\overline{X}$ is what Anderson calls soliton (or Coleman function). If y is 'fractional part' of ax,

$$g_x^{(N+1)}(\xi_a) = 1 + \binom{y}{q^N} = \prod_{n \in A_N+} (1 + y/n).$$

Thus on graph of $N + 1$-th Frobenius power it gives N-th term of gamma product.

For $x \in K_\infty = \mathbb{F}_q((1/t))$ and integers $N \geq 0$, let us define $\langle x \rangle_N$ to be 1, if the fractional part power series of x starts with $(1/t)^{n+1}$, and 0 otherwise. Then for bracket defined in 4.12, we have $\langle x \rangle = \sum_{N=0}^\infty \langle x \rangle_N$.

Theorem 8.4.8 *For $x \in f^{-1}A - A$, we have equality of divisors of $\overline{X}_f$:*

$$(g_x) = -\frac{1}{q-1} \infty X_f + \sum_{a \in A_{<\deg f}, (a,f)=1} \sum_{N=0} \langle ax \rangle_N \xi_a^{(N)}.$$

Proof. Note that there are only finitely many terms on the right and call the divisor on right D. Then $\deg D = -|(A/f)^*|/(q-1) + \sum \langle ax \rangle = 0$. Fix $a \in A$ prime to f such that $\langle ax \rangle = 1$. Let b be the unique element of A_+ such that $aa_0 \equiv b \mod f$ and $\deg b < \deg f$. Put $N := \deg f - \deg b - 1$ noting that $\langle b/f \rangle_N = 1$. We have

$$g_x(\xi_a^{(N)}) = 1 - \sum \mathbf{e}_*(a_i/f) C_{a_0 b_i}(\mathbf{e}(ax))^{q^N} = 1 - \sum_{i=1}^{\deg f} \mathbf{e}_*(a_i/f)\mathbf{e}(b_i b/f)^{q^N} = 0,$$

the last equality by the second part of the previous Theorem. Hence g_x has as many zeros in $\overline{U}_f$ as claimed, and since it has no singularities at infinity worse than simple poles, $(g_x) - D$ is effective of degree zero and hence is zero. ☐

The divisor D being the Stickelberger divisor of the cyclotomic theory, you get through the partial zeta function corresponding to a/f, we get soliton (specialization) as Stickelberger element, generalizing Coleman's constructions [And] corresponding to few denominators f.

The function g_x, called a "soliton" by Anderson, is thus a function on the cyclotomic curve cross itself and specializes at the graph of the N-th power of the Frobenius to essentially the N-th term in the product representing the gamma value at a fraction. Specializing the solitons at appropriate geometric points, Anderson has proved [And] two dimensional version of

Stickelberger's theorem. For connection with the arithmetic of zeta values and theta functions, see [And94; Tha92b]. The reason for the name soliton is that the way it arises in the theory of Drinfeld modules or Shtukas, when dealing with the projective line with some points identified, is analogous to the way the soliton solutions occur in Krichever's theory of algebro-geometric solutions of differential equations, as explained in [Mum78, pa. 130, 145] and [And; And94].

8.5   Another approach to solitons

Let us start again and get an algebraic equation [Tha99a] for a soliton $g_{a/f}$, which we will denote now by ϕ. For $f \in A_+$, let $\zeta_f := e(\tilde{\pi}/f)$. If we take t and T to be two independent variables, with the obvious abuse of notation,

$$L_f := \mathbb{F}_q(t, \zeta_{f(t)}, T, \zeta_{f(T)})$$

can be thought of as the function field of the product of two copies of X over $\mathbb{F}_q$. If f is understood, we put $\zeta = \zeta_{f(t)}$ and $Z = \zeta_{f(T)}$. Then $\phi \in L_f$ (or even a subfield) by the results of last section, but we will be content here to get an algebraic equation over $L_1 = \mathbb{F}_q(t, T)$.

Let us write $d := d_f$. Let $C_{f(t)} = \sum_{i=0}^d f_i(t)\tau^i$. The relation $C_t C_f = C_f C_t$ gives the recursion formula

$$f_0(t) = f(t), \quad f_i(t) = \frac{f_{i-1}^q(t) - f_{i-1}(t)}{t^{q^i} - t}.$$

For $i \geq 0$, we define ϕ_i recursively via

$$\phi_0 := \phi, \quad \phi_i := \frac{\phi_{i-1}^q - \phi_{i-1}}{T^{q^{i-1}} - t}.$$

We define $c_i \in L_1$ via

$$\sum_{i=0}^d c_i \phi^{q^i} = \sum_{i=0}^d f_{d-i} \phi_i^{q^{d-i}}.$$

Theorem 8.5.1   *With this notation, ϕ satisfies $\sum_{i=0}^d c_i \phi^{q^i} = 0$.*

Proof.   Now ϕ is characterized by its specializations at the graph of N-th power of Frobenius being $\{ {}_{q^N-1}^{a/f} \}$, as in 8.4.7 (2). We will see below that ϕ_i is characterized by its specializations at the graph of N-th power of Frobenius being $\{ {}_{q^{N+i-1}}^{a/f} \}$.

Let $\Phi_{a/f} := \sum_{i=0}^{\infty}\{\begin{smallmatrix}a/f\\q^i\end{smallmatrix}\}\tau^i$. Then

$$\Phi_{a/f}(z) = \sum\{\begin{smallmatrix}a/f\\q^i\end{smallmatrix}\}z^{q^i} = e(al(z)/f).$$

But $\rho_f(e(al(z)/f)) = \rho_a(z)$, for small z, so $\rho_f\Phi_{a/f} = \rho_a$. Equating coefficients of τ^{d+k}, we get $\sum f_{d-i}\{\begin{smallmatrix}a/f\\q^{i+k}\end{smallmatrix}\}q^{d-i} = 0$ for each non-negative k. These infinitely many equations satisfied by the specializations $\{\begin{smallmatrix}a/f\\q^i\end{smallmatrix}\}$'s of ϕ show that (note that the union of graphs of Frobenius powers is Zariski dense) ϕ satisfies $\sum c_i\phi^{q^i} = 0$, except that we have to justify that the ϕ_i characterized by its specializations as above is given by the recursion defined above.

Now even though $B\{\tau\}$ is a non-commutative ring, Drinfeld A-module provides a way to get its commutative subring isomorphic to A: $\rho_{a_1}\rho_{a_2} = \rho_{a_2}\rho_{a_1}$ for all $a_1, a_2 \in A$. Hence for $c \in A$,

$$\rho_f\rho_c\Phi_{a/f} = \rho_c\rho_f\Phi_{a/f} = \rho_c\rho_a = \rho_a\rho_c = \rho_f\Phi_{a/f}\rho_c,$$

which by cancellation of ρ_f gives

$$\rho_c\Phi_{a/f} = \Phi_{a/f}\rho_c, \quad c \in A.$$

Applying this to $c = t$ and comparing coefficients of τ^j, we get

$$t\{\begin{smallmatrix}a/f\\q^j\end{smallmatrix}\} + \{\begin{smallmatrix}a/f\\q^{j-1}\end{smallmatrix}\}^q = t^{q^j}\{\begin{smallmatrix}a/f\\q^j\end{smallmatrix}\} + \{\begin{smallmatrix}a/f\\q^{j-1}\end{smallmatrix}\}.$$

(Note that this equation is equivalent to the functional equation (37) of [And] and also that it gives a simple recursion formula for the terms in the product for gamma value.)

If now one puts $j = N+i-1$, then since $t^{q^j} = (t^{q^N})^{q^{i-1}}$ the characterization of ϕ_i stated in the beginning of the proof gets reconciled with the recursive definition for it. $\square$

Remarks 8.5.2 (a) In fact, what we have shown is that any element ϕ in a finite extension of L_1 (without assuming that it belongs specifically to L_f) which specializes on the graph of N-th power of Frobenius to $\prod_{n\in AN+}(1 + a/fn)$ (we will denote such an element by $\phi_{a/f}$) satisfies equation $\sum c_i\phi^{q^i} = 0$. Hence different solutions correspond to the different a's (not necessarily prime to f) with $d_a < d_f$. There is always the trivial solution $\phi = 0$ corresponding to $a = 0$. We will call the (non-trivial) solutions solitons. The solutions $\phi_{a/f}$, with a not prime to f, have lower 'conductor'.

(b) Note that ϕ_i defined by the recursion above has another description: ϕ_i is obtained by acting on ϕ by i-th power of partial Frobenius in the 'T variable'. This description makes sense for negative i also. Similarly, we define $\hat{\phi}_i$ to be the function defined by acting on ϕ by i-th power of partial Frobenius in the 't variable'. Then, clearly, $\phi_{-i}^{q^i} = \hat{\phi}_i$.

(c) Our proof, in fact, gives infinite hierarchy of equations: $\sum f_{d-i}\phi_{i+m}^{q^{d-i}} = 0$, one for each integer m, and m can be negative: One has to just use specializations of ϕ_{i+m} at the graphs of sufficiently high powers of Frobenius. For the same reason, the recursion formula for ϕ_i given before the Theorem is now justified for all integers i, not just the positive integers.

(d) Since the completion of $K(\zeta_f)$ at an infinite place is $\mathbb{F}_q((1/\zeta_t))$. (Note $\zeta_t^{q-1} = -t$.) Hence we can represent ϕ also by a power series in ζ_t, ζ_T over $\mathbb{F}_q$ and then the Theorem gives a recursion relation for the coefficients.

(e) In addition to the equation in the Theorem, we can also get, by the same method, algebraic recursion equations: For example, from the last displayed equation in the proof, we get

$$T\phi_{a/ft}^q - t\phi_{a/ft} = (T - t)(\phi_{a/f})_1.$$

Examples 8.5.3   (I) Let $f(t) = t$, then the Theorem gives

$$\phi^q + t\phi_1 = \phi^q + t\frac{\phi^q - \phi}{T - t} = 0.$$

Throwing out the trivial solution $\phi = 0$, we see that $\phi = (t/T)^{1/q-1} = \zeta_t/\zeta_T$. This is nothing but 4.11.1 [Tha91a, 9.1.1], where the additivity of the solitons (as functions of a/f) was established using Moore determinants and this formula was used (essentially the method of partial fractions) to get formulas for (what is now understood as) the solitons when f was a product of distinct linear factors.

(II) Let $f(t) = t^2$, then the Theorem gives

$$\phi^{q^2} + (t + t^q)\frac{\phi^{q^2} - \phi^q}{(T - t)^q} + t^2\frac{\frac{\phi^{q^2} - \phi^q}{(T-t)^q} - \frac{\phi^q - \phi}{T-t}}{T^q - t} = 0.$$

To get a rational expression, it is easier to use (10), say with $f = t$: With $\phi = (\zeta_t/\zeta_T)\psi$, (10) becomes $\psi^q - \psi = 1/T - 1/t$, hence $\psi = \zeta/\zeta_t - Z/\zeta_T$ gives an obvious solution. So $\phi = (\zeta C_T(Z) - ZC_t(\zeta))/C_T(Z)^2$. Full space of solutions of the displayed equation is then seen to be $\phi\mathbb{F}_q + (\zeta_t/\zeta_T)\mathbb{F}_q$.

These give formulas for the product of the monic polynomials in $\mathbb{F}_q[t]$ of degree i in a given congruence class (essentially the same as asking for a formula for the term in the definition of the geometric gamma function at a fraction), generalizing the formula for D_i in 2.5: The first example shows (4.11.1) that

$$\left(\prod_{\substack{a \in A_{i+} \\ a \equiv 1 \mod t}} a \right) / (t^{q^{i-1}} D_{i-1}) = 1 - \frac{\zeta_t}{\zeta_t^{q^i}} = 1 - (-t)^{-(q^i-1)/(q-1)},$$

and the second example shows that

$$\left(\prod_{\substack{a \in A_{i+} \\ a \equiv 1 \mod t^2}} a \right) / ((t^2)^{q^{i-2}} D_{i-2}) = 1 - \frac{\zeta_{t^2}(-t)^{q^{i-1}/(q-1)} - \zeta_{t^2}^{q^{i-1}}(-t)^{1/(q-1)}}{(-t)^{2q^{i-1}/(q-1)}}.$$

For several additional comments on irreducible factors and Galois action, field of rationality, divisor computations, explicit version of Anderson's earlier construction, occurrence of q as a parameter etc. from this view point, we refer to [Tha99a]. For Anderson's original approach, using theta and tau functions, symmetric function theory, see [And; And94]. Anderson's approach we described above is superior, but we still offered alternate approach in a belief that more ways of looking at important things are better.

Example 8.5.4 When $f = t^d$, we have $\phi = \sum_{i=0}^{d-1} C_{t^{d-i-1}}(\zeta)\overline{M}_{d-i-1}$, where

$$\overline{M}_0 := C_{T^{d-1}}(Z)^{-1}, \quad \overline{M}_i := -\overline{M}_0 \sum_{k=0}^{i-1} C_{T^{d+k-i-1}}(Z)\overline{M}_k.$$

Remarks 8.5.5 These explicit formulas reveal some interesting phenomena. First of all, q occurs as a parameter in the exponents, just as in the two examples above. Secondly, the formulas can be just thought of as specializations from universal formulas, e.g., for $f(t) \in \mathbb{Z}[t]$ if $q = p$, and hence their form is independent of questions such as whether f is a prime in A, though the properties of course depend on such questions. The dependence of ϕ on f_0 is implicitly through ζ and Z only.

8.6   Analog of Gross-Koblitz for Geometric Gamma: $\mathbb{F}_q[t]$ case

Now we show that using Anderson's analog of Stickelberger, the proof of Gross-Koblitz analog generalizes from the simplest case considered in 4.11.3 to the general $\mathbb{F}_q[t]$ case by a proof exactly as in previous cases:

Theorem 8.6.1   *Let $A = \mathbb{F}_q[t]$. Let v be a monic prime of degree d such that f (of positive degree) divides $v - 1$. Let Γ_v be the corresponding function of section 4.9. Let a be an element of degree less than that of f. Then $\Gamma_v(a/f)$ is (explicit) rational multiple of the a/f-Stickelberger element applied to v. In particular, it is algebraic.*

Proof.   (Compare 4.8.1, 4.11.3, 8.3.4). We define w_j and x_j as in 4.11.3, but with a/f replacing η/t. Then $\Pi_v(a/f)$ is the v-adic limit as $m \to \infty$ of $\prod_{j=0}^m w_j$, and as in 4.11.3, for $j \leq d$, we have $w_j = x_j/x_{j-d}v^{k_j}$, where k_j is such that this is a unit. The difference between w_j and x_j when $j < d$ makes up the explicit rational function, we ignore. The other product telescopes as in 4.11.3. As $j \to \infty$ within a fixed congruence class modulo d, x_j tends v-adically to $g_{a/f}$ evaluated at a $\overline{\mathbb{F}_q}$-valued point above v, and since the Galois orbit of these points constitutes v, we just get Stickelberger element applied to v by Theorem 8.4.8, and remarks after it.   □

Corollary 8.6.2   *Let $A = \mathbb{F}_q[t]$. If $\underline{a}$ consists of Frob_v-orbits, up to translation by elements in A, then $\Gamma_v(\underline{a})$ is algebraic.*

Proof.   Since the gamma value at a proper fraction get multiplied by an algebraic element when we change the fraction by an integer, it is enough to consider $\underline{a}$ which is a single Frobenius orbit. In the proof above, when v does not divide $f - 1$, dividing by v changes congruence classes, but now since the whole Frobenius orbit is there, the same proof works.   □

Remarks 8.6.3   (1) To prove just the algebraicity of these values, we do not need 8.4.8 above, but we can use weaker theorems 8.4.4 or 8.5.1 instead which prove algebraicity without giving much more immediate information on divisors.

8.7   What is known or expected in general case?

The Gross-Koblitz theorem (see 4.8 (a)) expresses Gauss sum, which is an algebraic (cyclotomic) number giving a discrete analog of the (transcendental) Gamma function, essentially as a value at a fraction of p-adic interpolation of the Gamma function.

So far, we have proved two analogs of the Gross-Koblitz theorem for $\mathbb{F}_q[t]$: The first in 4.8 (a) connects the Gauss sums of 4.1 to the arithmetic gamma interpolations in 4.5 (a), and the second in 8.6 (or 4.11 in special cases) connects the Stickelberger elements corresponding to geometric cyclotomic fields coming from the Carlitz module to geometric gamma interpolations in 4.9.

For general A, as we have seen in 4.3, the Gauss sums of 4.1 and Stickelberger elements for constant field extensions differ, but there is still a Gross-Koblitz formula 8.3 (a) connecting Gauss sums of 4.1 to the Gamma function interpolation of 8.3.

We expect that there is a Gross-Koblitz formula connecting the Stickelberger elements of the constant field extensions for general A case to the interpolations in 4.5 (b) of the arithmetic gamma function. This has been proved only for the four class number one A's of 3.1.2, using results of 4.15:

We want a meromorphic function Δ on the Hilbert cover of X cross itself whose specializations at $\overline{\mathbb{F}}_q$-points are 'Jacobi sums' corresponding to Stickelberger elements and at $\xi^{(i)}$ are D_{i+g}/D_{i+g-1}^q. When $g = d_\infty = 1$, Theorems 4.15.3, 4.15.4, 4.15.5 give a formula for Δ, proving, in particular, the algebraicity of the values of $\wp$-adic interpolations at appropriate fractions, but to check Gross-Koblitz one has to compare the value with the Stickelberger elements, and so far this has only been checked for the four class number one examples.

Example 8.7.1 For the example 3.1.2 (iii), we have

$$\Delta = \frac{((x - \overline{x})^2 + x - \overline{x} + 1 - y\overline{y})(y^3 - \overline{y} - y^3(x^3 - \overline{x}))}{(x^3 - \overline{x} + 1)(y^3\overline{y} - 1 - (x^3 - \overline{x})^2 - x^3 + \overline{x})}.$$

It is easy to verify directly from this formula that the divisor of $\Delta|_\theta$ is $(\theta^{(-1)}) + 3(\theta^{(1)}) - 3(\theta) - (\infty)$. A straightforward calculation with partial zeta functions shows that this divisor agrees with the divisor corresponding to 'Jacobi sum' associated to the Stickelberger elements.

What is the situation with Stickelberger elements for arithmetic or geometric extensions for general A? It is clear that the proof would be the the same once we have right interpolation function. When $d_\infty = 1$ (and even in full generality with more work), it should be possible to get such interpolations generalizing Greg Anderson's work in 8.4, but this has not been worked out. In [Tha99a] we show that interpolating functions do exist for curly brackets binomials and the connection in 4.15 shows that it is true also for round brackets, at least when $g = d_\infty = 1$. So we do get algebraicity again, but we do not have any good control of the field of rationality or divisors as in Anderson's theorem in 8.4 on $\mathbb{F}_q[t]$ case, so it is difficult to deduce the general theorem this way.

Remarks 8.7.2 (1) We have seen in Chapters 1 and 3 that Stickelberger elements, were shown to exist in the general situation by Tate and Deligne and Hayes [Hay85] provided an explicit construction. Interesting feature of his construction is that 'Gauss sum for a prime $\wp$' occurs as a torsion point for some rank one Drinfeld-A-module, where the infinite place for A lies above $\wp$. Hence an infinity-adic formula coming from the exponential for a torsion point is a'$\wp$-adic' limit formula for Gauss sums. This together with reflection formula 4.10.1 and product formula for $\tilde{\pi}$ such as 4.7.5 express 'Gauss sums' in terms of values of '$\wp$-adic gamma functions' at fractions. In this sense, this can be considered as an analog of the Gross-Koblitz formula, the main difference being that in this case the gamma function is constructed by uniform procedure at all places, instead of being interpolated from a fixed gamma function. The latter gives much closer analogy.

(2) For abelian extensions which are compositums of arithmetic and geometric extensions, the Stickelberger elements are most probably connected via the two variable gamma function of Goss in 4.12.

(3) We saw arithmetic gamma function for $\mathbb{F}_q[t]$ generalizes in two ways for general A, namely to those of 4.5 (b) and 8.3 (a), because D_i and d_i differ in general. Both give interesting theory. Similar relation holds between binomial coefficients as explained in 4.14-4.15. So in [Tha99a] gamma function $z^{-1} \prod_{i=0}^{\infty} (1 + \{ {z \atop q^i} \})^{-1}$ made up from curly bracket binomials of Section 4.15 was proposed, in addition to geometric gamma for A other than $\mathbb{F}_q[t]$. Natural generalization with Anderson's approach would be based on interpolations of $\binom{x}{q^i}$ (but it has not been achieved yet), whereas our approach in [Tha99a] deals with the other candidate $\{ {x \atop q^i} \}$. We refer to this paper for details of how interpolating function like ϕ can be obtained for $\{ {a/f \atop q^i} \}$. The equations obtained in general seem to be complicated enough to be of use in calculations as in the last section and better techniques are needed to understand the divisor and other properties. We do not yet understand special values of interpolations of this gamma function at any place.

8.8 Gamma values to Periods connection via solitons: Sketch

Let us use notation of 8.4 and let d be the degree of $f \in A_+$. Note that $[K(\zeta_f) : K] = \phi(f) = q^d - 1$ and $[K(\zeta_f)^+ : K] = \phi(f)/(q-1)$, where ϕ is analog (see 1.1) of Euler's ϕ function. Let $I := \{a \in A : (a, f) = 1, \deg(a) < d\}$ and $I_+ := I \cap A_+$. Note $|I| = \phi(f)$ and $|I_+| = \phi(f)/(q-1)$. By Theorem 8.4.8, the divisor of the soliton function $g = g_x$ is

$$(g) = W^{(1)} - W + E - \infty,$$

where

$$W := \sum_{j=0}^{d-2} \sum_{a \in I_+, \deg(a) \leq j} [a] \mathrm{Frob}^{d-j-2}(\xi), \quad E := \sum_{a \in I_+} [a]\xi$$

are effective divisors on $\overline{X_f}$.

Consider $M_f := \Gamma(\overline{U_f}, \Omega_{X_f}(W))$ as $\overline{K}[t, \tau]$-module by left multiplication action of $\overline{K}[t]$ and by

$$\tau m := g m^{(1)}, \quad m \in M_f.$$

In his University of Minnesota thesis of 1995, Sinha proved [Sin97b]:

Theorem 8.8.1 *With the notation as above, M_f is a uniformizable abelian t-motive over $\overline{K}$ of dimension $\phi(f)/(q-1)$ and rank $\phi(f)$. In fact, the corresponding t-module is HBD module with multiplications by $A[\zeta_f]_+$ and is of CM type with complex multiplication by $A[\zeta_f]$. Its period lattice is free rank one over $A[\zeta_f]$, with the a-th co-ordinate (for $a \in I_+$) of any non-zero period being (with appropriate explicitly given $C_\infty\{\tau\}$-basis of M to give co-ordinates) the (explicit non-zero algebraic multiple of) value $\Gamma(a/f)$.*

Brownawell and Papanikolas [BP02] proved

Theorem 8.8.2 *The co-ordinates of periods and quasi-periods of M_f (in the same co-ordinates as above) are exactly the (explicit non-zero algebraic multiples of) values $\Gamma(a/f)$, with $a \in I$.*

We refer to [Sin97b; BP02; ABP] respectively for very nice and clean treatment of these issues and applications (which will be mentioned in Chapter 10) to transcendence, transcendence plus linear independence and algebraic independence respectively, of the gamma values.

We will now sketch, briefly and informally, how this is achieved, by connecting to what we have seen before:

Note first that we have proved similar results in special cases: See 4.5 (a) for unramified cases and 4.11 and 6.4 (b) for the simplest ramified case $f = t$ which can be achieved using Drinfeld modules, as $d = 1$.

More generally, Shtuka (see 7.7) is a function f on $\overline{X}$ with divisor of the form $V^{(1)} - V + \xi - \infty$ and corresponds to a normalized rank one Drinfeld A-module. For example, as in 7.7.7, the Carlitz module for $A = \mathbb{F}_q[t]$ with $T := t|_\xi$ corresponds to the Shtuka $f = t - T$, with empty V, so that $f^{(i)} = t - T^{q^i}$ and $f^{(i)}|_{\xi^{(n)}} = t^{q^n} - t^{q^i}$ and the period of the Carlitz module is the residue at ξ of $(\prod_{i=0}^{\infty} f^{(i)})^{-1}$, as we have seen in 2.5 and in the discussion following 7.4.6.

How does all this generalize to soliton level? We are trying to generalize from this absolute (unramified) case to the relative (ramified with conductor the monic polynomial f: the notation unfortunately clashes with the Shtuka notation!) case of the f-th cyclotomic cover. (This is where the theta technology for the generalized Jacobian for conductor f shows up in Anderson's papers [And; And94].) The integral closure $A[\zeta_f]$ of A, which has several places above infinity now enters the picture. Allowing the twist by more than one zeros and poles in the twisting carries one to higher dimensional generalizations of Drinfeld modules with complex multiplication. Once one defines soliton $g = \phi$ by interpolations and calculates its divisor, it has the right form as above with several zeros and infinities, and thus gives rise to a (generalized) Shtuka. The corresponding rank one (over $A[\zeta_f]$) t-motive M_f, has period (essentially) $(\prod \phi^{(i)})^{-1}$, thus gives the gamma value as above. This rank one multiplication (complex multiplication) by cyclotomic integers is the reason for analogy with Fermat motives.

Let us see in more detail how (if not why) the period connection works: Write $c_g^{-1} := \prod_{i=0}^{\infty} g^{(i)}$. Choose a suitable natural $C_\infty\{\tau\}$-basis $\{n_a\}$ of M_f giving co-ordinates for the tangent space, indexed by $a \in I_+$. Then the period lattice is image of $A[\zeta_f]c_g$ in the tangent space given as follows: For $b \in A[\zeta_f]$, the a-th co-ordinate of the image of bc_g is b^{σ_a} times the residue of $c_g n_a$ at E or, what is the same, at ξ_a. Hence a calculation using (2) of 8.4.7 shows that it is a non-zero algebraic multiple of $\Gamma(a/f)$.

Hence the sign condition on a coming in Sinha's theorem is because of the 'brackets' corresponding to sign condition in the divisor calculation 8.4.8, which is responsible for sign condition in divisor E. In other words, all points ξ_a above $t = T$ (residues at these give periods and quasi-periods, as we have seen in 7.4 for periods and in examples of 4.11.3 and 6.4.7 for quasi-periods and periods in a complex multiplication case) do not contribute to poles of c_g. Anderson later realized that since the Carlitz module C has zero at $t = T$, all $\Gamma(a/f)$, $a \in I$ occur as periods of $C \otimes M_f$, which has dimension $(q^d - 1)q/(q - 1)$ and the same rank as M_f, as g gets replaced by $g(t - T)$ now. We will see applications in Chapter 10.

8.9 Log-algebraicity, Cyclotomic module and Vandiver conjecture

We have looked at analogs of cyclotomic units, connection with class groups and connection between p-divisibilities of orders of groups arising from them for $\wp$-th cyclotomic fields and $\wp$ divisibilities of the zeta values. This mixture of p and $\wp$ is not satisfactory analog. Also, we see that naive analogs of the Kummer's theorem that 'p does not divide $h(\mathbb{Q}(\zeta_p))$ implies p does not divide $h(\mathbb{Q}(\zeta_p)^+)$' and of Vandiver conjecture that 'p does not divide $h(\mathbb{Q}(\zeta_p)^+)$' both fail for simple examples. The p divisibilities or even Norm($\wp$) divisibilities do not work. To get good analog [Tha94] so that we can talk about $\wp$ divisibilities, we want A-modules and not abelian groups which are just $\mathbb{Z}$-modules.

For cyclotomic units groups, such an analog was provided by Anderson. Since exponential and logarithm are inverse functions, we have identity $\exp(-\sum e^{2\pi imnx} z^n/n) = 1 - (e^{2\pi ix})^m z$, which for $z = 1$ and x a proper fraction gives 'one minus a root of unity'. Now as in 4.15 and 5.9, we can think of harmonic series or $\zeta(1)$ instead of the logarithm. Mixing the Carlitz action by $a \in A = \mathbb{F}_q[T]$ and usual multiplicative action by $m \in \mathbb{Z}$, Anderson considered, for $A = \mathbb{F}_q[T]$,

$$S_m(t, z) := \sum_{i=0}^{\infty} \frac{1}{d_i} \sum_{a \in A_+} \left(\frac{C_a(t)^m}{a}\right)^{q^i} z^{q^{i+\deg a}} \in K[[t, z]]$$

and proved 'log-algebraicity' theorem:

Theorem 8.9.1 *(Anderson [And96b]) Let $A = \mathbb{F}_q[T]$. We have $S_m(t, z) \in A[t, z]$ and has degree $\leq q^{\lfloor (m-1)/q-1)\rfloor}$ in z.*

His first proof [And94], which works in more generality of any A with $d_\infty = 1$, uses his soliton ideas of deformation theory to get explicit interpolating functions on powers of the cyclotomic cover similar to what we have seen for solitons, and then showing directly by analysis of divisors and zeros that the coefficients for large enough power of z are zero. His second proof [And96b] is more elementary using the Dwork's v-adic trick (for all finite v) to deduce integrality of coefficients and then using analytic theory to get degree estimates to show that coefficients tend to zero eventually, so that they are zero from some point on-wards and the power series is really a polynomial. In the next section, we give yet another proof of a slightly weaker result but also providing interesting formulas for the polynomials you thus get and showing how q occurs as a 'parameter' suggesting more structure to the objects than we know of.

Using this Theorem Anderson then defines **cyclotomic module** (analog of group of the cyclotomic units in $\mathbb{Q}(\zeta_p)$) C to be A-sub-module of $\mathcal{O} := \mathcal{O}_{K(\Lambda_\wp)}$ (under the Carlitz action) generated by $S_m(e_C(\tilde{\pi}b/\wp), 1)$'s.

Theorem 8.9.2   *The cyclotomic module C is Galois stable module of rank* $(Norm(\wp) - 1)(1 - 1/(q - 1))$.

Now Kummer-Vandiver conjecture that p does not divide the class number of $\mathbb{Q}(\zeta_p)^+$ can be rephrased as the canonical map taking the cyclotomic units mod the p-th power to all units modulo p-th power is injective. Replacing $\mathcal{O}^*$ which are integral points for the multiplicative group by $\mathcal{O}$ which are integral points for the additive group (with Carlitz action), analog of Anderson for Vandiver conjecture would be that map $C/\wp C \to \mathcal{O}/\wp \mathcal{O}$ is injective. It would be nice to settle this.

As an analog of the Kummer's theorem that p does not divide h (see Theorem 5.3.8) implies p does not divide h^+, Anderson proves

Theorem 8.9.3   *The map above is injective, if $\wp$ does not divide $\zeta(1-i)$ for $1 \le i \le Norm(\wp)$, and $1 - i$ 'odd'.*

Remarks 8.9.4   (1) Using the remark after 5.3.1, the analog above of Kummer-Vandiver conjecture follows from the Theorem for $\wp = T$ (or more generally any degree one prime) or for $q = 2$.

(2) The rank should be compared with classical $(p - 1)(1 - 1/2) - 1$, where the difference of 1 has to do with having no pole of zeta at $s = 1$.

(3) If $q = 2$, C just consists of $\wp$-torsion points together with the rational torsion $1, T, T + 1$. For general q, but if $\wp = T$, then $C = \{C_a(\lambda_T)^m : 0 \le m < q - 1, a \in \mathbb{F}_q[T]\}$. For example, if $q = 3$, this just consists of $C_a(1)$'s together with T-torsion.

(4) If $\sqrt{C}$ is the divisible closure of C in $\mathcal{O}$ thought of as $\mathcal{O}$-points of the Carlitz module, then since the units are divisible closure of cyclotomic units, the analogies suggest that analog of the class number h^+ is the class polynomial which the (Fitting) index of A-module $\sqrt{C}/C$. (In both cases

$q = 2$ or $q = 3$, $\wp = T$ in (3), it is one.) General understanding of such 'class numbers' and their arithmetic significance is still lacking.

(5) In this paper, Anderson also defined from L-values some root numbers which are related [Fen97; Zha97] to Gauss sums of Chapter 4.

Using the analogy with cyclotomic units and the relation between class numbers and index of cyclotomic units, we thus get a polynomial as some kind of A-module version of class number in this special case. How about getting A-module 'class group' concept in general?

In this connection, Poonen has suggested trying something along the lines of the folklore fact (proved below) that the class group of a number field is the Tate-Shafarevich group of the units, by for example, trying to get a local analog of Anderson's module:

Let K be a global field. As usual, $\mathcal{O}_K$, I_K, P_K denote ring of integers (with some places chosen at infinity), group of its fractional ideals and the subgroup of principal ideals. If L is a finite Galois extension with Galois group G, then the exact sequence $0 \to \mathcal{O}_L^* \to L^* \to P_L \to 0$ and Hilbert's theorem 90 gives exact sequence $0 \to P_K \to (P_L)^G \to H^1(G, \mathcal{O}_L^*) \to 0$. Now I_L^G/I_K is isomorphic to the product over primes v of $\mathcal{O}_K$ of $\mathbb{Z}/e_v\mathbb{Z}$, where e_v is the ramification index in L. So locally, where all ideals are principal, $H^1(G, \mathcal{O}_{L_v}^*)$ is isomorphic to $\mathbb{Z}/e_v\mathbb{Z}$. So the kernel of $H^1(G, \mathcal{O}_L^*)$ mapping to the product of corresponding local things is $(P_L^G \cap I_K)/P_K$, i.e. the group of ideal classes of K which become trivial in L. Since every ideal class dies in a separable extension (eg. Hilbert class field), letting L go to separable algebraic closure, we get the claimed isomorphism

$$Cl_K = I_K/P_K \equiv \mathrm{Ker}\, H^1(G_K, \mathcal{O}_{K^{sep}}^*) \to \prod_v H^1(G_{K,v}, \mathcal{O}_{K_v^{sep}}^*).$$

First try may be to replace the multiplicative group (whose $\mathcal{O}_K$ points are the units) by the additive group, but the Tate-Shafarevich (TS) group of the additive group is zero for global field: Let K, L, G and $\mathcal{O}_L$ (the ring of S-integers in L for some S coming from K) be as above. The exact sequence $0 \to \mathcal{O}_L \to L \to L/\mathcal{O}_L$ gives

$$0 \to K/\mathcal{O}_K \to (L/\mathcal{O}_L)^G \to H^1(G, \mathcal{O}_L) \to 0.$$

Let A_K be the restricted direct product of K_v's for v not in S and O_K be the product of the completions O_v of $\mathcal{O}_K$ at those places. Then we get exact sequence

$$0 \to A_K/O_K \to (A_L/O_L)^G \to H^1(G, \mathcal{O}_L) \to 0.$$

Applying five-lemma to three vertical arrows connecting the last two exact sequences gives $\rightarrow A \rightarrow$ TS group $\rightarrow B \rightarrow$, where $A = ((L \cap \mathcal{O}_L)/\mathcal{O}_L)^G$ and $B = A_K/(K + \mathcal{O}_K)$. Now A is zero, because an element of K is in $\mathcal{O}_L$ if and only if it is in O_v for all places v of K not in S. Also B is zero: If $a \in A_K$, there is $0 \neq b \in \mathcal{O}_K$ with $ba \in \mathcal{O}_K$ and by the Chinese remainder theorem there is $c \in \mathcal{O}_K$ such that $c - ba \in b\mathcal{O}_K$. Then $c/b - a \in \mathcal{O}_K$ and $a = c/b + (a - c/b) \in K + \mathcal{O}_K$. (I learned this from John Tate.)

So it is unclear, how to proceed along this direction. It would be of great interest if we can still define some good A-modules to play a role of class groups.

8.10   Explicit Log-Algebraicity formulas

We give another proof, in $\mathbb{F}_q[T]$ case, of Anderson's theorem on log-algebraicity and special polynomials in new cyclotomic theory, using results of Carlitz and Lee. This provides formulas, gives bounds on degrees and shows that q occurs as a parameter.

Let us recall the notation from Chapters 2 and 4, but with T replacing t: Let $A = \mathbb{F}_q[T]$, $K = \mathbb{F}_q(T)$ be of characteristic p. Let $[n] = T^{q^n} - T$, $d_0 = l_0 = 1$, and $d_i = [i]d_{i-1}^q$, $l_i = -[i]l_{i-1}$, for $i > 0$ and $\{a, j\} = \sum_{k=0}^{j} a^{q^k}/(d_k l_{j-k}^{q^k})$, where we have changed Carlitz notation, from vertical to horizontal (and getting rid of base q and just keeping the exponent j), for typographical reason.

$$C_a(t) = \sum_{j=0}^{\deg a} \{a, j\} t^{q^j} = \sum_{j=0}^{\deg a} t^{q^j} \sum_{k=0}^{j} \frac{a^{q^k}}{d_k l_{j-k}^{q^k}} = \sum_{k=0}^{\deg a} \frac{a^{q^k}}{d_k} \sum_{j=k}^{\deg a} \frac{t^{q^j}}{l_{j-k}^{q^k}}. \qquad (I)$$

We will now prove the main part of Anderson's log-algebraicity theorem.

From the definition it follows that we can write $S_m(t, z)$ as a power series $\sum w_n z^{q^n}$, with $w_n \in K[t]$.

Before giving the proof, we illustrate the method by first proving the theorem (and providing a nice formula) in the case when m is sum of less than q, q-powers.

Let $S_i(k) = \sum_{a \in A_{i+}} a^k$. Given a positive integer k, let $l(k)$ denote the sum of its digits base q. Here are some facts (see 5.6) we need:
$\{a, j\} \in A$, if $a \in A$; $\{a, j\} = 0$, if $j > \deg a$; $\{a, \deg a\} = 1$, if $a \in A_+$. (II)
  If $i > l(k)/(q-1)$, then $S_i(k) = 0$. (III)

$$S_i(q^{k+i} - 1) = d_{i+k}/l_i d_k^{q^i}.\tag{IV}$$

First let $m = q^\mu$. Then w_n is

$$\sum_{i=0}^{n} \frac{1}{d_i} \Big(\sum_{a \in A_{(n-i)+}} \frac{C_a(t)^{q^\mu}}{a} \Big)^{q^i} = \sum_{i=0}^{n} \frac{1}{d_i} \Big(\sum_{j=0}^{n-i} t^{q^{j+\mu}} \sum_{k=0}^{j} \frac{1}{d_k^{q^\mu} l_{j-k}^{q^{k+\mu}}} S_{n-i}(q^{k+\mu} - 1) \Big)^{q^i}$$

by I, and since by III, the power sum can be nonzero only when $k \geq n-i-\mu$, when it can be evaluated by IV, we see that w_n is

$$\sum_{i=0}^{n} \frac{1}{d_i l_{n-i}^{q^i}} \Big[\sum_{j=0}^{n-i} t^{q^{j+\mu+i}} \sum_{k=n-i-\mu}^{j} \frac{d_{k+\mu}^{q^i}}{d_k^{q^{\mu+i}} l_{j-k}^{q^{k+\mu+i}} d_{k+\mu-(n-i)}^{q^n}} \Big].$$

Put $u = j+i+\mu-n$ and $v = k+i+\mu-n$, then $0 \leq u \leq \mu$ and $0 \leq v \leq u$, so that the last quantity inside the square brackets is

$$\sum_{u=0}^{\mu} t^{q^{n+u}} \sum_{v=0}^{u} \frac{[n-i+v]^{q^i} [n-i+v-1]^{q^{i+1}} \cdots [n-i+v-\mu+1]^{q^{i+\mu-1}}}{(d_v l_{u-v}^{q^v})^{q^n}}.$$

The numerator for the inside summand is

$$\prod_{\theta=0}^{\mu-1} (T^{q^{n+v}} - (T^{q^i})^{q^\theta}) = \sum_{l=0}^{\mu} (T^{q^{n+v}})^l (-1)^{\mu-l} \sum_{0 \leq j_1 < \cdots j_{\mu-l} \leq \mu-1} (T^{q^i})^{q^{j_1} + \cdots + q^{j_{\mu-l}}}.$$

So using the definition of $\{a, j\}$ for the sum over i as well as sum over v, we see that

$$w_n = \sum_{u=0}^{\mu} t^{q^{n+u}} \sum_{l=0}^{\mu} \{T^l, u\}^{q^n} (-1)^{\mu-l} \sum \{T^{q^{j_1} + \cdots + q^{j_{\mu-l}}}, n\}.$$

By II, we now see that $w_n \in A[t]$ (in fact, in $F_p[T, t]$ in general, as can be seen by $\text{Gal}(\mathbb{F}_q/\mathbb{F}_p)$-invariance in the definition) and $w_n = 0$ if $n > q^{\mu-1} + \cdots + 1 = (m-1)/(q-1)$ proving the claim. We further see that w_n can contain powers of t between q^n and $q^{n+\mu}$ only. In fact, the formula simplifies further: By I and II, we have $w_n = \sum_{l=0}^{\mu} \{P_{\mu,l}(T), n\} C_{T^l}(t)^{q^n}$, where $P_{\mu,l}(T) = (-1)^{\mu-l} \sum T^{q^{j_1} + \cdots + q^{j_{\mu-l}}}$. So

$$S_{q^\mu}(t, z) = \sum_{l=0}^{\mu} C_{P_{\mu,l}(T)}(z C_{T^l}(t)).$$

Now let $m = \sum_{e=0}^{k} q^{\mu_e}$, with $k < q$. We have

$$S_i(m-1) = (\prod [\mu_e]_i)/l_i, \text{ where } [h]_i = d_h/d_{h-i}^{q^i}.\tag{V}$$

Then w_n is

$$\sum_{i=0}^{n} \frac{1}{d_i} \Big(\sum_{a \in A_{(n-i)+}} \frac{C_a(t)^{q^{\mu_1}+\cdots+q^{\mu_k}}}{a} \Big)^{q^i}$$

which can be rewritten as before, but now with product of k similar sums with $0 \le j_e \le n-i$ and $0 \le k_e \le j_e$ multiplied by $S_{n-i}((\sum q^{k_e+\mu_e})-1)$. As before, by III, the power sum can be nonzero only when $k_e \ge n-i-\mu_e$, when it can be evaluated by V. Put $u_e = j_e + i + \mu_e - n$ and $v_e = k_e + i + \mu_e - n$, then $0 \le u_e \le \mu_e$ and $0 \le v_e \le u_e$, and the corresponding quantity becomes

$$\prod_{e=1}^{k} \sum_{u_e=0}^{\mu_e} t^{q^{n+u_e}} \sum_{v_e=0}^{u_e} \frac{[n-i+v_e]^{q^i}[n-i+v_e-1]^{q^{i+1}} \cdots [n-i+v_e-\mu_e+1]^{q^{i+\mu_e-1}}}{(d_{v_e} l_{u_e-v_e}^{q^{v_e}})^{q^n}}.$$

Expanding and simplifying as before by using the definition of $\{a,j\}$ for the sum over i as well as the sums over v_e's, and finally the sum over n's, we see that

$$S_m(t,z) = \sum_{l_1=0}^{\mu_1} \cdots \sum_{l_k=0}^{\mu_k} C_{P_{(\mu_j,l_j)}(T)} \Big(z \prod_{e=1}^{k} C_{T^{l_e}}(t) \Big)$$

where

$$P_{(\mu_j,l_j)}(T) = (-1)^{\sum(\mu_e-l_e)} \sum T^{q^{j_{1,1}}+\cdots+q^{j_{1,\mu_1}-l_1}+\cdots+q^{j_{k,1}}+\cdots+q^{j_{k,\mu_k}-l_k}}$$

with $0 \le j_{e,1} < j_{e,2} < \cdots j_{e,\mu_e-l_e} \le \mu_e - 1$, for $1 \le e \le k$.

Remarks 8.10.1   (1) As examples, we have $S_1(t,z) = tz$, $S_q = t^q z - t^q z^q$. For $q > 2$, we have $S_{q+1} = t^{q+1}z - t^{2q}z^q$ and $S_{2q} = t^{2q}z - ((T^q - T)t^{2q} + 2t^{q(q+1)})z^q + t^{2q^2}z^{q^2}$. One can also see that q occurs as a parameter (in the exponents) in the sense of the formula above, but if one expands as a polynomial in t and z, the number of terms can grow with q. For example, for $m = q^2$, the number of terms seems to be $\lfloor (m-1)/(q-1) \rfloor + 1$, as a polynomial in z.

(2) When m is as above, if we write $n = \lfloor (m-1)/(q-1) \rfloor$, then we see that the contribution to w_n can only come from terms with all $l_e = 0$ (because of II) and it is $(-1)^{\sum \mu_e} C_1(t)^{kq^n} = (-1)^n t^{kq^n}$. Hence Anderson's conjecture [And96b] about the leading term is verified for these infinite number of examples.

For the general case, we need a little more: Let $[k, i] = d_k/(d_i l_{k-i}^{q^i})$.

$$S_i(k) = \frac{d_i}{l_i} \text{ times the coefficient of } u^{k+1-q^i} \text{ in } (1+\sum_{j=0}^{i-1}[i,j]u^{q^i-q^j}-d_i u^{q^i})^{-1}.(VI)$$

Now we come to the general case, so that m is now any positive integer. We will first show that w_n is a sum of terms, each of which is zero, if $n > (m-1)/(q-1)$. This will imply that $S_m(t,z) \in K[t,z]$ with the degree bound as in Theorem 8.9.1.

If we let $u_k = (\sum_{j=k}^{n-i} t^{q^j}/l_{j-k}^{q^k})/d_k$, by I we get

$$w_n = \sum_{i=0}^{n} \frac{1}{d_i}(\sum_{a \in A_{(n-i)+}} \frac{1}{a}(\sum_{k=0}^{n-i} u_k a^{q^k})^m)^{q^i}$$

$$= \sum_{i=0}^{n} \frac{1}{d_i}(\sum_{k_1,\cdots,k_m} u_{k_1}\cdots u_{k_m} S_{n-i}(q^{k_1}+\cdots+q^{k_m}-1))^{q^i}.$$

If $\bar{k}$ denotes the least of the k_j's, then $(m-1)+(q-1)\bar{k} \geq$ the sum of the digits base q in the exponent $\geq (q-1)\bar{k}$, hence by III, the power sum can be non-zero only when $n - i \geq k_j \geq (n-i) - (m-1)/(q-1)$. (Note that this implies that l_r's occurring in u_k's have $r \leq (m-1)/(q-1)$.)

So the quantity inside the round bracket in the last expression for w_n is the sum of the terms of the form $u_{k_1}\cdots u_{k_m} S_{n-i}(q^{k_1}+\cdots+q^{k_m}-1)$, which by VI, are again sums of the terms of the form $\pm u_{k_1}\cdots u_{k_m}\frac{d_{n-i}}{l_{n-i}}d_{n-i}^g \prod_h [n-i, n-i-\beta_h]$, where $\sum q^{n-i-\beta_h}(q^{\beta_h}-1) = q^{k_1}+\cdots+q^{k_m}-(g+1)q^{n-i}$.

We ignore the signs and take out $l_{n-i}^{q^i}$, as in the special case above. The rest is the sum of the terms of the form

$$\frac{d_{n-i}^{g+1}}{d_{k_1}\cdots d_{k_m}l_{j_1-k_1}^{q^{k_1}}\cdots l_{j_m-k_m}^{q^{k_m}}}\prod\frac{d_{n-i}}{d_{n-i-\beta_h}l_{\beta_h}^{q^{n-i-\beta_h}}}.$$

Ignoring the l_j's (which occur with bounded j's) for the moment we are dealing with the ratio of the form $d_{n-i}^r/\prod d_{u_b}$, where $\sum q^{u_b} = rq^{n-i}$, with each $u_b \leq n - i$. Counting the exponents of primes (essentially 'integrality of binomial coefficient'), we see that this ratio is a polynomial and hence is a product of the terms of the form $[n - i - s]$. So raising to power q^i, and summing over i, by using II, we again see that the curly brackets we thus get are zero, if n is more than the maximum number of $[n - i - s]$'s. But counting the powers of T's that divide d_j's, we see that T divides the ratio to the power $r(q^{n-i}-1)/(q-1)-\sum(q^{u_b}-1)/(q-1) = (-r+\sum 1)/(q-1) =$

$(m-1-g)/(q-1)$. Now T divides each $[y]$ to exactly the first power. Hence we have proved that $w_n = 0$, if $n > (m-1)/(q-1)$.

To prove that $w_n \in A[t]$, we need to use l_j's and get everything in terms of curly brackets, as in the special case, and use II for integrality. This remains to be done.

Chapter 9

Diophantine approximation

Diophantine approximation is one more of the few areas where the number field situation is better understood, and the function field situation is not understood well, even conjecturally.

Still, in the function field case, the different tools, such as the Frobenius (or rather p-th power map) and differentiation, give many interesting results.

The main problem here is how 'well' we can approximate the algebraic irrational real numbers (Laurent series respectively) α by rational numbers (functions respectively) p/q. Since the rationals are dense in irrationals, by good approximations we really mean those which are close enough to α in comparison to their complexity traditionally measured by the size $|q|$ of their denominators. This can be taken to mean, either those which are closer than any with lower complexity, or those with error much smaller in comparison to the complexity. Both these notions lead directly to the approximations given by truncations of the continued fraction expansions of α, as we will see. Function field continued fraction study began with Emil Artin's thesis.

After exploring, in the first two sections, some basic results on diophantine approximation and continued fractions in classical case and their analogs in function field case, we concentrate on failure of Roth's theorem analog in function field case and study the range of distribution of diophantine approximation exponent in the (finite characteristic) function field case (it is two in number field case by Roth's theorem) in the next two sections. Then we study relation between deformation theory and diophantine approximation theory and present some results and speculations.

The diophantine approximation results and in particular, the improvements on the Liouville bound have interesting implications for the study of related diophantine equations, as the work of Thue, Siegel, Baker, Voloch, Vojta, Faltings etc. has shown. In the appendix to this chapter we quickly explore these connections.

In this chapter, by a function field we will mean function field over any field F, but in the second section F will be of finite characteristic p. As is traditional, we also use p for the numerator and q for the denominator of a fraction. This use should not be confused with use of p and q as prime and prime power.

9.1   Approximation exponents

We start with a basic result, due to Lagrange (for classical material, see [Bak90; Sch80; Sch91; Cas72]), who proved it using continued fractions (namely (1) of Section 2 below). The simple proof here is due to Dirichlet (to Mahler [Mah49] in the function field case).

Theorem 9.1.1   *(Lagrange-Dirichlet) Given any irrational α there are infinitely many approximations p/q satisfying $|\alpha - p/q| < 1/|q|^2$.*

Proof.   Denote the fractional part of a real number θ (the absolute value of the part of a Laurent series θ in $1/t$ obtained by removing the polynomial in t part, respectively) θ by $\{\theta\}$, so that $0 \le \{\theta\} < 1$. Let s be a positive integer (a polynomial, respectively) and let i run through $0 \le i < s$ (polynomials of lower degree than s, respectively). Then $\{i\alpha\}, 1$ are $|s| + 1$ real numbers in $|s| + 1$ boxes $[0, 1/(|s| + 1)), \cdots, [|s|/(|s| + 1), 1]$ and hence at least one of the box contains two of them. So, for some non-zero $|q| \le |s|$ and some p, we have $|q\alpha - p| = \{q\alpha\} \le 1/(|s| + 1)$. When α is irrational, letting $|s|$ larger and larger, we get the infinitely many approximations we want. $\square$

For almost all numbers (in the sense of measure one), the exponent 2 in the Dirichlet theorem can not be improved to exponent $\mu > 2$.

In fact, we have in the real number case,

Theorem 9.1.2   *(Khintchine) If $\psi(q)$ is a positive decreasing function of q, then if $\sum \psi(q)$ converges, $|\alpha - p/q| < \psi(q)/q$ has finitely many solutions for almost all α and if the sum diverges, the inequality has infinitely many solutions for almost all α.*

Proof.   The first part is easy and follows from the observation that

$$S := \{x \in [0, 1] : |x - p/q| < \psi(q)/q \text{ has infinitely many solutions}\}$$

is covered by union of intervals of length $2\psi(q)/q$ centered at i/q ($i = 0$ to q for $q > q_0$) of total length $\sum_{q>q_0} \psi(q)$ and hence is of measure zero.

For the second part (which is in the spirit of Borel's zero-one law in probability theory) and much more, see [Cas72]. □

We leave the statement and the proof of the function field case, which is parallel, to the reader, with the hint that one has to replace $[0, 1]$ by $\mathcal{O}_\infty$ with Haar measure and q by $|q|$.

On the other hand, it is easy to construct (uncountably many) numbers (called Liouville numbers) for which there are infinitely many solutions for any exponent μ: For example, the truncation of $\alpha = \sum 1/10^{n!}$ at $n = N$ approximates it with exponent greater than N, if N is large enough. But the Liouville's theorem below shows that such numbers can not be algebraic. In fact, these were the first explicit examples of non-algebraic numbers, which were known earlier to exist only by Cantor's counting arguments.

From now on we concentrate on algebraic irrational α and we denote its algebraic degree as $d(\alpha)$.

Theorem 9.1.3 (*Liouville*) *For α algebraic of algebraic degree $d = d(\alpha)$, $|\alpha - p/q| > c(\alpha)/|q|^d$ (for $\alpha \neq p/q$ which is automatic, if $d > 1$), with positive constant $c(\alpha)$ which is effective.*

Proof. The proof (real case) follows by the mean value theorem: Let f be the minimal polynomial for α, then for p/q close to but not equal to α, we have

$$|\alpha - p/q| = \frac{|f(\alpha) - f(p/q)|}{|f'(\beta)|} > c(\alpha)|f(p/q)| \geq c(\alpha)/|q|^d$$

by bounding (effectively) the derivative in the neighborhood of α

The function field case (over any base) is basically the same [Mah49]: Instead of the derivative f' of f, we deal directly with the polynomial $f(x)/(x - \alpha)$. □

Any improvement on Liouville bound has interesting diophantine applications as we will see in the last section.

From here on, the story diverges. Let us concentrate on exponents.

Definition 9.1.4 Let F be a field. For α an algebraic irrational real number (an element of $F((t^{-1}))$ algebraic irrational over $F(t)$ respectively), define its diophantine approximation exponent $E(\alpha)$ by

$$E(\alpha) := \limsup(-\frac{\log|\alpha - P/Q|}{\log|Q|})$$

where P and Q run over integers (polynomials in $F[t]$ respectively), the absolute value is the usual one in each case and the limit is taken as $|Q|$ grows.

Hence theorems of Dirichlet and Liouville show that $2 \leq E(\alpha) \leq d(\alpha)$.

For the real number case (characteristic zero function field case resp.), the well-known theorem of Roth (its analog by Uchiyama resp.) says that $E(\alpha) = 2$, for any algebraic α.

Mahler [Mah49] observed that, if q is a power of the characteristic $p > 0$ of F, then $E(\alpha) = d(\alpha) = q$ for $\alpha = \sum t^{-q^i}$. This is seen by a straight estimate of approximations by truncations of this series and the telescoping manipulation: $\alpha^q - \alpha = -t^{-1}$. Hence, the Liouville bound is the best possible in finite characteristic.

Contradicting the assertions in the literature that such a phenomenon is 'wild' and only occurs when the degree is divisible by p (this was raised as a question by Mahler), Osgood [Osg73] and Baum and Sweet [BS76; BS77] gave many examples in various degrees.

This story is continued after the next section, which is of independent interest. Only the equation (2) is needed from the next section, for that.

9.2 Good approximations: Continued fractions

What are all these promised good approximations? This is where the continued fractions come in.

Representation of a ratio a/b by a simple continued fraction is much more intrinsic notion than say by a decimal. In the latter case, we compare the magnitude with a unit, then the left-over with the tenth of a unit and so on. In forming a continued fraction, we would find how many b's fit in a, how many left-overs would fit in b and so on, so it is quite a natural way to compare two quantities a and b. We are led to the expansion $a_0 + 1/(a_1 + 1/(a_2 + \cdots))$ for which we use the short-form $[a_0, a_1, \cdots]$. In the function field case, we use 'polynomial part' in place of 'integral part'.

Let us fix an irrational α and let us quickly recall some standard basic notation and facts:

We write $\alpha_n = [a_n, a_{n+1}, \cdots]$, so that $\alpha = \alpha_0$.

With $p_0 = a_0$, $q_0 = 1$, $p_1 = a_1 a_0 + 1$, $q_1 = a_1$, $p_n = a_n p_{n-1} + p_{n-2}$ and $q_n = a_n q_{n-1} + q_{n-2}$ for $n > 1$, we have $p_n q_{n-1} - p_{n-1} q_n = (-1)^{n-1}$, so that p_n and q_n are relatively prime and $[a_0, \cdots, a_n] = p_n/q_n$. These truncation approximations $c_n := p_n/q_n$ are called the **convergents** to α.

The a_i's (α_i's respectively) are called **partial (complete respectively) quotients.**

In the real case, for $i > 0$, we have $a_i > 0$ and so q_i is positive and increases with i. In the function field case, for $i > 0$, a_i is any non-constant polynomial and so the degree of q_i increases with i, but q_i need not be monic. In fact, the formal identity $f[a_0, a_1, \cdots] = [fa_0, f^{-1}a_1, fa_2, f^{-1}a_3, \cdots]$ is very useful in the function field situation, for a non-zero constant f, since then f^{-1} is also integral and it gives valid simple continued fraction, in contrast to say $f = -1$ in the real case.

Since $\alpha = (\alpha_{n+1}p_n + p_{n-1})/(\alpha_{n+1}q_n + q_{n-1})$, we have the basic approximation formula

$$\alpha - p_n/q_n = (-1)^n/((\alpha_{n+1} + q_{n-1}/q_n)q_n^2) = (-1)^n/(q_n q_{n+1}) \qquad (1)$$

and $q_{n-1}/q_n = [0, a_n, a_{n-1}, \cdots, a_1]$. Hence we have $1/((a_{n+1} + 2)q_n^2) < |\alpha - p_n/q_n| < 1/(a_{n+1}q_n^2)$ in the real case, whereas

$$|\alpha - p_n/q_n| = 1/(|a_{n+1}||q_n|^2) \qquad (2)$$

in the function field case, because of the non-archimedean absolute value.

We immediately see the connection between the size of partial quotients a_n's and approximation exponent of α.

Remarks 9.2.1 (1) Classically, the result of Hurwitz says that for all irrational α there are infinitely many solutions to $|\alpha - p/q| < 1/(\sqrt{5}q^2)$, $\sqrt{5}$ being the best for the α's with all partial quotients 1 eventually. Otherwise it can be improved to $\sqrt{8}$, and so on, leading to Markoff chain and Markoff-Lagrange spectrum [Cas72]. The facts that the absolute value now jumps discretely and $|a'_{n+1}| = |a_{n+1}|$ reduces analogs of these interesting theorems in the function field case to simple observations: Analog of Hurwitz $\sqrt{5}q^2$ is clearly $|t|q^2$, but already uncountably many irrationals achieve it, as you can see by allowing all possible sequences of t and $t+1$ as partial quotients. Also, as $|a_{n+1}|$ can take all possible absolute values greater than 1, for all such absolute values c, limiting error $1/c|q|^2$ does occur, even with suitable quadratic α.

(2) The question of what errors $1/c|q|^2$ can occur for a given individual α is fully answered in Theorem 9.2.3 below.

(3) Another contrast is that by having $a_{n+1} = 1$ and a_{n+2} very large, the convergent approximation error can be very close to $1/|q|^2$, classically, whereas in our case, it would be always not worse than $1/|t||q|^2$.

(4) Lehmer's conjecture, an important unsolved problem in number field case, says that there is $\alpha > 1$ such that the Mahler measure $M(P) := |a_n| \prod \max(|\theta_i|, 1)$ of a polynomial $P(x) = \sum_{i=0}^{n} a_i x^i \in \mathbb{Z}[x]$ with roots θ_i (and $a_n \neq 0$) satisfies $M(P) > 1$, then $M(P) > \alpha$. This is also trivially true in our case of discrete absolute value in K_∞, when we replace $\mathbb{Z}$ by A. If $M(P) < q$, then $|a_n| = 1$ and the coefficients being elementary symmetric functions of roots, the strong non-archimedean triangle inequality gives $|a_i| \leq M(P) < q$ implying $a_i \in \mathbb{F}_q$, so that $\theta_i \in \overline{\mathbb{F}_q}$ are roots of unity, parallel to the classical result of Kummer that $M(P) = 1$ implies that the roots of P are roots of unity. But note that classically the weaker triangle inequality gives $|a_i| \leq \binom{n}{i} M(P)$, and we get a lower bound for each degree only. Dubrowski produced much better bound.

For more refined purposes one studies the **intermediate convergents**

$$c_{n,r} := (r p_{n-1} + p_{n-2})/(r q_{n-1} + q_{n-2}),$$

where $0 < r < a_n$ in the real case and $0 \leq \deg r \leq \deg a_n$, $0 \neq r \neq a_n$, in the function field case. Note that the denominator sizes of these are between $|q_{n-1}|$ and $|q_n|$, as there is no cancellation.

Remark 9.2.2   In the function field case, there are many subtleties. Firstly there is no order, so that we can not argue by saying take this intermediate convergent between these two convergents etc. We have eg., $|\alpha| = |p_n/q_n|$. There are many q's (even monic) of the same size, unlike natural order on positive q's in the real case.

We have,

$$\alpha - c_{n+1,r} = \pm \frac{\alpha_{n+1} - r}{(\alpha_{n+1} q_n + q_{n-1})(r q_n + q_{n-1})}.$$

With these preliminaries, we are now ready to compare the situation with the classical case. To avoid special cases, in what follows, we assume that $n > 1$, $|q| > |a_1|$ and that p, q are relatively prime.

Traditionally [Cas72], the approximation p/q to α is called **best** (**good** respectively), if $\{q\alpha\} = |q\alpha - p|$ and $\{q\alpha\} < \{q'\alpha\}$ ($|\alpha - p/q| < |\alpha - p'/q'|$, unless $p/q = p'/q'$ respectively) for $0 < |q'| \leq |q|$. In both cases, the best approximations are exactly those given by the convergents.

After that there are subtle differences in the situation. In the function field case, it makes quite a difference where the inequalities used are strict or not. Let us say that p/q is **good in a weak sense** if $|\alpha - p/q| \leq$

$|\alpha - p'/q'|$ for $|q'| \leq |q|$, unless $p/q = p'/q'$. Let us say that p/q is **fair** if $|\alpha - p/q| < |\alpha - p'/q'|$ for all $|q'| < |q|$.

A priori, 'best' implies 'good', 'good' implies 'good in weak sense' and 'fair'.

In the real case [Cas72], clearly, 'good' equals 'fair' equals 'good in weak sense'. But 'best' is stronger. In fact, p/q is best if and only if it is a convergent. It is good if and only if it is a convergent or an intermediate convergent $c_{n,r}$ either with $r > a_n/2$ or with $r = a_n/2$ and $[a_n, a_{n-1}, \cdots, a_1] > \alpha_n$. Also, $|\alpha - p/q| < 1/2q^2$ implies that p/q is a convergent and the coefficient 2 is best possible. In fact, $|\alpha - p/q| < 1/q^2$ implies that p/q is a convergent or an intermediate convergent $c_{n,r}$ with $r = 1$ or $r = a_n - 1$ and in fact it is a 'good approximation from its side'.

Theorem 9.2.3 *In the function field case, we have*

(1) The properties 'best', 'good', 'good in weak sense', 'convergent', 'approximation p/q with error less than $1/|q|^2$' (i.e., $\leq 1/|tq^2|$) are all equivalent.

(2) The approximation p/q is 'fair, but not good' if and only if it is an intermediate convergent $c_{n+1,r}$'s with $|a_{n+1} - r| < |a_{n+1}|$ (in particular, $|r| = |a_{n+1}|$).

(3) We have $|\alpha - p/q| = 1/|q|^2$ if and only if for some n, p/q is the intermediate convergent $c_{n,r}$ with $r \in F^*$.

Proof. We will divide the proof in several small steps.

(a) If $|\alpha - p/q| < 1/|q|^2$, then p/q is good. In particular, all convergents are good: If $|\alpha - p'/q'| \leq |\alpha - p/q|$ and $p/q \neq p'/q'$, then by triangle inequality, we have $1/|qq'| \leq |p/q - p'/q'| \leq \max(|\alpha - p/q|, |\alpha - p'/q'|) = |\alpha - p/q| < 1/|q|^2$. This implies $|q'| > |q|$ as needed.

(b) If $|\alpha - p/q| \leq |\alpha - p_n/q_n|$, $p/q \neq p_n/q_n$ and $|q| \leq |q_{n+1}|$, then $|q| = |q_{n+1}|$: By argument as above, we have $1/|qq_n| \leq |\alpha - p_n/q_n| = 1/|q_{n+1}q_n|$.

(c) If p/q is good or good in weak sense, then it is convergent: For some n, we have $|q_n| \leq |q| < |q_{n+1}|$, so that we will have $|\alpha - p/q| \leq |\alpha - p_n/q_n|$. Hence by (b), $p/q = p_n/q_n$. (In particular, it is good by (a).)

(d) If p/q is good, then $|\alpha - p/q| < 1/|q|^2$: This follows from (c), but we give a more direct proof: Write $pa - qb = 1$, where we can assume, without loss of generality, that $\deg a < \deg q$. Put $q' = q + a$ and $p' = p + b$. Then $|q| = |q'|$

and $pq' - qp' = 1$, so that $1/|q|^2 = |p/q - p'/q'| = \max(|\alpha - p/q|, |\alpha - p'/q'|) = |\alpha - p'/q'| > |\alpha - p/q|$ as needed.

(e) Convergents are best: In fact, if $|q\alpha - p| \leq |q_n\alpha - p_n|$ and $|q| < |q_{n+1}|$, then $|\alpha - p/q| \leq 1/|qq_{n+1}| < 1/|q|^2$, so that by (a) and (c), it is a convergent and hence $p/q = p_n/q_n$, by the inequalities assumed.

(1) follows from (a)-(e).

(f) If p/q is fair, q is a denominator of a convergent: Without loss of generality, $|q_n| < |q| \leq |q_{n+1}|$, then by (b), we have $|q| = |q_{n+1}|$.

(g) The intermediate convergent $c_{n+1,r}$ is fair, if and only if writing $r = a_{n+1} + s$, we have $|s| < |a_{n+1}|$: By (f), $|r| = |a_{n+1}|$. By the displayed formula for the difference, the error is $|s|/|q_{n+1}|^2$, which is better than error in approximating by p_n/q_n if and only if $|s| < |a_{n+1}|$. This, together with (b) and the definition of 'fair' proves the claim. In fact, by reducing the degree of s one by one, we see that errors go through all the possible values between $1/|a_{n+1}q_n^2|$ (namely, the error for p_n/q_n) to $1/|a_{n+1}^2 q_n^2| = 1/|q_{n+1}^2|$. Error smaller than that occurs, by (1), only at the convergent p_{n+1}/q_{n+1}, which gives error $1/|a_{n+2}^2 q_{n+1}^2|$.

(h) If p/q is fair, but not good, then it is an intermediate convergent: Without loss of generality, assume that $|q_n| < |q| \leq |q_{n+1}|$. Then by (f), $|q| = |q_{n+1}|$. Since $p_n q_{n-1} - q_n p_{n-1} = \pm 1$, we can write $q = aq_n + bq_{n-1}$ and $p = ap_n + bp_{n-1}$, with integral (i.e., polynomial) a and b. By proof of (g), there is r such that $|r| = |a_{n+1}|$ and $\alpha - c_{n+1,r} = 1/|q_{n+1}^2|$. We have $(rp_n + p_{n-1})(aq_n + bq_{n-1}) - (rq_n + q_{n-1})(ap_n + bp_{n-1}) = \pm(rb - a)$. Writing $a = rb + m$, we then have $|m/q_{n+1}^2| = |p/q - c_{n+1,r}| \leq \max(|\alpha - p/q|, |\alpha - c_{n+1,r}|) < 1/|q_n q_{n+1}|$. Hence $|m| < |a_{n+1}|$. So $|a_{n+1}q_n| = |q| = |(rb + m)q_n + bq_{n-1}| = |rbq_n| = |b||a_{n+1}q_n|$, which implies that $b \in \mathbb{F}_q^*$ and so $p/q = c_{n+1,a/b}$.

(2) follows from (g) and (h).

(i) Let $|q_n| \leq |q| < |q_{n+1}|$, then $|\alpha - p/q| = 1/|q|^2$ if and only if $p/q = c_{n+1,r}$ with $r \in F^*$: Write $p = ap_n + bp_{n-1}$ and $q = aq_n + bq_{n-1}$ as above. We have $1/|qq_n| \leq |p/q - p_n/q_n| \leq \max(|\alpha - p/q|, |\alpha - p_n/q_n|) = |\alpha - p/q| = 1/|q|^2$, hence all the inequalities are equalities. In particular, $1 = |pq_n - qp_n| = |b|$. Hence p/q is intermediate convergent. But then $|q| = |q_n|$ implies $a \in F^*$ as needed. Converse follows by substitution in the displayed formula for the difference.

This implies (3) and finishes the proof of the theorem. $\qquad\qquad$ □

Remarks 9.2.4 (1) The part that the notions of 'best approximations', 'convergents' and 'those with error $< 1/|q|^2$' agree was proved in [dM70] and (3) can be obtained from easy modification of the proof of [BS76] for $q = 2$.

(2) As is the case classically, the tails of the continued fraction expansion are the same if and only if the numbers are related by integral Mobius transformation of determinant ± 1. Eventually periodic continued fraction expansion for α immediately implies that α is its own Mobius transformation with integral coefficients, so quadratic. The converse is true in real case and for function fields over finite fields. For other function fields, we have analog of Abel's theorem. So the analogies are stronger for function fields over finite fields. See [Sch00; dM70] for a discussion and development of the basics.

Euler proved that

$$e = [2, 1, 2, 1, 1, 4, 1, \cdots, 1, 2n, 1, \cdots],$$

which you can think of as consisting eventually of three arithmetic progressions, two with common difference zero and one with common difference two. Hurwitz [Per50] proved that integral linear fractional transformation of a non-zero determinant of a continued fraction expansion consisting eventually of some arithmetic progressions (eg., of $e^{2/n}$) has continued fraction with the same property.

On the other hand, nobody knows much about the pattern for the continued fraction for π, γ or even e^3.

Interestingly, we again have nice but completely different patterns [Tha92a; Tha96b; Tha97] for analogs of e and these Hurwitz numbers, for the Carlitz module. For example, for $e = e(1)$ for the exponential of Carlitz module for $A = \mathbb{F}_2[t]$, we have, with $[k] = t^{2^k} - t$, as before,

$$e = [1, \underbrace{[1]}, [2], \underbrace{[1]}, [3], [1], [2], [1], [4], [1], [2], [1], [3], [1], [2], [1], [5], \cdots].$$

For any integral linear fractional transformation of $e(1/a)$, with $a \in A = \mathbb{F}_q[t]$, we have fully proved patterns which follow interesting inductive scheme of mixture of block repetition and reversals.

It seems that the right analog of 2 in $e^{2/n}$ is θ, with $\theta \in \mathbb{F}_q^*$ for $q > 2$, and $\theta = 1, t, t{+}1$ or $t^2{+}t$ for $q = 2$. The analogy here is that the n-th roots of unity are rational exactly for $n = 1$ or 2, just as the Carlitz a-torsion is rational exactly for $a = \theta$. Why there are patterns exactly for these analogs, even when the proofs and patterns are so different, and why this analogy with the rational torsion comes up is a mystery.

9.3   Range of exponents : Frobenius

In contrast to the quadratic case, the continued fraction expansion pattern is not known even for a single algebraic real number of degree more than two. It is not even known whether the sequence of partial quotients is bounded or not for such a number. (Because of the numerical evidence and a belief that algebraic numbers are like most numbers in this respect, it is often conjectured that the sequence is unbounded.) It is hard to get such expansions for algebraic numbers, because the effect of basic algebraic operations (except for adding an integer or more generally an integral Mobius transformation of unit determinant), such as addition or multiplication or even multiple or power, is not at all transparent on the continued fraction expansions.

In finite characteristic p, the algebraic operation of taking p-th power has a very transparent effect: If $\alpha = [a_0, a_1, \cdots]$, then $\alpha^p = [a_0^p, a_1^p, \cdots]$.

Let F be a field of characteristic p and let q be a power of p. If $A_i(t) \in F[t]$ are any non-constant polynomials, the remark above shows that

$$\alpha := [A_1, \cdots, A_k, A_1^q, \cdots, A_k^q, A_1^{q^2}, \cdots] \tag{3}$$

is algebraic over $F(t)$ because it satisfies the algebraic equation

$$\alpha = [A_1, \cdots, A_k, \alpha^q].$$

So one gets a variety of explicit continued fractions with explicit equations. By integral Mobius transformations with determinant ± 1 (which keeps the tail of the expansion the same), one can get many more such explicit examples. One can let the determinant be any $f \in F^*$ by using $f[a, b, c, d, \cdots] = [af, b/f, cf, d/f, \cdots]$.

Since continued fractions give best diophantine approximations, let us analyze what we get for α's as above.

Theorem 9.3.1 *(1) Let the degree of A_i be d_i, and put $r_i := d_i/((d_1 + \cdots + d_{i-1})q + d_i + \cdots + d_k)$. Then for α as in equation (3), we have*

$$E(\alpha) = 2 + (q-1)MAX(r_1, \cdots, r_k).$$

(2) Given any rational μ between $q^{1/k} + 1$ (which tends to 2 as k tends to infinity) and $q + 1$, we can find a family of α's as in equation (3), with $E(\alpha) = \mu$ and $d(\alpha) \leq q + 1$.

Proof.   We have seen that the convergents p_n/q_n of the continued fractions give best approximations, so that, in the definition of the exponent, one needs to use P/Q coming from the truncations of the continued fraction only. By equation (2), we have $E(\alpha) = 2 + \limsup(\deg a_{n+1}/\deg q_n)$.

By the usual recursion formula for q_n, we see that the degree of q_n is the sum of the degrees of $a_1, \cdots, a_n$. By (3), $\deg a_n = d_{j+1}q^{(n-j)/k}$, where j is the smallest non-negative residue of n modulo k. Hence a straight calculation (but see Remark 9.3.5 (2) below) proves (1). (In more detail, the summation of the resulting geometric series shows that $\deg q_{n-1} = (d_1 + \cdots + d_k)(q^{(n-j)/k} - 1)/(q-1) + (d_1 + \cdots + d_j)q^{(n-j)/k} - d_1$. Hence $\deg a_n/\deg q_{n-1}$ tends to r_{j+1}, as n (congruent to j modulo k) tends to infinity.) This proves (1). Then (2) follows by straight-forward manipulations (see [Tha99b] for details) showing that the formula just obtained for the exponent takes all rational values in the given range. $\square$

Remarks 9.3.2   (1) It seems thus reasonable to guess that the set of exponents of algebraic α's of degree d, which is a priori just a countable subset of interval $2 \leq x \leq d$, is just the set of rational numbers in this Dirichlet-Liouville range. That the set can not contain any irrational number is known for degree 3, as every α of degree 3 has the property that α^q is Mobius integral transformation of α. Numbers with this property are called numbers of class I. For such numbers, the rationality of the exponent is a result of de Mathan [dM]. So for $q = 2$ and $d = 3$, the guess is correct. In general, to show that all rationals occur, we need to control the exact degrees. Since we can always take k large, so that there is huge choice in choosing A_i's of given degree, it might be reasonable to expect that the equation above is irreducible (for example) for some choice. This has been done only in a few cases. To settle whether this happens for each degree (not necessarily of the form $q + 1$) may require more complicated combinatorics.

(2) The theorem and easy operation of scalar multiplication by constants, takes care of explicit continued fractions as well as exponents for all α's whose some q-th power is integral linear fractional (i.e., Mobius) transformation of α of constant determinant (called α's of class Ia). Since the exponent is invariant with respect to integral Mobius transformations of non-zero determinants, we get exponents for such, but not the explicit continued fractions for these or for class I elements in general. The method of automata/transducers [MR86] generates such expansions, but it has not led to direct description of the patterns yet.

In retrospect, in our example $\alpha = [A_1, \cdots, A_k, A_1^q, \cdots]$ the Frobenius 'shifts' the continued fraction giving the equation $\alpha^q =$ Mobius transformation of α, whereas in Mahler's example $\alpha = \sum t^{-q^i}$ it shifts the sum giving the equation $\alpha^q - \alpha = -t^{-1}$ and in Osgood's example [Osg73; Osg75] (which is essentially) $\alpha = \prod t^{-q^i}$ it shifts the product giving the equation $\alpha^q / \alpha = t$.

Voloch [Vol88] looked at the variants of Osgood and Mahler examples by replacing $1/t$ by $f(1/t)$ for a polynomial f:

(1) Let f be a polynomial over F with highest degree m and lowest degree $n > 0$, with $nq/m > q^{1/2} + 1$. If $\alpha = \sum f(1/t)^{q^i}$, so that $\alpha^q - \alpha = -f(1/t)$, then $E(\alpha) = nq/m$.

(2) Let f be a polynomial over F with highest degree m and lowest degree appearing in $f - 1$ be $n > 0$ and let d be a divisor of $q - 1$ with $nd/m > q^{1/2} + 1$. If $\alpha = \prod f(1/t)^{-q^i(q-1)/d}$, so that $\alpha^d = f(1/t)$, then $E(\alpha) = nd/m$.

The proof follows, as in Mahler case, by comparing the approximation given by the truncations, but now there is a subtlety: For continued fractions, we have shown how to read off the exponent. In the Mahler and Osgood cases, the truncations give the best possible (Liouville bound) approximations, so we get the exponent. But in Voloch's more general examples, the truncations give strong approximations, but as we can not guarantee that all good approximations that may influence the exponent are in these truncations, in contrast to the continued fraction case, Voloch proved and used the following useful Lemma:

Lemma 9.3.3 *(Voloch): If $a_n/b_n \to \alpha$, with relatively prime polynomials a_n and b_n, satisfying*

$$\limsup \deg(b_{n+1})/\deg(b_n) = b, \quad \log|\alpha - a_n/b_n|/\log|b_n| \to a, \quad a > b^{1/2} + 1,$$

then $E(\alpha) = a$.

We refer to [Vol88] for the nice proof using just triangle inequalities for our non-archimedean absolute value.

The truncations of our continued fraction examples above give sequences satisfying the hypotheses of the lemma above, and show that the inequality in the hypothesis is the best possible: When you use truncations at i modulo k stages, the ratio of the degrees is asymptotically q, whereas error exponent is $s_i + 1$. Since the product of $s_i > 1$'s is q, we have $s_i \leq \sqrt{q}$ for the s_i giving next to the optimum, but we can have it as close to $\sqrt{q}$ as we wish in the examples.

Theorem 9.3.4 *(Voloch) For α as in 9.3.1, $E(R(\alpha)) = E(\alpha)/\deg(R)$, if $R(x) \in K(x)$ has non-zero derivative and $E(\alpha) > d(q^{1/2} + 1)$.*

Proof. This follows from the Lemma together with the mean value theorem applied to approximations $R(r_n)$ to $R(\alpha)$, where r_n are those approximations obtained by truncations of continued fractions of α which lead to exponent approximation, so that $b = q$.   □

Remarks 9.3.5 (1) Voloch [Vol95] stated this for more general $\alpha \in K_\infty$ satisfying $\alpha^q = (a\alpha + b)/(c\alpha + d)$ where $ad - bc \neq 0$ (our examples correspond to $ad - bc = \pm 1$), but the existence of approximations r_n used there by referring to Theorem 2 of [Vol88] is not clear (because of n in the theorem and inequalities rather than equalities needed for application). So the more general statement seems not fully justified, as Voloch confirmed to the author.

(2) In the straight-forward calculation mentioned in the proof of the Theorem 9.3.1, I had made a simple arithmetic mistake at first, leading to a wrong formula for the exponent, which reduced the problem to $k = 2$ and part (2) of the Theorem then gives the bound $q^{1/2} + 1$ seen above a few times, so I checked no further. Only after I saw the preprint of Schmidt [Sch00], who came to the same examples independently with a different route, I realized the mistake. But in retrospect, the mistake was beneficial, since to get the exponent near 2, I then tried another family, using block repetition and reversal patterns I had learned about for e (see previous section), which is transcendental. The same kind of patterns worked, this time for algebraic quantities! And as a bonus, some kind of degeneration of the family led to algebraic examples with bounded sequence of partial quotients. The same method produced many non-Riccati families (see the next section for definition and significance of this) and also many with bounded partial quotient families. Some isolated examples of bounded partial quotient fam-

ilies, but of Riccati or rather class I type, were known, starting with first break-through by Baum and Sweet. See the references in [Tha99b].

In retrospect, the difficulty in giving examples in earlier literature was that earlier explicit examples were found by large computer searches of continued fractions starting with equations for algebraic laurent series to find examples where the patterns can be guessed and proved by ad hoc methods. We start with nice continued fractions, and then writing equations is easy.

The new examples, as are the pattern for e mentioned in the previous section, are based on the following simple lemma, due to Mendes France, which has been rediscovered many times.

Lemma 9.3.6   *Let* $[a_0, a_1, \cdots, a_n] = p_n/q_n$, *with* p_n, q_n *normalized as usual (i.e., as in 9.2), then* $[a_0, \cdots, a_n, y, -a_n, \cdots, -a_1] = p_n/q_n + (-1)^n/yq_n^2$.

We will refer to this pattern as 'a signed block reversal pattern with the new term y'.

Now, if $\alpha := \sum f_i t^{-n_i} \in F((1/t))$ (where n_i is an increasing sequence of integers) is algebraic over $F(t)$ and satisfies $n_{i+1} > 2n_i$, for $i \geq i_0$ say, then the repeated applications of the lemma starting with the continued fraction of the rational function obtained by truncating at i_0-th power gives complete continued fraction of α consisting of signed block reversals, with the new y's being $t^{n_{i+1}-2n_i}$ (up to signs which are easy to calculate from the lemma). As before, the exponent calculation and its range determination is routine.

First we give the main examples [Tha03] : By taking linear combinations of Mahler's example above, we know that any

$$\alpha = \sum_{i=1}^{k} f_i \left(\sum_{j=0}^{\infty} t^{-m_i q^j + b_i} \right)$$

(where $m_i \geq 0$ and b_i are rational numbers so that the exponents are integers) is algebraic. (With integral coefficients, we can write the exponent as $a_i q^j + b_i (q^j - 1)/(q-1) + c_i$.) And it is easy to write down conditions on the coefficients to satisfy $n_{i+1} > 2n_i$ for large i. For example, $m_{i+1} > 2m_i$ for $1 \leq i < k$ and $qm_1 > 2m_k$ is clearly sufficient, but not necessary. With this condition, as in Theorem 2 of [Tha99b], we see that $E(\alpha) = \mathrm{MAX}(m_2/m_1, \cdots, m_k/m_{k-1}, qm_1/m_k)$ and that it takes any rational value between $q/2^{k-1}$ and $q^{1/k}$, if further that $q > 2^k$.

The algebraic equation for each term (corresponding to a fixed i) is immediate, since it is just a multiple of Mahler's example. So the polynomial equation satisfied by α follows, for example, by the usual elimination method using resultants. For example, when $k = 2$, we use

$$\text{Resultant}(x^q + ax + b, x^q + cx + d, x) = (d - b)^q + (ad - bc)\sum_{i+j=q-1} a^i c^j.$$

The flexibility in the choice of number and coefficients m_i and b_i can be used to produce many families of α's not satisfying rational Riccati equation. We leave the manipulation details to the interested reader and just give some examples below.

Most (all, in odd characteristic, as we will see below) families we thus construct have unbounded sequence of partial quotients, and, in fact, have exponent greater than two. But we can also produce many explicit continued fraction families with bounded sequence of partial quotients in characteristic two: For example, any α as above with $q = 2^k$, $m_i = 2^{i-1}$ and $b_i > b_{i+1}/2$, for i modulo k will do.

Let us show that most of these do not satisfy the rational Riccati equation (and so are of degree more than 3): Take $f_i = 1$ for simplicity, and write α_i for the i-th term of the sum expression for the α above. Then $\alpha = \sum_{i=1}^{k} \alpha_i$, and $\alpha_i = \alpha_1^{2^{i-1}} p_i$ with $p_i := t^{b_i - 2^{i-1}b_1}$. Again for simplicity, take b_1 odd and other b_i's even, so that $\alpha' = \alpha_1/t + t^{b_1-2}$. If α were to satisfy rational Riccati equation $\alpha' = a\alpha^2 + b\alpha + c$, then we would have

$$\alpha_1/t + t^{b_1-2} = a(\alpha_1^2 + \alpha_1^4 p_1^2 + \cdots + \alpha_1^{2^k} p_k^2) + b(\alpha_1 + \alpha_1^2 p_1 + \cdots + \alpha_1^{2^{k-1}} p_k) + c.$$

But by the degree comparison, this equation has to be the Mahler type irreducible equation $\alpha_1^{2^k} = t^{b_1(q-1)}\alpha_1 + t^{qb_1-1}$, which is clearly impossible for most choices of p_i for $k > 2$. The same construction in characteristic $p > 2$, with say exponent p, gives examples (now $k > 1$ is fine) which are non-Riccati (in fact not of the form α' equals polynomial of degree $\leq p$).

Remarks 9.3.7 (1) Since we are allowed to modify finitely many terms of the series for α, we can construct examples with almost arbitrary partial quotients occurring infinitely often.

(2) We can relax $n_{i+1} > n_i$ to $n_{i+1} \geq n_i$ by changing from resulting degenerate expansion to a proper one, as explained in [Tha99b].

(3) If we just assume the inequality above for infinitely many i rather than for all large i, we do not get full continued fraction, but the resulting continued fraction has infinitely many signed block reversal places and we

thus get a lower bound on its exponent.

(4) There is one more flexibility in the method, which allows for additional families, not satisfying our conditions: Usually we can substitute everywhere some polynomial $P(t)$ for each occurrence of t. But here, satisfying certain mild conditions, you can make different substitutions $P_i(t)$ for t for different i in the formula for α. Rather than giving general conditions, we will just write down a very simple illustrative example, for $q > 8$:

$$\sum (t(t+1))^{-q^j} + (t^2(t+1))^{-2*q^j}.$$

By the lemma, we again get the complete continued fraction, with block reversals at each stage with the new partial quotients y's being (up to signs) powers of t and powers of t times those of $t+1$ mixed alternately.

(5) Choosing suitable m_i and b_i, we can clearly construct explicit non-quadratic elements whose sums or various rational multiples are also explicit having signed block reversal patterns.

Then arises the question of the classification of all algebraic α's satisfying the conditions above. For F a finite field, this has been done by computer scientists!:

(i) By Theorem 11.1.3, the expansion of α is automatic.

(ii) By Theorem 11.2.2, our condition of exponential growth implies the sparsest possible automatic sequences (excluding the finite ones which correspond to just rational functions).

(iii) By the main result of [SYZS92] (we give this as a convenient reference, but for our special case, there are earlier references given in this paper and Shallit tells me that he later found even earlier references in the computer science literature) we see that sequence with such a density is a finite union of regular expressions of the form xy^*z (i.e., with y repeated a few times) for strings x, y, z in the base q alphabet. Translated in our language of numbers, it means that such a sequence is union of sequences $n_i := a_j p^{id} + b_j(p^{id}-1)/(p^d-1) + c_j$ over finitely many j. (Several d's can be combined by least common multiple, by elementary manipulation.)

Putting these together, we see that our examples take care of all the examples satisfying the growth conditions.

Next, we show that if α of the type above has bounded sequence of partial quotients, then the characteristic p is two:

Let n_i be a p-automatic sequence of positive integers. By [Eil74, Cha. V, cor. 4.2] (which is our 11.2.2), there are rational $a > 0$ and b and a power q of p such that $aq^m + b = n_{i_m}$ for all m and a subsequence i_m. Now assume that $c > n_{i+1} - 2n_i > 0$ for all i. If we fix $i = i_{m_0}$ and let l_m run through values so that $i + l_m = i_m$, then $(2^{l_m} - 1)n_i < a(q^m - q^{m_0}) < (2^{l_m} - 1)(n_i + c)$. Hence for all large enough m, we have

$$\log_2(n_i/a) < m \log_2(q) - l_m < \log_2((n_i + c)/a).$$

Now for large enough i, the two extremes of this inequality are sufficiently close, whereas if p were odd, then $\log_2(q)$ would be an irrational and the fractional part of $m \log_2(q)$ would be dense in the interval $(0, 1)$, leading to a contradiction.

Another immediate implication of this classification is that any of our examples with exponent two has bounded sequence of partial quotients.

We are not only interested in the range of the exponents, but in their distribution. This is partially addressed in the next two sections. We just note here that at least in our families of 9.3.1, 'most' numbers (in the sense of asymptotics of heights) seem to have exponent near two. A precise distribution is being worked out.

9.4 Range of exponents: Differentiation

In addition to Frobenius in finite characteristic, another well-known tool that we have for function field of any characteristic is differentiation. Maillet, Kolchin and Osgood [Osg73; Osg75] used it to get interesting results for diophantine approximation.

Kolchin's idea was to use Liouville argument replacing the minimal polynomial of α with 'small' differential polynomial that kills α. Frequently this gives a smaller exponent and we get good effective bounds by more refined work of Osgood. Hence, even in characteristic zero, where the optimal exponent is known, you get improvement because of the effective bounds. We will only concentrate on the exponents.

We denote the m-th derivative of y with respect to t by $y^{(m)}$ and also write y' for $y^{(1)}$ following the usual practice. Note that differentiating the minimal polynomial $P(x)$ for α we get the equation $\alpha' P_x(\alpha) + P'(\alpha) = 0$. Simplifying, we get $\alpha' = \sum_{j=0}^{w} a_j(t)\alpha^j$, with $w < \deg \alpha$.

For a vector $\bar{e} = (e_0, \cdots, e_k)$ of non-negative integers, let us write $y^{\bar{e}}$ as a short-form for the differential monomial $y^{e_0}(y')^{e_1} \cdots (y^{(k)})^{e_k}$. Consider a differential polynomial $P(y) = \sum p_{\bar{e}} y^{\bar{e}}$. Note that j-th derivative of a/b has power b^{j+1} in the denominator. So define the denomination $\bar{d}(P)$ to be the maximum of $\sum_{j=0}^{k}(j+1)e_j$ corresponding to $\bar{e}$ such that $p_{\bar{e}} \neq 0$. So, if $P(a/b) \neq 0$, then $|P(a/b)| \geq 1/|b|^{\bar{d}(P)}$, hence $\bar{d}(P)$ replaces the degree in the Liouville argument. Define $\bar{d}(\alpha)$ to be the smallest possible $\bar{d}(P)$ for P satisfying $P(\alpha) = 0$. Here α can be differentially algebraic and if, in fact, it is also algebraic, then clearly $\bar{d}(\alpha) \leq \deg(\alpha)$.

Kolchin's analog of Liouville's theorem says that

Theorem 9.4.1   *Given irrational α which is differentially algebraic over a characteristic zero function field, there is a constant $c > 0$ such that $|\alpha - a/b| > c/|b|^{\bar{d}(\alpha)}$.*

The proof is the same as Liouville argument, except the catch is that the differential minimal polynomial P has, in general, infinitely many zeros and some a/b can be among those. Kolchin shows that this is impossible by showing that other zeros can not come close to α by estimation of Wronskians using '$c_1|a| \leq |a'| \leq c_2|a|$, for some positive constants c_i', which is true for characteristic zero function field, but fails in characteristic p, for example, for non-zero p-th power a.

How small can we get the denomination for α's of algebraic degree n? Since any $n+1$ elements of $K(t, \alpha)$ are dependent over $K(t)$, a simple count shows that for large n, as there are more than $n+1$ differential monomials $y^{\bar{e}}$ satisfying $\sum(j+1)e_j \leq (\log n)^2$, we can achieve $\bar{d}(\alpha) \leq (\log n)^2$ for large n, in characteristic zero. In characteristic p, since $y^{(p)} = 0$, we can only choose $\bar{e}$ with $k < p$, so for a fixed p the denomination can not be improved to order smaller than $n^{1/p}$, for large n. (For p large compared to n, we can reach near the characteristic zero bound.) Plus small denomination is not useful, unless there is a corresponding non-vanishing.

We get smallest denomination 2 for irrational α satisfying the Riccati equation $y' = ay^2 + by + c$ with rational functions as coefficients. So for such elements (for example, any element of degree 3 or any irrational n-th root of a rational) we have (effective) Roth estimate (there is no ϵ even). Since we did not give the full proof of Kolchin's theorem, let us see how Osgood proves this special case:

We only need to show that other roots β of the Riccati equation can not come arbitrarily close to the root α. Fix two other roots γ, δ. We use the well-known fact, easy to verify, that cross-ratio of any 4 roots of the Riccati equation is a constant function, to deduce that the cross ratio $(\alpha - \gamma)(\delta - \gamma)^{-1}(1 + (\alpha - \delta)(\beta - \alpha)^{-1})$ is a constant function. But this implies that $|\beta - \alpha|$ can not come arbitrarily close to zero, as required.

In contrast, in characteristic p, α which is a rational Mobius transformation of its p^n-th power, satisfies Riccati equation, and we saw in the last section that in this case the Riccati examples can have any rational exponent within Dirichlet and Liouville bounds (at least for some degrees).

In fact, Osgood proved [Osg73; Osg75] very interesting theorem:

Theorem 9.4.2   *In finite characteristic, the exponent bound can be reduced from Liouville bound to Thue bound $E(\alpha) \leq \lfloor \deg(\alpha)/2 \rfloor + 1$ for all non-Riccati α's.*

The relevance of Riccati to this question is clearly brought out by the theorem of Osgood and Schmidt [Sch76]

Theorem 9.4.3   *If $y'B(y) + A(y) = 0$, where A and B are coprime polynomials with coefficients integral (i.e., polynomial in t), then all its rational solutions have height bounded in terms of those of A and B, as long as the equation is not Riccati (i.e., we do not have $\deg(B) = 0$ and $\deg(A) \leq 2$).*

So the close enough rational approximations will not be the roots and hence the Liouville-Thue type argument goes through applied to $y'B(y) + A(y)$.

More precisely, for any $0 \leq d < n = \deg(\alpha)$, the $n+1$ elements $\alpha', \alpha'\alpha, \cdots, \alpha'\alpha^d, 1, \alpha, \cdots, \alpha^{n-d-1}$ being linearly dependent over $K(t)$, we get A, B, with $\deg(A) \leq n - d - 1$ and $\deg(B) \leq d$ such that $P = y'B + A$ vanishes at α. Since P has denomination $\max(d+2, n-d-1)$, we get the Thue bound for optimum $d = \lfloor (n-2)/2 \rfloor$.

Solutions $1/t^k$ (respectively $1/(t^k-1)$) to the Riccati equation $ty' = -ky$ (respectively $ty' = -ky(y+1)$), where k is a natural number, so that $|k| = 1$ in function field absolute value, shows that analogous statement fails for Riccati equations, situation being even worse in characteristic p, since then $1/t^{k+np}$ are also solutions to the first equation, for any n.

Since we did not prove Schmidt theorem, let us prove, following Osgood, an easier result, that for non-Riccati α of degree n, the exponent bound n can be improved to $n - 1$, without using Schmidt theorem:

If P is the monic minimal polynomial for α, then α satisfies two differential polynomials of denomination at most $n - 1$: namely $y' - \sum_{j=0}^{m} a_j y^j$, with $m > 2$, obtained by differentiating, and $(P - y^n) + a_m^{-1} y^{n-m}(y' - \sum_{j=0}^{m-1} a_j y^j)$. Since any approximation vanishing for both these differential polynomials has to satisfy P, which has only finitely many roots, we get the result.

Finally, a rational Riccati equation in characteristic $p > 2$ (but not for $p = 2$) has infinitely many rational solutions, if it has a non-quadratic solution. (See [LdM99] for a nice direct calculation proving the $p = 3$ case.

The general case can be easily extracted from classification in [vdP96, pa. 369-370] as van der Put explained to the author.)

9.5   Connection with deformation theory

Following on the initial result 9.4.2 of Osgood, we study [KTV00] the influence of differential equations on diophantine approximation properties. We start with an observation of Voloch [Vol95] that the Riccati condition on β is equivalent to the vanishing of Kodaira-Spencer (we write KS in short-form) class of projective line minus conjugates of β:

Let $\beta \in F((1/t))$ be algebraic (and so automatically separable) of degree d over $F(t)$, with $\beta^d + b_{d-1}\beta^{d-1} + \cdots = 0$. Differentiation with respect to t gives

$$\beta' = a_n\beta^n + \cdots + a_0, a_i \in F(t), a_n \neq 0, n \leq d-1.$$

Then $n = 2$, i.e., β satisfies the rational Riccati equation $\beta' = a\beta^2 + b\beta + c$, with $a, b, c \in F(t)$ if and only if [KTV00] the cross ratio of any four conjugates of β has zero derivative. This is equivalent to having no infinitesimal deformations, i.e., vanishing of KS.

Hence, it might be possible to successively improve on Osgood's bound, if we throw out some further classes of differential equations coming from the conditions that some corresponding Kodaira-Spencer map (or say the vector space generated by derivatives of the cross-ratios of conjugates of β) has rank not more than some integer. It should also be noted that even though the KS connection holds in characteristic zero, analog of Roth's theorem holds in the complex function field case. The Osgood bound still holds (conjectured in [Vol88], proved in [LdM96; LdM99]) by throwing out only a subclass given by 'Frobenius' equation $\beta^q = (a\beta + b)/(c\beta + d)$. This might be the best one can get. Similarly, the differential equation hierarchy suggested above might have some corresponding more refined Frobenius hierarchy.

In this Section, we will associate certain curves over function fields to given algebraic power series and show that bounds on the rank of Kodaira-Spencer map of this curves imply bounds on the diophantine approximation exponents of the power series, with more 'generic' curves (in the deformation sense) giving lower exponents. If we transport Vojta's conjecture on height inequality to finite characteristic by modifying it by adding suitable deformation theoretic condition, then we see that the numbers giving rise to general curves approach Roth's bound. We also prove a hierarchy of exponent bounds for approximation by algebraic quantities of bounded degree.

(a)   Height inequalities for algebraic points

Though we have not succeeded yet in improving the Osgood bound unconditionally, by throwing out more classes of numbers (we want good conditions like Osgood rather than a trivial way of obtaining this by throwing solutions of $y' = P_k(y)$, where P_k is a polynomial of degree k, for $j \leq k \leq d$ to force exponents to be less than j via the Kolchin theorem [Osg73] on denomination, combined with 9.4.3), we prove existence of hierarchies, given by deformation theoretic conditions, of bounds, using results of [Kim97], which we now recall: Let X be a smooth projective surface over a perfect field k. Assume that X admits a map $f : X \to S$ to a smooth projective curve S defined over k, with function field L in such a way that the fibers of f are geometrically connected curves and the generic fiber X_L is smooth of genus $g \geq 2$. Consider algebraic points $P : T \to X$ of X_L, where T is a smooth projective curve mapping to S (such that the triangle commutes). Define the canonical height of P to be $h(P) := \deg P^*\omega/[T : S] = \langle P(T).\omega\rangle/[K(P(T)) : L]$, where $\omega = \omega_X := K_X \otimes f^*K_S^{-1}$ denotes the relative dualising sheaf for $X \to S$. This is a representative for the class of height functions on $X_L(\overline{L})$ associated to the canonical sheaf K_{X_L}. Define the relative discriminant to be $d(P) := (2g(T) - 2)/[T : S] = (2g(P(T)) - 2)/[K(P(T)) : L]$.

The Kodaira-Spencer map is constructed on any open set $U \subset S$ over which f is smooth from the exact sequence $0 \to f^*\Omega^1_U \to \Omega^1_{X_U} \to \Omega^1_{X_U/U} \to 0$, by taking the coboundary map $KS : f_*(\Omega^1_{X_U/U}) \to \Omega^1_U \otimes R^1f_*(\mathcal{O}_{X_U})$.

Theorem 9.5.1   *(Kim [Kim97]) (1) Suppose the KS map of X/S (defined on some open subset of S) is non-zero. Then $h(P) \leq (2g - 2)d(P) + O(h(P)^{1/2})$ if $g > 2$. If $g=2$, then $h(P) \leq (2 + \epsilon)d(P) + O(1)$.*

(2) Suppose the KS map of X/S has maximal rank, then $h(P) \leq (2 + \epsilon)d(P) + O(1)$.

Remarks 9.5.2   (i) The inequality in (2) was proved [Voj91] in the characteristic 0 function field analog, without any hypothesis, by Vojta, who also conjectured [Voj87] the stronger inequality with 2 replaced by 1 in the number field (and presumably also in the characteristic 0 function field) case.

(ii) Modifying the proof in [Kim97] of Theorem 2, we get [KTV00] the following hierarchy of bounds: If the rank of the kernel of the KS map is $\leq i$, then we have

$$h(P) \leq (Max((2g-2)/(g-i), 2) + \epsilon)d(P) + O(1), \quad (0 \leq i < g).$$

(Note that the maximum is 2 only for $i = 0, 1$.)

(iii) For examples of explicit calculations of KS matrix and formulas for rank conditions in cases we consider, see [KTV00].

(b) Exponent hierarchy

We now apply this theorem to get bounds on the exponents for the diophantine approximation situation by associating to $\beta = \beta(t)$ some curves X over $F(t)$ and associating to its approximations some algebraic points P on them.

Let $f(x) = \sum_{i=0}^{d} f_i x^i$ be an irreducible polynomial with $\beta = \beta(t)$ as a root and with $f_i \in F[t]$ being relatively prime. Let $F(x, y) = y^d f(x/y)$ be its homogenization (there should be no confusion with the field F).

Assume p does not divide d. Let X be the projective Thue curve with its affine equation $F(x, y) = 1$. Given a rational approximation x/y (reduced in the sense that $x, y \in F[t]$ are relatively prime) to β, with $F(x, y) = m(t)$, we associate the algebraic point $P = (x/m^{1/d}, y/m^{1/d})$ of X. We have $[K(P(T)) : L] = d$. By Example 1.2.1, X is non-singular with $2g - 2 = d^2 - 3d$ and Theorem 2 can be applied. Note that $g \geq 2$ implies that $d \geq 4$, which implies $g \geq 3$.

Now the naive height of P is $\deg(y)$ (if $\deg(x)$ is bigger than $\deg(y)$, it differs by the fixed $\deg(\beta)$, so the difference does not matter below, similarly we ignore ϵ's which do not matter at the end). Hence $h(P) + O(h(P)^{1/2}) = (2g - 2) \deg(y)/d = (d - 3) \deg(y)$.

Let $e = E(\beta)$. We want upper bounds for e. If x/y is an approximation approaching the exponent bound, then degree of the polynomial $m(t)$ is asymptotically (as $\deg(y)$ tends to infinity) $(d - e) \deg(y)$. Since $K(P)$ over L is totally ramified at zeros of m and at infinity, by the Riemann-Hurwitz formula, we have (since p does not divide d) $2g(P) - 2 \leq -2d + ((d - e) \deg(y) + 1)(d - 1)$, which is asymptotic to $(d - e)(d - 1) \deg(y)$.

Hence under the maximal rank hypothesis, the Theorem gives us $d - 3 \leq 2(d - 1)(d - e)/d$. This simplifies to $e \leq d/2 + d/(d - 1)$, which is slightly worse than Osgood bound, but approaches it for large even d.

Also note that if Vojta's conjectured inequality is assumed to hold in characteristic p under the maximal rank hypothesis (this may be reasonable to do, taking into account parallel results (Remark 9.5.2 (i) above) in the two cases), we get $e \leq 2d/(d-1)$, which tends to the Roth bound 2 as d tends to infinity.

Let X have affine equation $y^k = f(x)$, with k relatively prime with p and d. Corresponding to a (reduced) approximation x/z, let $P = (x/z, (m/z^d)^{1/k})$, where $m = F(x,z)$. By Example 1.2.1, $2g - 2 = (d-1)(k-1) - 2$, and $[K(P(T)) : L] = k$. We assume that $g > 1$. The naive height is $\deg(z)$ ($\deg(x)$ differs by an additive constant, so it does not matter which is bigger). Hence the $h(P) + O(h(P)^{1/2})$ is $((d-1)(k-1) - 2)/k$ times that. (In the asymptotic, it does not matter that we computed the height on this model, singular at infinity.)

Now the zeros of m and z can ramify totally, so that Riemann-Hurwitz (as p does not divide k) gives (for approximation approaching the exponent bound) $2g(P) - 2 \leq -2k + (1+d-e)(k-1)\deg(z)$, which is $(1+d-e)(k-1)\deg(z)$ asymptotically. Hence under the maximal rank hypothesis, the Theorem gives $(d-1)(k-1) - 2 \leq 2(1+d-e)(k-1)$, which simplifies to $e \leq (d+3)/2 + 1/(k-1)$. This is again worse than, but asymptotic to, Osgood bound.

In this case, if we assume Vojta's bound under the maximal rank hypothesis, then we get $e \leq 2 + 2/(k-1)$ again approaching Roth bound, this time with k approaching infinity. So we can say that $e = 2$ under the maximal rank hypothesis and assuming the corresponding modification of Vojta's conjecture.

(c) Approximation by algebraic functions

Let β be of degree d over $F(t)$. Now we want to see how close β can be to α of degree $r < d$ over $F(t)$ in the spirit of Wirsing's theorem [Sch80].

Let α have height H and be such that $-\log |\beta - \alpha|/H$ is close to e. We again use the curves $y^k = f(x)$ and the point $P = (\alpha, f(\alpha)^{1/k})$ and $[F(t)(P) : F(t)] = kr$ now. Then $h(P) = ((d-1)(k-1) - 2)H/kr$ by the same calculation as before. Also the ramification of $K(P)$ over $K(\alpha)$ (where $K = F(t)$) is bounded by $(1+d-e)(k-1)H$. So by Riemann-Hurwitz $2g(P) - 2 < k(2g(\alpha) - 2) + (1+d-e)(k-1)H$, where $g(\alpha)$ is the genus of $K(\alpha)/F$. To bound $g(\alpha)$ apply the Castelnuovo inequality to $K(\alpha)$ viewed as the compositum of $F(t)$ and $F(\alpha)$ both function fields of genus 0 together with $[K(\alpha) : F(t)] = r$, $[K(\alpha) : F(\alpha)] = H$. So $g(\alpha) \leq (H-1)(r-1)$ and $2g(P) - 2 < 2krH + (1+d-e)(k-1)H$. Finally, we can apply our Theorem, provided the Kodaira-Spencer map is of maximal rank and get

$$((d-1)(k-1)-2)H/kr < (2+\epsilon)(2k(r-1)(H-1)+(1+d-e)(k-1)H)/kr+O(1).$$

Now divide by H and make H big and ϵ small, obtaining

$$((d-1)(k-1)-2)/kr \le 2(2k(r-1)+(1+d-e)(k-1))/kr.$$

The last inequality gives a bound for e in terms of d, r and k. If we can take k arbitrarily large it gives $e \le 2r + (d-1)/2$. Of course this is only interesting when $r < (d+1)/2$ but it seems that other methods that yield improvements on Liouville's inequality, such as Osgood's, do not give anything in this setting.

In this case, our finite characteristic version of Vojta's conjecture gives (under the maximal rank condition) $e \le 2r$. This is what was conjectured earlier in number fields case and proved via extension of Roth's methods, but Schmidt then improved it to $r + 1$.

Remarks 9.5.3   (1) Interestingly, Dirichlet type result is conjectured but not known in either case, whereas Roth's type results are better understood in this situation of approximations by algebraic quantities.

(2) There are many **open questions** on issues we discussed: It is unclear what the differential, Frobenius, Kodaira-Spencer hierarchy conditions and corresponding best exponent bounds would be. How can we characterize α's having the exponent 2, or bounded partial quotients or those with 'no ϵ' in exponent? If we order the algebraics, say by heights in each degree, do we have exponent 2 and unbounded partial quotients for most (density one)? How close to random behavior of quotients do we get for 'general' algebraic elements? In Class Ia elements, clearly almost all have exponent 2, by 9.3.1, as generic element has d_i large generic. For $q = 2$, $d = 3$, all elements are of Class I and Mobius transforms preserve exponent. This case is being worked on.

9.6   Note on connection with Diophantine equations

Let us recall some simple connections between diophantine approximation questions and the diophantine geometry. As Thue noticed, any improvement on the exponent upper bound of d has interesting consequences to the question of integral points on curves: Consider the homogenized version $P(x, y) = y^d p(x/y)$ of the minimal polynomial $p(x)$ of α with integral coefficients. Then for a given constant c, the affine curve $P(x, y) = c$ can have only finitely many integral points, because as $p'(\alpha) \neq 0$, by the mean value theorem

$$|P(x,y)| = |y|^d |p(\alpha) - p(x/y)| = |y|^d |\alpha - p/q| |p'(\beta)|$$

then tends to infinity with y, and it is easy to see that y can not stay bounded.

Siegel using his improvement on exponent bounds showed finiteness of integral points on all affine curves of genus more than 0, defined over number fields. Finally, Vojta generalized the work of Dyson and Roth in a wider context to give another proof of Mordell conjecture, which was by then Faltings' theorem that curves of genus at least two, defined over number fields can have only finitely many rational points. In fact, Faltings' proof was influenced by function field analog proof by Parshin and Arakelov theory developed to carry over function field analogies.

Not only Vojta first proved function field (characteristic zero) version of Mordell conjecture by these ideas, before he worked out the number field case, but Vojta and Lang's conjectures in arithmetic geometry were influenced by analogies between diophantine approximation theorems on algebraic numbers such as Roth's theorem and function theoretic approximation theorems in Nevanlinna theory. We refer to work of Vojta and Julie Wang etc. on this.

The famous *abc*-conjecture with wonderful diophantine applications, also has origins in works of Stothers and Mason on function fields, and main progress on it is through the diophantine approximation results on linear forms in logarithms. In fact, qualitative finiteness results about it follow from finiteness of solutions of S-unit equation $ax + by = 1$ in $x, y \in \mathcal{O}_S$, which itself is consequence of Siegel's exponent bound improvements mentioned above.

Diophantine geometry for function fields is a vast subject and we refer to surveys [Vol97; Tha02], a recent survey by Ulmer, and references therein. The p-th power map and differentiation has been used in diophantine geometry over function fields, in works of Manin, Voloch, Buium, Denis etc. There is also a recent work by Thomas Scanlon [Sca02] on Manin-Mumford conjecture in the setting of Drinfeld modules.

We finish by mentioning the work of Denis on analog due to Goss of Fermat equations. The usual Fermat equation $x^n + y^n = z^n$, which connects to the usual roots of unity and thus constant field extensions, is well-understood: If p does not divide $n > 2$, then since the genus is not zero, there are no parametric solutions, i.e., no non-trivial solutions in $A = \mathbb{F}_q[t]$, whereas eg. for $n = p^k$, we have $x^{p^k} + y^{p^k} = z^{p^k}$, whenever $x + y = z$.

Goss [Gos82] looked at analog $z^{q^d} = y^{q^d} C_\wp(x/y)$, for Carlitz module C, of the Fermat equation $z^p = y^p((x/y)^p - 1)$ and proved various analogous features of the theory and made conjectures about its solutions, saying there are none, except for some small exceptions as in the classical case. These were settled by Denis in a nice work [Den94]: Write the equation as $(z/y)^{q^d} = \sum a_i(x/y)^{q^i}$ and differentiate with respect to t, noting that $q = 0$ in characteristic p. Clearing out the denominators by multiplying by y^{q^d}, we see that y divides a power of x. But essentially by usual reductions, y could have been taken prime to x, and hence apart from some low cases and trivial solutions, there are no more!

Chapter 10

Transcendence results

We will give a short, incomplete and applications oriented survey of irrationality and transcendence results in the function field arithmetic, postponing such results obtained by automata techniques to the next chapter.

After some work in 1940's by Carlitz student Wade (and in 1960's by Geijsel) about the polynomial ring and the Carlitz module related objects, Jing Yu laid down the foundations and developed the transcendence theory of Drinfeld modules and Anderson's higher dimensional t-modules. Other important work then followed by Denis, Brownawell, Tubbs, Sinnou David, de Mathan, Cherif, Dammame, Thiery, Bertrand, Hellagouarch, Papanikolas, Anderson etc. For detailed discussions of transcendence theory aspects, we refer to the surveys [Gei79; Wal90; Yu; Bro98].

We start with techniques of the last chapter. In function field arithmetic, we seem to get good approximations easily compared to their number field counterparts, but as a compensation, we can not draw conclusions from them that we could have drawn in number fields case, where the Roth theorem holds, as discussed in the last chapter.

We will not even sketch the proofs of basic transcendence results, but will just state them, describe analogies or contrasts with the classical case and sometimes make comments of comparison of proof techniques for those familiar with the proofs of classical counterparts. On the other hand, we will sketch how to deduce the applications. We mention general, as well as special theorems, so that the reader can see which applications need results of which strength.

10.1 Approximation techniques and irrationality

Imitating Cantor's proof of existence of numbers in $\mathbb{C}$ (or $\mathbb{R}$) which are transcendental over $\mathbb{Q}$, we see that there are uncountably many numbers in C_∞ (or K_∞) which are transcendental over a function field K, since only countably many can be algebraic.

Just as Liouville produced explicit transcendental real numbers such as $\sum 10^{-n!}$, whose series converges so rapidly that its truncations give approximations not possible by his theorem for any algebraic number. We can do the same, using Mahler's analog Theorem 9.1.3 of Liouville's theorem.

How about numbers coming up naturally in the function field arithmetic? Consider $A = \mathbb{F}_q[t]$ first and let $e = e_C(1)$, where C is the Carlitz module. We have seen its continued fraction in 9.2. In particular, the continued fraction development is not bounded or ultimately periodic, so e is not rational or even quadratic irrational. In fact, the exponent [Tha99b] of e is q, hence Liouville theorem analog 9.1.3 implies that it is transcendental or algebraic of degree at least q. We can also imitate the standard irrationality proof using its fast converging series expansion $e = \sum 1/D_i$.

In fact, the same proof shows $\zeta(sp^m)$ is irrational, if $1 \le s < q$, using the formulas in 5.6, 5.9: Since $\zeta(sp^m) = \zeta(s)^{p^m}$ and both are in K_∞, without loss of generality $m = 0$. If $\zeta(s)$ were a rational with denominator of degree m, then $L_m^s \zeta(s)$ is integral, by the interpretation in 2.5 of L_m as the least common multiple of monic polynomials of degree m. On the other hand, by Theorem 5.9.1, it is integer plus $\pm 1/[m+1]^s$ plus lower degree terms by the straight multiplication on expansion, leading to a contradiction. This result can be pushed further [Tha87] to $s \le q^2$ by using the formulas in 5.6.

Note that $\zeta(1)$ makes sense here and can be thought of as analog of Euler's γ because of analogies described in 4.9.

Similar approximation proofs [Tha86; Tha87] show irrationality of the ratio $(\zeta(1)/\tilde{\pi})^2$ when $q = 3$. This is at 'odd' s. For 'even' s, we know by Theorem 5.2.1 that the ratio $\zeta(s)/\tilde{\pi}^s$ is rational. But $\tilde{\pi}$ is not 'real' (if $q > 2$) (i.e., $\tilde{\pi}^{q-1} \in K_\infty$, but $\tilde{\pi} \notin K_\infty$, if $q > 2$), let alone rational, that is why we took square to get a non-trivial result. Let us now look at 'real' variants of $\tilde{\pi}$ described in 2.5:

Let $\pi_0 := \prod_{i=1}^\infty (1 - [i-1]/[i])$, and $\pi := \prod_{i=1}^\infty (1 - 1/t^{q^i-1})^{-1}$.

Comparing the size of denominator of m-th truncation approximation with size of error for either, shows that they are transcendental of algebraic of degree at least $q - 1$, by Liouville analog 9.1.3.

With a little more work, the approximations obtained with zeta formula 5.9.1 similarly show [Tha87] that $\zeta(s)/\pi_0^s$ is transcendental or irrational of algebraic degree $\geq \lfloor (q-2)/s \rfloor + 1$, for $1 \leq s \leq q - 2$.

Similar argument applying Mahler-Liouville inequality to approximations by truncations, now applied in K_v instead of K_∞, show that if v is a monic prime of degree 1 of $\mathbb{F}_q[t]$, then (see remark after 5.9.1) $\zeta_v(1) = \log_v(v)/v$ can not be algebraic of degree $< q - 1$. This can be generalized to v of degree 2, using $\zeta_v(1) = \log_v(C_{v-1}(1))/v$, but concluding only irrationality, when $q > 2$.

Most of these approximations would have been good enough to imply transcendence, if the Roth analog were true.

These results can also be pushed a little more by these methods. But, as we will see, much more is known by better methods. Bounds for measure of irrationality of several of these quantities were given by de Mathan, Cherif [dM95; CdM93].

10.2 Transcendence results on Drinfeld modules

The exponential, logarithm, because of their connection with algebraic groups, have well-known transcendence properties [Bak90] proved in theorems of Hermite, Lindemann, Gelfond, Schneider, Baker etc.

The transcendence of $\tilde{\pi}$ and e was proved by Wade for $A = \mathbb{F}_q[t]$ case and generalized by Yu to an arbitrary A.

Wade [Wad41; Wad43; Wad44; Wad46a; Wad46b] found that the usual trick to show transcendence, by assuming algebraicity and then breaking up, using the arithmetic information, such as good series expansion for the quantity, the resulting zero equated to polynomial satisfied by it in integral part and a part smaller than one to get a contradiction, works especially well, when you use $\mathbb{F}_q$-linear polynomial satisfied by the quantities coming up naturally in Carlitz module arithmetic. Wade proves analogs of Hermite-Lindemann, Gelfond-Schneider theorems in the setting of the Carlitz module. In [Tha86; Tha87] it was noted that using explicit formulas for zeta sums as in 5.6, this method shows the transcendence of zeta values for $s = p^m$, $1 \leq s \leq q$. In [DH88], this was extended by using Wade's methods again, to $1 \leq s \leq q^2$, and then in [DH91] to all positive values, giving another proof of the first part of Jing Yu's theorem 10.5.1 below.

Entire function $f(z) = \sum_{n=0}^{\infty} f_n z^n/n!$ is called E-function, if all f_n's lie in a number field, for any $\epsilon > 0$, the maximum modulus of conjugates of f_n is $O(n^{\epsilon n})$ as $n \to \infty$ and there exists a sequence of natural numbers $q_n = O(n^{\epsilon n})$, such that for all n, $q_n c_j$ are algebraic integers for $0 \le j \le n$.

Examples are polynomials with algebraic coefficients, e^z, $\sin(z)$, $\cos(z)$, generalized hypergeometric functions with rational parameters. They often arise naturally as solutions of algebraic differential equations. The ring of E-functions is closed under differentiation, integration from 0 to z and scaling of variable z by a non-zero algebraic number.

Roughly, E_q-functions [Yu86] with respect to a finite extension L of function field K over $\mathbb{F}_q$ are similarly defined: entire functions, additive, with control on growth, height and denominators of Taylor expansion coefficients which are supposed to lie in L (eg. exponentials of Drinfeld modules with algebraic coefficients). They satisfy functional equations which are Frobenius equations analogs of differential equations.

Yu [Yu86] by generalizing E-functions techniques of Gelfond-Schneider-Siegel to E_q-functions proves analog of Schneider-Lang theorem that there are only finitely many simultaneous algebraic values of algebraically independent E-functions.

Theorem 10.2.1 *Let L be a finite extension of K and f_1, f_2 be E_q functions with respect to L and algebraically independent over $\overline{K}$. Then there are only finitely many points at which they simultaneously assume values in L.*

As applications of this analog of Schneider-Lang theorem, Yu proves: Hermite-Lindemann analog (apply to $e(z)$ and z):

Theorem 10.2.2 *If ρ is a Drinfeld A-module over $\overline{K}$, then $e_\rho(\alpha)$ is transcendental for all non-zero $\alpha \in \overline{K}$. In particular, the non-zero periods of e_ρ are transcendental.*

Schneider's theorem analog (apply to $e(z)$ and $e(\alpha z)$):

Theorem 10.2.3 *If ρ is a Drinfeld A-module over $\overline{K}$, with corresponding lattice Λ, then $e_\rho(\alpha\lambda)$ is transcendental, for non-zero $\lambda \in \Lambda$ and $\alpha \notin K_\Lambda$, the field of multiplication of M. In particular, if there is no complex multiplication, and the period ratio is irrational then it is transcendental.*

Schneider's theorem analog (apply to two exponentials, see also [Yu90]):

Theorem 10.2.4 *If ρ_1 and ρ_2 are non-isogenous Drinfeld A-modules over $\overline{K}$ and $u_1, u_2 \in \overline{K_\infty}^*$ are such that $e_{\rho_i}(u_i) \in \overline{K}$, then u_1/u_2 is transcendental. In particular, the ratio of any non-zero periods of non-isogenous Drinfeld A-modules over $\overline{K}$ is transcendental.*

Gelfond-Schneider theorem analog (apply to $e(z)$ and $e(\beta z)$, where $\log(\alpha_2) = \beta \log(\alpha_1)$):

Theorem 10.2.5 *If ρ is a Drinfeld A-module over $\overline{K}$ and α_1, α_2 are non-zero algebraic, then if $\log(\alpha_i)$ are linearly independent over K_Λ, then they are linearly independent over $\overline{K}$.*

Schneider's theorem analog:

Theorem 10.2.6 *For algebraic, non-real, non-quadratic α over K, we have $j(\alpha)$ is transcendental.*

Proof. If it is algebraic, and the corresponding rank 2 ρ has period ratio α, then $e(\alpha z)$, $e(z)$ must be algebraically dependent, so that $\alpha \in K_\Lambda$, which is quadratic. $\square$

On the other hand, Gekeler [Gek83] showed that for non-real quadratic α on the other hand $j(\alpha)$ is algebraic integer as in the complex multiplication theory.

Yu [Yu89] proves qualitative baker linear forms in logarithms analog, generalizing Gelfond-Schneider (separability assumption in this paper is removed in [Yu97]).

Theorem 10.2.7 *For Drinfeld A-module ρ over $\overline{K}$, if $\{\alpha_1, \cdots, \alpha_n\} \subset \overline{K_\infty}$ are linearly independent over K_ρ with $e_\rho(\alpha_i) \in \overline{K}$, then $\{1, \alpha_1, \cdots, \alpha_n\}$ are linearly independent over $\overline{K}$.*

In [Yu90] Yu proves quasi-period transcendence as in Schneider's work on elliptic integrals of the second kind:

Theorem 10.2.8 *Let ρ be a Drinfeld A-module over $\overline{K}$ and F_δ be quasi-periodic function corresponding to non-inner bi-derivation δ. Let $u \in \overline{K_\infty}^*$ and $e_\rho(u) \in \overline{K}$. Then $F_\delta(u)$ is transcendental. In particular, $\int_w \delta$ is transcendental for all non-zero periods w of ρ.*

Thiery [Thi] proves analog of Chudnovsky's result giving algebraic independence of periods and quasi-periods of rank 2 Drinfeld modules over $\mathbb{F}_q[t]$:

Theorem 10.2.9 *Let λ and ω be non-zero periods of rank 2 Drinfeld $\mathbb{F}_q[t]$-module over $\mathbb{F}_q(t)$ and η denote a bi-derivation with corresponding quasi-periodic function F_η as in Chapter 6. Then $F_\eta(\omega) - \omega F_\eta(\lambda)/\lambda$ and $F_\eta(\lambda)/\lambda$ are algebraically independent over $\mathbb{F}_q(t)$.*

10.3   Application to Zeta and Gamma values

Our results on connections of the special zeta and gamma values with $\tilde{\pi}$, combined with transcendence of $\tilde{\pi}$ shown by Wade and Yu, immediately gives transcendence results for those gamma and zeta values. For example, applying Theorems on transcendence of periods and 10.2.9 in the previous section to Theorems 4.7.5, 4.10, 4.11.2, 6.4.7 we get

Theorem 10.3.1   *Let A be general.*

For the arithmetic gamma, $\Gamma(i - a/(q-1)), \Gamma(n)$ are transcendental, where a, i, n are integers, $0 < a < q - 1$ and $n \leq 0$.

For the geometric gamma, if $q = 2$, all values at proper fraction are transcendental.

For $A = \mathbb{F}_q[t]$, $(a/(1 - q^2))!$, $0 < a < q^2 - 1$ and $\Gamma(1/t)$ are transcendental. If, further $q = 3$, $\Gamma(-1/t)$ is also transcendental.

Remark 10.3.2   Let us compare with the results on Euler's gamma: The most well-known result is : $\Gamma(1/2) = (-1/2)! = \sqrt{\pi}$, parallel to our first result with denominator $q - 1$, as 2 and $q - 1$ are cardinalities of $\mathbb{Z}^*$ and A^* respectively. Chudnovsky [Chu84] showed that Γ values at proper fractions with denominator 4 or 6 are transcendental (this is basically all that is known about the transcendence of the individual gamma values at proper fractions), parallel to our third result with denominator $q^2 - 1$ (in arithmetic case) and t (or degree one prime, in geometric case): Whereas 4 and 6 being possible number of roots of unity in (full cyclotomic) imaginary quadratic fields (and enter via Chowla-Selberg results on periods of elliptic curves with complex multiplication by such fields), these are number of roots of unity in $\mathbb{F}_{q^2}(t)$ and $\mathbb{F}_3(\sqrt{-t})$, (full cyclotomic) quadratic extensions with only one infinite place (and enter similarly via Chowla-Selberg analogs).

For the zeta values, as an application of Theorems 5.2.1, 5.2.6 we get

Theorem 10.3.3   *For even positive m, $\zeta(m)$ is transcendental. In particular, for $q = 2$, $\zeta(m)$ is transcendental, for positive integer m.*

For relative zeta, we can get similar applications from Theorem 5.2.8.

For values at odd positive integers, we can do a lot using Jing Yu's higher dimensional theorems in the next section, but even with Drinfeld modules, we have simple applications eg. of Theorem 5.9.1 and 5.9.2:

Theorem 10.3.4   *For $A = \mathbb{F}_q[t]$, $\zeta(p^n)$ are transcendental.*

For A as in (iii) of 3.1.2, $\zeta(3^n)$ and $\zeta(3^n)/\tilde{\pi}^{3^n}$ are transcendental for all nonnegative integers n.

Proof. Hermite-Lindemann analog Theorem 10.2.2, and Theorems 5.9.1 and 5.9.2 connecting the zeta value to logarithm, imply that $\zeta(1)$ is transcendental. Let $\alpha = \zeta(1)/\tilde{\pi}$. By Theorem 5.9.2, $e(\alpha\tilde{\pi})$ is algebraic, hence by Gelfond-Schneider analog Theorem 10.2.5, it is enough $\alpha \notin K$. In fact, we claim that α is not in K_∞: By the non-archimedean nature of K_∞, $\zeta(1) \in K_\infty$ has degree 0. Hence 5.2.7 implies that $2\deg \tilde{\pi} = \deg d_1 = 3\deg x - \deg x_1 = -3$, as we see by comparing the coefficients of z^q in $e(az) = \rho_a(e(z))$ with $a = x$. Hence $\tilde{\pi}$ is not in K_∞. This proves the claim. We have shown that $\zeta(1)$ and $\zeta(1)/\tilde{\pi}$ are transcendental. On the other hand, $\zeta(kp^n) = \zeta(k)^{p^n}$. $\qquad\square$

Remarks 10.3.5 (1) There are many other such applications, using results of [Tha92b; And94]. As mentioned above, $\zeta(1)$ can be thought of as analog of Euler's gamma constant.

(2) Let $q = 2$. Then we have seen that special values at fractions of geometric Γ or Γ_v, special values of zeta or multizeta at positive integers are all algebraic, once divided by appropriate power of the fundamental period. The special values of zeta and multizeta at negative integers are all zero (5.10.19 (2)) leading to identically zero interpolations at all primes. As seen in 3.7, the moduli approach is also easy, and as seen in 8.9.4, the cyclotomic module is torsion and corresponding class number is trivial for any prime. This is all related to the fact that $\mathbb{F}_q^*$ is then trivial, but it should have ramification in whatever theory of extension of motives in this context. Also note that we do not see such trivial behavior for arithmetic gamma or hypergeometric functions, in $q = 2$ case.

10.4 Transcendence results in higher dimensions

Using techniques of 10.2.1 generalized to $n + 1$ functions of n variables, Jing Yu [Yu89] proved analog of multi-variable version of Schneider-Lang theorem:

Theorem 10.4.1 *Let L be a finite extension of K. Let $f_1, \cdots, f_{d+1}$ be E_q-functions of d variables with respect to L (see [Yu89] for precise definition), which are algebraically independent over L. If Λ is an A-submodule of $\overline{K_\infty}^d$ containing d elements linearly independent over $\overline{K_\infty}$, then not all values $f_i(\lambda)$ $(1 \le i \le d+1, \lambda \in \Lambda)$ can lie in L.*

Applying this to d co-ordinate functions of exponential of the d-dimensional G below and a fixed co-ordinate function (these are algebraically independent E_q functions) and to $\Lambda = \mathcal{O}_+ u$ for u below, Jing Yu proved [Yu89]

Theorem 10.4.2 *If G is HBD module over $\overline{K}$ with multiplications by $\mathcal{O}_+$. Let e be corresponding exponential map with respect to a fixed co-ordinate system normalized so that chosen basis for $Lie(G)$ consists of eigenvectors for $\mathcal{O}_+$-action. Then every co-ordinate component of nonzero $u \in C_\infty^d$ (eg. a non-zero period) such that $e(u) \in \overline{K}^d$ is transcendental.*

Using this, and adapting Bertrand-Masser method to finite characteristic, analog of Baker's theorem was proved in [Yu89; Yu97]:

Theorem 10.4.3 *Let ρ be a Drinfeld A-module over $\overline{K}$, with the corresponding exponential $e(z)$. Let $\alpha_1, \cdots, \alpha_n$ be elements of $\overline{K_\infty}$ such that $e(\alpha_i)$ are all algebraic over K. If α_i are linearly independent over the field of multiplications K_ρ, then $1, \alpha_1, \cdots \alpha_n$ are linearly independent over its algebraic closure.*

Using these, Jing Yu proved analog of Hermite-Lindemann for t-modules [Yu91; Yu97]:

Theorem 10.4.4 *If G is a simple, non-trivial t-module over $\overline{K}$ and if $X \in LieG(\overline{K_\infty})$ is a non-zero point with $\exp_G(X) \in G(\overline{K})$, then the last co-ordinate (where the co-ordinates are chosen so that at the lie level t-action is θ plus a nilpotent upper triangular) of X is transcendental.*

(In 1997 paper, Jing Yu had assumed that G is regular, but has told the author that it can be dropped.)

It follows that X and the corresponding algebraic point are uniquely determined by this last co-ordinate. If we drop simplicity assumption, we can guarantee transcendence of at least one co-ordinate. Jing Yu [Yu91] also proved $\wp$-adic version of this theorem as well as analog (together with its $\wp$-adic version) of Baker's qualitative theorem on linear independence of logarithms of algebraic point to last co-ordinates of logarithms of algebraic points in this case. More generally, reasonable linear relations satisfied by a logarithmic vector of an algebraic point must come from algebraic relations inside the t-module. This analog of Wustholz's theorem for commutative algebraic groups is Yu's theorem [Yu97]:

Theorem 10.4.5 *Let G be a regular (see 7.2, eg. abelian) t-module defined over $\overline{K}$ and X be a point in $Lie(G)(\overline{K_\infty})$ such that $\exp_G(X) \in G(\overline{K})$. Then the smallest vector subspace in $Lie(G)(\overline{K_\infty})$ defined over $\overline{K}$ which is* *invariant under d[t]$_G$ and contains X is the tangent space at the origin of a t-submodule of G.*

(According to Jing Yu, in all of his theorems in this chapter, C_∞ can be used in place of $\overline{K}_\infty$, as the same can be done throughout the proofs.)

Anderson, Brownawell, Papanikolas [ABP] proved a linear independence criterion (see [ABP] for remarks on comparison with Yu's result above) directly applicable in Shtuka language:

Let $K = \mathbb{F}_q(T)$ with ∞ being the usual place at infinity. Let t be a variable independent of T. Given $f = \sum a_i t^i \in C_\infty[[t]]$, put $f^{(n)} := \sum a_i^{q^n} t^i$, and extend this operation entry-wise to matrices. Let $\mathcal{E} \subset \overline{K}[[t]]$ be the subring of power series $\sum a_i t^i$ such that all a_i's are contained in a finite extension of K_∞, and $\sqrt[i]{|a_i|_\infty}$ tending to zero, as $i \to \infty$.

Theorem 10.4.6   *With notation as above, if $\Phi(t) \in Mat_{\ell \times \ell}(\overline{K}[t])$ and $\psi(t) \in Mat_{\ell \times 1}(\mathcal{E})$ are such that $Det(\Phi(t))$ is a polynomial in t vanishing (if at all) only at $t = T$ and $\psi^{(-1)}(t) = \Phi(t)\psi(t)$. Then for every $\rho \in Mat_{1 \times \ell}(\overline{K})$ such that $\rho\psi(T) = 0$, there exists $P(t) \in Mat_{1 \times \ell}(\overline{K}[t])$ such that $P(T) = \rho$ and $P(t)\psi(t) = 0$.*

Thus, in the situation of this theorem, every $\overline{K}$-linear relation among entries of the specialization $\psi(T)$ is explained by a $\overline{K}[t]$-linear relation among entries of $\psi(t)$ itself.

10.5   Application to Zeta and Gamma values

Theorem 8.1.1 expresses $\zeta(n)$ (and its $\wp$-adic counterparts) as last coordinate logarithm of an algebraic point for $C^{\otimes n}$. When n is 'odd', $\zeta(n)/\tilde{\pi}^n$ is not in K, as it is not even in K_∞, so $\tilde{\pi}^n$ which is also such last-coordinate logarithm is linearly independent over K to the above. Thus Jing Yu's results in 10.4 in case of $G = C^{\otimes n}$ combined with 8.1.1 give

Theorem 10.5.1   *Let ζ be the Carlitz zeta function associated to $\mathbb{F}_q[t]$. Then $\zeta(n)$ is transcendental for all positive integers n.*

Further, $\zeta(n)/\tilde{\pi}^n$ as well as $\zeta_\wp(n)$ are transcendental for 'odd' (i.e., not divisible by $q - 1$) positive integers n.

All linear relations over $\overline{K}$ between zeta values at positive integers and positive powers of Carlitz period are those coming from Carlitz result 5.2.1 for 'even' values.

Remarks 10.5.2   (1) For n 'even', the first part follows from Carlitz' 5.2.1 and Wade's proof of transcendence of Carlitz' $\tilde{\pi}$. For the Riemann zeta function, we only have Apery and Rioval's results that there are infinitely many odd n ($n = 3$ being the only known value) such that $\zeta(n)$ is irrational. The values in the second part are rational and zero respectively, if n is 'even'.

(2) We deduce from 10.4.3 that point Z_n in 8.1.1 is a torsion point, when n is 'even', since then, by 5.2.1, $\zeta(n)/\tilde{\pi}^n \in K$.

(3) After Jing Yu's proofs, the first part was also proved by Dammame-Hellegouarch [DH91] applying Wade's method and by de Mathan [dM94] by giving another transcendence criterion.

Now we give an application to the study of multi-zetas in 5.9 by giving the **proof of Theorem 5.10.2**:

Proof.   Let $q = 2$. By 5.9.1,

$$\zeta(2,1) - \zeta_d(2,1) - \zeta_d(1,2) = \sum_{d=0}^{\infty} \left(\sum_{n_1 \neq n_2, d_1 = d_2 = d} \frac{1}{n_1^2 n_2} - \sum_n \frac{1}{n^3} \right)$$

$$= \sum_{d=0}^{\infty} (S_d(-2) S_d(-1) - S_d(-3))$$

$$= \left(\sum_{d=0}^{\infty} \frac{1}{l_d^3} \right) - \zeta(3).$$

Since $\zeta(2,1)$ and $\zeta(3)$ are eulerian, if both the ζ_d's are eulerian, then $\sum 1/l_d^3$ is eulerian. But that would mean that $(0,0,1)$ is a torsion point for $C^{\otimes 3}$ for $\mathbb{F}_2[t]$. We now show that this is not true.

One way to proceed is to calculate the canonical height of this point and show that it is non-zero. As we have not developed this theory, we will proceed directly.

The action of t on $C^{\otimes 3}$ for $\mathbb{F}_2[t]$ takes vector (a,b,c) to $(ta+b, tb+c, a^2+tc)$. Write degree of the three entries of $t^n(0,0,1)$ as a_n, b_n, c_n respectively. Once we achieve the condition '$b_n > a_n + 1$, $c_n > b_n + 1$ and $2a_n > c_n + 1$', then we have $a_{n+1} = b_n$, $b_{n+1} = c_n$ and $c_{n+1} = 2a_n$ and it follows that the condition (and hence these equalities) continues to hold with n replaced by $n + 1$. Now straight calculation shows that the condition holds for $n = 11$ and so for $n \geq 11$. Hence if $n > m \geq 11$, the t^n-th multiple of our point will never be the same as t^m-th multiple, because degrees blow up. If our point had order $f(t)$, since by Fermat's little theorem $f(t)$ divides some $t^n - t^m$ with $n > m$ as above, we get a contradiction. This finishes the first part of the proof.

Now let $q = 3$. The second part is proved similarly by showing that $(0, 0, 0, 1)$ is not a torsion point on $C^{\otimes 4}$ for $\mathbb{F}_3[t]$, which implies that $\sum 1/l_d^4$ is not eulerian. But $2\zeta_d(2, 2) + \sum 1/l_d^4 = \zeta(2)^2$. □

Now we turn to gamma values. It can be easily deduced [Yu, pa. 261] from 10.4.4 that $\tilde{\pi}_{q_1}^{n_1}/\tilde{\pi}_{q_2}^{n_2}$ is transcendental, if $q_1 \neq q_2$, $n_i \in \mathbb{Z}_{>0}$, and where $\tilde{\pi}_q$ denotes the period of Carliz module for $\mathbb{F}_q[t]$. This (together with 4.7.3 and formulas immediately after it) gives another proof of transcendence of values of arithmetic gamma for $\mathbb{F}_q[t]$ at proper fractions with denominator $q^2 - 1$, deduced in 10.3.1 from Thiery's result, at least for some numerators including 1 and q. We would get all proper fractions with denominator $q^2 - 1$, if we knew transcendence of products $\tilde{\pi}_{q_1}^{n_1} \tilde{\pi}_{q_2}^{n_2}$ for $n_i \in \mathbb{Z}_{>0}$. This can be attacked via Jing Yu's theorem by analyzing submodules of corresponding tensor products, but this has not been carried out.

Applying Theorem 10.4.2 to the period vector u of Theorem 8.8.1, we get that its a-th co-ordinate z_a, $a \in I_+$ is transcendental, so that we get transcendence of values $\Gamma(a/f)$, $a \in I_+$, and if we apply the same to $C \otimes M_f$ mentioned in 8.8, we get transcendence of $\Gamma(a/f)$, $a \in I$, and hence of all geometric gamma values for $\mathbb{F}_q[t]$ at proper fractions.

The transcendence of all values of geometric gamma function for $\mathbb{F}_q[t]$ at proper fractions was proved [BP02] before, by applying Theorem 10.4.4 to Theorem 8.8.2: In fact, Brownawell and Papanikolas [BP02] developed complex multiplication theory and theory of quasi-periods for general t-motives, and analyzing the connection between the CM types of soliton t-motives of 8.8 showed, using 10.4.4, that only $\overline{K}$- linear relations among the gamma values at fractions are those explained by the bracket criterion in 4.12, as they mirror relations coming via CM type relations.

More recently, Anderson, Brownawell and Papanikolas [ABP] using tensor powers to handle all monomials (do not need quasi-periods then, see 8.8) have proved very complete result showing that known algebraic relations (note that linear independence of monomials corresponds to algebraic independence) are 'all' algebraic relations, proving analog of Lang-Rohrlich's conjecture (also implication of Grothendieck conjecture on period relations coming from motivic relations and Deligne's [DMOS82] work):

Theorem 10.5.3 *Let* Γ *stand for geometric gamma function for* $A = \mathbb{F}_q[T]$. *By a* Γ-*monomial, we mean an element of the subgroup of* C_∞^* *generated by* $\tilde{\pi}$ *and* Γ-*values at proper fractions in* K.

*Then a set of Γ-monomials is $\overline{K}$-linearly dependent exactly when some pair of Γ-monomials is, and pairwise $\overline{K}$-linear dependence is entirely decided by bracket criterion of 4.12 (**H1**).*

In particular, for any $a \in A_+$ of positive degree, the extension of $\overline{K}$ generated by $\tilde{\pi}$ and $\Gamma(x)$ with x ranging through proper fractions with denominator (not necessarily reduced) a, is of transcendence degree $1 + (q-2)|(A/a)^*|/(q-1)$ over $\overline{K}$.

The reader should read very nice and almost self-contained exposition in [ABP]. Their independence criterion of the last section giving 'functional relations' from relations between specializations, applied to solitons specializing to gamma products is very well suited to show that algebraic relations between special values come from functional relations criterion above same type. As we saw in 8.8, the gamma special values are attached to Shtukas made up from solitons and as mentioned above the triviality of CM type is seen by bracket criterion in 4.12.

Remarks 10.5.4 (1) There are several other transcendence results (eg., by Becker, Brownawell, Tubbs, Denis) we have not mentioned. For generalization of many techniques and results to σ-transcendence concept of Hellegouarch, where some results work in zero characteristic also, we refer to his article in [G+92]. In spirit, this fits in Drinfeld dictionary of Chapter 7, which also works with general automorphism σ of infinite order.

(2) There are many natural **open questions**: What are results for gamma, zeta values and algebraic independence results in v-adic, arithmetic, two variable gamma, or general A setting? What is the nature of $\Gamma(0)$ for the gamma of 8.3 (b)? What are the transcendence results for the hypergeometric functions?

Chapter 11

Automata and algebraicity: Applications

We recognize rationals from their decimal (or p-adic or Laurent series expansions in the function field case) expansions from the eventual periodicity of the digits. We may hope for a good criterion for recognizing which Laurent series in $1/t$ over $\mathbb{F}_q$ are algebraic over $\mathbb{F}_q(t)$, because there is no carry over of digits in the expansion when we manipulate them and since the residue field is finite.

A nice criterion was indeed found by Christol, based on earlier work by Furstenberg. It makes connection, via results of Cobham, to work done on finite automata by computer scientists. We will describe this and then turn to applications to the function field arithmetic, especially to the gamma values, to the periods of Tate elliptic curve and to the q-expansions of modular forms.

We saw in the last chapter that transcendence results on the values of the arithmetic gamma obtained by relating to periods were exactly parallel to known classical counterparts. The automata technique on the other hand gives complete results, at least for $A = \mathbb{F}_q[t]$ case, going beyond this. In particular, we know that the monomials proved to be algebraic earlier are the only algebraic ones.

Finally, we propose a finer classification of finite characteristic numbers, extending these ideas and tools. In some cases, we then generalize transcendence results by showing that some important numbers are 'almost algebraic'.

11.1 Automata and algebraicity

Interesting quantities of number theory or geometry are often defined by analytic processes such as integration or infinite series or product expansions.

A central question which transcendence theory addresses is whether they can also be described by simpler ('finite') algebraic methods; in particular, whether they turn out to be algebraic. For example, the most common analytic description of a number is by its infinite decimal or p-adic expansion and that of a function is by its power series expansion. It is well-known that any expansion that is eventually periodic represents a rational number (or a function). On the other hand, rational numbers and rational functions over a finite field give expansions that are eventually periodic.

More generally, we can ask for a similar, simple characterization of digit patterns for algebraic numbers or functions. We begin by describing an automata-theoretic criterion of discrete mathematics for this, and will prove three transcendence results in number theory as applications.

The first major advance in answering this general question was made by Furstenberg [Fur67]:

For $r = \sum r_{n_1, \cdots, n_k} x_1^{n_1} \cdots x_k^{n_k}$, define the 'diagonal' $Dr := \sum r_{n, \cdots, n} x^n$.

Theorem 11.1.1 *For $k = \mathbb{C}$ or $\mathbb{F}_q$, the set of algebraic power series $f(x)$ over $k(x)$ is the same as the set of diagonals Dr of two-variable rational functions $r(x_1, x_2)$. The diagonal of a rational function of many variables over $\mathbb{F}_q$ (but not over $\mathbb{C}$) is algebraic.*

Proof. Over $\mathbb{C}$, for small ϵ and $|x|$, we have

$$Dr(x) = \frac{1}{2\pi i} \int_{|z|=\epsilon} r\left(z, \frac{x}{z}\right) \frac{dz}{z}.$$

Evaluating by residues gives an algebraic function. On the other hand, suppose f is algebraic, satisfying a polynomial equation $P(x, f(x)) = 0$. Also assume $f(0) = 0$ and that 0 is an isolated root of $P(0, w) = 0$, then expressing

$$f(x) = \int_\gamma w \frac{\partial P}{\partial w}(x, w) / P(x, w) dw$$

as an integral above, we see that f is a diagonal of a rational function. Also, if we have more than two variables, the resulting contour integration of an algebraic function of two variables is in general transcendental. Though this proof does not work over $\mathbb{F}_q$, the resulting formulas can be directly checked. □

Deligne [Del84] generalized the last statement of the Theorem to algebraic functions of many variables. For proof of this involving automata ideas, see University of Bordeaux thesis of 1995 of Koskas.

From now on, we focus on our usual function fields by restricting to a coefficient field $\mathbb{F}_q$ of characteristic p.

Corollary 11.1.2 *In the function field case, (1) Algebraic power series are closed under Hadamard (term-by-term) product.*

(2) $\sum a_n x^n$ is algebraic if and only if $\sum_{a_n=a} x^n$ is algebraic for each $a \in \mathbb{F}_q$.

Proof. If $\sum a_n x^n = Dr_1(x_1, x_2)$, $\sum b_n x^n = Dr_2(x_1, x_2)$, then $\sum a_n b_n x^n = D(r_1(x_1, x_2) r_2(x_3, x_4))$. This implies (1). Then (2) follows from (1) by expanding out $\sum_{a_n=a} x^n = \sum (1 - (a_n - a)^{q-1}) x^n$. □

By the Corollary, we can focus on characteristic sequences on subsets of natural numbers, i.e., on the series $\sum x^{n_i}$, if we wish. The main theorem giving the automata-theoretic criterion for algebraicity is the following theorem due to Christol [Chr79; CKMR80]. In fact, the equivalence of the last two conditions (as well as another description in terms of substitutions) of the Theorem is due to Cobham [Cob72]. We sketch the proof later.

Theorem 11.1.3 *(i) $\sum f_n x^n$ is algebraic over $\mathbb{F}_q(x)$ if and only if (ii) $f_n \in \mathbb{F}_q$ is produced by a q-automaton if and only if (iii) there are only finitely many subsequences of the form $f_{q^k n + r}$ with $0 \leq r < q^k$.*

Here, an m-automaton (we shall usually use $m = q$, a prime power in the applications) consists of a finite set S of states, a table of how the digits base m operate on S, and a map Out from S to $\mathbb{F}_q$ (or some alphabet in general). For a given input n, fed in digit by digit from the left, each digit changing the state by the rule provided by the table, the output is $\mathrm{Out}(n\alpha)$ where α is some chosen initial state. For more details, see [All87; AS03]. So instead of our ideal Turing machine which has infinite tape and no restriction on input size (and still is a good approximation to computers because of enormous memories available these days) the finite automata has a restricted memory, so integer has to be fed in digit by digit, with the machine retaining no memory of previous digits fed except through its changed states.

Remarks 11.1.4 (1) Change of view point makes some hard things easy and vice versa. For example, if $\sum a_n x^n$ and $\sum b_n x^n$ are algebraic, the theorem implies that there are automata which produce a_n (respectively b_n) on (digit by digit) input n. Running them in parallel (i.e. direct product), we can get both a_n and b_n and thus their product as output implying (1) of the corollary above. We leave deduction of (2) as an exercise. So the closure of algebraic power series under Hadamard product is trivial by automata approach, easy from Furstenberg theorem, but challenging by usual algebraic techniques, whereas closure under usual product is immediate from the usual point of view, whereas challenging to deduce from these criteria.

(2) Here is another direct nice consequence (pointed out to me by Allouche): The series $\sum a_i t^{p^i} \in \mathbb{F}_p[[t]]$ is algebraic over $\mathbb{F}_p(t)$ if and only if $\sum a_i t^i \in \mathbb{F}_p(t)$. This is since both the statements are equivalent to the fact that a_i is an eventually periodic sequence.

Examples 11.1.5 (1) Consider the 2-automata given by the following table with $\alpha = s_1$, $\text{Out}(s_1) = \text{Out}(s_2) = 1$ and $\text{Out}(s_3) = \text{Out}(s_4) = 0$.

	s_1	s_2	s_3	s_4
0	s_1	s_2	s_2	s_4
1	s_2	s_3	s_2	s_4

If the corresponding power series is $g = \sum_{m \in M} x^m$, then we see that $0, 1 \in M$ and $m \in M$ if and only if $4m, 4m + 1 \in M$, because their expansions are obtained by appending 00 and 01 to the expansion of m respectively. Hence $g = \sum t^m = (\sum t^{4m})(1 + t) = f^4(1 + t)$, so that $g = (1 + t)^{-1/3}$.

(2) The following table, where $2 \leq i < p$, together with $\alpha = s_1$, $\text{Out}(s_1) = \text{Out}(s_2) = 0$ and $\text{Out}(s_3) = 1$ defines a p-automaton whose output f_n is the characteristic sequence of $\{p^m\}$, i.e., the numbers of the form $1000\cdots$ base p.

	s_1	s_2	s_3
0	s_1	s_2	s_3
1	s_2	s_3	s_3
i	s_3	s_3	s_3

It is easy to see directly that $f := f(x) := \sum x^{p^n}$ satisfies $f^p - f + x = 0$, in accordance with Theorem 2. In fact, this f is Mahler's famous counterexample to the analogue of Roth's theorem in characteristic p. The partial sums to f approximate f with Liouville bounds.

From the proof of the first theorem, we find that $f = D(x_1/(1 - (x_1^{p-1} + x_2)))$. We leave to the reader the interesting exercise of verifying this equation directly from the definition by expanding the right hand side via a geometric series, and proving the necessary divisibility properties of the binomial coefficients thus arising.

As a warm-up to our applications, we now prove that f is transcendental in finite characteristic $\ell \neq p$. Clearly, there are infinitely many k's such that $0 < p^m - \ell^k \mu < \ell^k$, for some m and some $0 < \mu < \ell$. So the subsequence $f_{\ell^k n + (p^m - \ell^k \mu)}$ assumes the value 1, for $n = \mu < \ell$. The next n for which it is 1 corresponds to $\ell^k n + (p^m - \ell^k \mu) = p^{m+w}$, with $w > 0$ the least such. So ℓ^k divides $p^w - 1$ and hence $n = \mu + p^m(p^w - 1)/\ell^k > p^m \to \infty$ as k tends to ∞. Hence there are infinitely many such subsequences, and (iii) of Theorem 11.1.3 finishes the proof.

In fact, this is a special case of a very general result of Cobham [Cob69]:

Theorem 11.1.6 *Non-periodic sequences produced by m-automata cannot be produced by n-automata, if m and n are multiplicatively independent.*

We do not sketch the proof. To quote Eilenberg [Eil74], "The proof is correct, long and hard. It is a challenge to find a more reasonable proof of this fine theorem". See [BHMV94] for survey of other proofs based on logic.

Together with Theorem 11.1.3, this implies

Corollary 11.1.7 *If $\sum x^{n_i}$ is irrational and algebraic in one finite characteristic, then it is transcendental in all other finite characteristics.*

The natural question is whether corresponding real numbers, e.g. the decimal $\sum 10^{-n_i}$, are transcendental. For our example $f(x)$ above, instead of using the algebraicity equation $f(x)^p - f(x) + x = 0$, which is special to characteristic p, Mahler used the characteristic-free functional equation $f(x^p) - f(x) + x = 0$ and established (for $p = 2$) the transcendence of various values of $f(x)$. It implies the transcendence of real numbers $f(1/k) = \sum k^{-2^n}$, for any integral base $k > 1$. See [LvdP77] for a nice exposition of the proof of a more general result.

Loxton and van der Poorten generalized Mahler's method, but it should be noted here that the proof of the often quoted result of Loxton and van der Poorten that under the hypothesis of the Corollary, the real number $\sum 10^{-n_i}$ is transcendental, has a gap, as van der Poorten has mentioned to the author. Thus, this statement and similar p-adic statement remain as challenging open questions.

We now present an example due to Richie [Rit63] of interesting non-automatic sequence. We shall use it later in applications.

Theorem 11.1.8 *The characteristic sequence of the set of squares is not produced by any 2-automaton.*

Proof. Consider the 2-automaton given by the following table and $\mathrm{Out}(f) = 1$, $\mathrm{Out}(\text{rest}) = 0$.

	α	s_1	s_2	s_3	s_4	f	n
0	n	s_3	s_4	s_4	s_3	n	n
1	s_1	s_2	s_1	n	f	n	n

A straight entry chase in the table shows that this produces the characteristic function χ_A of the set $A = \{1^n 0^m 1 : n, m > 0, n + m \text{ odd}\}$. It is easy to see that the intersection of A with the set of squares is $B = \{1^n 0^{n+1} 1 : n > 0\}$ and also that in general, the intersection corresponds to the direct product of automata or the Hadamard product of series.

But χ_B can not be produced by a 2-automaton: As there are only a finite number of states, $1^\ell \alpha = 1^{\ell+m} \alpha$, for some ℓ, m. But $1^{\ell+m} 0^{\ell+1} 1$ is not in B whereas $1^\ell 0^{\ell+1} 1 \in B$. $\qquad\square$

Now let us give a sketch of the proof of Theorem 11.1.3. For more details, see [CKMR80; All87; AS03], see also [Chr79] for original approach using Furstenberg's result.

Proof. (ii) implies (iii): There are only finitely many possible maps $\beta : S \to S$ and any $f_{q^k n + r}$ is of the form $\mathrm{Out}(\beta(n\alpha))$.

(iii) implies (i): Let V be the vector space over $\mathbb{F}_q(x)$ generated by monomials in $\sum f_{q^k n + r} x^n$. Then V is finite dimensional with $fV \subset V$, so f satisfies its characteristic polynomial.

(i) implies (iii): For $0 \leq r < q$, define C_r (twisted Cartier operators) by $C_r(\sum f_n x^n) = \sum f_{qn+r} x^n$. Considering the vector space over $\mathbb{F}_q$ generated by the roots of the polynomial satisfied by f, we can assume that $\sum_{i=0}^{k} a_i f^{q^i} = 0$, with $a_0 \neq 0$. Using $g = \sum_{r=0}^{q-1} x^r (C_r(g))^q$ and $C_r(g^q h) = g C_r(h)$, we see that

$$\{h \in \mathbb{F}_q((x)) : h = \sum_{i=0}^{k} h_i (f/a_0)^{q^i}, h_i \in \mathbb{F}_q[x], \deg h_i \leq \max(\deg a_0, \deg a_i a_0^{q^{i-2}})\}$$

is a finite set containing f and stable under C_r's.

(iii) implies (ii): If there are m subsequences $f_n^{(i)}$ with $f_n^{(1)} = f_n$ say, put $S := \{\alpha := \alpha_1, \cdots, \alpha_m\}$. Define a digit action by, $r\alpha_i := \alpha_k$ if $f_{qn+r}^{(i)} = f_n^{(k)}$. Define $\mathrm{Out}(\alpha_i) := f_n$, if $n^- \alpha_1 = \alpha_i$ with n^- being the base q expansion of n written in the reverse order. $\qquad\square$

11.2 Some useful automata tools

We now discuss some more properties of the sequences produced by automata that have been proved by computer scientists. We will focus on those, which have turned out to be helpful so far in number theoretic applications. Other properties, such as equivalences with fixed points of uniform substitutions, with non-deterministic automata, with q-definability in logic, with regular grammars in linguistic point of view, might turn out to be useful in other applications. For these properties, see for example, textbooks on computation by Sipser, Papadmitrou or [HU79; Sal85; AS03].

The criterion (iii) of Theorem 11.1.3 is already such a useful tool, which is necessary and sufficient. But for transcendence applications, we just need simple necessary conditions.

By Corollary 11.1.2, we can restrict to power series of type $\sum x^{n_i}$ or equivalently to subsets $\{n_i\}$ of the set of natural numbers, which we describe for convenience by strictly increasing sequence n_i (ignoring the finite sets). This restriction to the characteristic sequences allows us to consider only two valued output functions, i.e., some states are accepted (output 1) and others are rejected (output 0). We thus call a subset of the set of natural numbers m-automatic, if there is an m-automata whose output is 1 exactly on this subset. We visualize the members as written in base m digit expansions. So these are just 'words' in the base m alphabet.

Here is a useful tool, whose variants we will encounter in the last section also.

Theorem 11.2.1 *(Pumping lemma for finite automata) Let S be an m-automatic set. Then there is N such that any word in S of length at least N can be written as juxtraposition xyz, with y being a non-empty word and length of xy not more than N and so that for all $i \geq 0$, the juxtraposition $xy^i z$ is also in S.*

That is subword y can be pumped as many times without leaving the automatic set. In fact, for linguists learning formal languages, the digits base m or what we called alphabet are words and what we called words are sentences. Automatic sets thus consist of language (i.e. collection of sentences) produced by 'regular' grammar. The conclusion of the pumping lemma then reflects the fact common to many languages that subclauses such as 'son of' in 'A is son of B' can be pumped many times giving still a grammatical sentence.

Proof.   The number N is just (greater than) the number of states in the corresponding automata. As you keep inputing the digits from left, we will get a repeat of states: the portion between the repeat is y, and can be clearly pumped.   □

Note that end of the proof of Richie's theorem uses essentially the same argument.

Here are some results [Cob72; Eil74; All87; AS03] about restrictions on densities, gaps and asymptotic behavior of automatic sequences:

Theorem 11.2.2   *Let $S = \{n_i\}$ be an m-automatic set. Define its maximum growth rate to be $\limsup n_{i+1}/n_i$, its natural density to be $\lim s_n/n$, where $s_x := |\{n_i \leq x\}|$, and its logarithmic density to be $\lim(\sum 1/n_i)/\log(n)$. Then*

(1) The maximum growth rate is finite.

(2) Either the maximum growth rate is more than one or $\limsup(n_{i+1} - n_i) < \infty$ and these are mutually exclusive.

(3) The logarithmic density exists and if the natural density exists, then both are the same and rational. If the natural density is zero, then either there is an integer $d \geq 1$ and a real number $0 < s < 1$ such that

$$0 < \liminf s_x/(x^s \log^{d-1} x) < \limsup s_x/(x^s \log^{d-1} x) < \infty$$

or there are integers $d \geq 1$, $m \geq 2$ and rational $c > 0$ such that s_x is asymptotic to $c(\log_m x)^{d-1}$.

(4) There is c such that the number of n_i's in S of given length n is at most cn.

(5) There is a subsequence n_i' of n_i such that $n_{i+1}'/n_i' \to m^d$. More precisely, there are non-negative integers a, $b > 0$, c, $d >$ such that $am^{nd} + b(m^{nd} - 1)/(m^d - 1) + c \in S$, for all positive integers n.

Remark 11.2.3   Part (2) immediately implies that n^k is not q-automatic for any q, vastly generalizing Richie's theorem mentioned above.

11.3   Applications to transcendence of gamma values and monomials

Now we shall present some transcendence results on quantities naturally occurring in our setting such as gamma values, zeta values, periods. These quantities are not usually naturally presented as power series, but are convertible to manageable power series. The first such result was the automata style proof by Allouche [All90] of Wade's result establishing transcendence of Carlitz $\tilde{\pi}$.

In this section, we deal with values of the arithmetic gamma function of 4.5 (a).

In this case, it turns out that while the methods of the last Chapter based on transcendence of periods and Chowla-Selberg type formulas give weak results (see 10.3) parallel to those known in the number fields case, the automata method settles completely [Tha96c; All96] the question of which monomials in gamma values at fractions are algebraic and which are transcendental.

Just as in 4.6, where the functional equations reduce to combinatorics of p-adic digits, this shows that algebraicity questions are also dealt efficiently by such combinatorics handled by automata.

Let us compare the situation for the classical case and arithmetic gamma case for $\mathbb{F}_q[t]$. In the notation and framework of 4.12, we have

Theorem 11.3.1 *(Usual Gamma) If $m(\underline{f}^{(\sigma)}) = 0$ for all σ relatively prime to N, then $\Gamma(\underline{f})$ is algebraic.*

The way we have presented it, this was conjectured (together with Galois action) by Deligne [Del79] (proved in [DMOS82]). But using the ideas of Lang and Kubert on distributions, it was shown in the appendix by Koblitz and Ogus to [Del79] that the algebraicity also follows by taking the correct combinations of multiplication and reflection formulas. The converse is not known, but is conjectured, because it follows from the general belief that functional equations force all the relations and also from conjectures [DMOS82] in algebraic geometry.

Theorem 11.3.2 *(Arithmetic gamma) Let $A = \mathbb{F}_q[t]$. $m(\underline{f}^{(\sigma)}) = 0$ for all $\sigma = q^j$ if and only if $\Gamma(\underline{f})$ is algebraic.*

In this case, the monomials are not obtained by combinations of naive analogues of multiplication and reflection formulas, but the 'only if' part is proved in Theorem 4.6.4 by showing that in the multiplicative basis of factorials of $1/(1-q^i)$ our monomial turns out to be a trivial monomial. Automata theory takes care of the converse, as we will see. Mendes-France and Yao [MFY97] generalized to p-adic values from the fractional values and simplified further and we present below an account based on their method:

Lemma 11.3.3 *For positive integers a, b, c; $q^c - 1$ divides $q^a(q^b - 2) + 1$ if and only if c divides (a, b), the greatest common divisor of a and b.*

Proof. If the remainder obtained by dividing x by c is written as x_c; then since $q^r \equiv q^{r_c} \mod (q^c - 1)$, the first condition is equivalent to $q^{(a+b)_c} - 2q^{a_c} + 1 \equiv 0$, i.e., $(a+b)_c = a_c = 0$, which is equivalent to the second condition. $\qquad\square$

Theorem 11.3.4 *If the sequence $n_j \in \mathbb{F}_q$ is not ultimately zero, then $\sum_{j=1}^{\infty} n_j/(t^{q^j} - t) \in K_\infty$ is transcendental over K.*

Proof. We have $\sum t\, n_j/(t^{q^j} - t) = \sum c(m)t^{-m}$, where

$$c(m) = \sum_{l(q^j-1)=m} n_j = \sum_{(q^j-1)|m} n_j.$$

Consider the subsequences $c_t(m) := c(q^t m + 1)$, as there are infinitely many non-zero n_j's, it is enough to show, by Christol's theorem 11.1.3, that $c_a \neq c_b$, for any $a > b$ such that n_a and n_b are non-zero.

Let h be the least positive integer s dividing a, but not dividing b and with $n_s \neq 0$. (Note $s = a$ satisfies the three conditions, so h exists.) By the lemma,

$$c_a(q^h - 2) - c_b(q^h - 2) = \sum_{(q^l-1)|(q^a(q^h-2)+1)} n_l - \sum_{(q^l-1)|(q^b(q^h-2)+1)} n_l$$

$$= \sum_{l|\gcd(a,h)} n_l - \sum_{l|\gcd(b,h)} n_l$$

$$= n_h \neq 0.$$

Hence $c_a \neq c_b$ and the theorem follows. $\qquad\square$

Theorem 11.3.5 *Let $A = \mathbb{F}_q[t]$. If $n \in \mathbb{Z}_p$ is not a non-negative integer, then $n! = \Gamma(n+1)$ is transcendental over K. In particular, the values of the gamma function at the proper fractions and at non-positive integers are transcendental.*

Proof. If a power series f is algebraic, so is its derivative f' (we can see this by differentiating the algebraic relation, but in finite characteristic this might give back the trivial relation. We can also deduce this by Hadamard product Corollary 11.1.2 or 11.1.3), and hence also the logarithmic derivative f'/f. In other words, transcendence of the logarithmic derivative implies the transcendence. This is a nice tool to turn products into sums, sometimes simplifying the job further because now exponents matter modulo p only.

Write the base q expansion $n = \sum n_j q^j$ as usual, so that $n! = \prod d_j^{n_j}$, and hence

$$\frac{n!'}{n!} = \sum n_j \frac{d'_j}{d_j} = -\sum \frac{n_j}{t^{q^j} - t}.$$

Now if all sufficiently large digits n_j are divisible by p^k, then modifying the first few digits (which does not affect transcendence) we can arrange that all are divisible by p^k and then take p^k-th root (which does not affect transcendence). So without loss of generality, we can assume that the sequence n_j is not ultimately zero (modulo p). Hence the previous theorem applies and finishes the proof.   □

Now we describe the **proof of 11.3.2**:

Proof. By taking a common divisor and using Fermat's little theorem, any monomial of gamma values at proper fractions can be expressed as a rational function times monomial in $(q^j/(1 - q^d))!$'s, where d is fixed and $0 \leq j < d$. It was shown in the proof of Theorem 4.6.4 that the condition says that this monomial is non-trivial. So again taking p-powers out as necessary, as in the proof of the previous theorem, we can assume that the exponents are not all divisible by p. But the exponents matter only modulo p, when we take the logarithmic derivative. So this logarithmic derivative is logarithmic derivative of some gamma value also, and hence is transcendental by the previous theorem.   □

Remark 11.3.6 We cannot expect to have analogous result for the classical gamma function, because its domain and range are archimedean, and continuity is quite a strong condition in the classical case. In other words, a non-constant continuous real valued function on an interval cannot fail to take on algebraic values. Since, as we have seen above, the values at proper fractions are expected to be transcendental, many values at irrationals should be algebraic.

Morita's p-adic gamma function has domain and range $\mathbb{Z}_p$, which being non-archimedean is closer to our situation. Let us now look at interpolation (see 4.5) of $\Pi(n)$ at a finite prime v of $A = \mathbb{F}_q[t]$:

We have proved, as a corollary to analog 4.8.1 of Gross-Koblitz theorem that if d is the degree of v, then $\Pi_v(q^j/(1 - q^d))$ $(0 \leq j < d)$ is algebraic. The straight manipulation with digits then shows that $\Pi_v(n)$ is algebraic, if the digits n_j are ultimately periodic of period d. The converse, in a case n is a fraction, is a question raised earlier whether (H2) of 4.12 is best possible.

In analogy with the Theorem above, Yao has conjectured the converse for $n \in \mathbb{Z}_p$.

Things become quite simple [Tha98] when v is of degree one, so that without loss of generality we can assume that $v = t$. Yao used [MFY97] result to simplify and generalize again.

Theorem 11.3.7 *Let $A = \mathbb{F}_q[t]$. If v is a prime of degree 1, then $\Pi_v(n)$ is transcendental if and only if the digits n_j of n are not ultimately constant.*

Proof. Using the automorphism $t \to t + \theta$ for $\theta \in \mathbb{F}_q$ of A, we can assume without loss of generality that $v = t$. Then $\Pi_v(n) = \prod(-D_{j,v})^{n_j}$, for $n = \sum n_j q^j$. Since $q = 0$ in characteristic p, when we take logarithmic derivative, it greatly simplifies to give

$$t \frac{\Pi_v(n)'}{\Pi_v(n)} = \sum n_j \left(\frac{t^{q^j}}{1 - t^{q^j - 1}} - \frac{t^{q^{j-1}}}{1 - t^{q^{j-1}-1}} \right) = \sum \frac{n_j - n_{j+1}}{(1/t)^{q^j} - (1/t)}.$$

This power series in $\mathbb{F}_q((t))$ is transcendental over $\mathbb{F}_q(t) = \mathbb{F}_q(1/t)$ by Theorem 11.3.4, by just replacing t by $1/t$, because the hypothesis implies that $n_j - n_{j+1}$ is not ultimately zero. Then the theorem follows as in 11.3.5. □

Remarks 11.3.8 (1) What should be the implications for the Morita's p-adic gamma function? As explained in [Tha], the close connection to cyclotomy leads us to think that the situation for values at proper fractions should be parallel. But then this implies that the algebraic values in the image not taken at fractions (conjecturally (see [Tha]) the only algebraic values at fractions arise at fractions with denominators dividing $p - 1$, and we know these values by the Gross-Koblitz theorem and functional equations) should be taken at irrational p-adic integers. Thus we do not expect a Mendès France-Yao type result for Morita's p-adic gamma function, but it may be possible to have such a result for Π_v's. This breakdown of analogies seems to be due to an important difference: in the function field situation, the range is a 'huge' finite characteristic field of Laurent series over a finite field, and the resulting big difference in the function theory prevents analogies being as strong for non-fractions.

(2) Another proof of gamma values transcendence has been given by Hellegouarch [Hel95], using de Mathan's criterion [dM94] instead. On the other hand, Koskas, in his thesis referred above, has given an automata-theoretic proof of de Mathan's criterion. See [FKdM00] for these and effective version of Christol's theorem.

(3) For automata style alternate proofs of many transcendence results

for Carlitz zeta values mentioned in the last chapter, see eg. [Ber92; Ber93; Ber94; Ber95].

11.4   Applications to transcendence: periods and modular functions

First, let $p = 2$. Then

$$a_4 = a_6 = \sum_{n \text{ odd} \geq 1} q^n/(1 - q^n) = \sum_{n \text{ odd} \geq 1} \sum_{k=0}^{\infty} q^{kn} = \sum_{m=1}^{\infty} d_o(m) q^m,$$

where $d_o(m) = $ number of odd positive divisors of m. Hence, if $m = 2^k \prod p_i^{m_i}$, then $d_o(m) = \prod (m_i + 1)$. So $d_o(m)$ is odd if and only if $m = n^2$ or $2n^2$. Hence, with $f := \sum q^{n^2}$ (essentially theta), we have

$$a_4 = \sum_{n=1}^{\infty} (q^{n^2} + q^{2n^2}) = f + f^2.$$

Now Theorems 11.1.3 and 11.1.8 imply that f and so $a_4 = a_6 = f + f^2$ is transcendental over $k(q)$, i.e., q is transcendental over $K = k(a_4)$, finishing the proof when $p = 2$.

Using the techniques of automata theory, we give another proof of the function field analogue of Mahler-Manin conjecture and prove transcendence results for the power series associated to higher divisor functions $\sigma_k(n) = \sum_{d|n} d^k$.

Let p be a prime number, and k be an algebraic closure of $\mathbb{F}_p$. Let q be a variable and consider $a_4, a_6 \in k[[q]]$ defined by

$$a_4 := \sum_{n \geq 1} \frac{-5n^3 q^n}{1 - q^n}, \qquad a_6 := \sum_{n \geq 1} \frac{-(7n^5 + 5n^3) q^n}{12(1 - q^n)}.$$

Theorem 11.4.1   *The period q of the Tate elliptic curve $y^2 + xy = x^3 + a_4 x + a_6$ over $K := k(a_4, a_6)$ is transcendental over K.*

Proof.   It suffices to prove that a_4 (resp. a_6) is transcendental over $k(q)$, if $p \neq 5$ (resp. $p = 5$). Namely the Hasse invariant of the Tate elliptic curve, i.e., the coefficient of x^{p-1} in $(x^3 + x^2/4 + a_4 x + a_6)^{(p-1)/2}$ for $p > 3$, is equal to one, which shows (essentially first noticed in [S-D 73]) that a_4 and a_6 are algebraically dependent, for $p > 3$. (See the first part of the remarks below for the case $p \leq 3$.) In fact, we prove

Proposition 11.4.2 *If $h := (p-1)/\gcd(u, p-1)$ is even (e.g., if $p > 2$ and u is odd), then $\sum_{n \geq 1} n^u q^n / (1 - q^n)$ is transcendental over $\mathbb{F}_p(q)$.*

Proof. With $\sigma_u(\ell) := \sum_{d | \ell} d^u$, we have

$$\sum_{n \geq 1} \frac{n^u q^n}{1 - q^n} = \sum_{n \geq 1} n^u \sum_{k \geq 1} q^{kn} = \sum_{\ell \geq 1} \sigma_u(\ell) q^\ell.$$

It is known that [Ran61], since h is even, for some $A > 0$, we have

$$\#\{n \leq x, \ \sigma_u(n) \neq 0 \bmod p\} \sim \frac{Ax}{(\log x)^{1/h}}.$$

Now define $b_n = 0$ if $\sigma_u(n) = 0 \bmod p$, and $b_n = 1$ otherwise. If $\sum_{n \geq 1} \sigma_u(n) q^n$ were algebraic over $k(q)$, then the sequence $(\sigma_u(n) \bmod p)_n$ would be p-automatic, so would be the sequence b_n, and this asymptotic contradicts the restriction in part (3) of Theorem 11.2.2. $\quad\square$

Remarks 11.4.3 (1) This (11.4.1) function field analogue of Mahler-Manin conjecture was proved by Voloch [Vol96b], by approximating q by algebraic quantities and getting a contradiction by analyzing the Galois action using Igusa's theorem. Soon after-wards, the original conjecture was proved. See [Wal97] for the history and account of the proof. We just mention an interesting application that '$\log_p q$' appearing in the p-adic Birch-Swinnerton-Dyer conjectures of Mazur, Tate and Teitelbaum (Theorem of Stevens/Greenberg) does not then vanish, so that the order of vanishing is exactly as predicted in the conjectures.

(2) In [Tha96a], another proof of Theorem 11.4.1 was given by reducing the question to the transcendence of the theta function and using the theory of modular forms to show algebraic dependence between the theta function and a_4, a_6, which are related to Eisenstein series. The present proof [AT99] avoids this modular forms technology, by directly establishing the transcendence of a_4 or a_6. For $p \leq 7$, the proof is simpler in that we do not need the fact on Hasse invariant either then: it is easy to see that $a_4 = a_6$ if $p = 2$, $a_4 = 0$ if $p = 5$, $a_4 = 5a_6$ if $p = 7$ and (see [Tha96a]) $a_6 + a_4 = a_4^2$ if $p = 3$.

(3) For a non-negative integer u, let $S_u := \sum_{n \geq 1} \sigma_u(n) q^n \in \mathbb{F}_p[[q]]$. One might ask in general whether S_u is transcendental over $k(q)$ and when S_u and S_v are algebraically dependent over $\mathbb{F}_p$ or more generally what the transcendence degree of the field generated by all S_u's over $\mathbb{F}_p$ is. It is easy to see that $S_u = S_{u+k(p-1)}$, if $u, k > 0$ and $S_{p-1} = S_0 - S_0^p$, so that the

answers depend only on u, v modulo $p - 1$ and the case $p = 2$ has been completely settled. The situation of algebraic dependence, in general, is unclear, when u or v is even. See [AT99] for more discussion.

(4) See also [ADR99] for analog of 11.4.1 for rank 2 Drinfeld modules over $\mathbb{F}_q[t]$.

As for the transcendence question, here is another simple, but conditional, application of the automata tools:

Theorem 11.4.4 *If $c_p := \zeta(p)/\zeta(p - 1)$ is an irrational real number (here ζ is the Riemann zeta function), and if $p - 1$ divides u, then S_u is transcendental over $k(q)$.*

Proof. We again use Rankin's result [Ran61] for the case where the number h defined above is odd. Here $h = 1$, and Rankin's result implies that, for some positive rational r,

$$\#\{n \leq x, \ \sigma_u(n) \neq 0 \bmod p\} \sim (c_p r)x.$$

But from Theorem 11.2.2, if the sequence b_n is p-automatic and if the set $\{n, \ b_n = \alpha\}$ has a natural density, then this density must be a rational number. $\square$

11.5 Classifying finite characteristic numbers

Now we look at [BT98] a computational classification of finite characteristic numbers (Laurent series with coefficients in a finite field) and prove that some classes have good algebraic properties. This provides tools from the theories of computation, formal languages and formal logic for finer study of transcendence and algebraic independence questions. Using them, we place some well-known transcendental numbers occurring in number theory in the computational hierarchy.

Existence of or lack of patterns in the digit sequences of naturally occurring real numbers is a natural question. Rational numbers (and only rational numbers) have eventually periodic digit sequences. But the question has not been studied much for irrational real numbers, except for statistical studies on normality and randomness of digits for general numbers as well as special numbers such as π. Apart from the fact that no interesting patterns are found in general, the other reason for the lack of such studies is that irrationals usually are not naturally presented by their digit expansions (say decimals) at least in theoretical studies. Also, since it is hard to control carry-overs well, when we add or multiply, it is usually hard to manipulate the formulas to get good control on digit expansions of sums and products.

The situation seems to be much better for finite characteristic numbers: The Laurent series representation is widely used: There are no carry-over difficulties and many times the expressions can be manipulated to find the expansions. Also, as we have seen in 11.1, the algebraic Laurent series are precisely those whose digit patterns are recognizable by a finite automaton, which is a very simple and weak model of a computer.

The notion of pattern is closely linked with the notion of computation: the stronger (easier) patterns can be produced by weaker (easier) machines. Most Laurent series (and real numbers) arising naturally in number theory / geometry are computable in the sense that they can be produced (see below for more precise description of how) by Turing machine, which is the strongest theoretical model of the computer. There are in-between categories (hierarchies) of complexity studied in computer science, formal language theory and logic and remarkably really diverse viewpoints have converged to the same notions. This is at least well-known for the notion of computability: Recursive function theory, Church's lambda calculus, Turing machines, Post systems, generative grammars, Cellular automata etc. all lead to the same notion. Similarly, the notion of finite automata corresponds to regular languages, images of fixed points of uniform substitutions (uniform tag sequences), definability (in logic) and algebraicity (in the context of finite characteristic numbers).

Hence it seems natural to attempt a finer classification of transcendental Laurent series arising in number theory, by finding out where they fit in the hierarchy. Then we can use techniques of all these diverse fields for further study. Basically, we classify numbers $\sum a_i t^i \in \mathbb{F}_q((t))$ by looking at what kind of machines can produce a_i, given i (say positive), expanded in base q, as the input. (This is different from the study of (time or space) complexity of producing the first i terms of a power series.) Since there are only countably many computable numbers, we are really attempting a finer classification of this small class of numbers, which nonetheless contains many naturally occurring important numbers. For such a classification to be useful, it is necessary to relate closure properties of language classes to algebraic properties of the sets of numbers representable in the class. Making use of such good algebraic (as well as differential) closure properties of the classes, we can manipulate from numbers we are interested in to the numbers which easily yield to the standard tools of language theory, such as pumping lemmas, for instance. We give such examples of classification for some important Laurent series such as analogues of π, e and q-expansions of modular forms.

11.6 Computational classes and basic tools

In this section, we collect some basic information on computational classes and the corresponding language classes.

The most well-known hierarchy is the Chomsky hierarchy consisting of: Finite automata (regular languages), pushdown automata (context-free languages), linear bounded automata (context-sensitive languages) and Turing machines (languages generated by an unrestricted grammar). See [HU79; Sal85] for precise definitions and many equivalent descriptions. Informally, Turing machines have infinite two directional writing tape, whereas linear bounded automata uses tape size linear in the input size (but note that we can erase and use the same tape several times) and finite automata corresponds to zero or constant size working tape. For more on finite automata, especially in connections with number theory, the reader may look at [All87; CKMR80; Eil74; AS03; Tha98]. We will also consider some refined categories such as various space and time classes.

We give a brief summary of properties of the language classes we treat. These will help us later to establish whether a given power series belongs to the class or not. The main tools to show that something does not belong to a given class are various closure properties and the pumping lemma. We include, for completeness, some properties which we do not actually use (but which may be useful in further investigations). Unless noted otherwise, this material is from [HU79], where many other closure properties that we do not discuss are proved.

Fact 11.6.1   The regular sets are closed under union, concatenation, complementation, intersection.

Here is a pumping lemma for context free languages (CFL's).

Fact 11.6.2   Let L be a context-free language. Then there is a constant n_0, such that for any $z \in L$ with length $|z| \geq n_0$, z is a concatenation $uvwxy$ such that $vx \neq \epsilon$, $|vwx| \leq n_0$, and for all $n \geq 0$, $uv^n wx^n y \in L$. (Here and below ϵ denotes the empty string.)

Fact 11.6.3   Let L be a context-free language. Then there is a constant

n_0, such that for any $z \in L$ with n_0 or more positions of z marked as distinguished, z is a concatenation $uvwxy$ such that v and x together have at least one distinguished position, vwx has at most n_0 distinguished positions, and for all $n \geq 0$, $uv^n wx^n y \in L$.

Here the substrings v and x are said to be "pumped". Also note that this fact clearly implies the previous fact.

Fact 11.6.4   Context-free languages are closed under union, concatenation and intersection with regular sets. Context-free languages are not closed under complementation or intersection.

CFL's may be equivalently defined as those languages generated by context-free grammars, or those languages accepted by pushdown automata (PDA i.e., nondeterministic machines with a finite state control and a stack). An important subclass of CFL's is the class of deterministic context-free languages (DCFL's), which are accepted by deterministic pushdown automata (DPDA). The class of deterministic finite automata is the same, as far as its computational power is concerned, as that of non-deterministic finite automata.

Fact 11.6.5   DCFL's are closed under complementation and intersection with a regular set. DCFL's are not closed under union or concatenation. There exist DCFL's L_1 and L_2 such that $L_1 \cap L_2$ is not a CFL (and therefore is not a DCFL).

Now we turn our attention to space complexity. The class $NSPACE(S(n))$ $(DSPACE(S(n))$ respectively) contains exactly those languages which are accepted by nondeterministic (deterministic respectively) machine which uses $O(S(n))$ tape cells on inputs of length n. The most well-known space complexity class is the class of context-sensitive languages (CSL's), which are generated by context-sensitive grammars (the left side of a production may contain more than one symbol, but the right side must be at least as long). Equivalently, the CSL is $NSPACE(n)$.

Clearly, deterministic space classes are closed under complementation.

Fact 11.6.6   For $S(n) \geq \log n$, $NSPACE(S(n))$ is closed under complementation.

Fact 11.6.7   Let $SPACE$ stand for $DSPACE$ or $NSPACE$. Then, for $\epsilon > 0$ and $r \geq 0$, $SPACE(n^r)$ is a proper subset of $SPACE(n^{r+\epsilon})$.

We define the polynomial space class as follows: $PSPACE :=$ $\bigcup DSPACE(n^i) = \bigcup NSPACE(n^i)$. The well-known class P (NP respectively) of polynomial time (non-deterministic polynomial time respectively) complexity is a (conjecturally proper) subset of $PSPACE$. But unlike the separation theorems above for the space classes, it is not even known (though conjectured) that $LOGSPACE := DSPACE(\log(n))$ is properly contained in P. So it is not known whether CFL is contained in $LOGSPACE$. On the other hand, it is easy to see that CFL does not contain $LOGSPACE$, because $\{a^n b^n c^n : n \geq 1\}$ is clearly in the latter, but well-known [HU79] to be not CFL. It is known that

$$DTIME(S) \subseteq NTIME(S) \subseteq DSPACE(S) \subseteq NSPACE(S) \subseteq \bigcup_{c>0} DTIME(c^S)$$

$$LOGSPACE \subseteq NSPACE(\log n) \subseteq P \subseteq NP \subseteq PSPACE = NPSPACE$$

with at least one strict inclusion in each line. All the inclusions here are conjectured to be strict.

For ease of exposition, when considering space classes, we will consider machines which, on input N, compute the value of the t^N coefficient. Let L_c be the language over the alphabet $\Sigma_q := \{0, 1, \cdots q-1\}$ consisting of the base q representations of those N for which the coefficient of t^N is c. Note that if all of the languages L_c, for $c \in \mathbb{F}_q$, are in $NSPACE(S(n))$ or $DSPACE(S(n))$, then a nondeterministic or, respectively, deterministic machine can calculate the t^N coefficient in space $S(n)$, where n is the length of N. For a nondeterministic machine to compute a function, we mean that the machine halts with either the correct value of the function, or with an honest failure report. For any input, it must be the case that some computation path exists in which the machine halts with the correct function value.

Note that in any base, there are infinitely many representations of any number obtained by prepending zeroes. Of course, we do not want to have two different representations of the same number as elements of different L_c's. We simply require that numbers be written without leading zeroes. Since the set of strings of digits with no leading zeroes is a regular set, and all the classes we consider are closed under intersections with regular sets, we may henceforth ignore the problem of leading zeroes. That is, we simply do not care whether any strings with leading zeroes are in any of the L_c's, since they may be removed with no increase in complexity by an intersection with a regular set.

If L is a language over an alphabet Σ, we denote by $\overline{L}$ the complement $\Sigma^* \setminus L$.

11.7 Algebraic properties of computational classes

In this section, we divide finite characteristic numbers (i.e. the Laurent series in $\mathbb{F}_q((t))$, where we fix $\mathbb{F}_q$) into computational classes and explore which classes have good algebraic properties.

Given a finite characteristic numbers $\sum a_i t^i$ we look at a machine which can, for each positive integer i, produce $a_i \in \mathbb{F}_q$ as output when given the base q-expansion of i as input. For example, we will say that $\sum a_i t^i$ is context-free, if there is a pushdown automaton which accomplishes this task. For finer details on comparison with complexity of computing first i digits or language recognition, we refer to [BT98].

For language class $\mathcal{C}$, the corresponding Laurent series set will be denoted by $\mathbb{F}_q((t))_{\mathcal{C}}$.

For convenience, we recall our starting point of the chapter:

Fact 11.7.1 Automatic Laurent series in $\mathbb{F}_q((t))$ are exactly the Laurent series algebraic over $\mathbb{F}_q(t)$, so that $\mathbb{F}_q((t))_{Aut}$ is a field algebraically closed in Laurent series.

In contrast to such nice properties for the automata, as we shall see, the next class in the Chomsky hierarchy of the context-free languages has only very weak algebraic properties:

Theorem 11.7.2 *The class of context-free (or deterministic context-free) Laurent series is closed under addition of algebraic Laurent series, but is not closed under addition or multiplication in general.*

Proof. The first statement follows because a PDA (DPDA respectively) can simulate a PDA (DPDA respectively) and a finite state automaton in parallel, and add the coefficients obtained at the end.

For $q = 2$, this boils down to closure under the symmetric difference with a regular set:

$$\sum_{n \in L} t^n + \sum_{n \in R} t^n = \sum_{n \in (L \cup R) \setminus (L \cap R)} t^n.$$

Recall that we require both L and the complement $\overline{L}$ to be context free, so the symmetric difference is a union $(L \cap \overline{R}) \cup (\overline{L} \cap R)$ of context free languages, as is the complement of the symmetric difference.

There are (see 6.4 of [HU79]) deterministic context-free languages L_1, L_2, whose intersection is not context-free. Let $p \neq 2$ and $q = p^k$. Then

$$\sum_{n \in L_1} t^n + \sum_{n \in L_2} t^n = \sum_{n \in L_1 \cap L_2} 2t^n + \sum_{n \in (L_1 \cup L_2)-(L_1 \cap L_2)} t^n.$$

So the sum is not context-free. For more on this and multiplication claim, see [BT98]. □

Next, we turn to the properties of the space classes. Let S be a function from positive integers to positive integers. We say that a finite characteristic number is in $DSPACE(S)$ (respectively $NSPACE(S)$) if the deterministic (respectively nondeterministic) space complexity of computing the Nth coefficient is bounded above by $S(n)$, where n is the encoding length of N.

Theorem 11.7.3 *Let a and b be finite characteristic numbers in $DSPACE(S)$ (respectively $NSPACE(S)$). Then $a + b$ is also in $DSPACE(S)$ (respectively $NSPACE(S)$).*

Proof. We describe the machine. On input N, of length n, the machine computes a_N using space $S(n)$. Since $a_N \in \mathbb{F}_q$, it can be stored in constant space. Re-using the $S(n)$ space, it computes b_N and the sum $a_N + b_N$. □

Theorem 11.7.4 *Let $S(n) \geq n$ for all n. Let a and b be finite characteristic numbers in $DSPACE(S)$ (respectively $NSPACE(S)$). Then ab is also in $DSPACE(S)$ (respectively $NSPACE(S)$).*

Proof. Let N be an input of length n. The Nth coefficient of ab is $\sum a_i b_{N-i}$. To compute this, a machine can run through all possible values of i, using space $O(n)$ to store both i and $j = N - i$. For each pair i, j, the machine computes a_i and b_j, using space $S(n)$. This space is re-used for each a_i and each b_j. The running total of the products $a_i b_j$ is kept using constant space. So the total space is $O(S(n))$. □

Theorem 11.7.5 *Let $S(n) \geq n$ for all n. Let a be a finite characteristic number in $DSPACE(S)$ (respectively $NSPACE(S)$), with $a \neq 0$. Then a^{-1} is also in $DSPACE(S)$ (respectively $NSPACE(S)$).*

Proof. We assume without loss of generality that $a \in 1 + t\mathbb{F}_q[[t]]$. Let $\alpha = 1-a$. Let p be the characteristic, and let $\beta = 1+\alpha+\ldots+\alpha^{p-1}$. By Theorems 11.7.3 and 11.7.4, β is in $DSPACE(S)$ (respectively $NSPACE(S)$). We have $a^{-1} = 1 + \alpha + \alpha^2 + \ldots = \beta \beta^p \beta^{p^2} \ldots$.

Let $\beta = \sum_i b_i t^i$ (note $b_0 = 1$). Since p is the characteristic, $\beta^{p^k} = \sum_i (b_i)^{p^k} t^{ip^k}$. Let $\Omega = \Omega(N, n) = \{\overline{N} := (N_0, N_1, N_2, \ldots, N_{n-1}) \mid N_0 + pN_1 + \ldots + p^{n-1}N_{n-1} = N\}$, where as usual n is the length of the encoding of N. Then the Nth coefficient of a^{-1} is

$$\sum_{\overline{N} \in \Omega} \prod_{i=0}^{n-1} (b_{N_i})^{p^i}.$$

We describe a space-efficient divide and conquer strategy to compute this sum. Basically, the same space $S(n)$ is re-used to compute all of the b_{N_i}, and the difficultly lies in keeping track of $N_0, N_1, \ldots, N_n$. Naively, this would take space n^2, since half of these n numbers are of length at least $n/2$. Let $f_k(m)$ be the t^m coefficient of $\beta^{(1+p+p^2+\ldots+p^{k-1})}$, i.e.,

$$f_k(m) = \sum_{\overline{N} \in \Omega(m,k)} \prod_{i=0}^{k-1} (b_{N_i})^{p^i}.$$

We use the recurrence

$$f_k(m) = \sum_{i+p^{\lfloor k/2 \rfloor} j = m} f_{\lfloor k/2 \rfloor}(i) f_{\lceil k/2 \rceil}(j)^{p^{\lfloor k/2 \rfloor}}.$$

So, to compute $f_k(m)$, we run through all possible values of i and j, recursively computing $f_{\lfloor k/2 \rfloor}(i)$ and $f_{\lceil k/2 \rceil}(j)$ (re-using space). This leads to a $S(n) + n \log n$ algorithm if we simply write down i, compute $f_{\lfloor k/2 \rfloor}(i)$, then write down j (re-using the space where i was written), and compute $f_{\lceil k/2 \rceil}(j)$, since the same $S(n)$ space is used for all computations of coefficients of β (the $\log n$ factor comes from the fact that the depth of the recursion is $\log n$, since each recursive call involves cutting k in half).

We improve on this by observing that i and j only require half the space as m to write down. This is because j simply is a number with half the digits of m, while i, which may be as long as m, must agree with m in the rightmost $\lfloor k/2 \rfloor$ places. So each of i, j can be written in half the space as m, and the total space is therefore n plus the $S(n)$ that we use for computing coefficients of β. $\qquad \square$

Theorem 11.7.6 *Let r be a Laurent series which is a root of $f(x) \in \mathbb{F}_q((t))[x]$. Suppose that the coefficients of f are in $DSPACE(S)$ (respectively $NSPACE(S)$), for some $S(n) \geq n$. Then r is in $DSPACE(nS(n))$ (respectively $NSPACE(nS(n))$).*

Proof. Let K be the field generated by the coefficients of f. By Theorems 11.7.3, 11.7.4, and 11.7.5, everything in K has space complexity $S(n)$. So we can assume without loss of generality that f is irreducible over K (otherwise consider the minimal polynomial of r), and that r is separable over K (otherwise consider r^p, r^{p^2}, etc. all of which have the same complexity).

Let $a, b \in K[x]$ be such that $af + bf' = 1$. Let $g(x) = x - b(x)f(x)$. Suppose that $\alpha \equiv r \bmod (t^i)$. Then $g(\alpha) \equiv r \bmod (t^{2i-c})$, for some constant c depending only on f, since $g(r) = r$ and $g'(r) = 0$.

Let $\alpha_0 \in \mathbb{F}_q$ be congruent to $r \bmod (t^{(c+1)})$. For $i \geq 0$ let $\alpha_{i+1} = g(\alpha_i)$. By induction we have that $\alpha_i \equiv r \bmod (t^{2^i})$. Therefore, to compute the Nth coefficient of r, it suffices to compute the Nth coefficient of α_k for some k with $2^k > N$. Let n be the encoding length of N. Then for some $k = O(n)$, we have $2^k > N$.

Using the technique of Theorems 11.7.3 and 11.7.4, we may build a machine which, using work space $S(n)$, computes the coefficients of α_{i+1} given an oracle for the coefficients of α_i. To compute the Nth coefficient of α_k it suffices to simulate k of these machines hooked together, which takes space $kS(n)$, which is $O(nS(n))$ as desired. $\square$

Theorem 11.7.7 *(1) If $S(n) \geq n$, then the class of Laurent series corresponding to deterministic (non-deterministic resp.) space class $S(n)$ form a field. In particular, context-sensitive Laurent series form a field.*

(2) The Laurent series in $PSPACE$ form a field ($\mathbb{F}_q((t))_{PSPACE}$ in our notation) algebraically closed in the field $\mathbb{F}_q((t))$ of all Laurent series. More generally, any space class of the form $\bigcup DSPACE(n^i S(n))$ or $\bigcup NSPACE(n^i S(n))$ (such as the class corresponding to exponential space or the Turing machines), has the same property.

Proof. The first part follows from Theorems 11.7.3, 11.7.4, 11.7.5. For the second part, we also use Theorem 11.7.6: If g is in the space class for S and r is algebraically dependent on g, with dependency relation $P(g, r) = 0$, then f is just P developed as a polynomial in the second variable. $\square$

Remarks 11.7.8 (1) Theorem 11.7.6 can be used as a tool to prove algebraic independence of two given numbers by comparing their relative complexity.

(2) More generally, Fact 11.6.7 together with Theorem 11.7.6 implies, for example, that $SPACE(n^{r+1+\epsilon})$ contains some element transcendental over $SPACE(n^r)$.

(3) To compute $ia_i \in \mathbb{F}_q$, one needs to only look at the last digit of i, so all classes (including automata, context-free and space classes) are

closed under derivatives. So by Theorems 11.7.4, 11.7.5 the space classes for $S(n) \geq n$ are closed under derivative and logarithmic derivative. The same is true for automata. This is quite useful tool to show non-membership in a class, because logarithmic derivatives turn products into (often simpler) sums and derivatives kill all powers which are multiples of the characteristic.

(4) In the first section, we saw closure under Hadamard products (i.e. term-wise product) for algebraic power series (so for automata by Christol's theorem). At the automata or regular languages level, it corresponds to direct product or intersection respectively and the property is very transparent and natural in that viewpoint. From computational classes perspective, it is clearly valid for all space classes as well. It does not hold for CFL's: let f_1 and f_2 be power series with only $0, 1$ coefficients. Then the Hadamard product corresponds to taking the intersection of the $L'_1 s$, but CFL's are not closed under intersection. (Indeed, the intersection of two DCFL's need not be context-free.)

(5) One can use these algebraic (and differential) properties to manipulate from numbers we are interested in to numbers whose Laurent series expansion yields more easily to computational analysis by various computational/ language theoretic tools such as pumping lemma. This is illustrated in the examples in the next section.

(6) The class $PSPACE$ has independent characterizations: A problem is in $PSPACE$ if and only if it is describable in first-order logic with the addition of the partial fixed point operator if and only if it is describable in second-order logic with a transitive closure operator. See [Imm99] and references there for the terminology and many such equivalences, which also may lead to more tools.

11.8   Applications to refined transcendence

In this section, we show where several naturally occurring finite characteristic numbers in number theory lie in our classification.

The theory of Drinfeld modules gives analogues of e, $2\pi i$, gamma and zeta values. Also, one encounters periods of elliptic curves and Fourier expansions of modular forms in finite characteristic. We have seen that many of these are non-automatic (i.e. transcendental). Most of them are easily seen to be computable or even in PSPACE (this will be clear from the formulas below: For modular form expansions, this follows from their algebraic dependence on say theta and eisenstein series). Here we attempt to pin some of them down more accurately.

First we consider analog of $2\pi i$, namely Carlitz' $\tilde{\pi}$ for $\mathbb{F}_q[t]$. Note $\tilde{\pi}^{q-1} \in \mathbb{F}_q((1/t))$ just as $(2\pi i)^2 \in \mathbb{R}$. Put $\pi := t^{-q/(q-1)}\tilde{\pi}$ (see discussion at the end of 2.5). For $q = 2$, $\pi = \tilde{\pi}/t^2$.

Theorem 11.8.1 (1) $1/\pi$ is in LOGSPACE, π is in linear space. In particular, $\tilde{\pi}^{q-1}$ (which is just $\tilde{\pi}$ for $q = 2$) is context-sensitive. On the other hand, $1/\pi$ is not context-free.

(2) Let w be the logarithmic derivative of π with respect to t. Then w is context-sensitive, but not context-free.

Proof. The product formula for π in 2.5 shows [All90] $1/\pi = \sum a_n t^{-n}$, with $a_n = 0$ if n can not be represented as a sum of distinct $q^j - 1$'s ($j > 0$) and with $a_n = (-1)^k$ if n is sum of k such numbers (k is then uniquely defined). Note that this occurs exactly when for some m, the following hold:

(1) The base q representation of m has only zeroes and ones.
(2) k is the number of ones in the base q representation of m.
(3) $m - k = n$.

Since the encoding length of k is logarithmic in the encoding length of n, a machine can run through all possible values of k in LOGSPACE. For each value of k, the machine checks that $n + k$ has only zeroes and ones, and that k of the digits are one. Therefore, $1/\pi$ is in LOGSPACE. By Theorem 11.7.5, π is in linear space, i.e., it is context-sensitive. Theorem 11.7.4 and remarks before the Theorem now finish the proof of the first part of (1).

Suppose the language of n's with $a_n = -1$ is context-free. Then so is its intersection, say L, with the regular set $1^a 0^b$. Now let r be large and $f := q^r$. Then $n := (q^{r+f-1} - 1) + (q^{r+f-2} - 1) + \cdots + (q^r - 1)$ has base q expansion $1^{f-1} 0^{r+1}$, and thus $n \in L$. Then by marking some 1's as distinguished, the Fact 11.6.3 implies that for some $b > 0$ and c, we have $n_i := 1^{f-1+bi} 0^{r+1+ci} \in L$, for all i. But this easily leads to contradiction, by looking for possibilities for $m_i -$ (sum of digits of m_i) $= n_i$ as follows: Since $\sum_{j=0}^{l-1}(q^j - 1) < q^l - l$, the digits of m_i up to exponent $l - 1$ can not influence the l-th digit of n_i. So m_i is forced to have the expansion of the form $1^{f-1+bi} a_{r+1+ci} \cdots a_1$, where a_j are 0 or 1. So in order to get the expression for n_i, with all those zeros at the end, the $f - 1 + bi$ of the -1's occurring in $\sum(q^j - 1)$ from the 1^{f-1+bi} part have to cancel with $q^j - 1$'s from the next digits, i.e., $f - 1 + bi = \sum(q^{a_j(j-1)} - 1)$. Now Fact 11.6.3 implies that b, c are absolutely bounded, independently of r. So one can choose large enough r, and i so that $q^r - r < bi < q^r - 1$. Then $q^r - 1 < f - 1 + bi < q^{r+1} - 1$. So when we express $f - 1 + bi$ as sum of $q^j - 1$'s, one term has to be $q^r - 1 = f - 1$, but by the inequalities above, bi is not a sum of this type. This contradiction finishes the proof of (1).

Now we use formula [All96; MFY97] from 11.3, namely $w = \sum c_n t^{-n-1}$ with $c_n = \sum_{q^j - 1 | n} 1 \in \mathbb{F}_q$. Just dividing n by $q^j - 1$'s one at a time in linear space (and reusing the space), we can see that w is context-sensitive. (Alternately, we can conclude this combining part (1) above, part (1) of Theorem 11.7.7, and the Remark (3) of the last section.) Suppose it is also context-free, then its intersection with the regular set $\{q^u - 1\}$ (which consist of numbers with all digits $q - 1$) is also context-free. But $c_{q^u - 1} = d(u) = \prod(u_i + 1)$, where $u = \prod p_i^{u_i}$ is the prime factorization of u. Now the subset of these $q^n - 1$'s where $c_{q^n - 1} = d(n) = 2$ is also context-free. But since all these numbers have $q - 1$ as the only digit, we can only pump it. This implies that if $d(n) = 2$, then for some $h > 0$ (pump-length), $d(n + hi)$ is also 2 for all i. Now for some m, the number $(nh)m + 1$ is a prime, so $d(n + h(mn^2)) = d(n)d((nh)m + 1) = 2d(n)$. This gives a contradiction if the characteristic is not two. On the other hand if the characteristic is two, we look at the subset where $d(n)$ is odd, i.e., n is a square. By pumping the context-free assumption implies $d(n + hi)$ are all odd, which means $n + hi$ is a square for all i, which is a contradiction again. This completes the proof. Note that we have generalized the proof of non-automaticity of w (and hence of transcendence of π), by using pumping lemma. □

Next we look at an analogue of e, namely Carlitz' $e = \sum D_i^{-1} \in \mathbb{F}_q((1/t))$ for $\mathbb{F}_q[t]$, where $D_i = \prod_{0 \le j < i} (t^{q^i} - t^{q^j})$ as usual.

Theorem 11.8.2 *e is in $DSPACE(n)$ and so is context-sensitive.*

Proof. Since $D_i^{-1} \in t^{-iq^i} \mathbb{F}_q[[1/t]]$, to compute the Nth coefficient of e it suffices to consider D_i^{-1} for $i \le \log N$. So, using $\log n$ space (where n is the length of N), a machine can enumerate values of i such that $iq^i < \log N$. Since N is written in base q, once i is written down, using $\log n$ space for scrap work, the machine has access to the digits of $N - iq^i$. So it suffices to describe a machine which, in linear space, computes the Nth coefficient of $(t^{-iq^i} D_i)^{-1}$ given N and i. Since

$$(t^{-iq^i} D_i)^{-1} = \prod_{j=0}^{i} (1 - t^{-(q^i - q^j)})^{-1} = \prod_{j=0}^{i} (1 + t^{-(q^i - q^j)} + t^{-2(q^i - q^j)} + t^{-3(q^i - q^j)} \ldots),$$

it follows that the Nth coefficient is the number of ways to write N as

$\sum_{j<i} \ell_j(q^i - q^j)$ with non-negative integers ℓ_j. Call this number $f_i(N)$; we have the recurrence

$$f_i(N) = \sum_{q^{\lfloor i/2 \rfloor} a+b=N} f_{\lceil i/2 \rceil}(a) + f_{\lfloor i/2 \rfloor - 1}(b).$$

As in the proof of Theorem 11.7.5, the machine can effect this recursion in linear space; either a or b can be written using only half as much space as n, since a is a number of half the length, and b agrees with N in the rightmost $n/2$ positions. □

Next we look at the sequence of squares (essentially the expansion of the theta function) when $q = 2$.

Theorem 11.8.3 *The set L of squares (and hence $\sum t^{n^2} \in \mathbb{F}_q[[t]]$) is context-sensitive, and under the generalized Riemann hypothesis (GRH) it is even in LOGSPACE. But for $q = 2$ it is not context-free.*

Proof. L is easily recognizable (just check whether $a^2 = n$, reusing the same space for a with $a^2 \leq n$) in deterministic linear space. So, in particular, it is context-sensitive. In fact, one can do better: Under GRH, if n is not square, it is a quadratic non-residue for a prime smaller than $c(\log n)^{2+\epsilon}$ and this can be checked in *LOGSPACE*: A machine need only check, for each $m < c(\log n)^3$, that n is a quadratic residue modulo m. Each such m can be written in $O(\log \log n)$ space (this space is re-used for each m), which is logarithmic in the input length. Also, computing n modulo m, and enumerating the quadratic residues modulo m, can all be done in the same space.

Since CFL's are closed under intersections with regular sets, it suffices to show that assumption that $L_5 = L \cap \{x \in \{0,1\}^* \mid x$ has exactly 5 ones$\}$ is context free leads to a contradiction. This is shown in [BT98] (with help from Poonen) by elementary number theoretic /combinatoric argument on the conclusion of pumping lemma for the context-free languages. □

We expect that the second assertion of the theorem generalizes to any q, but have not proved it. For connections with q-expansions of modular forms, we refer to [Tha98; AT99].

The tools seem to be successful to place some natural numbers in Chomsky hierarchy, and one might hope to do the same for some gamma and zeta values in function field setting. But to get a finer classification, tools need to be developed to show that certain languages are not in the given space class.

See [BT98] for references for the facts in the following remarks.

Remarks 11.8.4

(I) The algebraicity result 11.1.3 of Christol is quite different from Chomsky's well-known result on algebraic power series and context-free languages.

(II) What can one say about digit expansions of real numbers in some base or of p-adic or λ-adic numbers? There is a classical result of Rice which showed that recursive real numbers form a field algebraically closed in the field of real numbers. Though there is a lot of work done on complexity of the basic operations, it seems that no other such good class is known. The class of automatic real numbers is closed under addition and under multiplication by rationals but it is not closed under multiplication or reciprocation.

(III) On the other hand there is well-known classification of complex numbers z due to Mahler (see also [Mah71]) into A, S, T, U numbers by optimizing diophantine approximation properties of values of all polynomials evaluated at z. The A numbers turn out to be algebraic numbers. The other classes also have the property that algebraically dependent (transcendental) numbers belong to the same class. But these classes (or simple modifications by taking unions) are not closed under addition or multiplication. Analogues of this classification have been studied also for p-adic case, finite characteristic Laurent series, general Laurent series. In each case, there are four classes (subdivided often into finer classes, but the finer classes usually do not have the nice algebraic property mentioned above) with similar properties. But, except in the complex case the A numbers are now algebraic by definition rather than by intrinsic diophantine approximation characterization. In all these situations, almost all numbers are S numbers and Liouville numbers are U numbers. So there are uncountably many S and U numbers. In the complex case, Schmidt proved the existence of (uncountably many) T-numbers.

We can also consider disjoint classes: eg. Aut, $PSPACE \setminus Aut$, $Comp \setminus PSPACE$ and $\mathbb{F}_q((t)) \setminus Comp$ with the same algebraic dependence (and more) properties. (We can consider even EXP class, for example.) But except for $\mathbb{F}_q((t)) \setminus Comp$ class, our classes are countable. Also, since complex numbers form a two-dimensional space over the real numbers, it is easy to make computational classes for complex numbers (as in Mahler's classification), but the finite characteristic complex numbers form an infinite dimensional space over the finite characteristic real numbers. On the other hand, at least in Carlitz modules theory, most of the interesting numbers lie in the finite extension $\mathbb{F}_q((u))$, with $u = t^{-1/(q-1)}$.

(IV) Now let us point out some **open questions** which naturally arise: Can our classes be characterized by some diophantine approximation properties? Can one provide a class of interesting numbers which generate $\mathbb{F}_q((t))_C$ or generate a field whose algebraic closure in the Laurent series is $\mathbb{F}_q((t))_C$, say for $C = PSPACE$ or *Comp*? Does the well-known result of Cobham, which says that if m and n are multiplicatively independent (eg. powers of distinct primes), then a non-periodic sequence of integers given in base m or n can not both be recognized by automata, generalize to other low complexity computational classes, say $LOGSPACE$? (Recently, an interesting algorithm to compute N-th digit base 16 of π, without computing all the previous digits has been discovered. Unfortunately, its time complexity is worse than the time complexity of fastest known algorithm to generate first N digits, and also its space complexity is of order n^k, probably too large for Cobham type expectation.) What are the closure properties with respect to solutions of some interesting class of differential equations? What are the good time complexity, mixed complexity, quantum or randomized complexity classes with good algebraic properties? (It is easy to see that they are closed under addition, but even for multiplication, the naive approach leads to time exponential in the size of input.) What are the algebraic properties of say $PSPACE$ for real or p-adic numbers?

Are **periods** (obtained by integrating algebraic functions) in PSPACE? How about for periods in real or complex setting?

In [KZ01] Kontsevich and Zagier raise a question of producing a non-periods. From numerical computation using definition in [KZ01], it seems quite plausible, and experts seem to believe it, but I have not nailed down references yet, that periods are in PSPACE or at least in exp-space. If that is the case, any algorithm, coded as computable number, out of exp-space will be an answer to their question.

Note on the Notation

Main warning: Same notation is used for the same concepts in general, specific or analogous situations. eg., Π (or !), Γ or ζ is used for classical as well as different function field factorial, gamma or zeta functions respectively, for general A or for $A = \mathbb{F}_q[t]$ etc. Many quantities such as $\tilde{\pi}$ (see 2.1, 2.5, 4.7), d_i, l_i depend on A. There should be no confusion, as they are clearly identified.

In addition to usual usage of ∞, we also often use it for a place at infinity (chosen) for a function field.

We use p, q for numerator, denominator in chapter 9 as well as usual usage as prime and prime power. We also use q in q-expansion.

We use t for a variable as well as for standard substitution $t = q^{-s}$.

$A = \mathbb{F}_q[t]$ case usually means Carlitz module C_a, unless stated otherwise.

M is used both for motives and matrices.

For integral parts of x, we use Knuth's floor notation $\lfloor x \rfloor$. The cardinality of the set S is denoted by $|S|$ or by $\#S$.

As usual, $\mathbb{Z}$, $\mathbb{Z}_p$, $\mathbb{Q}$, $\mathbb{Q}_p$, $\mathbb{R}$, $\mathbb{C}$ stand respectively for the ring of integers, of p-adic integers, the field of rational numbers, of p-adic rationals, of real numbers and of complex numbers.

We use standard notations End, Hom, Aut, Gal, Spec, Ker, Pic, mod etc. and also Tr for trace, det or Det for determinant, Re for the real part, Res for residue, $\mathbb{G}_a$ for additive group and $\mathbb{G}_m$ for multiplicative group.

Here are places where some basic notation is fixed. (Local notation and definitions used only in the section where it is introduced are not repeated here):

See 1.1 for ϕ (Euler's, except in 8.5 it is soliton), sgn, 'even', 'odd'.

See 2.2 for $\mathbb{F}_q, X, K, \infty, d_\infty, A, K_\infty, C_\infty, \overline{K}, \overline{K_\infty}, \mathbb{F}_\infty, A_v, g, h, h_A, \iota$

See 2.3 for $\rho[I]$, ρ_I.

See 2.5 for $A_d, A_{<d}, A_+, A_{i+}, A_{\leq d}, \mathcal{A}_d, \mathbb{Z}_{>0}, [i], D_i, L_i$ etc. (also 4.5, 5.1 for more general sign convention).

See 1.4, 3.3, 3.4 for $H = H_A, H_1$.

See 4.6, 4.12 for $\underline{a}, m(\underline{a}), n(\underline{a}), \underline{a}^\sigma$ etc.

See 4.14, 4.15 for binomial coefficient notations.

See 2.5, 4.15, 7.7, 8.3 for different notions of D_i, d_i, L_i, l_i.

See 5.6 for $S_d(k)$ and $l(k)$.

Bibliography

[ABP]   Greg Anderson, Dale Brownawell, and Matthew Papanikolas. Determination of the algebraic relations among special gamma values in positive characteristic. *To appear in Ann. Math.*

[ADR99]   M. Ably, L. Denis, and F. Recher. Transcendance de l'invariant modulaire en caractéristique finie. *Math. Z.*, 231(1):75–89, 1999.

[All87]   Jean-Paul Allouche. Automates finis en théorie des nombres. *Exposition. Math.*, 5(3):239–266, 1987.

[All90]   Jean-Paul Allouche. Sur la transcendance de la série formelle Π. *Sém. Théor. Nombres Bordeaux (2)*, 2(1):103–117, 1990.

[All96]   J.-P. Allouche. Transcendence of the Carlitz-Goss gamma function at rational arguments. *J. Number Theory*, 60(2):318–328, 1996.

[And]   G. W. Anderson. A two-dimensional analogue of Stickelberger's theorem. In *[G⁺92]*, pages 51–73.

[And86]   G. Anderson. *t*-motives. *Duke Math. J.*, 53(2):457–502, 1986.

[And94]   Greg W. Anderson. Rank one elliptic *A*-modules and *A*-harmonic series. *Duke Math. J.*, 73(3):491–542, 1994.

[And96a] Greg W. Anderson. Another look at the index formulas of cyclotomic number theory. *J. Number Theory*, 60(1):142–164, 1996.

[And96b] Greg W. Anderson. Log-algebraicity of twisted *A*-harmonic series and special values of *L*-series in characteristic *p*. *J. Number Theory*, 60(1):165–209, 1996.

[And02]   Greg W. Anderson. Kronecker-Weber plus epsilon. *Duke Math. J.*, 114(3):439–475, 2002.

[Ang97]   Bruno Angles. On some characteristic polynomials attached to finite Drinfeld modules. *Manuscripta Math.*, 93(3):369–379, 1997.

[Ang01]   Bruno Anglès. On Gekeler's conjecture for function fields. *J. Number Theory*, 87(2):242–252, 2001.

[Art64]   Emil Artin. *The gamma function*. Translated by Michael Butler. Holt, Rinehart and Winston, New York, 1964.

[Art65]   Emil Artin. *Collected papers*. Springer-Verlag, New York, 1965.

[Art67]   Emil Artin. *Algebraic numbers and algebraic functions*. Gordon and Breach Science Publishers, New York, 1967.

[Art03]   James Arthur. The principle of functoriality. *Bull. Amer. Math. Soc.*, 40(1):39–53, 2003.

[AS03]   J.-P. Allouche and J. Shallit. *Automatic sequences.* Cambridge University Press, Cambridge, 2003.

[AT68]   E. Artin and J. Tate. *Class field theory.* W. A. Benjamin, Inc., New York-Amsterdam, 1968.

[AT90]   G. Anderson and D. Thakur. Tensor powers of the Carlitz module and zeta values. *Ann. of Math. (2)*, 132(1):159–191, 1990.

[AT99]   Jean-Paul Allouche and Dinesh S. Thakur. Automata and transcendence of the Tate period in finite characteristic. *Proc. Amer. Math. Soc.*, 127(5):1309–1312, 1999.

[Bak90]   Alan Baker. *Transcendental number theory.* Cambridge University Press, Cambridge, second edition, 1990.

[BCH$^+$66]   A. Borel, S. Chowla, C. Herz, K. Iwasawa, and J. Serre. *Seminar on complex multiplication.* Springer-Verlag, Berlin, 1966.

[Ber92]   Valérie Berthé. De nouvelles preuves "automatiques" de transcendance pour la fonction zêta de Carlitz. *Astérisque*, (209):13, 159–168, 1992. Journées Arithmétiques, 1991 (Geneva).

[Ber93]   Valérie Berthé. Fonction ζ de Carlitz et automates. *J. Théor. Nombres Bordeaux*, 5(1):53–77, 1993.

[Ber94]   Valérie Berthé. Automates et valeurs de transcendance du logarithme de Carlitz. *Acta Arith.*, 66(4):369–390, 1994.

[Ber95]   Valérie Berthé. Combinaisons linéaires de $\zeta(s)/\Pi^s$ sur $\mathbf{F}_q(x)$, pour $1 \leq s \leq q - 2$. *J. Number Theory*, 53(2):272–299, 1995.

[BGKY03]   S. Bae, E.-U. Gekeler, P. Kang, and L. Yin. Anderson's double complex and gamma monomials for rational function fields. *Trans. Amer. Math. Soc.*, 355(9):3463–3474, 2003.

[BGY01]   Sunghan Bae, Ernst-U. Gekeler, and Linsheng Yin. Distributions and Γ-monomials. *Math. Ann.*, 321(3):463–478, 2001.

[Bha97]   Manjul Bhargava. P-orderings and polynomial functions on arbitrary subsets of Dedekind rings. *J. Reine Angew. Math.*, 490:101–127, 1997.

[Bha00]   Manjul Bhargava. The factorial function and generalizations. *Amer. Math. Monthly*, 107(9):783–799, 2000.

[BHMV94]   Véronique Bruyère, Georges Hansel, Christian Michaux, and Roger Villemaire. Logic and p-recognizable sets of integers. *Bull. Belg. Math. Soc. Simon Stevin*, 1(2):191–238, 1994.

[BK99]   M. Bhargava and K. Kedlaya. Continuous functions on compact subsets of local fields. *Acta Arith.*, 91(3):191–198, 1999.

[Böc02]   Gebhard Böckle. Global L-functions over function fields. *Math. Ann.*, 323(4):737–795, 2002.

[BP89]   Armand Borel and Gopal Prasad. Finiteness theorems for discrete subgroups of bounded covolume in semi-simple groups. *Inst. Hautes Études Sci. Publ. Math.*, (69):119–171, 1989.

[BP02]   W. Dale Brownawell and Matthew A. Papanikolas. Linear independence of gamma values in positive characteristic. *J. Reine Angew. Math.*, 549:91–148, 2002.

[Bro98]   W. Dale Brownawell. Transcendence in positive characteristic. In *Number theory (Tiruchirapalli, 1996)*, volume 210 of *Contemp. Math.*, pages 317–332. Amer. Math. Soc., Providence, RI, 1998.

[BS76]   L. Baum and M. Sweet. Continued fractions of algebraic power series in characteristic 2. *Ann. of Math. (2)*, 103(3):593–610, 1976.

[BS77]   L. Baum and M. Sweet. Badly approximable power series in characteristic 2. *Ann. of Math. (2)*, 105(3):573–580, 1977.

[BS96]   Eric Bach and Jeffrey Shallit. *Algorithmic number theory. Vol. 1.* MIT Press, Cambridge, MA, 1996.

[BS97]   A. Blum and U. Stuhler. Drinfeld modules and elliptic sheaves. In *Vector bundles on curves—new directions (Cetraro, 1995)*, volume 1649 of *Lecture Notes*, pages 110–193. Springer, 1997.

[BT98]   Robert M. Beals and Dinesh S. Thakur. Computational classification of numbers and algebraic properties. *Internat. Math. Res. Notices*, (15):799–818, 1998.

[Car35]   L. Carlitz. On certain functions connected with polynomials in a galois field. *Duke Math. J.*, 1:137–168, 1935.

[Car40a]   L. Carlitz. An analogue of the Staudt-Clausen theorem. *Duke Math. J.*, 7:62–67, 1940.

[Car40b]   L. Carlitz. A set of polynomials. *Duke Math. J.*, 6:486–504, 1940.

[Car41]   L. Carlitz. An analogue of the Bernoulli polynomials. *Duke Math. J.*, 8:405–412, 1941.

[Car60]   L. Carlitz. Some special functions over $GF(q, x)$. *Duke Math. J.*, 27:139–158, 1960.

[Cas72]   J. W. S. Cassels. *An introduction to Diophantine approximation.* Hafner Publishing Co., New York, 1972.

[CdM93]   H. Chérif and B. de Mathan. Irrationality measures of Carlitz zeta values in characteristic p. *J. Number Theory*, 44(3):260–272, 1993.

[CF67]   J. W. S. Cassels and A. Fröhlich, editors. *Algebraic number theory*, London, 1967. Academic Press Inc.

[Che63]   Claude Chevalley. *Introduction to the theory of algebraic functions of one variable.* Amer. Math. Soc., Providence, R.I., 1963.

[Che76]   Greg Cherlin. *Model theoretic algebra—selected topics.* Springer-Verlag, Berlin, 1976. Lecture Notes in Mathematics, Vol. 521.

[Chr79]   G. Christol. Ensembles presque periodiques k-reconnaissables. *Theoret. Comput. Sci.*, 9(1):141–145, 1979.

[Chu84]   Gregory V. Chudnovsky. *Contributions to the theory of transcendental numbers.* Amer. Math. Soc, Providence, RI, 1984.

[CKMR80]   G. Christol, T. Kamae, M. Mendèsfrance, and G. Rauzy. Suites algébriques, automates et substitutions. *Bull. Soc. Math. France*, 108(4):401–419, 1980.

[CL01]   Wen-Chen Chi and Anly Li. Kummer theory of division points over Drinfeld modules of rank one. *J. Pure Appl. Algebra*, 156(2-3):171–185, 2001.

[Cob69]   Alan Cobham. On the base-dependence of sets of numbers recognizable by finite automata. *Math. Systems Theory*, 3:186–192, 1969.

[Cob72]   A. Cobham. Uniform tag sequences. *Math. Systems Theory*, 6:164–192, 1972.

[Con00]   Keith Conrad. The digit principle. *J. Number Theory*, 84(2):230–257, 2000.

[Dav01]   Chantal David. Frobenius distributions of Drinfeld modules of any rank. *J. Number Theory*, 90(2):329–340, 2001.

[Del79]   P. Deligne. Valeurs de fonctions *L* et périodes d'intégrales. In *Automorphic forms, representations and L-functions, Part 2*, pages 313–346. Amer. Math. Soc., Providence, R.I., 1979.

[Del84]   P. Deligne. Intégration sur un cycle évanescent. *Invent. Math.*, 76(1):129–143, 1984.

[Den94]   Laurent Denis. Le théorème de Fermat-Goss. *Trans. Amer. Math. Soc.*, 343(2):713–726, 1994.

[Deu73]   Max Deuring. *Lectures on the theory of algebraic functions of one variable*. Springer-Verlag, Berlin, 1973. Lecture Notes in Mathematics, Vol. 314.

[DH87]   Pierre Deligne and Dale Husemoller. Survey of Drinfel'd modules. In *Current trends in arithmetical algebraic geometry*, pages 25–91. Amer. Math. Soc., Providence, RI, 1987.

[DH88]   Gilles Damamme and Yves Hellegouarch. Propriétés de transcendance des valeurs de la fonction zéta de Carlitz. *C. R. Acad. Sci. Paris Sér. I Math.*, 307(12):635–637, 1988.

[DH91]   G. Damamme and Y. Hellegouarch. Transcendence of the values of the Carlitz zeta function by Wade's method. *J. Number Theory*, 39(3):257–278, 1991.

[dM70]   Bernard de Mathan. Approximations diophantiennes dans un corps local. *Bull. Soc. Math. France Suppl. Mém.*, 21:93, 1970.

[dM94]   Bernard de Mathan. Un critère de transcendance en caractéristique positive. *C. R. Acad. Sci. Paris Sér. I Math.*, 319(5):427–432, 1994.

[dM95]   Bernard de Mathan. Irrationality measures and transcendence in positive characteristic. *J. Number Theory*, 54(1):93–112, 1995.

[DMOS82]   Pierre Deligne, James S. Milne, Arthur Ogus, and Kuang-yen Shih. *Hodge cycles, motives, and Shimura varieties*, volume 900 of *Lecture Notes in Mathematics*. Springer-Verlag, Berlin, 1982.

[Dor91]   David R. Dorman. On singular moduli for rank 2 Drinfel'd modules. *Compositio Math.*, 80(3):235–256, 1991.

[Dri74]   V. G. Drinfel'd. Elliptic modules. *Mat. Sb. (N.S.)*, 94(136):594–627, 656, 1974.

[Dri77a]   V. G. Drinfel'd. Commutative subrings of certain noncommutative rings. *Funkcional. Anal. i Priložen.*, 11(1):11–14, 96, 1977.

[Dri77b]   V. G. Drinfel'd.   Elliptic modules. II.   *Mat. Sb. (N.S.)*, 102(144)(2):182–194, 325, 1977.

[Dri77c]   V. G. Drinfel'd. A proof of Petersson's conjecture for function fields. *Uspehi Mat. Nauk*, 32(2 (194)):209–210, 1977.

[Dri87]   V. G. Drinfel'd. Moduli varieties of *F*-sheaves. *Funktsional. Anal. i Prilozhen.*, 21(2):23–41, 1987.

[DV96]   Javier Diaz-Vargas. Riemann hypothesis for $F_p[T]$. *J. Number Theory*, 59(2):313–318, 1996.

[Dwo90]   Bernard Dwork. *Generalized hypergeometric functions.* The Clarendon Press Oxford University Press, New York, 1990.

[EF99]   I. Efrat and I. Fesenko. Fields Galois-equivalent to a local field of positive characteristic. *Math. Res. Lett.*, 6(3-4):345–356, 1999.

[Eic66]   Martin Eichler. *Introduction to the theory of algebraic numbers and functions.* Academic Press, New York, 1966.

[Eil74]   Samuel Eilenberg. *Automata, languages, and machines. Vol. A.* Academic Press, New York, 1974.

[Elk99]   Noam D. Elkies. Linearized algebra and finite groups of Lie type I. In *Applications of curves over finite fields (Seattle, WA, 1997)*, pages 77–107. Amer. Math. Soc., 1999.

[Elk01]   Noam D. Elkies. Explicit towers of Drinfeld modular curves. In *European Congress of Mathematics, Vol. II (Barcelona, 2000)*, pages 189–198. Birkhäuser, Basel, 2001.

[Fen97]   Keqin Feng. Anderson's root numbers and Thakur's Gauss sums. *J. Number Theory*, 65(2):279–294, 1997.

[FKdM00]   Jean Fresnel, Michel Koskas, and Bernard de Mathan. Automata and transcendence in positive characteristic. *J. Number Theory*, 80(1):1–24, 2000.

[Fur67]   Harry Furstenberg. Algebraic functions over finite fields. *J. Algebra*, 7:271–277, 1967.

[FW87]   Peter G. O. Freund and Edward Witten. Adelic string amplitudes. *Phys. Lett. B*, 199(2):191–194, 1987.

[G+92]   David Goss et al., editors. *The arithmetic of function fields*, volume 2 of *Ohio State University Mathematical Research Institute Publications*, Berlin, 1992. Walter de Gruyter & Co.

[G+97]   E.-U. Gekeler et al., editors. *Drinfeld modules, modular schemes and applications*, River Edge, NJ, 1997. World Scientific Publishing Co. Inc.

[Gei79]   J. M. Geijsel. *Transcendence in fields of positive characteristic*, volume 91 of *Mathematical Centre Tracts*. Mathematisch Centrum, Amsterdam, 1979.

[Gek83]   Ernst-Ulrich Gekeler. Zur Arithmetik von Drinfel'd-Moduln. *Math. Ann.*, 262(2):167–182, 1983.

[Gek85a]   E.-U. Gekeler. Automorphe Formen über $F_q(T)$ mit kleinem Führer. *Abh. Math. Sem. Univ. Hamburg*, 55:111–146, 1985.

[Gek85b]   Ernst-Ulrich Gekeler. A product expansion for the discriminant function of Drinfel'd modules of rank two. *J. Number Theory*,

21(2):135–140, 1985.

[Gek86]   Ernst-Ulrich Gekeler. *Drinfel'd modular curves*, volume 1231 of *Lecture Notes in Mathematics*. Springer-Verlag, Berlin, 1986.

[Gek88]   Ernst-Ulrich Gekeler. On the coefficients of Drinfel'd modular forms. *Invent. Math.*, 93(3):667–700, 1988.

[Gek89a] Ernst-Ulrich Gekeler. On the de Rham isomorphism for Drinfel'd modules. *J. Reine Angew. Math.*, 401:188–208, 1989.

[Gek89b] Ernst-Ulrich Gekeler. Quasi-periodic functions and Drinfel'd modular forms. *Compositio Math.*, 69(3):277–293, 1989.

[Gek89c] Ernst-Ulrich Gekeler. Some new identities for Bernoulli-Carlitz numbers. *J. Number Theory*, 33(2):209–219, 1989.

[Gek90]   Ernst-Ulrich Gekeler. On regularity of small primes in function fields. *J. Number Theory*, 34(1):114–127, 1990.

[Gek91]   Ernst-Ulrich Gekeler. On finite Drinfel'd modules. *J. Algebra*, 141(1):187–203, 1991.

[Gek99]   Ernst-Ulrich Gekeler. A survey on Drinfeld modular forms. *Turkish J. Math.*, 23(4):485–518, 1999.

[Gek01]   Ernst-Ulrich Gekeler. Finite modular forms. *Finite Fields Appl.*, 7(4):553–572, 2001.

[GGPS90] I. M. Gel'fand, M. I. Graev, and I. I. Pyatetskii-Shapiro. *Representation theory and automorphic functions*. Academic Press Inc., Boston, MA, 1990.

[GK79]   Benedict H. Gross and Neal Koblitz. Gauss sums and the p-adic Γ-function. *Ann. of Math. (2)*, 109(3):569–581, 1979.

[GK88]   R. Gold and H. Kisilevsky. On geometric $\mathbf{Z}_p$-extensions of function fields. *Manuscripta Math.*, 62(2):145–161, 1988.

[Gol03]   David M. Goldschmidt. *Algebraic functions and projective curves*. Springer-Verlag, New York, 2003.

[Gos]   David Goss. L-series of t-motives and Drinfel'd modules. In *[G$^+$92])*, pages 313–402.

[Gos78]   D. Goss. von Staudt for $\mathbf{F}_q[T]$. *Duke Math. J.*, 45(4):885–910, 1978.

[Gos79]   David Goss. v-adic zeta functions, L-series and measures for function fields. *Invent. Math.*, 55(2):107–119, 1979.

[Gos80a] David Goss. The algebraist's upper half-plane. *Bull. Amer. Math. Soc. (N.S.)*, 2(3):391–415, 1980.

[Gos80b] David Goss. The Γ-ideal and special zeta-values. *Duke Math. J.*, 47(2):345–364, 1980.

[Gos80c] David Goss. Modular forms for $\mathbf{F}_r[T]$. *J. Reine Angew. Math.*, 317:16–39, 1980.

[Gos80d] David Goss. π-adic Eisenstein series for function fields. *Compositio Math.*, 41(1):3–38, 1980.

[Gos82]   D. Goss. On a Fermat equation arising in the arithmetic theory of function fields. *Math. Ann.*, 261(3):269–289, 1982.

[Gos87]   David Goss. Analogies between global fields. In *Number theory*, pages 83–114. Amer. Math. Soc., Providence, RI, 1987.

[Gos88] David Goss. The Γ-function in the arithmetic of function fields. *Duke Math. J.*, 56(1):163–191, 1988.

[Gos89] David Goss. Fourier series, measures and divided power series in the theory of function fields. *K-Theory*, 2(4):533–555, 1989.

[Gos94] David Goss. Drinfel'd modules: cohomology and special functions. In *Motives (Seattle, WA, 1991)*, volume 55 of *Proc. Sympos. Pure Math.*, pages 309–362. Amer. Math. Soc., 1994.

[Gos96] David Goss. *Basic structures of function field arithmetic.* Springer-Verlag, Berlin, 1996.

[Gos00] David Goss. A Riemann hypothesis for characteristic *p* L-functions. *J. Number Theory*, 82(2):299–322, 2000.

[Gos03] David Goss. The impact of the infinite primes on the riemann hypothesis for characteristic *p* valued *l*-series. In *Algebra, Arithmetic and Geometry with Applications*, pages 357–380. Springer, Berlin, NY, 2003.

[GR81a] Steven Galovich and Michael Rosen. The class number of cyclotomic function fields. *J. Number Theory*, 13(3):363–375, 1981.

[GR81b] Steven Galovich and Michael Rosen. Distributions on rational function fields. *Math. Ann.*, 256(4):549–560, 1981.

[GR82] S. Galovich and M. Rosen. Units and class groups in cyclotomic function fields. *J. Number Theory*, 14(2):156–184, 1982.

[GR96] E.-U. Gekeler and M. Reversat. Jacobians of Drinfeld modular curves. *J. Reine Angew. Math.*, 476:27–93, 1996.

[GS85] David Goss and Warren Sinnott. Class-groups of function fields. *Duke Math. J.*, 52(2):507–516, 1985.

[GS95] Arnaldo García and Henning Stichtenoth. A tower of Artin-Schreier extensions of function fields attaining the Drinfel'd-Vlăduţ bound. *Invent. Math.*, 121(1):211–222, 1995.

[GV87] Arnaldo García and J. F. Voloch. Wronskians and linear independence in fields of prime characteristic. *Manuscripta Math.*, 59(4):457–469, 1987.

[Ham93] Y. Hamahata. Tensor products of Drinfel'd modules and *v*-adic representations. *Manuscripta Math.*, 79(3-4):307–327, 1993.

[Ham02] Y. Hamahata. On a product expansion for the Drinfeld discriminant function. *J. Ramanujan Math. Soc.*, 17(3):173–185, 2002.

[Has49] Helmut Hasse. *Number theory.* Akademie-Verlag, Berlin, 1949. Translated 1979 by Horst Günter Zimmer.

[Hay] David R. Hayes. A brief introduction to Drinfel'd modules. In *[G+92]*, pages 1–32.

[Hay74] D. R. Hayes. Explicit class field theory for rational function fields. *Trans. Amer. Math. Soc.*, 189:77–91, 1974.

[Hay79] David R. Hayes. Explicit class field theory in global function fields. In *Studies in algebra and number theory*, pages 173–217. Academic Press, New York, 1979.

[Hay85] David R. Hayes. Stickelberger elements in function fields. *Compositio Math.*, 55(2):209–239, 1985.

[Hay91]   David R. Hayes. On the reduction of rank-one Drinfel'd modules. *Math. Comp.*, 57(195):339–349, 1991.

[Hay93]   David R. Hayes. Hecke characters and Eisenstein reciprocity in function fields. *J. Number Theory*, 43(3):251–292, 1993.

[Hel92]   Yves Hellegouarch. Modules de Drinfel'd généralisés. In *Approximations diophantiennes et nombres transcendants (Luminy, 1990)*, pages 123–164. de Gruyter, Berlin, 1992.

[Hel95]   Yves Hellegouarch. Une généralisation d'un critère de de Mathan. *C. R. Acad. Sci. Paris Sér. I Math.*, 321(6):677–680, 1995.

[HU79]   John E. Hopcroft and Jeffrey D. Ullman. *Introduction to automata theory, languages, and computation.* Addison-Wesley Publishing Co., Reading, Mass., 1979.

[HY98]   L. Hsia and J. Yu. On singular moduli of Drinfeld modules in characteristic two. *J. Number Theory*, 69(1):80–97, 1998.

[HY00]   Liang-Chung Hsia and Jing Yu. On characteristic polynomials of geometric Frobenius associated to Drinfeld modules. *Compositio Math.*, 122(3):261–280, 2000.

[HY01]   Chih-Nung Hsu and Jing Yu. On Artin's conjecture for rank one Drinfeld modules. *J. Number Theory*, 88(1):157–174, 2001.

[Iha81]   Yasutaka Ihara. Some remarks on the number of rational points of algebraic curves over finite fields. *J. Fac. Sci. Univ. Tokyo Sect. IA Math.*, 28(3):721–724 (1982), 1981.

[Imm99]   N. Immerman. *Descriptive complexity.* Springer, N. Y., 1999.

[Iwa69]   Kenkichi Iwasawa. Analogies between number fields and function fields. In *Some Recent Advances in the Basic Sciences, Vol. 2*, pages 203–208. Belfer Graduate School of Science, Yeshiva Univ., New York, 1969.

[Iwa72]   Kenkichi Iwasawa. *Lectures on p-adic L-functions.* Princeton University Press, Princeton, N.J., 1972.

[Iwa86]   Kenkichi Iwasawa. *Local class field theory.* The Clarendon Press Oxford University Press, New York, 1986.

[Iwa93]   Kenkichi Iwasawa. *Algebraic functions.* Amer. Math. Soc, 1993.

[Jar96]   Moshe Jarden. Infinite Galois theory. In *Handbook of algebra, Vol. 1*, pages 269–319. North-Holland, Amsterdam, 1996.

[Jeo00a] Sangtae Jeong. A comparison of the Carlitz and digit derivative bases in function field arithmetic. *J. Number Theory*, 84(2):258–275, 2000.

[Jeo00b] Sangtae Jeong. Continuous linear endomorphisms and difference equations over the completion of $\mathbf{F}_q[T]$. *J. Number Theory*, 84(2):276–291, 2000.

[Jeo01]   Sangtae Jeong. Hyperdifferential operators and continuous functions on function fields. *J. Number Theory*, 89(1):165–178, 2001.

[Kap95]   M. Kapranov. A higher-dimensional generalization of the Goss zeta function. *J. Number Theory*, 50(2):363–375, 1995.

[Kat75]   Nicholas M. Katz. *p*-adic *L*-functions via moduli of elliptic curves. In *Algebraic geometry (Proc. Sympos. Pure Math., Vol. 29*, pages 479–506. Amer. Math. Soc., Providence, R. I., 1975.

[KC02]   Victor Kac and Pokman Cheung. *Quantum calculus.* Universitext. Springer-Verlag, New York, 2002.

[Ked01]   Kiran S. Kedlaya. The algebraic closure of the power series field in positive characteristic. *Proc. Amer. Math. Soc.*, 129(12):3461–3470 (electronic), 2001.

[Kim97]   Minhyong Kim. Geometric height inequalities and the Kodaira-Spencer map. *Compositio Math.*, 105(1):43–54, 1997.

[Kis93]   H. Kisilevsky. Multiplicative independence in function fields. *J. Number Theory*, 44(3):352–355, 1993.

[Knu69]   Donald E. Knuth. *The art of computer programming. Vol. 2: Seminumerical algorithms.* Addison-Wesley Pub, 1969.

[Kob77]   Neal Koblitz. *p-adic numbers, p-adic analysis, and zeta-functions.* Springer-Verlag, New York, 1977.

[Kob80]   Neal Koblitz. *p-adic analysis: a short course on recent work.* Cambridge University Press, Cambridge, 1980.

[Koc67]   Helmut Koch. Über die Galoissche Gruppe der algebraischen Abschliessung eines Potenzreihenkörpers mit endlichem Konstantenkörper. *Math. Nachr.*, 35:323–327, 1967.

[Koc96]   Anatoly N. Kochubei. *p*-adic commutation relations. *J. Phys. A*, 29(19):6375–6378, 1996.

[Koc98]   Anatoly N. Kochubei. Harmonic oscillator in characteristic *p*. *Lett. Math. Phys.*, 45(1):11–20, 1998.

[Koc99]   Anatoly N. Kochubei. F_q-linear calculus over function fields. *J. Number Theory*, 76(2):281–300, 1999.

[Koc00]   Anatoly N. Kochubei. Differential equations for F_q-linear functions. *J. Number Theory*, 83(1):137–154, 2000.

[KS99]   Nicholas M. Katz and Peter Sarnak. *Random matrices, Frobenius eigenvalues, and monodromy.* Amer. Math. Soc, 1999.

[KTV00]   M. Kim, D. Thakur, and J. F. Voloch. Diophantine approximation and deformation. *Bull. Soc. Math. France*, 128(4):585–598, 2000.

[Kuh01]   Franz-Viktor Kuhlmann. Elementary properties of power series fields over finite fields. *J. Symbolic Logic*, 66(2):771–791, 2001.

[KZ01]   Maxim Kontsevich and Don Zagier. Periods. In *Mathematics unlimited—2001 and beyond*, pages 771–808. Springer, 2001.

[Laf02]   Laurent Lafforgue. Chtoucas de Drinfeld et correspondance de Langlands. *Invent. Math.*, 147(1):1–241, 2002.

[Lan90]   Serge Lang. *Cyclotomic fields I and II.* Springer-Verlag, New York, second edition, 1990.

[Lan94]   Serge Lang. *Algebraic number theory.* Springer-Verlag, New York, second edition, 1994.

[Lau02]   Gérard Laumon. La correspondance de Langlands sur les corps de fonctions (d'après Laurent Lafforgue). *Astérisque*, (276):207–265, 2002. Séminaire Bourbaki, Vol. 1999/2000.

[LdM96] A. Lasjaunias and B. de Mathan. Thue's theorem in positive characteristic. *J. Reine Angew. Math.*, 473:195–206, 1996.

[LdM99] A. Lasjaunias and B. de Mathan. Differential equations and Diophantine approximation in positive characteristic. *Monatsh. Math.*, 128(1):1–6, 1999.

[LMQ75] James R. C. Leitzel, Manohar L. Madan, and Clifford S. Queen. Algebraic function fields with small class number. *J. Number Theory*, 7:11–27, 1975.

[Lor96] Dino Lorenzini. *An invitation to arithmetic geometry*, volume 9 of *Graduate Studies in Mathematics*. American Mathematical Society, Providence, RI, 1996.

[LV00] Gilles Lachaud and Serge Vlăduţ. Gauss problem for function fields. *J. Number Theory*, 85(2):109–129, 2000.

[LvdP77] J. H. Loxton and A. J. van der Poorten. Transcendence and algebraic independence by a method of Mahler. In *Transcendence theory: advances and applications*, pages 211–226. Academic Press, London, 1977.

[LZ98] Chaoqun Li and Jianqiang Zhao. Class number growth of a family of $\mathbf{Z}_p$-extensions over global function fields. *J. Algebra*, 200(1):141–154, 1998.

[Mad69] Manohar L. Madan. Class number relations in fields of algebraic functions. *J. Reine Angew. Math.*, 238:89–92, 1969.

[Mad70] Manohar L. Madan. On class numbers in fields of algebraic functions. *Arch. Math. (Basel)*, 21:167–171, 1970.

[Mah49] K. Mahler. On a theorem of Liouville in fields of positive characteristic. *Canadian J. Math.*, 1:397–400, 1949.

[Mah58] K. Mahler. An interpolation series for continuous functions of a p-adic variable. *J. Reine Angew. Math.*, 199:23–34, 1958.

[Mah71] K. Mahler. On the order function of a transcendental number. *Acta Arith.*, 18:63–76, 1971.

[MFY97] M. Mendès France and J. Yao. Transcendence and the Carlitz-Goss gamma function. *J. Number Theory*, 63(2):396–402, 1997.

[MM74] B. Mazur and William Messing. *Universal extensions and one dimensional crystalline cohomology*. Springer-Verlag, Berlin, 1974.

[MM80] M. Madan and D. Madden. On the theory of congruence function fields. *Comm. Algebra*, 8(17):1687–1697, 1980.

[Mor75] Yasuo Morita. A p-adic analogue of the Γ-function. *J. Fac. Sci. Univ. Tokyo Sect. IA Math.*, 22(2):255–266, 1975.

[Mor91] Carlos Moreno. *Algebraic curves over finite fields*. Cambridge University Press, Cambridge, 1991.

[MR86] W. Mills and D. Robbins. Continued fractions for certain algebraic power series. *J. Number Theory*, 23(3):388–404, 1986.

[Mul95] S. B. Mulay. On integer-valued polynomials. In *Zero-dimensional commutative rings (Knoxville, TN, 1994)*, pages 331–345. Dekker, New York, 1995.

[Mul99] S. B. Mulay. Integer-valued polynomials in several variables.

Comm. Algebra, 27(5):2409–2423, 1999.

[Mum78] D. Mumford. An algebro-geometric construction of commuting operators and of solutions to the Toda lattice equation, Korteweg deVries equation and related nonlinear equation. In *Proceedings of the International Symposium on Algebraic Geometry*, pages 115–153, Tokyo, 1978. Kinokuniya Book Store.

[Neu86] Jürgen Neukirch. *Class field theory*. Springer, 1986.

[NSW00] Jürgen Neukirch, Alexander Schmidt, and Kay Wingberg. *Cohomology of number fields*. Springer-Verlag, Berlin, 2000.

[Oka91] Shozo Okada. Kummer's theory for function fields. *J. Number Theory*, 38(2):212–215, 1991.

[Ore33] Oystein Ore. On a special class of polynomials. *Trans. Amer. Math. Soc.*, 35(3):559–584, 1933.

[Osg73] Charles F. Osgood. An effective lower bound on the "diophantine approximation" of algebraic functions by rational functions. *Mathematika*, 20:4–15, 1973.

[Osg75] C. Osgood. Effective bounds on the "Diophantine approximation" of algebraic functions over fields of arbitrary characteristic and applications to differential equations. *Nederl. Akad. Wetensch. Proc. Ser. A* **78**=*Indag. Math.*, 37:105–119, 1975.

[Ove52] Gordon Overholtzer. Sum functions in elementary p-adic analysis. *Amer. J. Math.*, 74:332–346, 1952.

[Per50] Oskar Perron. *Die Lehre von den Kettenbrüchen*. Chelsea Publishing Co., New York, N. Y., 1950. 2d ed.

[Pin97] Richard Pink. The Mumford-Tate conjecture for Drinfeld-modules. *Publ. Res. Inst. Math. Sci.*, 33(3):393–425, 1997.

[Poo95] Bjorn Poonen. Local height functions and Mordell-Weil theorem for Drinfel'd modules. *Compositio Math.*, 97(3):349–368, 1995.

[Poo96] Bjorn Poonen. Fractional power series and pairings on Drinfeld modules. *J. Amer. Math. Soc.*, 9(3):783–812, 1996.

[Poo97] Bjorn Poonen. Torsion in rank 1 Drinfeld modules and uniform boundedness conjecture. *Math. Ann.*, 308(4):571–586, 1997.

[Poo98] Bjorn Poonen. Drinfeld modules with no supersingular primes. *Internat. Math. Res. Notices*, (3):151–159, 1998.

[Pot99] Igor Yu. Potemine. Drinfeld-Anderson motives and multicomponent KP hierarchy. In *Recent progress in algebra*, pages 213–227. Amer. Math. Soc., Providence, RI, 1999.

[PR03] Matthew A. Papanikolas and Niranjan Ramachandran. A Weil-Barsotti formula for Drinfeld modules. *J. Number Theory*, 98(2):407–431, 2003.

[Ran61] R. A. Rankin. The divisibility of divisor functions. *Proc. Glasgow Math. Assoc.*, 5:35–40 (1961), 1961.

[Rit63] Robert W. Ritchie. Finite automata and the set of squares. *J. Assoc. Comput. Mach.*, 10:528–531, 1963.

[Ros87] Michael Rosen. The Hilbert class field in function fields. *Exposition. Math.*, 5(4):365–378, 1987.

[Ros02]   Michael Rosen. *Number theory in function fields.* Springer-Verlag, New York, 2002.

[RV99]   Dinakar Ramakrishnan and Robert J. Valenza. *Fourier analysis on number fields.* Springer-Verlag, New York, 1999.

[Sal85]   Arto Salomaa. *Computation and automata,* volume 25 of *Encyclopedia of Mathematics and its Applications.* Cambridge University Press, Cambridge, 1985.

[Sca02]   Thomas Scanlon. Diophantine geometry of the torsion of a Drinfeld module. *J. Number Theory,* 97(1):10–25, 2002.

[Sch76]   W. Schmidt. On Osgood's effective Thue theorem for algebraic functions. *Comm. Pure Appl. Math.,* 29(6):749–763, 1976.

[Sch80]   Wolfgang M. Schmidt. *Diophantine approximation,* volume 785 of *Lecture Notes in Mathematics.* Springer, Berlin, 1980.

[Sch90]   Fred Schultheis. Carlitz-Kummer function fields. *J. Number Theory,* 36(2):133–144, 1990.

[Sch91]   W. Schmidt. *Diophantine approximations and Diophantine equations,* volume 1467 of *Lecture Notes in Math.* Springer, 1991.

[Sch00]   W. Schmidt. On continued fractions and Diophantine approximation in power series fields. *Acta Arith.,* 95(2):139–166, 2000.

[Ser79]   Jean-Pierre Serre. *Local fields.* Springer-Verlag, New York, 1979.

[Ser88]   Jean-Pierre Serre. *Algebraic groups and class fields.* Springer-Verlag, New York, 1988.

[She98]   Jeffrey T. Sheats. The Riemann hypothesis for the Goss zeta function for $F_q[T]$. *J. Number Theory,* 71(1):121–157, 1998.

[Shu94a] Linghsueh Shu. Class number formulas over global function fields. *J. Number Theory,* 48(2):133–161, 1994.

[Shu94b] Linghsueh Shu. Kummer's criterion over global function fields. *J. Number Theory,* 49(3):319–359, 1994.

[Sin97a] Samarendra K. Sinha. Deligne's reciprocity for function fields. *J. Number Theory,* 63(1):65–88, 1997.

[Sin97b] Samarendra K. Sinha. Periods of t-motives and transcendence. *Duke Math. J.,* 88(3):465–535, 1997.

[Sla66]   Lucy Joan Slater. *Generalized hypergeometric functions.* Cambridge University Press, Cambridge, 1966.

[ST68]   Jean-Pierre Serre and John Tate. Good reduction of abelian varieties. *Ann. of Math. (2),* 88:492–517, 1968.

[Sta87]   H. M. Stark. Modular forms and related objects. In *Number theory (Montreal, Que., 1985),* volume 7 of *CMS Conf. Proc.,* pages 421–455. Amer. Math. Soc., Providence, RI, 1987.

[Sti93]   Henning Stichtenoth. *Algebraic function fields and codes.* Universitext. Springer-Verlag, Berlin, 1993.

[SV86]   K.-O. Stöhr and J.F. Voloch. Weierstrass points and curves over finite fields. *Proc. London Math. Soc. (3),* 52(1):1–19, 1986.

[SW85]   G. Segal and G. Wilson. Loop groups and equations of KdV type. *Inst. Hautes Études Sci. Publ. Math.,* (61):5–65, 1985.

[SYZS92] Andrew Szilard, Sheng Yu, Kaizhong Zhang, and Jeffrey Shallit. Characterizing regular languages with polynomial densities. In *Mathematical foundations of computer science 1992 (Prague, 1992)*, pages 494–503. Springer, Berlin, 1992.

[Tag91] Yuichiro Taguchi. Semisimplicity of the Galois representations attached to Drinfel'd modules over fields of "finite characteristics". *Duke Math. J.*, 62(3):593–599, 1991.

[Tag93] Yuichiro Taguchi. Semi-simplicity of the Galois representations attached to Drinfel'd modules over fields of "infinite characteristics". *J. Number Theory*, 44(3):292–314, 1993.

[Tag95] Yuichiro Taguchi. The Tate conjecture for *t*-motives. *Proc. Amer. Math. Soc.*, 123(11):3285–3287, 1995.

[Tag96] Yuichiro Taguchi. On ϕ-modules. *J. Number Theory*, 60(1):124–141, 1996.

[Tag99] Yuichiro Taguchi. Finiteness of an isogeny class of Drinfeld modules. Correction to a previous paper: "Ramifications arising from Drinfel'd modules" [in *the arithmetic of function fields (columbus, oh, 1991)*, 171–187, de Gruyter, Berlin, 1992; MR 94b:11049]. *J. Number Theory*, 74(2):337–348, 1999.

[Tak82] Toyofumi Takahashi. Good reduction of elliptic modules. *J. Math. Soc. Japan*, 34(3):475–487, 1982.

[Tam94] Akio Tamagawa. Generalization of Anderson's *t*-motives and Tate conjecture. *Sūrikaisekikenkyūsho Kōkyūroku*, (884):154–159, 1994.

[Tat84] John Tate. *Les conjectures de Stark sur les fonctions L d'Artin en s = 0*. Birkhäuser Boston Inc., Boston, MA, 1984.

[Tat99] Koichi Tateyama. Continuous functions on discrete valuation rings. *J. Number Theory*, 75(1):23–33, 1999.

[Tei] Jeremy Teitelbaum. Rigid analytic modular forms: an integral transform approach. In *[G⁺92]*, pages 189–207.

[Tha] D. S. Thakur. On gamma functions for function fields. In *[G⁺92]*, pages 75–86.

[Tha86] D. S. Thakur. Number fields and function fields (zeta and gamma functions at all primes). In *Proceedings of the conference on p-adic analysis (Houthalen, 1987)*, pages 149–157, Brussels, 1986. Vrije Univ. Brussel.

[Tha87] Dinesh S. Thakur. Gauss sums and gamma functions for function fields and periods of drinfeld modules. *Thesis, Harvard University.*, 1987.

[Tha88] Dinesh S. Thakur. Gauss sums for $\mathbf{F}_q[T]$. *Invent. Math.*, 94(1):105–112, 1988.

[Tha90] Dinesh S. Thakur. Zeta measure associated to $\mathbf{F}_q[T]$. *J. Number Theory*, 35(1):1–17, 1990.

[Tha91a] Dinesh S. Thakur. Gamma functions for function fields and Drinfel'd modules. *Ann. of Math. (2)*, 134(1):25–64, 1991.

[Tha91b] Dinesh S. Thakur. Gauss sums for function fields. *J. Number Theory*, 37(2):242–252, 1991.

[Tha92a] Dinesh S. Thakur. Continued fraction for the exponential for $\mathbf{F}_q[T]$. *J. Number Theory*, 41(2):150–155, 1992.

[Tha92b] Dinesh S. Thakur. Drinfel'd modules and arithmetic in the function fields. *Internat. Math. Res. Notices*, (9):185–197, 1992.

[Tha93a] Dinesh S. Thakur. Behaviour of function field Gauss sums at infinity. *Bull. London Math. Soc.*, 25(5):417–426, 1993. With an appendix by José Felipe Voloch.

[Tha93b] Dinesh S. Thakur. Shtukas and Jacobi sums. *Invent. Math.*, 111(3):557–570, 1993.

[Tha94] Dinesh S. Thakur. Iwasawa theory and cyclotomic function fields. In *Arithmetic geometry (Tempe, AZ, 1993)*, volume 174 of *Contemp. Math.*, pages 157–165. Amer. Math. Soc., 1994.

[Tha95a] Dinesh S. Thakur. Hypergeometric functions for function fields. *Finite Fields Appl.*, 1(2):219–231, 1995. Special issue dedicated to Leonard Carlitz.

[Tha95b] Dinesh S. Thakur. On characteristic p zeta functions. *Compositio Math.*, 99(3):231–247, 1995.

[Tha96a] Dinesh S. Thakur. Automata-style proof of Voloch's result on transcendence. *J. Number Theory*, 58(1):60–63, 1996.

[Tha96b] Dinesh S. Thakur. Exponential and continued fractions. *J. Number Theory*, 59(2):248–261, 1996.

[Tha96c] Dinesh S. Thakur. Transcendence of gamma values for $\mathbf{F}_q[T]$. *Ann. of Math. (2)*, 144(1):181–188, 1996.

[Tha97] Dinesh S. Thakur. Patterns of continued fractions for the analogues of e and related numbers in the function field case. *J. Number Theory*, 66(1):129–147, 1997.

[Tha98] Dinesh S. Thakur. Automata and transcendence. In *Number theory (Tiruchirapalli, 1996)*, volume 210 of *Contemp. Math.*, pages 387–399. Amer. Math. Soc., Providence, RI, 1998.

[Tha99a] Dinesh S. Thakur. An alternate approach to solitons for $\mathbf{F}_q[t]$. *J. Number Theory*, 76(2):301–319, 1999.

[Tha99b] Dinesh S. Thakur. Diophantine approximation exponents and continued fractions for algebraic power series. *J. Number Theory*, 79(2):284–291, 1999.

[Tha00] Dinesh S. Thakur. Hypergeometric functions for function fields. II. *J. Ramanujan Math. Soc.*, 15(1):43–52, 2000.

[Tha01] Dinesh S. Thakur. Integrable systems and number theory in finite characteristic. *Phys. D*, 152/153:1–8, 2001. Advances in nonlinear mathematics and science.

[Tha02] Dinesh S. Thakur. Elliptic curves in function field arithmetic. In *Currents trends in number theory (Allahabad, 2000)*, pages 215–238. Hindustan Book Agency, New Delhi, 2002.

[Tha03] Dinesh S. Thakur. Diophantine approximation in finite characteristic. In *Algebra, Arithmetic and Geometry with Applications*, pages 757–765. Springer, Berlin, NY, 2003.

[Thi] Alain Thiery. Indépendance algébrique des périodes et quasipériodes d'un module de Drinfel'd. In *[G+ 92]*, pages 265–284.

[TVZ82] M. A. Tsfasman, S. G. Vlăduţ, and Th. Zink. Modular curves, Shimura curves, and Goppa codes, better than Varshamov-Gilbert bound. *Math. Nachr.*, 109:21–28, 1982.

[TW96] Y. Taguchi and D. Wan. *L*-functions of ϕ-sheaves and Drinfeld modules. *J. Amer. Math. Soc.*, 9(3):755–781, 1996.

[TW97] Yuichiro Taguchi and Daqing Wan. Entireness of *L*-functions of ϕ-sheaves on affine complete intersections. *J. Number Theory*, 63(1):170–179, 1997.

[VD83] S. G. Vlèduts and V. G. Drinfel'd. The number of points of an algebraic curve. *Funktsional. Anal. i Prilozhen.*, 17(1):68–69, 1983.

[vdP96] Marius van der Put. Reduction modulo *p* of differential equations. *Indag. Math. (N.S.)*, 7(3):367–387, 1996.

[Voj87] Paul Vojta. *Diophantine approximations and value distribution theory*. Springer-Verlag, Berlin, 1987.

[Voj91] Paul Vojta. On algebraic points on curves. *Compositio Math.*, 78(1):29–36, 1991.

[Vol88] J. F. Voloch. Diophantine approximation in positive characteristic. *Period. Math. Hungar.*, 19(3):217–225, 1988.

[Vol95] José Felipe Voloch. Diophantine approximation in characteristic *p*. *Monatsh. Math.*, 119(4):321–325, 1995.

[Völ96a] H. Völklein. *Groups as Galois groups*, volume 53 of *Cambridge Studies in Adv. Math.*. Cambridge Univ. Press, 1996.

[Vol96b] José Felipe Voloch. Transcendence of elliptic modular functions in characteristic *p*. *J. Number Theory*, 58(1):55–59, 1996.

[Vol97] José Felipe Voloch. Diophantine geometry in characteristic *p*: a survey. In *Arithmetic geometry (Cortona, 1994)*, Sympos. Math., XXXVII, pages 260–278. Cambridge Univ. Press, 1997.

[Vol98] José Felipe Voloch. Differential operators and interpolation series in power series fields. *J. Number Theory*, 71(I):106–108, 1998.

[Wad41] L. I. Wade. Certain quantities transcendental over $GF(p^n, x)$. *Duke Math. J.*, 8:701–720, 1941.

[Wad43] L. I. Wade. Certain quantities transcendental over $GF(p^n, x)$. II. *Duke Math. J.*, 10:587–594, 1943.

[Wad44] L. I. Wade. Two types of function field transcendental numbers. *Duke Math. J.*, 11:755–758, 1944.

[Wad46a] L. I. Wade. Remarks on the Carlitz ψ-functions. *Duke Math. J.*, 13:71–78, 1946.

[Wad46b] L. I. Wade. Transcendence properties of the Carlitz ψ-functions. *Duke Math. J.*, 13:79–85, 1946.

[Wag71a] Carl G. Wagner. Interpolation series for continuous functions on π-adic completions of GF(*q*, *x*).. *Acta Arith.*, 17:389–406, 1970/1971.

[Wag71b] Carl G. Wagner. Linear operators in local fields of prime characteristic. *J. Reine Angew. Math.*, 251:153–160, 1971.

[Wag74] Carl G. Wagner. Differentiability in local fields of prime characteristic. *Duke Math. J.*, 41:285–290, 1974.

[Wal90] M. Waldschmidt. Transcendence problems connected with Drinfel'd modules. *İstanbul Üniv. Fen Fak. Mat. Derg.*, 49:57–75 (1993), 1990.

[Wal97] Michel Waldschmidt. Sur la nature arithmétique des valeurs de fonctions modulaires. *Astérisque*, (245):Exp. No. 824, 3, 105–140, 1997. Séminaire Bourbaki, Vol. 1996/97.

[Wan93] Daqing Wan. Newton polygons of zeta functions and *L* functions. *Ann. of Math. (2)*, 137(2):249–293, 1993.

[Wan96a] Daqing Wan. Meromorphic continuation of *L*-functions of *p*-adic representations. *Ann. of Math. (2)*, 143(3):469–498, 1996.

[Wan96b] Daqing Wan. On the Riemann hypothesis for the characteristic *p* zeta function. *J. Number Theory*, 58(1):196–212, 1996.

[Was97] Lawrence C. Washington. *Introduction to cyclotomic fields.* Springer-Verlag, New York, second edition, 1997.

[Wei73] André Weil. *Basic number theory.* Springer-Verlag, Berlin, 1973.

[Wei79] André Weil. *Scientific works. Collected papers. Vol. I-III.* Springer-Verlag, New York, 1979.

[Yan98] Zifeng Yang. Locally analytic functions over completions of $\mathbf{F}_r[U]$. *J. Number Theory*, 73(2):451–458, 1998.

[Yan01] Zifeng Yang. A note on zeta measures over function fields. *J. Number Theory*, 90(1):89–112, 2001.

[Yin00] Linsheng Yin. Distributions on a global field. *J. Number Theory*, 80(1):154–167, 2000.

[Yu] Jing Yu. Transcendence in finite characteristic. In *[G+92]*, pages 253–264.

[Yu86] Jing Yu. Transcendence and Drinfel'd modules. *Invent. Math.*, 83(3):507–517, 1986.

[Yu89] Jing Yu. Transcendence and Drinfeld modules: several variables. *Duke Math. J.*, 58(3):559–575, 1989.

[Yu90] Jing Yu. On periods and quasi-periods of Drinfel'd modules. *Compositio Math.*, 74(3):235–245, 1990.

[Yu91] Jing Yu. Transcendence and special zeta values in characteristic *p*. *Ann. of Math. (2)*, 134(1):1–23, 1991.

[Yu95] Jiu-Kang Yu. Isogenies of Drinfel'd modules over finite fields. *J. Number Theory*, 54(1):161–171, 1995.

[Yu97] Jing Yu. Analytic homomorphisms into Drinfeld modules. *Ann. of Math. (2)*, 145(2):215–233, 1997.

[Zha97] J. Zhao. On root numbers connected with special values of *L*-functions over $\mathbf{F}_q(T)$. *J. Number Theory*, 62(2):307–321, 1997.

www.ingramcontent.com/pod-product-compliance
Lightning Source LLC
Chambersburg PA
CBHW061615220326
41598CB00026BA/3776